P9-DXN-471

THE YEASTS

Volume 3

THE YEASTS

Edited by

ANTHONY H. ROSE

School of Biological Sciences, Bath University,
Claverton Down, Bath, England.

AND

J. S. HARRISON

The Distillers Co. (Yeast) Ltd., Great Burgh, Epsom,
Surrey, England.

Volume 3

YEAST TECHNOLOGY

SEP 1 8 1970

1970

ACADEMIC PRESS · LONDON and NEW YORK

ACADEMIC PRESS INC. (LONDON) LTD.
BERKELEY SQUARE HOUSE
BERKELEY SQUARE
LONDON, WIX 6BA

U.S. Edition published by

ACADEMIC PRESS INC.
111 FIFTH AVENUE
NEW YORK, NEW YORK 10003

Copyright © 1970 by ACADEMIC PRESS INC. (LONDON) LTD.

All Rights Reserved

No part of this book may be reproduced in any form by photostat, microfilm or any
other means, without written permission from the publishers.

Library of Congress Catalog Card Number : 72–85465

SBN : 12–596403–X

LIBRARY

SEP 16 1970

UNIVERSITY OF THE PACIFIC

220977

Printed in Great Britain by
William Clowes and Sons, Limited, London and Beccles

Contributors

AMERINE, M. A., *Department of Viticulture and Enology, University of California, Davis, California, U.S.A.*

AYRES, J. C., *Department of Food Science, The University of Georgia, Athens, Georgia, U.S.A.*

BEECH, F. W., *University of Bristol, Research Station, Long Ashton, Bristol, England.*

BURROWS, S., *Distillers Company Ltd., Glenochil Technical Centre Menstrie, Clackmannanshire, Scotland*

DAVENPORT, R. R., *University of Bristol, Research Station, Long Ashton, Bristol, England.*

GRAHAM, J. C. J., *Research Department, Distillers Company (Yeast) Ltd., Epsom, Surrey, England.*

HARRISON, J. S., *Research Department, Distillers Company (Yeast) Ltd., Epsom, Surrey, England.*

KODAMA, K., *Kodama Brewing Co. Ltd., Japan.*

KUNKEE, R. E., *Department of Viticulture and Enology, University of California, Davis, California, U.S.A.*

PEPPLER, H. J., *Universal Foods Corporation, Milwaukee, Wisconsin, U.S.A.*

RAINBOW, C., *Bass Charrington Ltd., Burton-on-Trent, England,*

ROSE, A. H., *School of Biological Sciences, Bath University, Bath, England.*

WALKER, H. W., *Department of Dairy and Food Industry, Iowa State University, Ames, Iowa, U.S.A.*

Preface to Volume I

It is often said that yeasts have been more intimately associated than any other group of micro-organisms with the progress and well-being of mankind. Many would base this claim largely on the capacity of certain yeasts to bring about a rapid and effective alcoholic fermentation, an activity which has been exploited for centuries in the manufacture of alcoholic beverages and in the leavening of bread. Others may prefer to acknowledge the contribution which yeasts have made in the elucidation of the basic metabolic processes of living cells. Few, however, would deny that yeast species, although less numerous than those of other major groups of micro-organisms, are sufficiently important that information on them needs to be fully documented so that scholars can continue to exploit their characteristic qualities. Over the years, many excellent texts have been published that deal comprehensively with yeasts and their activities. The present set of three volumes aims to bring the literature on this subject up to date.

So much has been written on yeasts that it is now impossible to encompass our knowledge of their microbiology in one volume. Moreover, it is beyond the capacity of any one person to write such a review, and so recourse must be made to a team of specialists each reviewing his or her own field of interest.

The first of the present volumes deals with the biology of yeasts, the second with yeast physiology and biochemistry, and the third with yeast technology. Together, these volumes present an up-to-date account of knowledge on yeasts and their activities. At the same time, each volume is sufficiently complete in itself to be perused separately.

We are extremely grateful to the authors of the chapters, whose labours have made this effort possible. Their forbearance, often under conditions of pressure and always with a deadline looming ahead, is very greatly appreciated. We also thankfully acknowledge permission to use the original drawings of apple blossom at various stages of development, which are the property of Messrs Farbenfabriken Bayer AG. Our hope is that readers of these volumes will consider the venture to have been worthwhile.

Anthony H. Rose
J. S. Harrison

July 1969

Contents

Contents of Volume 1

Biology of Yeasts

Contents of Volume 2

Physiology and Biochemistry of Yeasts

Abbreviations

The following abbreviations are used for names of yeast genera:

Aureobasidium	*A.*	*Lipomyces*	*L.*
Ashbya	*Ash.*	*Metschnikowia*	*M.*
Bullera	*B.*	*Nematospora*	*N.*
Brettanomyces	*Br.*	*Oospora*	*O.*
Candida	*C.*	*Pichia*	*Pi.*
Citeromyces	*Cit.*	*Pityrosporum*	*Pit.*
Coccidiascus	*Co.*	*Pullularia*	*Pull.*
Cryptococcus	*Cr.*	*Rhodotorula*	*Rh.*
Debaryomyces	*D.*	*Saccharomyces*	*Sacch.*
Endomycopsis	*E.*	*Saccharomycodes*	*S.*
Eremascus	*Erem.*	*Schizosaccharomyces*	*Schizosacch*
Geotrichum	*G.*	*Schwanniomyces*	*Schw.*
Hansenula	*H.*	*Sporobolomyces*	*Sp.*
Hanseniaspora	*Ha.*	*Torulopsis*	*T.*
Itersonilia	*I.*	*Trichosporon*	*Trich.*
Kloeckera	*Kl.*	*Trigonopsis*	*Trig.*
Kluyveromyces	*Kluyv.*	*Zygosaccharomyces*	*Zygosacch.*

The abbreviations used for chemical compounds are those recommended by the *Biochemical Journal* (1967; **102**, 1). Enzymes are referred to by the trivial names recommended by the "Report of the Commission on Enzymes of the International Union of Biochemistry" (1961, Pergamon Press, Oxford). All temperatures recorded in this book are in degrees Centigrade.

Chapter 1

Introduction

J. S. HARRISON AND A. H. ROSE

*Research Department, Distillers Company (Yeast) Ltd., Epsom, England,
and School of Biological Sciences, Bath University of Technology, Bath,
England*

The yeasts are, without doubt, both quantitatively and economically the most important group of micro-organisms commercially exploited by Man. The third volume of this treatise is concerned with their industrial applications, the chief of which are baking, feeding of humans and animals, and the production of ethanol for potable, power, solvent and synthetic uses. The technology is approached primarily from the viewpoint of the yeast as a living organism.

The total amount of yeast produced annually, including that formed during brewing and in distillery practice, is of the order of a million tons (World-Wide Survey of Fermentation Industries, 1966), and ethanol made for all purposes by fermentation processes involving yeast totals at least one and a half million tons. The benefit to national exchequers is counted in thousands of millions of dollars. In the United Kingdom alone, the excise duty amounts to more than four hundred million pounds sterling per annum (Anon., 1968a, b).

The industrial uses of yeast can be classified into three groups according to the relation of the product to the biochemistry of the organism, namely: (*a*) cell constituents; (*b*) excretion products; (*c*) compounds produced by the action of cellular enzymes on specific substrates. The first group can be subdivided into: (i) dry whole cells, which have a composition broadly similar to all other living matter and may therefore be used as a food supplement; (ii) lipids and macromolecular constituents such as proteins, enzymes and nucleic acids; (iii) extractable compounds, including coenzymes and vitamins; (iv) breakdown products, for example amino acids formed by the hydrolysis of proteins, and purines and pyrimidines from nucleic acids. Most commercial applications fall in Group *b*, comprising excreted compounds such as ethanol, glycerol and carbon dioxide, which is used both as a saleable pure product in gaseous, liquid or solid form and as the active agent in baking. Typical examples from Group *c* are the synthesis of ephedrine via benzaldehyde, and the formation of thiamine from its thiazole and pyrimidine moieties (Chapter 10, p. 529).

The extensive use still made of yeasts for large-scale manufacturing operations is particularly interesting since other microbiological processes, such as the production of acetone and butanol by species of *Clostridium* (Wilkinson and Rose, 1963), have given place to synthetic chemical methods. The reasons for this are two-fold: (i) the efficiency of the process by which ethanol can be produced from cheap or waste raw materials such as sulphite liquor from the paper-making industry; and (ii) the widespread insistence on fermentation spirit for pharmaceutical and perfumery applications. The only field in which yeast is being displaced is the production of industrial spirit, although fermentation methods are still used in countries where suitable raw materials are readily available, while the technological climate is not suitable for the operation of complex chemical plant.

The two main yeast genera of commercial interest are *Saccharomyces* and *Candida*, especially *Saccharomyces cerevisiae* and closely related species, which are intimately involved in baking and alcohol production, and have the special virtue of possessing particularly efficient aerobic and anaerobic metabolic capabilities. So many strains of the same species have been selected, and more recently bred for these purposes, that taxonomy in the classical sense becomes almost meaningless. Strain differentiation is in many cases more important than species classification, and has become a matter of extremely accurate quantitative measurement of a number of parameters, as can be seen by the characterizations listed in modern patents protecting new hybrids. The technical problems met with in this field are well illustrated in the chapter dealing with baker's yeast (p. 349). By careful crossing of suitably chosen parent strains of the same or different species, a few hybrids with small gains in one of more desired functions are obtained. Among the remaining hybrids there is a wide range of lower activities, although all may have the same parents.

As these yeasts are usually maintained as pure strains by vegetative reproduction the mass of the culture, or clone, of which each cell is similar, can be considered at all stages as a continuously growing individual, rather than as a species or race. Mutants, whether naturally occurring or produced intentionally, add to the confusion. The custom has therefore arisen of identifying these new yeast strains by a sequence of letters and numbers which is usually derived from the place of origin followed by the number in the laboratory records. The particular species to which any such yeast belongs would need to be determined by the accepted tests; in the case especially of mutants, it is conceivable that the progeny would be classified differently from the parents (see Kreger-van Rij, Chapter 2, Vol. 1, p. 5). This confusing situation is likely to become increasingly acute with the expansion of research in the field of genetics, and a workable scheme of classification, involving quantitative values of a large number of accurately defined parameters, is badly needed (Wiles, 1954).

In the present volume, a contrast will be seen between the patterns of nomenclature of yeasts used in different industries. Many species, with comparatively few well-defined strains, contribute to the production of wines and cider, for example, whereas most baker's, brewer's and *saké* yeasts in modern use are highly developed strains of a limited number of species. The variation in presentation in the following chapters is a reflection of the essentially different character of the various industries. The production of baker's and food yeasts by highly aerobic methods, and the anaerobic manufacture of beer, *saké* and industrial spirit, have largely lost their early empirical nature and developed into sophisticated systems more akin to chemical engineering processes. The recent introduction of continuous processes has tended to accelerate this change.

While much research has been carried out on other alcoholic beverages such as wine, cider and many of the potable spirits, the complexities of the problems involved are such that the biochemical mechanisms, and therefore the precise methods of control of the production processes, have not been fully worked out. While the proportions of water and ethanol present are vastly greater than those of other compounds, the value and acceptability of the products depend, in the case of almost all alcoholic beverages, on the small residual percentage of other volatiles, sometimes comprising hundreds of compounds each present in extremely low concentration (see Chapter 6, p. 283), mostly produced directly or indirectly through the action of yeast. The organoleptic properties of mixtures of these compounds are still far from completely understood, but active research, often based on gas–liquid chromatographic techniques, is being pursued in many centres. Various aspects of these congeners are discussed in each of the chapters dealing with alcoholic products, and some details of their biosynthesis are given.

The economic importance of yeast in the food and drink industries so outweighs its other applications that the discussion of examples of literally hundreds of other uses which have been recorded in the literature occupies only a small proportion of the present volume. An attempt has been made, however, to indicate the scope of this subject and the theories behind the various ways in which yeasts can usefully be employed. Although many of these uses are still largely of academic interest, the possibility remains that wider industrial applications will be developed in the future. The potential large-scale use of hydrocarbons as the substrate for yeast metabolism is an interesting example of this possibility.

On the reverse side of the picture, yeasts occupy a special position in commercial technology as spoilage organisms. Many species of yeast are involved in this destructive capacity (Chapter 9, p. 463); in certain fields the control of these undesirable types is of major industrial importance.

In many respects this volume, with its accent on small but subtle differences largely within single species, differs from Volumes 1 and 2 which, in the main, deal with a wide variety of species and with general cytological, genetical and biochemical principles and mechanisms. It can be concluded that the competitive application of yeasts in modern technology brings into play a wide range of scientific disciplines, and relies on highly developed and quantitatively precise techniques for research and process control. From a glance at the chapter titles it is evident that the world would be a different, and perhaps unhappier, place without the yeasts; there would be no bread in the usually accepted form, and no wine or its various alternatives.

References

Anon. (1968a). "Board of Trade Report on the Census of Production 1963: 18. Brewing and Malting". Her Majesty's Stationery Office, London.

Anon. (1968b). "Board of Trade Report on the Census of Production 1963: 19. Spirit Distilling and Compounding". Her Majesty's Stationery Office, London.

Wiles, A. B. (1954). *Wallerstein Labs Commun.* **17**, 259–281.

Wilkinson, J. F. and Rose, A. H. (1963). *In* "Biochemistry of Industrial Microorganisms" (C. Rainbow and A. H. Rose, eds.), pp. 379–414. Academic Press, London.

World-Wide Survey of Fermentation Industries 1963. (1966). *Pure appl. Chem.* **13**, 405–417.

Chapter 2

Yeasts in Wine-making

RALPH E. KUNKEE AND MAYNARD A. AMERINE

*Department of Viticulture and Enology, University of California,
Davis, California, U.S.A.*

I. History of Studies of Yeast in Wine-making

Wine-making is one of the oldest examples of a food production industry; the intentional fermentation of grape juice seems to be as ancient as Man's earliest written records. Even though van Leeuwenhoek (1687, pp. 972–973) seems to have seen yeasts in the sediment of wines (Fig. 1), the role of yeasts in wine-making was not recognized until long after this discovery. Prior to the findings of Pasteur, processes using yeasts were empirical. The scientific theories at that time recognized the presence of yeasts in alcoholic fermentation, but the yeasts were considered complex and lifeless chemical catalysts. Pasteur (1866, 1873) noted that Fabroni in 1785, Chaptal in 1799, Thenard in

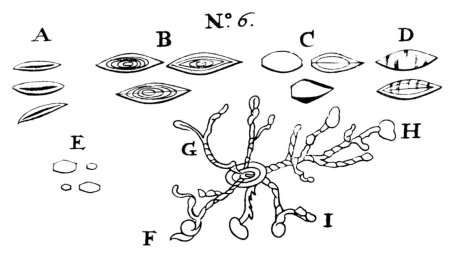

FIG. 1. Drawing by van Leeuwenhoek (1687, Fig. 6) of microscopic view of Hogmer [Hochheimer] wine. Schanderl (1959, p. 14) suggested that *E* are yeast cells, and that *A-D* are tartrate crystals and *F-I* are hyphae of *Botrytis cinerea*.

1803 and Cagniard de Latour and Berzelius in 1836 had suggested vitalistic theories of fermentation, and their suggestions may have contributed to Pasteur's own perception of the importance of yeasts as living organisms in fermentation. Bersch (1878, p. 114) pointed out that Erxleben, as early as 1818, deserves credit for associating yeasts with fermentation.

Pasteur proved yeasts to be the agents of alcoholic fermentation and of certain types of wine spoilage. Pasteur's (1866, 1873) illustrations clearly showed true wine yeasts (Fig. 2); and his text and illustrations indicated that he differentiated the wine yeasts from film-forming yeasts, from acetic acid bacteria and from lactic acid bacteria; the latter he called "new yeasts" (Pasteur, 1858). Pasteur's studies with yeasts in wine have a most important place in the history of microbiology. His demonstration in 1878 (cf. Duclaux, 1896; Dubos, 1960) of the absence of spontaneous fermentation in juices of grapes grown under conditions which prevented the development of yeasts on the skins was dramatic proof of *omnis cellula e cellula*.

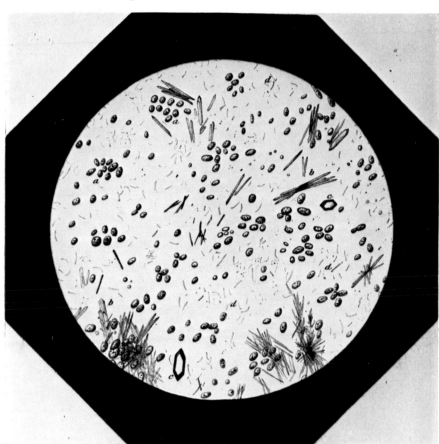

FIG. 2. Drawing by Pasteur (1873, Fig. 10) of microscopic view of spoiled wine, showing yeast of alcoholic fermentation, *a*, as well as tartrate crystals, *b* and *c*. *d* are spoilage bacteria, undoubtedly *Lactobacillus* spp.

In his review of the history of yeasts in wine-making, Schanderl (1959) declined to date precisely the beginning of modern yeast studies as applied in the wine industry. Possibly Pasteur's "Études sur les Vin" in 1866 deserves the honour; although there were a number of important studies before his work, and Pasteur had published some of this work in other forms by 1863. Some would credit Emil Christian

Hansen with the beginning of modern yeast studies. He noted the great differences between yeasts (Hansen, 1881) and introduced the use of pure cultures for beer production (Hansen, 1886, 1888). This marked the final triumph of the recognition of the essential nature of yeasts as fermentation organisms. It is not usually noted that Hansen not only showed that fermenting wort contained many types of yeasts of varying value for beer production, but that he also made useful observations on wine yeasts (Hansen, 1881). He showed that wine yeasts were invariably present in the soil of vineyards. Castelli (1960) considers the work of Pasteur and Hansen to constitute the "first period" of the history of yeasts in fermented beverages.

Hansen's use of pure culture technique in brewing was introduced into the wine industry by Müller-Thurgau between 1880 and 1890 and, accordingly, Castelli (1960) dated this as the beginning of the "second period". However, Bersch (1878, p. 123) used a kind of yeast culture as early as 1878; sound grapes were crushed several days before the main harvest and the juice was allowed to ferment naturally in a warm room. This then, was used as a starter culture. Böhringer (1960) considered Müller-Thurgau, Wortmann, Meissner and Aderhold to be the leaders in the field of the microbiology of wines at the turn of the century. Schanderl (1959, p. 17) specifically dates Müller-Thurgau's report to the wine congress at Worms in 1890 as the beginning of the application of modern yeast studies to the wine industry (see Müller-Thurgau, 1889, for the substance of this report). The first distribution of pure yeast cultures for wine-making was apparently made from Geisenheim-am-Rhein. The famous strain Steinberg 92 was released by Müller-Thurgau in 1890 according to Schmitthenner (1937). This was followed by numerous releases from the same laboratory by Julius Wortmann, starting in 1892 (see also Wortmann's text, 1905). In France, Dubourg (1897) reported that Gayon used pure yeast cultures in Bordeaux as early as 1894. Jacquemin (1893) and Jacquemin and Alliot (1903) also distributed pure yeast cultures in France shortly thereafter. Jacquemin believed that the bouquet of the wine was *solely* due to yeasts, and in particular to those yeasts isolated from premium quality vineyards. To develop special bouquet he used grape leaves and these pure yeast cultures. His cultures were widely distributed from the two institutes he established. At least two of the wine yeast cultures still maintained at the University of California (Champagne and Burgundy strains) were obtained from Jacquemin's Institut la Claire.

One of the objectives of the modern oenologist is to suppress the activity of the undesirable micro-organisms in the fermenting must (i.e. crushed grapes or grape juice) and to encourage the growth of desirable organisms. In a normal fermentation the growth of almost all the undesirable micro-organisms is suppressed, or at least retarded, when the ethanol content reaches 4%. Another factor favouring the oenologists' objectives is the now general practice of the addition of sulphur dioxide prior to fermentation. As early as 1889, Müller-Thurgau in Germany (and later Matinand, 1909, and others in France; and Cruess, 1912, in California) showed that sulphur dioxide severely limited the growth of undesirable micro-organisms—the so-called "wild yeasts". The suppression was particularly noticeable after the initial stages of fermentation when ethanol had been formed. The wild yeasts are primarily of the genera *Candida*, *Brettanomyces*, *Hansenula*, *Kloeckera* and *Pichia*.

Another factor which led to better fermentations in the period 1870 to 1900 was correction of the musts by addition of sugar, reduction of total acidity and increase in pH level. This was particularly important under cold climatic conditions of viticulture, such as occur in Germany, where the pH value of the must often was below 3·0 and the sugar content so low that the new wines had only 7–9% ethanol.

Some of the information supposedly obtained on yeasts during this period around the turn of the century was actually on bacteria; Kunkee (1967) discussed the controversy between Müller-Thurgau, A. Koch and Seifert on the one hand, and Kulisch, Amthor, Wortmann, Schukow and Mestrezat on the other, concerning the causative agents of micro-biological reduction of acidity in wine during secondary fermentations.

The "third period" is represented by studies on the ecology of yeasts and their natural distribution. Castelli (1960) dates this from the work of De'Rossi which started in 1933. However, ecological studies were common prior to this date. In France, Algeria, Tunisia and Romania, the yeast flora had been examined by Descoffre (1904), Ventre (1913), Niteseu (1915) and Musso (1932), and later by Béraud (1937) and Renaud (1941). Both Holm (1908) and Cruess (1918) made extensive isolations in California, and this work was continued by Mrak and McClung (1940). It was during the third period that the fermentative and oxidative biochemical pathways of breakdown of sugars by yeasts were elucidated by enzymologists (cf. Nord and Weiss, 1958). Also important for their time were the discussions on the practical aspects of yeasts in wine-making by Antunes (1916), Meissner (1920), von Babo and Mach (1922–27), Mensio and Forti (1928) and Ventre (1930, 1934).

II. Sources of Information

For important reviews of wine microbiology from the modern point of view see Jörgensen and Hansen (1956), Schanderl (1959), Castelli

(1960), Verona and Florenzano (1956) and Amerine and Kunkee (1968). Other useful information is found in texts on yeasts and wine-making: Amerine (1951), Schanderl (1957), Cook (1958), Böhringer (1960), Ribéreau-Gayon and Peynaud (1960, 1961), Phaff et al. (1966) and Amerine et al. (1967). A popular monograph on the yeasts in wine-making has also appeared (Barnett, 1965).

III. Wine Yeast Species and Strains

A. CHARACTERISTICS

Wine yeasts are those yeasts which function well at the relatively high acidity of grape juice, which are resistant enough to allow the formation of greater than 10% ethanol, and which can adapt to low concentrations of sulphur dioxide added as an antiseptic. It was early established that the primary agent of alcoholic fermentation in wine was *Saccharomyces cerevisiae* Hansen var. *ellipsoideus* (Hansen) Dekker. (Yeast taxonomists now often prefer to classify this yeast simply by its specific name, *Saccharomyces cerevisiae*. In Europe it is frequently classified as *Sacch. vini* or as *Sacch. ellipsoideus*. In this chapter we generally conform to the taxonomic nomenclature used by Lodder and Kreger-van Rij, 1952.) *Saccharomyces cerevisiae* is distinguished from the closely related beer yeast *Sacch. carlsbergensis* by the inability of the former to ferment the sugar melibiose. During fermentation most *Sacch. cerevisiae* var. *ellipsoideus* strains rise to the surface of the fermenting must and thus function as "top yeasts", in contrast to most strains of *Sacch. carlsbergensis* which form no yeast head. *Saccharomyces willianus* is separated from *Sacch. cerevisiae* on the basis of cell size and shape. Other species of *Saccharomyces* which have often been justifiably classified as wine yeasts include *Sacch. fermentati, Sacch. oviformis* and *Sacch. bayanus*.

Böhringer (1960) estimated that more than 1,000 isolates of *Sacch. cerevisiae* var. *ellipsoideus* have been made. Lodder and Kreger-van Rij (1952) reported three groups of the variety based on cell size. The variety can also be divided into strains which are differentiated by a number of other characteristics. For example, strains were recognized by Schanderl (1959) and Böhringer (1960) on the basis of the following characteristics: high ethanol producing ability (18–20% v/v); cold resistance (fermenting at 4°); resistance to sulphate; resistance to ethanol (i.e. ability to start a new fermentation at 8–12% ethanol); resistance to tannin (for red wine production); ability to produce sparkling wine (i.e. ferment wine under pressure at low temperature to dryness and settle rapidly with the formation of no permanent deposit on the glass); film formation; resistance to high concentrations of sugar (i.e. ability to start fermentation at greater than 30% w/v); heat resistance (i.e. ability to ferment at 30–32°); and production of low volatile acidity (mostly accounted for as acetic acid).

Unfortunately, studies on yeast strains have often been conducted under variable and unspecified conditions of must composition. The number of yeast strains, as described above, might be fewer if must composition were held constant, particularly the content of amino acids and vitamins. Furthermore, many studies have been made under laboratory conditions where surface-to-volume ratios and conditions of aeration are quite different from those used in wineries.

Although clear physiological and morphological distinctions between strains of *Sacch. cerevisiae* var. *ellipsoideus* and consistent sensory differences between the products formed by them have not been established, there are many indications of real differences in behaviour of the various strains and in the composition of the wines produced by them. In addition to the possible differences listed above, there are numerous reports on strain variability in formation of fixed acids, esters, higher alcohols, hydrogen sulphide, glycerol and 2,3-butylene glycol. It is appropriate to discuss here a few of the many isolated strains of wine yeast, especially those described or studied recently.

Marques Gomes and Vaz de Oliviera (1963–64) made a detailed study of 14 strains of wine yeasts (all classified as *Sacch. cerevisiae* var. *ellipsoideus*) from the port-producing district of Portugal. Since port wines are fortified with spirits during early stages of fermentation, it would not be expected that the yeasts would have a marked effect on wine quality. This was essentially true. However, one strain produced more volatile acidity and one (recommended) strain produced an exceptionally adherent deposit which resulted in brilliantly clear wines. The differences in other components and characteristics were negligible as far as sensory detection was concerned.

Şeptilici *et al.* (1966) made a special study of various strains of yeasts to isolate those which utilized the greatest amount of organic acids. Besides species of *Schizosaccharomyces*, four strains of *Sacch. cerevisiae* var. *ellipsoideus* were found which fermented organic acids. Surprisingly the report praised species of *Schizosaccharomyces* as having good fermentative properties and of producing well-balanced normal-flavoured wines. Further discussion of yeasts, especially *Schizosaccharomyces* spp., and the utilization of organic acids, is given in Section VII.B (p. 44).

A number of experimenters have tried to use technological characteristics of yeasts (usually by means of fermentation tests) as an aid for strain classification. Minárik *et al.* (1965) and others failed to accomplish this objective. (In the beer industry, however, "tall-tube fermentation tests" are used to compare relative mechanical performances of strains.) As an example of this means of differentiation, one of three strains of *Sacch. cerevisiae* var. *ellipsoideus* contained a pectinase in the work of Corrao (1956). No strains of *Saccharomycodes ludwigii*, *Sacch. rosei*, *Kloeckera apiculata*, *Kl. magna* and *Hanseniaspora guilliermondi* tested had pectolytic activity. Osterwalder (1942a) found difference in the yield of yeast. Among six strains, the yield ranged from 9 to 19 g per litre; low yeast yields are normally preferable.

Gayon and Dubourg (1890) reported yeasts which preferred glucose to fructose, and *vice versa*. They did not identify the yeasts, but one was a "sorte de *Sacch. exiguus*". Peynaud and Domercq (1955) divided yeasts in their collection into three groups: in the first class were those which had fermented 75% of the glucose when only 50% of the fructose had fermented; comprising most species of *Saccharomyces*, *Saccharomycodes ludwigii*, *Kl. africana*, *Brettanomyces* spp. and *Pi. fermentans*. In the second class were those fermenting fructose without attacking more than 50% of the glucose; including *Sacch. acidifaciens*, *Sacch. elegans*, *Sacch. rouxii* and *T. bacillaris*. In the third class were those fermenting the two sugars at about equal rates; comprising *Kl. apiculata*, *Kl. magna*, *H. anomala*, *C. parapsilosis*, *Sacch. veronae*, *Sacch. rosei*, *Sacch. delbrueckii* and *Pi. membranaefaciens*.

The fructophiles were noted to be especially good in fermenting solutions containing large concentrations of sugar. Also, *Sacch. elegans* and *Sacch. acidifaciens* were very resistant to ethanol, and the latter to sulphur dioxide, acetic acid, and concentrated solutions of glycerol and of potassium nitrate. In a study by Requinyi and Soós (1935), two yeast strains, Tokaj 4 and 20, fermented sucrose but did not invert it (presumably because they fermented the inverted sugar as fast as it was formed). Normal yeasts completely invert sucrose in five to eight days, but fermentation sometimes follows much later. Soós and Ásvány (1952) reported numerous biochemical differences between strains which they isolated, most of which were fructophiles. Their strains, with two exceptions, produced hydrogen sulphide, but the amount of hydrogen sulphide produced varied from strain to strain. The strains were *Sacch. cerevisiae* var. *ellipsoideus*, except three which appeared to be *Sacch. oviformis*.

Rankine (1964) also studied hydrogen sulphide formation by ten different strains of wine yeasts. The amounts formed were extremely variable, especially in the presence of elemental sulphur where there was a ten-fold variation, depending on strain. Further discussion of hydrogen sulphide is given in Section VIII.D (p. 49). The first studies on the fermentation of over-sulphited musts dates from Kroemer's (1912a, b) reports. Fermentation of highly sulphited musts had also been studied by Mensio, Baragiola and Godet in Italy and Switzerland.

The place of *Saccharomycodes* in these fermentations was established by Krumbholz and coworkers (Kroemer and Krumbholz, 1931, 1932; Krumbholz, 1931a, b; Karamboloff and Krumbholz, 1932); in general, *Saccharomycodes* yeasts are sulphite-tolerant. Scardovi (1953) found that yeasts resistant to sulphur dioxide had a high glutathione content. Ásvány (1952–57) isolated several Hungarian strains of *Sacch. cerevisiae* var. *ellipsoideus* that were resistant to as much as 1,000 mg sulphur dioxide per litre at pH 3·45.

Rankine (1955a) made a detailed study of ethanol production by many strains of wine yeast. With two strains the highest ethanol yield was produced in aerated cultures with added sugar. In another strain, the highest yield was produced with aerated juice without added sugar. His conclusion was that with musts of high sugar content it was advantageous to aerate the musts. With three wine-grape varieties, two yeast strains gave consistently high ethanol yields in all juice samples, one gave consistently low ethanol yields, and two had varying ethanol yields. Juice samples with equivalent sugar content had varying ethanol yields. Rankine also reported that some yeast strains gave better wine clarity, faster onset of fermentation and improved flavour. Lescure (1956) also reported dependency of ethanol formation on yeast strain. Furthermore, Lescure (1956) found the colour stability of the wine depended to a certain extent on the strain of yeast used for fermentations.

Gray (1941, 1945) and Kunkee and Amerine (1968) reported differences in ethanol tolerance of various strains of wine yeast. Ferreira (1959) studied the temperature response of six strains of yeasts, two of *Sacch. cerevisiae* var. *ellipsoideus*, one each of *Sacch. oviformis*, *Sacch. marxianus*, *Sacch. fragilis* and *Schizosacch. pombe*. For all strains the best temperature for yeast cell growth was 30°. Ethanol production decreased at temperatures above 30°. The highest temperature tolerance was shown by *Schizosacch. pombe* and the least by *Sacch. fragilis*. Volatile acid production decreased at higher temperatures except for *Schizosacch. pombe*. The highest ethanol and lowest volatile acid production was with a Montrachet strain of *Sacch. cerevisiae* var. *ellipsoideus*.

A summary of the physiological characteristics of various yeasts was made by Lafon (1956). The interpretation was made on the basis of Genevois's (1936) equation on the equivalence of glycerol and other secondary products of alcoholic fermentation:

$$5s + 2a + b + 2m + h = \Sigma \leqslant g$$

where (in moles) s = succinic acid,
a = acetic acid,

$$b = \text{2,3-butylene glycol,}$$
$$m = \text{acetoin (acetyl methylcarbinol),}$$
$$h = \text{acetaldehyde,}$$
$$\text{and } g = \text{glycerol.}$$

Lafon determined the strain influence on the relationship between g and Σ. She correctly noted the necessity of always measuring the respiratory and fermentative characteristics of yeasts at the same stage of yeast development. The equation was valid for *Saccharomyces* and *Saccharomycodes* and for certain strains of *Torulospora* [*Saccharomyces*], even in aerated cultures.

Some yeast strains produced more glycerol, succinic acid, 2,3-butylene glycol or acetoin than predicted. Acetoin was only formed in aerated cultures, and 2,3-butylene glycol in non-aerated cultures. Thus, during the last part of a fermentation (when oxygen was nearly absent), acetoin tended to disappear and the 2,3-butylene glycol content increased. The yeast strains varied considerably in this respect. For example, with strains of *Sacch. acidifaciens*, *Sacch. rosei* and *Kl. africana* the content of acetoin and of 2,3-butylene glycol increased in the last stages of fermentation. With strains of *Pichia* spp., *Hansenula* spp., *Kl. apiculata*, *T. bacillaris* and *S. ludwigii* the acetoin content relative to the 2,3-butylene glycol content was higher at the end of the fermentation. However, except for *Kl. apiculata* and *S. ludwigii*, the final 2,3-butylene glycol content was higher than the acetoin at the end of the fermentation. *Brettanomyces vini* formed little of either compound. The intensity of respiration of the yeast strain apparently determined the amounts of these two compounds in the final product. Yeast strains with the highest fermentation capacity produced the most 2,3-butylene glycol, and those with the largest respiratory capacity produced the most acetoin. It is only in the last stages of fermentation that these differences were evident. However, production of acetoin was apparently not due to oxidation of 2,3-butylene glycol. The formulae of Genevois or of Rebelein (1957) are valid if various yeast strains produce the same relative amounts of by-products. Siegel *et al.* (1965) in one study found a low but relatively constant production of glycerol and 2,3-butylene glycol, while with another must there was considerable variation. The results of Siegel *et al.* (1965), given in Table I, are typical.

Rankine (1967a) found that several strains of *Sacch. cerevisiae* produced different amounts of pyruvic acid. This has some significance in practical winery operations, since pyruvic acid binds sulphur dioxide and thus decreases its antiseptic activity. Rankine (1967a) also reported consistent differences in the amount of pyruvic acid produced by strains of *Sacch. cerevisiae, Sacch. cerevisiae* var. *ellipsoideus, Sacch. oviformis* and *Sacch. carlsbergensis.*

In another recent report, Rankine (1967b) showed variation among wine yeasts in the amounts (as much as two-fold) and kinds of higher alcohols (*n*-propanol, isobutanol and amyl alcohols) formed during alcoholic fermentation. The variety of grapes also had an influence on the higher alcohol production, but this influence was not as great as that of the yeast strain. Peynaud and Guimberteau (1962) also reported small variations in fusel oil formation by three wine species of *Saccharomyces.*

TABLE I. *Comparison of End Product Formation by Various Strains of Yeast.*
Reproduced from Siegel *et al.* (1965) by permission of the publisher,
Springer-Verlag

Strain	Glycerol (g/litre)	2,3-Butylene glycol (g/litre)	Ethanol (g/litre)
K 2	3·20	0·423	113·0
Kosterneuberg Auslese	4·41	0·376	110·5
Ay Champagne	3·59	0·296	113·0
Gumpoldskirchner V 89	3·57	0·491	113·0
Stamm 5	3·54	0·441	114·5
Stamm 2	4·27	0·460	124·0
Tokay	2·95	0·394	112·0
Stamm 7	3·44	0·440	115·2
Obsthefe	3·95	0·400	124·7
Riesling	2·98	0·594	113·7
Sulfit	4·23	0·409	113·0
Sherry	4·66	0·455	121·5
Stamm 1	3·15	0·406	112·0
Stamm 3	3·33	0·415	114·5
Stamm 8	3·48	0·365	113·7

Ribéreau-Gayon and Peynaud (1961, p. 496) discussed the effects of various yeast strains on secondary fermentations by bacteria. The relative amounts of micronutrients taken up by the yeasts, depending on strain, and also the excretion of materials inhibitory or stimulatory (Challinor and Rose, 1954) to bacteria by the yeasts would influence the extent and rate of subsequent bacterial growth and fermentation. It may be well to mention here that the antibiotic effect of yeasts on *Escherichia coli* and *Micrococcus* [*Staphylococcus*] *pyogenes* var. *albus haem* was studied by Nyerges (1952–57); some inhibition of the bacteria was observed on the third and fourth day of fermentation. Red wines and Tokay wines of certain years also exhibited some inhibiting effect. Some fungi had an effect on wine yeasts (increasing yeast autolysis), but others had no effect. Very few yeasts had any inhibiting effect on fungi. Some species of *Acetobacter* also possess the ability to inhibit yeast (Marcilla Arrazola *et al.*, 1936; Gilliland and Lacey, 1966). The antagonistic effect of one yeast on another (Schanderl, 1959, pp. 277–282) has not received sufficient attention; however, the effect does exist, particularly between *C. mycoderma* and *Sacch. rouxii*. Zardetto de Toledo and Gonzalves Teixeira (1955) claimed lower volatile acid production when *Sacch. rosei* was used as a starter two to four days before *Sacch. cerevisiae* var. *ellipsoideus* was added.

Nyerges (1963) made a detailed study of 45 wine yeast cultures held for ten years under paraffin oil. She noted some differences in the shape of the cells and in their ability to ferment some rare sugars. However, the ethanol-forming capacity of the strains had not decreased, since the ability to ferment glucose and fructose had not changed.

The relationship of yeast strains to the formation of organic acids has been studied. Most alcoholic fermentations of *Sacch. cerevisiae* var. *ellipsoideus*, *Sacch. oviformis*, *Sacch. chevalieri*, *Sacch. carlsbergensis*, *Sacch. elegans*, *Sacch. rosei*, *S. ludwigii*, *Schizosacch. pombe*, *Pi. fermentans*, *Pi. membranaefaciens*, *H. anomala* or *Kl. apiculata* under anaerobic conditions produce primarily D(−)-lactic acid with only a small amount of the L(+)-isomer. Peynaud *et al.* (1966a, 1967) reported that *Sacch. veronae* produced mainly L(+)-lactic acid; of the 16 different cultures, five produced moderate amounts of L(+)-lactic acid and the remainder produced predominantly or exclusively L(+)-lactic acid. They recommended that this characteristic of *Sacch. veronae* be used as a diagnostic test for the species. Sai (1966) demonstrated differences in citramalic acid (α-methylmalic acid) formation with respiratory deficient strains of *Sacch. carlsbergensis* and *Sacch. cerevisiae*. The citramalic acid content increased parallel to glucose consumption. Normal strains (respiratory sufficient) did not accumulate citramalic acid under aerobic conditions even in the presence of respiratory inhibitors such as antimycin A or acriflavine.

Marques Gomes *et al.* (1961) reported strains of *Sacch. fructuum* that after several transfers on malt became strong maltose fermenters. Other strains did not undergo this transformation, or underwent it only when stored on a malt medium for several months. The fermentation characteristics for other sugars were not altered, so malt agar is acceptable for storing yeast cultures intended for wine-making.

Bashtannayà (1966) called attention to differences in foam production by different strains of yeast. Two strains (T-242 and Novotsimlyaanskei 3) produced little foam. Addition of 100 mg of sulphur dioxide per litre to musts gave moderate foam production independent of the strain of yeast. Previous to this, Amerine *et al.* (1942) noted the dearth of investigations of foam capacity and head retention on wines, in contrast to investigations on beers. They noted greater foam stability with low temperature fermentations, and that saponins did not seem to be important; the effect of yeast strain was not studied.

When considering individual strains of yeast, each of the characteristics mentioned probably has relevance to the wine-maker; either from the influence of these characteristics on the quality of the end product itself, or from their influence on the ease or speed of production. From the phylogenic point of view, Zsolt (1959) believed that the specific characteristics of the present "pure" yeast strains (see Section III.C, p. 20) now used for wine production are probably the result of conscious or unconscious selection by man. Zsolt (1959) reported a pure wine yeast which grew well on wort but produced an undrinkable beer! Ribéreau-Gayon and Peynaud (1960, p. 226) mentioned the isolation by Duclaux (Pasteur's student) in 1887 of a flocculent yeast from a wine of the Champagne region. When this yeast was used for fermentation in other regions, not only was a clear wine obtained, but one in which the special characteristics of wines from the Champagne region were imparted. The sensory evaluation could be questioned.

B. NATURAL FLORA AND ECOLOGY

Oenologists have long been interested in yeast ecology because of the obvious importance of the natural microflora on the extent and nature of the fermentation. Knowledge of yeast flora is also important for tracing the source of spoilage yeasts (see Section X, p. 51) and for providing important information on the relation of climatic conditions to the flora. As a result of these studies, interest has greatly increased in the possible role of mixed yeast cultures in flavour production (see Section III.C, p. 20).

Over 150 species and varieties of yeasts which have been reported on grapes or in musts and wines are listed in the Appendix. Some of these organisms, referred to as rare, were found only in a few instances and should not be considered as being of great importance.

In a table published elsewhere, Amerine and Kunkee (1968, Table I) have summarized the frequency of occurrence of the more common species of yeast found in must and wine. The table was compiled from results of extensive surveys in Italy (Castelli, 1960), Spain (Castelli and Iñigo Leal, 1958), Bordeaux (Domercq, 1956a, b) and Czechoslovakia (Minárik et al., 1960). The major problem of studies on yeast ecology results from the methods used for isolating the microflora. So many different techniques have been used that comparisons of frequency of occurrence must be made with caution. Nevertheless some general conclusions may be drawn.

In the period before full fruit maturity, relatively small numbers of *Saccharomyces* can be found on grapes, especially on grapes from very cool vineyards or from those in new oenological regions. Under very cool climatic conditions, such as occur in Japan, species of moulds tend to predominate (Shimatani and Nagata, 1967). Examination of grape

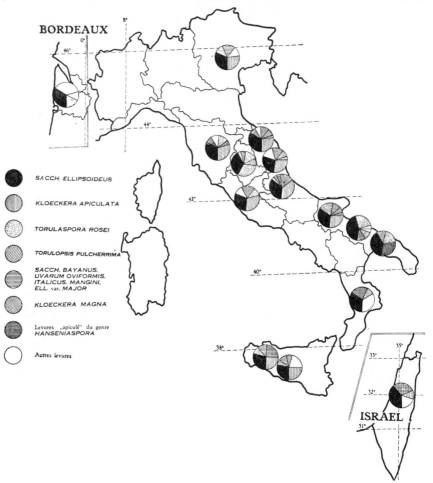

FIG. 3. Distribution of wine yeasts in Israel, Italy and Bordeaux. Reprinted from "Les Levures de la Fermentation du Vin en Israel" (Castelli, 1952) by permission of the publisher, Weizmann Science Press.

flowers (in Brazil, Verona and Zardetto de Toledo, 1954; in South Africa, Van Zyl and Du Plessis, 1961) and of green grapes (in New Zealand, Parle and Di Menna, 1966) revealed no *Saccharomyces*. As the grapes ripen, wine yeasts become more plentiful on the skins, the inoculation possibly coming from the soil via *Drosophila* spp. (Phaff *et al.*, 1966, pp. 115–116) as vectors. In musts, *Sacch. cerevisiae* var. *ellipsoideus* is found in virtually all samples (cf. Amerine and Kunkee, 1968), and it is clearly the predominant organism. *Kloeckera apiculata* seems to be the next most prevalent species. In warmer regions, such as Malta, Israel, Iraq, Spain and southern Italy (Castelli 1954, 1960, 1965, 1967), *Hanseniaspora* spp. are found in large amounts on grapes and in musts. In cooler regions *Kloeckera* spp. tend to predominate. *Saccharomyces pastorianus* also occurs mainly in cool regions. (Castelli 1967, recommended the latter species for sparkling wine fermentations to be conducted at low temperatures.) See Fig. 3 for a summary of yeast distribution in Bordeaux, Italy and Israel.

In addition to the information already mentioned, other reports on floras of grapes, musts and wines have appeared: Benda (1962a), Bréchot *et al.* (1962), Capriotti (1966), Engel (1950), Marcilla Arrazola *et al.* (1936), Minárik (1966, 1967a, b), Zájara Jiménez (1958) and Zardetto de Toledo *et al.* (1959). Castelli (1965) gave a list of over 150 wine yeasts in the collection at the University of Perugia. Other organisms besides yeasts are present on grapes and in musts; see Zhuravleva (1963), Amerine *et al.* (1967, pp. 151–176) and Amerine and Kunkee (1968) for information on moulds and bacteria present.

In studies of ecology of musts and fermentation, there is also the problem of how much of the flora originated on the grapes and how much came from the winery equipment. The studies of Engel (1950), Ribéreau-Gayon and Peynaud (1960, pp. 216–220), Peynaud and Domercq (1959) and Minárik (1967c) indicate that wineries and winery equipment have a rich flora of their own, which undoubtedly influences the microbial flora of musts and wines processed and stored in them. Besides yeasts this flora includes species of *Penicillium, Aspergillus, Oospora, Trichoderma, Citromyces, Oidium* and *Absidia.*

In spontaneous fermentations of must from established vineyards, where no yeasts are added, the usual course of ecological events (Ribéreau-Gayon and Peynaud, 1960, pp. 216–217; Peynaud and Domercq, 1959) is the early growth of and fermentation by *Kl. apiculata* with red grapes or *Kl. apiculata* and *T. bacillaris* with white grapes. As the ethanol concentration reaches 3% or 4% it inhibits the activities of these yeasts, and species of *Saccharomyces* complete the fermentation. Soon after the beginning of fermentation, conditions become more or less anaerobic and oxidative yeasts such as *Cryptococcus* spp. and *Rhodotorula* spp. are thus inhibited. The initial addition of sulphur dioxide inhibits the growth of most other organisms (*Saccharomycodes ludwigii* may be found in wines made from highly sulphited musts). The practice of settling of juice in the cold for white wine production also results in the removal of many organisms. In musts with initially high concentrations of sugar, the fermentation may be completed by *Sacch. oviformis*, many strains of which are more resistant to ethanol than *Sacch. cerevisiae* var. *ellipsoideus* (Peynaud and Domercq, 1959; Martini, 1960; Minárik, 1966).

C. PURE AND MIXED YEAST CULTURES

As already noted, the use of pure yeast cultures of strains of *Sacch. cerevisiae* var. *ellipsoideus* was greatly favoured by some early zymologists; these include, besides Müller-Thurgau and Wortmann, Jacquemin (1893), Jacquemin and Alliott (1903), Osterwalder (1942a), Jörgensen (1900) and Jörgensen and Hansen (1956). The relative value of the use of pure yeast cultures, as contrasted to the use of mixed cultures, either added to the must or naturally supplied from the grapes, is not without dispute. For example, Ribéreau-Gayon and Peynaud (1960, pp. 226 and 509) expressed two views: on the one hand, fine wines are those which they considered to be fermented naturally by yeasts of quality, and "On peut dire que la vinification dans les régions de vins fins n'a jamais bénéficié de l'intervention de levures étrangères au cru". On the other hand there are wines of average quality where one obtains "à l'aide de levures cultivées, de bon résultats ... et un parfum plus agréable de vin jeune".

The relative desirability of addition of yeast to must depends on the amounts and kinds of natural flora present on the grapes. In well established vineyards, especially those whose fruit has been used in production of high quality wines for many years, it might seem reasonable to rely either on the native flora as the agents of fermentation, or on the addition of cultured natural yeast which had been isolated from the vineyard. For example, Michel (1958) reported that such a culture gave better results with the grape varieties Gamay and Chardonnay than either no addition of yeast or addition of another pure yeast culture. However, with new plantings or in new areas, the wine-maker would probably wish to add to the musts yeast which has been shown to have oenologically desirable properties.

It should be mentioned here that the difference in the importance of pure yeast cultures to the beer industry, as compared to the wine industry, lies in the fact that the beer medium is sterilized by heating prior to fermentation and thus contains no viable yeasts. Moreover, the widespread use of sulphur dioxide by the wine industry suppresses the growth of most undesirable yeasts but does permit desirable yeasts to grow. Pure yeast cultures are, therefore, of less critical importance to the wine industry than to the beer industry. The beer industry in the United States is especially yeast conscious, and rigorously controls its use of pure cultures.

1. *Use of Pure Cultures*

The term "pure yeast culture", as used in wine-making, is not precise, since grape must is not sterile before addition of yeast starter. However, it is assumed that the use of sulphur dioxide and large amounts of yeast starter overwhelms the natural flora and results in a fermentation by essentially one kind of yeast.

Most oenologists would agree that the use of pure yeast cultures is generally advisable at the beginning of the harvest season or for special conditions. Schanderl (1959) and Böhringer (1960) indicated that addition of a pure culture of selected strains of *Saccharomyces* spp. is essential when: (i) mouldy fruit and high sulphur dioxide concentrations are used; (ii) fermentation proceeds too slowly or "sticks"; (iii) low fermentation temperatures are used; (iv) must temperature is lowered too rapidly; (v) preparing berry and fruit wines; (vi) sterilized musts are used; (vii) producing sparkling wine. They further noted that addition of pure yeast cultures is strongly recommended for: (i) cold red wine fermentations; (ii) completion of the fermentation of some wines; (iii) fermentation of frozen grapes; (iv) fermentation of fruit musts subject to oxidation and requiring high concentrations of sulphur dioxide. One can easily predict the use of specific pure yeast cultures for other desirable purposes such as: (i) fermentation of high sugar musts; (ii) low or high ethanol yield per gram of sugar fermented; (iii) decrease in production of non-volatile organic acids (see Section VII, p. 42); (iv) production of special flavours (see Section VIII, p. 45); (v) low or high resistance to ethanol; (vi) more or less resistance to sulphur dioxide (Porchet, 1931; Osterwalder, 1942b) or other permissible antiseptics; (vii) ability to ferment various sugars at the same or different rates; (viii) fermentation at high temperatures; (ix) control of fermentation by specific amino acid requirements. Special cultures are also desirable for sparkling wine production (see Section V, p. 38) and for Spanish-type sherry production (Section VI, p. 39).

Probably the original attraction of pure yeast cultures was that the growth of undesirable micro-organisms could be inhibited by conducting the fermentations under strictly anaerobic conditions, a practice directly derived from Pasteur's research. With this method the growth of undesirable film-forming and oxidative yeasts and of acetic acid bacteria was particularly suppressed, and the added yeast brought about a clean fermentation. Pure yeast-culture practices were well established in the best European cellars by 1900. The results of Llinca (1961) are typical: pure yeast cultures gave a much better yield of ethanol; and when selected yeasts from the vineyard were used, the results were better than when selected cultures of foreign yeasts were used.

Pure yeast cultures have immense value to the wine industry. Combined with the inhibitory effects of sulphur dioxide they have made wine spoilage rare. The practice is now common in many parts of the world. For example, Rankine (1955b, 1963) noted the widespread use of pure cultures in Australia. Amerine and Joslyn (1951) and Amerine et al. (1967) favoured pure yeast cultures for the conditions prevalent in California. Nearly all wineries in California use pure yeast cultures. Schulle (1954) favoured the use of pure cultures in Germany but noted that, even when used, the actual fermentation is only relatively pure. Minárik (1963) showed that of 72 yeast cultures distributed in Czechoslovakia, 48 were *Sacch. cerevisiae* var. *ellipsoideus* and 14 were *Sacch. oviformis*. The remainder were *Sacch. carlsbergensis*, *Sacch. heterogenicus*, *Sacch. chevalieri* and *Sacch. globosus* [*delbrueckii*].

An important reason for the use of pure cultures is to produce wine with the most desirable flavours and odours. Malan (1956) tested 127 strains of *Sacch. cerevisiae* var. *ellipsoideus*, 83 of *Kl. apiculata*, 13 of *Sacch. chevalieri*, 11 of *Sacch. oviformis*, six of *C. pulcherrima*, five of *Sacch. rosei* and one each of *Sacch. steineri*, *Sacch. uvarum* and *S. ludwigii* for odour and flavour production. In general, yeast strains producing the highest ethanol concentration resulted in less flavour in the wine. *Saccharomyces rosei*, however, in spite of good ethanol production, resulted in flavourful products but modified the grape flavour. While extensive sensory analysis was used, no statistical analysis of the sensory results was made.

Some other examples of the use of pure cultures can be given. Cold-resistant yeast was developed by Osterwalder (1934) and Porchet (1938). Osterwalder called his strain *Sacch. intermedius* var. *turicensis*. This was similar to his variety *valdensis*. Rzędowska and Rzędowski (1955) and Hulač (1955) recommended certain yeast strains because of their low production of volatile acidity, high production of desirable volatile esters and easier clarification of the product. Sarishvili (1963) reduced the time of fermenting sparkling wines at low temperature in tanks by adapting the yeast to a temperature of 6°. A larger number of yeast cells were produced by the adapted strain compared to the non-adapted, and this seemed to explain the results.

The effect of the size of the yeast inoculum on the quality of wine seems not to have been critically studied. Some of the objection to pure yeast cultures may be due to their being used in excessively large amounts and the resulting high temperatures produced by the too rapid fermentation rate (Saller, 1954).

It is known that, in beer production, strains of yeast suitable for batch fermentation may not be satisfactory for continuous systems. Egamberdiev (1967) reported that the strain Parkentskaya-1 was superior to the usual Rkatziteli-6 (both *Sacch. cerevisiae* var. *ellipsoideus*) for continuous fermentation of musts.

In this chapter we do not discuss brandy or other distilled products of wine (see Chapter 5, p. 283). Special strains, including amino acid-requiring mutants of yeast (cf. Guymon, 1966), have been considered for fermentation of must for subsequent distillation, but in practice ordinary wine yeasts are being used.

We are in agreement with Ribéreau-Gayon and Peynaud (1960, pp. 227–228) that the sensory characteristics imposed on wine by different genera or species of yeast may be extremely varied, but the wine made with the use of different strains of *Sacch. cerevisiae* var. *ellipsoideus* are similar in composition and flavour. It would seem that the variety, the condition of the grapes used as starting materials, and the conditions of the fermentation would be far more influential on quality of wine than the particular strain of *Sacch. cerevisiae* var. *ellipsoideus*.

2. *Use of Mixed Cultures*

Under natural conditions, musts often *seem* to be fermented by a succession of organisms (see Section III.B, p. 17); therefore, one might expect that the use of a proper mixture of yeasts would be beneficial. Saller (1954) suggested that the favourable results sometimes found with mixed cultures result from the slower fermentation rates and the lower temperatures rather than any specific contributions of the strains used.

In natural mixed-culture fermentations, the initial fermentation starts with *Kloeckera* spp. These yeasts have relatively slow growth and fermentation rates and low ethanol tolerances. Because of the slow fermentation and the presence of low concentrations of ethanol, the fermentating must is vulnerable to spoilage by oxidation of ethanol to acetic acid. Undesirable effects of apiculate yeasts in mixed fermentations have been reported (Ribéreau-Gayon and Peynaud, 1960, p. 510; Anon., 1967). However, the desirability of these yeasts has also been noted, particularly in Italy, by De'Rossi (1920), Mensio and Forti (1928), Tarantola (1945–46) and Picci (1955). Fischler (1930) and Schulle (1953a) also found that small amounts of apiculate yeasts may be desirable. Schulle (1953a) noted that the apiculate yeasts have a retarding effect on the growth of *Saccharomyces* when present in large amounts, and a stimulating effect when present in small quantities.

Mixed cultures have been used to minimize volatile acidity. Verona (1952) used *Sacch. cerevisiae* var. *ellipsoides* to start the fermentation, and five days later *Sacch. veronae* to complete it. High volatile acid production by strains of the former, as compared to low production by the combination of yeasts, was the reason for the procedure. Since *Sacch. rosei* produces little volatile acid, Castelli (1967) recommended it for white musts where a neutral white wine is desired. He recommended that, on the fourth or fifth day of fermentation, a pure culture of *Sacch. cerevisiae* var. *ellipsoides* should be added to complete the fermentation.

To achieve higher ethanol yields, mixed cultures were employed by Abramovich (1952). Schulle (1953b) showed that small amounts of *Kloeckeraspora* [*Hanseniaspora*] sp. accelerated the rate of fermentation and increased the ethanol yield, while large amounts retarded the fermentation rate. His strains of *Sacch. cerevisiae* var. *ellipsoideus* could only produce about 9% ethanol, which may account for the results.

Other studies also show favourable results with mixed cultures: Florenzano (1961) increased production of glycerol and 2,3-butylene glycol; Mosiashvili (1958) used ester-producing and pectic-decomposing *Sacch. bailii*, *Sacch. eupagicus* and *Hanseniaspora apiculata*; Kir'yalova and Shklyar (1953) reported improved quality of fruit wines with mixed cultures of *Saccharomyces* sp., *Torulopsis* sp. and *Sacch. cerevisiae* var. *ellipsoideus*; Mosiashvili (1955) used cold-resistant and heat-resistant acclimatized yeasts with normal yeasts to increase the rate of alcoholic fermentation with high-sugar musts and to improve flavour; Sulkhanishvili (1963) obtained high grade white wines by inoculation with *Sacch. oviformis* and *Sacch. chodati* [*steineri*]; Kir'yalova (1955) improved flavour with *Sacch. cerevisiae* var. *ellipsoideus*, *Sacch. apiculatus* (possibly *Candida mycoderma*) and *Torulopsis* sp., but with the notation that the first tended to crowd out the latter because of its ability to resist higher ethanol content. Later, Kir'yalova (1959) found that *Sacch. cerevisiae* var. *ellipsoideus* crowded out *Hanseniaspora apiculata* and *Torulopsis* sp. after 24 hours. With the former, suppression was delayed at lower temperatures. The significance of the sensory differences claimed is not often verifiable.

The problem of suppression of one culture of yeast by another is of major importance in the use of mixed cultures. In experiments involving inoculations with mixed cultures, we have found it extremely difficult, even under the most carefully controlled conditions, to prevent contamination by rapidly growing wild yeasts whenever large volumes of unclarified must were used.

A discussion of mixed cultures involving *Schizosaccharomyces* spp. for decrease of acidity of must is given in Section VII.B (p. 44).

3. *Pure Culture Availability*

In spite of the frowns of some purists, pure yeast cultures are widely available. No less than eight state agencies distribute them in West

Germany. Cultures are available from several institutes, world-wide, including l'Institut Pasteur and the American Type Culture Collection. Several firms in California also provide cultures. Generally the pure cultures are supplied on agar slants, and the yeast should be propagated in sterile medium by the wine-maker. Usually autoclaved or pasteurized grape juice is used, but Osterwalder (1942a) recommended grape juice plus sugar and ammonium sulphate and phosphate as the best practical medium for preparation of pure yeast cultures in Switzerland. Vigorous aeration is used to secure a large yield of yeast. Osterwalder (1942a) found that the yeast yield was greater with his strain at 14°, compared to 24°. De Soto (1955) described a large scale (200 hl) yeast propagator designed for use in a California winery.

The main objections to liquid yeast cultures is the time taken to pre-pare them during the busy vintage season. Usually 1–3% by volume of a liquid starter is required. It is also of concern that the culture be in the active stage of growth when required. Therefore preparation of yeasts before the vintage season has attracted the attention of oenolo-gists.

Storage of yeast in the dry state was recommended by Kir'yalova (1958). She noted that strains varied in their resistance to desiccation. *Torulopsis* sp. and *Hanseniaspora* sp. were less resistant to desiccation than *Saccharomyces* sp. The optimal drying temperature range was 25°–35°.

Freeze drying of yeasts under vacuum (lyophilization) has been recommended by Barini-Banchi (1956) and Ostroukhova (1961). Yeasts with residual moistures of about 2% gave preparations which retained their original properties for a year. Less desirable results were obtained by Pieri and Contini (1955). In Canada frozen yeast starters, as moist cakes or powders, have largely replaced liquid starters accord-ing to Adams (1961). In California, Castor (1953a) experimentally made compressed cake and dry granular preparations of pure yeasts. Thoukis *et al.* (1963) prepared large quantities of yeast under aerobic conditions, centrifuged and chilled the yeast to 4·4° and, by pressing, formed a crumbly cake. The cake was packaged in 1-lb and 25-lb cartons. The active-dry form of the Montrachet strain of *Sacch. cerevisiae* var. *ellipsoideus* and two Champaign strains of *Sacch. bayanus* are now commercially available in the United States. Care must be taken to reconstitute the active-dry yeast in tepid liquid to prevent physical loss of intracellular material before metabolic activity recommences. One big advantage of such products is that the number of viable yeast cells is approximately constant for a given weight. The wine-maker thus can easily control the amount of yeast. Too much yeast should not be used. Böhringer (1960) noted that dried or pressed yeast is not allowed in West Germany. Saller and Stefani (1962) found that the "dry" yeast (pressed baker's yeast) which they tested gave less ethanol and developed an off-flavour. Use of better strains of yeasts might overcome their objections.

Bertin (1927) noted that some French commercial "pure" yeast cultures were in fact mixed cultures with some bacteria present. Wine-makers obviously should inspect cultures obtained on slants before propagation. The pressed cake and active dry forms are not sterile either, but no difficulty in their use has occurred.

IV. Alcoholic Fermentation

A. UNIQUE CHARACTERISTICS IN WINE-MAKING

The biochemical pathways of catabolism of sugars to ethanol and carbon dioxide by yeast is familiar and need not be discussed here (see Volume 2, Chapter 7; also Wood, 1961, and Nord and Weiss, 1958). The metabolic mechanisms for control of the pathways are becoming clearer and better understood. A complete understanding of these mechanisms will undoubtedly have practical importance on the production of wine and other alcoholic beverages. As concerns the alcoholic fermentation, wine-making has some unique characteristics. The starting material contains high concentrations of sugar (glucose and fructose) and acid. Thus, the yeast has to contend with the inhibitory effects of high concentrations of sugar and of ethanol formed from the sugar, as well as from the acidity of the medium and from added antiseptics. In addition, good wine-making practices demand cool temperatures, which also retard fermentation. However, wine yeasts are characterized by their hardiness, and they grow and ferment relatively rapidly in spite of these conditions. One should realize that a great deal, if not most, of the alcoholic fermentation in wine-making occurs after the yeasts have completed their growth phase, and in which very likely their complement of enzymes is more or less established. Thus, the factors which influence growth and yield are as important as those directly affecting fermentation rate.

B. FACTORS INFLUENCING ALCOHOLIC FERMENTATION

1. *Ethanol*

The maximum ethanol concentration that can be produced by yeasts varies from 0 to about 19% (v/v). The latter figure was obtained in "syruped" fermentations, where small amounts of concentrated grape juice were added during the course of the fermentation (Cruess et al., 1916).

The amount of ethanol which can be produced per unit of sugar fermented is also important to the wine-maker. According to the Gay-

Lussac equation, under anaerobic conditions one molecule of sugar is converted to two molecules each of ethanol and carbon dioxide:

$$C_6H_{12}O_6 \rightarrow 2\,C_2H_5OH + 2\,CO_2$$

On a weight basis, 51·1% of the sugar is converted to ethanol and 48·9% to carbon dioxide. This theoretical yield cannot be achieved because of losses in by-products formed, entrainment of ethanol and the utilization of sugar for yeast growth and metabolism (see Chapter 6, p. 283). In practice, yields amount to 90–95% of the theoretical value. The yield varies with yeast strain, must composition, fermentation temperature, amount of mixing (stirring or pumping-over of the fermenting liquid) and the design of the fermenters (particularly the surface-to-volume ratio).

The yield of ethanol per gram of sugar fermented varies during the course of fermentation. According to Tyurina (1960) it is higher in the later stages because intermediates formed in the earlier stages of fermentation are then converted to ethanol. Also, the utilization of sugar for yeast growth has practically ceased.

Ethanol, of course, inhibits fermentation. The inhibition increases with concentration of ethanol and temperature. According to Schanderl (1959, p. 229), at zero concentration of ethanol 15% of the yeasts survive heating for one minute at 50°, whereas at 6% ethanol only 0·28% survive, and at 15% ethanol 0·05% of the cells survive.

Recently Amerine and Kunkee (1965) and Kunkee and Amerine (1968) have re-investigated the concept of Delle (1911) that wines can be stabilized by the proper balance of sugar and ethanol contents. Delle suggested that yeast inhibition could be achieved when the Delle Units (DU) = 80, where:

$$DU = a + 4·5c$$

a is the percentage of sugar (w/v), and

c is the percentage of ethanol (v/v).

The equation is based on the observation that alcoholic fermentation normally does not occur at 18% ethanol or at 80% sugar, thus ethanol has about 4·5 times the inhibitory power of sugar. Amerine and Kunkee (1965) verified that the DU value for stability was approximately 80, but ranged from 75 to 85 depending on when the spirits addition was made; less ethanol was required when added during the early stages of fermentation. Kunkee and Amerine (1968) substantiated this, and also reported that the Tokay strain of wine yeast gave the lowest DU value of the strains they tested. They also found that the clarity of the must, the amount of sulphur dioxide, and the use of fractional fortification had little effect on the minimum DU value. When stored properly, the wines prepared by the Delle procedure were stable with respect to bacteria as well as yeast, and received higher sensory scores than wines prepared by the usual procedure. Since the amount of spirits used is only about half that required in the conventional procedure, the procedure would appear to be economically attractive.

The maximum percentage of ethanol that can be produced by yeast was, as already noted, about 19% in the experiments of Cruess *et al.* (1916) and Cruess and Hohl (1937). These high yields were obtained by adding concentrated grape juice during fermentation and by controlling the temperature. In practice much lower yields are obtained.

The yield varies markedly with the strain of yeast, its adaptation to ethanol, and the phase of growth when it is transferred. Castan (1927) tested 17 strains, probably all *Sacch. cerevisiae* var. *ellipsoideus*, on a must containing 32% sugar. The maximum ethanol produced varied from 11% to 17%; this behaviour is typical.

2. *Sugar*

Almost all strains of wine yeast ferment glucose, fructose, maltose, sucrose and usually galactose. Raffinose is fermented in part and lactose, melibiose, pentoses, dextrins and starch are not fermented at all. Grapes contain primarily the sugars glucose and fructose, in about equal amounts at full maturity. Szabó and Rakcsányi (1937) found that glucose fermented more rapidly when grape musts contained 17–20% total reducing sugar; glucose and fructose fermented about equally readily between 20% and 25%, and fructose more rapidly at higher concentrations. At normal sugar concentration glucose is fermented more rapidly than fructose, and with the glucophilic yeast at a much more rapid rate. However, several fructophilic yeasts (*Sacch. elegans*, for example) are known which ferment fructose more rapidly. Gottschalk (1946) believed this was due to the great permeability of the cell membrane of fructophilic yeasts to fructose. To produce wines with residual sugar (such as Sauternes) and with the greatest intensity of sweetness, one should use glucophilic yeasts; fructose is considerably sweeter than glucose and it would be an advantage in this case to have fructose as the residual sugar.

As the sugar concentration is raised, the rate of fermentation and the maximum amount of alcohol produced decreases. Again there is considerable variation depending on the species and strain and the conditioning of the yeast to grow at high sugar concentrations. At higher sugar concentrations, there is an increased production of acetic acid (Schanderl, 1959, pp. 232–233). This may result from the inhibition of the fermentation rate by the high concentration of sugar. This in turn would make the ethanol present more vulnerable to oxidation to acetic acid (see the discussion on mixed cultures, Section III.C, p. 20).

Sugars are by no means the only carbon source for yeasts. Acetic acid, lactic acid and many other organic compounds, including ethanol, can be utilized by yeasts under aerobic conditions. Direct use of carbon dioxide is also possible.

3. Carbon Dioxide and Pressure

The inhibitory effect of carbon dioxide on yeast growth or fermentation is very small at atmospheric pressure. Ough and Amerine (1966b, 1967a) reported that removal of the major part of the carbon dioxide with nitrogen or air did not speed up the rate of fermentation. The slow fermentation of clarified musts has been attributed to the slower rate of removal of carbon dioxide owing to lack of surfaces on which to collect and escape. More likely, the slow fermentation is due to a diminished concentration of nutrients in clarified musts. The effect may also involve aerobiosis (see below).

However, at pressures greater than atmospheric, the inhibitory effect of carbon dioxide is striking. Schmitthenner (1950) reported that yeast growth ceases at 7·3 atm. pressure. All yeast species tested showed this effect. This is the basis of the Böhi process for preservation of grape juice by storage at 8 atm.

The use of carbon dioxide under pressure is also the basis of the Geiss (1952) gezügelte Gärung. In this system, musts are fermented in pressure tanks. In order to reduce the rate of fermentation (and control the rise in temperature) the tank is periodically closed and the pressure allowed to rise to about 8 atm. At this pressure the rate of fermentation is very slow. When the pressure is released the rate of fermentation increases. The pressure may be allowed to build up several times during the fermentation and thus control the rate of fermentation and the temperature. Another advantage of this system is that apparently less sugar is used in yeast multiplication and the ethanol yield per gram of sugar fermented is greater. Fermentation usually ceases with some residual sugar. The current demand for semi-dry white table wines thus favours this system. It is now widely used in Germany, Australia and South Africa for the production of white wines. Less favourable results have been obtained in California, possibly because of the higher pH values of the musts, which allows some bacterial contamination (see Amerine and Ough, 1957).

Although yeast multiplication is partly inhibited at a few atmospheres pressure, and completely inhibited at about 7 atm. of carbon dioxide (Schmitthenner 1950; Kunkee and Ough, 1966), fermentation does not cease until some higher pressure is reached (Kunkee and Ough, 1966). In practice it is important to note that inhibition of growth of bacteria requires much greater pressures. In fact, Schmitthenner (1950) reported increased lactic acid production, probably by lactic acid bacteria, under carbon dioxide pressure. We have observed similar results in practical laboratory studies. This probably accounts for the less favourable results in California noted in the preceding paragraph.

4. *Oxygen*

Alcoholic fermentation can be thought of as consisting of two different, but overlapping, phases: the aerobic growth phase (respiration) and the anaerobic stationary phase (fermentation). Although Pasteur established "fermentation" as *la vie sans air*, it is customary for oenologists and other industrial microbiologists to refer to all activities of micro-organisms, including respirations, as "fermentation".

In the initial growth, or exponential, phase of alcoholic fermentation, the effect of oxygen is to stimulate yeast growth since the metabolic pathways associated with respiration are also those which provide precursors of the cell constituents. As early as 1895, Wortmann showed that much higher yields of yeast were obtained in aerated cultures. Schanderl (1959, pp. 238–242) noted important differences between yeast species and strains in the effect of aeration. We have noticed that the presence of some air during the beginning of fermentation in nutritionally deficient medium is essential to obtain yeast cells capable of completing the fermentation if the medium contains high concentrations of sugar. Wikén and Richard (1955) showed that very small amounts of oxygen stimulated fermentation by young (growing) yeast cultures but not by old cultures. Ough and Amerine (1966a) reported a delay of two days in the growth of yeast in grape juice stripped of oxygen, but the growth rate after the cell population had reached 10^6/ml was practically the same in the two cases. Presumably the redox potential of the two liquids was the same after the yeast had grown to this concentration.

Musts from freshly-crushed grapes may be saturated with oxygen, but the redox potential of the juice falls rapidly after the beginning of fermentation (Ribéreau-Gayon and Peynaud, 1960, pp. 332–334; Ferenczi, 1961–62). Thus, the second phase of fermentation is the anaerobic conversion of sugar to ethanol and carbon dioxide. The two phases overlap; yeast multiplication continues until about halfway through the fermentation (Ough, 1964), but most of the sugar is converted to ethanol when anaerobiosis is established. Fermentations of must in large open tanks, as on a commercial scale, can be so rapid as to cause a tumultuous rolling agitation of the liquid. (Indeed, the derivation of the English word "yeast" is from the Greek word for "boiled"). This agitation tends to increase the rate of aeration and leads to more rapid fermentation. The stirring may also be important in removing fermentation products from the surface of the yeast cell and in supplying fresh nutrients to the cell.

In spite of the ultimate effect of oxygen in stimulation of rates of fermentation, it is to the wine-maker's advantage to maintain anaerobiosis during the secondary phase. The lack of oxygen will force the yeasts to use intermediates (*viz.* acetaldehyde) of the ethanol pathway as hydrogen acceptors and stimulate ethanol formation. The Pasteur effect (Pasteur, 1861; Dixon, 1937) is the inhibition of carbohydrate utilization by oxygen (or respiration). During the second phase of alcoholic fermentation, the lack of oxygen releases the inhibition and reverses the Pasteur effect. (The reversal of the Pasteur effect should not be confused with the *contre-effet Pasteur* or Crabtree (1929) effect where respiration, or oxygen utilization, is inhibited by high glucose concentration, or high rates of glycolysis, although the mechanisms of the two effects are intimately related.) The stimulation of the utilization of sugar, will in turn also tend to shift (at the level of pyruvic acid) the utilization of glucose toward ethanol formation (Holzer, 1961).

Other advantages of an anaerobic second phase are: (i) decreased production of by-products such as acetaldehyde and acetic acid; (ii) less loss of ethanol; (iii) more glycerol and lactic acid production; (iv) a lowering of the redox potential of the new wine. There is also less nitrogen loss from the must, particularly at low temperatures, in the absence of oxygen. This was true for total, amino and protein nitrogen in the experiments of Nilov and Valuĭko (1958). Some oenologists believe that some of the favourable effects of sulphur dioxide is due, at least in part, to its removal of oxygen.

5. *Acetic Acid*

The repressive effect of acetic acid on yeast growth and on alcoholic fermentation has been known since 1885, according to Schanderl (1959, pp. 264–266). Acetic acid occurs in grape musts only under unusual circumstances, mainly in insect- or mould-damaged grapes, but can be caused by yeasts or by bacteria. The inhibitory effect is noted at about 0·1% acetic acid, according to Böhringer (1960). At about 0·5% acetic acid, alcoholic fermentation may be delayed for several weeks. The inhibitory effect is particularly noted with *Saccharomyces* spp.; with *S. ludwigii* the inhibiting effect is much less. Butyric and propionic acids have an effect similar to acetic acid.

Acetic acid in wine is sometimes produced by spoilage yeasts (Section X, p. 51). There is a wide variation in the amount of acetic acid produced by various strains (Section III.A, p. 10), and many yeasts in the film stage (Section VI, p. 39) utilize acetic acid as a carbon source. Under anaerobic conditions *Sacch. cerevisiae* produces only small amounts of acetic acid. The mechanism for acetic acid formation may be a dismutation of acetaldehyde to give ethanol and acetic acid (Neuberg's third form of fermentation; cf. Wood, 1961), but this is usually expected to occur in alkaline media. *Saccharomyces acidifaciens*, a spoilage yeast (see Chapter 9, p. 463), is often distinguished from *Sacch. cerevisiae* by the ability of the former to produce large amounts of acetic acid. This is probably not a sufficient characteristic for differentiation since a hybrid of the two was found to be an acetate producer (Rosell *et al.*, 1968).

6. *Organic Acids and Hydrogen Ion Concentration*

The high acidity of grapes is a very important characteristic; the acidity inhibits growth of pathogenic and other undesirable organisms in must and wine, and acids are prominent flavour components of wine. The acidity is due mostly to D-tartaric and L-malic acids (Genevois, 1938; Amerine and Winkler, 1942; Kliewer *et al.*, 1967). Citric acid is present only at low concentrations unless it has been added as an acidulating agent; other organic acids are present in trace amounts.

Tartaric acid is practically microbiologically inert. Malic acid can be fermented by some *Schizosaccharomyces* yeasts, but usually only a small portion of malic acid is lost during alcohol fermentation with *Saccharomyces* spp. Further discussion on metabolism of organic acids by yeasts is given in Section VII (p. 42).

The pH value of grape must ordinarily lies between 3 and 4. In this range the pH value has a definite effect on both the growth rate of yeast and the fermentation rate. Ough (1966a) showed that the growth rate of *Sacch. cerevisiae* var. *ellipsoideus* in sulphited, sugared grape juice at pH 3·5 was three-fourths that at pH 4·0, while at pH 3·0 it was less than half the pH 4·0 value. There was also a longer lag period at the lower pH values, part of which was due to the greater inhibitory effect of sulphur dioxide under these conditions. The inhibition of yeast growth is reflected in the rate of fermentation of the whole culture, although the fermentation rate based on yeast concentration is unaffected (Franz, 1961).

7. *Nitrogenous Components*

Grape musts normally contain adequate concentrations of nitrogenous constituents in a readily assimilable form for yeast growth. Murolo (1966) reported that 63–78% of the total nitrogenous material of musts was readily assimilable by several strains of *Sacch. cerevisiae* var. *ellipsoideus* and by one strain of *Sacch. rosei*, although there was some variation between different musts. Numerous experiments have shown that the amount of nitrogen present is adequate for yeast development for several successive fermentations. Some oenologists have recommended addition of ammonium salts to sparkling wine cuvées. This is usually unnecessary, and may lead to very fine yeast deposits which are difficult to remove.

Yeast growth removes some nitrogen, and Mensio (1910) utilized this phenomenon for the development of the Asti *spumante* process for making sweet sparkling wines. In this process the yeasts are allowed to multiply, but before fermentation is fully developed the must is filtered. Successive growth of yeasts and filtration produces a wine of sufficiently low nitrogen content to inhibit yeast growth and fermentation. Amerine *et al.* (1967) suggested that a more rational procedure would be to develop mutant yeast strains which have a requirement for one or more essential amino acids, preferably those that are present in low amounts in musts.

Wine yeasts, apparently, can assimilate all of the common amino acids present in a synthetic mixture except proline, according to Peynaud and Lafon-Lafourcade (1962, 1963). Proline is utilized when present alone; thus its utilization is apparently spared when enough other amino acids are present. Thus, some of the proline (20–30%) in must can be assimilated. Peynaud and Lafon-Lafourcade (1962, 1963) confirmed earlier findings of the direct assimilation of amino acids (Thorne, 1949; Barton-Wright and Thorne, 1949; Thorne, 1950). Often they were able to enrich the intracellular concentration of certain amino acids when the amino acid was used as a nitrogen source. However, they also found that assimilation of an amino acid and its nutritive value, as measured by the weight of yeast formed, were not always the same. The yeast did not assimilate the amino acids according to the normal amino acid complement of the yeast for that particular amino acid. Again, when individual amino acids were used as the sole nitrogen source, the wine yeast acted similarly to brewing yeast (Walters and Thiselton, 1953); although they were able to use proline, they could not use lysine unless other amino acids were present (Peynaud and Lafon-Lafourcade, 1962, 1963).

The composition of by-products formed during anaerobic fermentation remained unchanged when single amino acids or mixtures of acids were used (Peynaud and Lafon-Lafourcade, 1962, 1963). However, the amounts of the various by-products varied over a wide range. Valine, for example, resulted in high acetic acid content, methionine in more succinic acid, arginine in increased 2,3-butylene glycol and acetaldehyde. The influence of branched-chain amino acids on higher alcohol formation is discussed below (Section VIII.B, p. 47). Peynaud and Lafon-Lafourcade (1962, 1963) confirmed that there is a large and specific excretion of amino acids from the yeast cells (completely independent of yeast autolysis).

In a series of papers, Ough and coworkers (Ough, 1964, 1966a, b; Ough and Amerine, 1966b; Ough and Kunkee, 1968) sought to predict fermentation rates from the constitution of grape juice. They found that about 84% of the rate variation could be accounted for by a regression equation which involved only the concentration of total nitrogen and biotin as variables. Bizeau (1963) also found that nitrogen played the most important role in the development of yeasts in must.

8. *Growth Factors*

It is well known that most strains of *Sacch. cerevisiae* var. *ellipsoideus*, *Sacch. oviformis* and other wine yeasts can produce nearly all the growth substances needed for cell division (Wikén and Richard, 1951, 1952). In some of these strains only biotin is required. Many growth substances are found, some in appreciable quantities, in the juice of grapes and other fruits. Castor (1953b) found changes in several B-group vitamins during fermentation. Thiamine, which may be found in concentrations as high as 1 mg/litre, has been the subject of several studies. The disappearance of thiamine during alcoholic fermentation is well known, and also that it is absorbed by yeasts. There are also reports that under anaerobic conditions some yeasts synthesize thiamine. Ournac and Flanzy (1967) reported that when wines were held on the lees, varying amounts of thiamine were released into the wine, sometimes as much or more than was originally present. By this means wines containing 0·2 to 0·5 mg thiamine/litre were produced. If 2 mg per day is the mean daily requirement for humans, one litre of wine represents 10–25% of the need.

Some experimenters have obtained a slight increase in ethanol production when thiamine is added to the must. (The report of da Silva, 1953, that riboflavin also increased the alcohol yield should be confirmed.) Lafon-Lafourcade and Peynaud (1965) considered that the liberation of pyruvic acid indicated that alcoholic fermentation occurs with a slight deficiency of thiamine (thiamine pyrophosphate is a cofactor in the enzymic conversion of pyruvic acid to ethanol). Paronetto (1955) considered thiamine to be a regulator, rather than an accelerator, of fermentation; he found higher amounts (200 mg/litre) decreased the rate of fermentation. Although sulphur dioxide destroys thiamine, Flanzy and Ournac (1963) were not able to re-activate fermentation of desulphited musts by addition of thiamine.

Ough and Kunkee (1968) found that fermentation rates could be related to the natural content of biotin in the must. Biotin is a required growth factor for the yeast, but it was present in much higher amounts than needed. The biotin concentration seemed to be correlated with the overall micronutrient content, and thus was a good index of subsequent fermentation rate. Joslyn (1951) noted that, in the absence of an essential nutrilite, yeasts after a long incubation period would grow slowly. For more rapid growth several nutrilites should be present. Nevertheless, conclusive evidence as to the value of adding growth factors under winery conditions is scarce, and the practical value of using them under commercial conditions has not been clearly demonstrated. Other yeasts may require addition of growth factors. Kusewicz (1965) found that *Schizosacch. pombe* and *Schizosacch.* [*pombe* var.] *acidodevoratus* required nicotinic acid and pantothenic acid and mesoinositol. Pantothenic acid is also required for some *flor* sherry yeast (Castor and Archer, 1957).

There have been some reports of favourable effects of proprietory activators (containing growth factors and nitrogenous constituents). Insufficient data are usually presented as to number of musts tested, the chemical composition of the wine, and especially as to its sensory quality. Kolek (1961) used a preparation of the mycelia of *Aspergillus niger*. It was reported to increase yeast growth and to decrease volatile acid formation. Further data on this subject would be desirable.

9. *Tannins*

There is a need for research on the influence of grape and fruit polyphenols on yeast growth and fermentation. In practice oenologists have recommended addition of tannins, primarily to decrease the growth of undesirable yeasts and bacteria. The different effects noted are probably due to the wide variation in composition of commercial tannins. In general, however, it is true that *Saccharomyces* spp. are relatively more resistant to tannins than *Pichia* spp., *Hanseniaspora* spp. or *Kloeckera* spp. (Schanderl, 1959, p. 263). Nevertheless, even red wines, which contain up to 0·3% tannin, readily support film growth by film-forming yeasts. It is also well known that yeast cells from red wine fermentations are coloured from adsorbed anthocyanins. Schanderl (1959, p. 264) suggested that the addition of small amounts (0·005–0·02%) of tannins to cuvées for sparkling wines was more to improve the character of the lees than for any antiseptic effect.

Šikovec (1966) reported the use of chlorogenic and isochlorogenic acids to stimulate, and gallic, ellagic and caffeic acids to inhibit fermentation. During ageing of wine chlorogenic acid is hydrolyzed to caffeic acid. She suggested that this might explain the cessation of fermentation (or difficulty of renewing fermentation) of bottle-fermented sparkling wines. She also reported that involution forms of yeasts were produced in the presence of tannin, and at the higher ethanol concentrations death of the yeast resulted. There was considerable variability between yeast strains in the extent of these effects.

10. *Particulate Material*

Grape musts for white wine making are often clarified before fermentation by centrifugation, filtration or by being "settled" by storage in the cold. The particulate material thus removed includes solid pieces of grape cell walls, micro-organisms, and elemental sulphur in cases where sulphur dusting has been recent. It is well known that the presence of particulate material increases the rate of fermentation (Schanderl, 1959, p. 234). In fact, musts which have been sterilized by filtration are so free of insoluble material that fermentation of them to completion is usually impossible. For experimental wines made from this kind of juice, Rankine (1963) added sterilized kieselguhr as suspended solids. The effect of the insoluble material may be the attachment of the particles to carbon dioxide gas bubbles. The resulting buoyancy of the solids would then impart an internal mixing of the liquid. We have already discussed the positive influence of agitation and aeration on fermentation rates (Section IV.B.4, p. 30).

11. *Metals*

Grapes and other fruits contain small but adequate amounts of the metals (calcium, cobalt, copper, iron, magnesium, potassium and zinc) needed for yeast growth and alcoholic fermentation. They also contain sufficient phosphorus, sulphur and iodine. Much more common than an inadequate amount of metal ions is the inhibiting effect of excessive amounts.

Schanderl (1938) concluded that excess iron accounted for failure of some sparkling wines to be fermented to completion, particularly when the cuvée contained elemental sulphur or hydrogen sulphide. If the excess iron is removed, and the wine is close-filtered, prompt resumption of fermentation occurs. The close-filtration removes the elemental sulphur and the sulphur-containing aged yeast cells.

Excess copper, at about 9 mg/litre, has a retarding effect on fermentation; this is far higher than normally found in wines. Aluminium likewise may inhibit fermentation (at 25 mg/litre or more). Only wines stored in aluminium containers (particularly if corroded) would contain this concentration. Böhringer (1960) classified iron, aluminium, lead and zinc as slight inhibitors of fermentation, and cadmium, copper, mercury, palladium, osmium and silver as strong retardants.

12. *Temperature*

Temperature is of importance to the oenologist both because of its direct effects on yeast activity and because of its indirect effects on the quality of wine, such as evaporative losses of alcohols and other aromatic materials. We will concern ourselves only with the direct effect on yeast, and restrict the discussion to temperatures ordinarily used in wine-making (10–33°).

The length of the lag period of yeast growth in must has been shown to be inversely proportional to the temperature, within the range given above (Ough and Amerine, 1966a). The coefficient of proportionality would need to be empirically determined for each must. There is surprisingly little information about the effect of temperature on growth rate of yeast in must, though one would expect from other metabolic studies with yeasts that the growth rate would approximately double for every increase of 10°. Ough (1966a) found an increase of 4·7 times in yeast growth rate between 10° and 21°, but an increase in fermentation rate of about two-fold. The relationship between temperature and fermentation rate, which is complicated by other factors such as the composition of the musts, has been the subject of study by Ough (1964, 1966a, b).

In practice, the wine-maker attempts to maintain a constant temperature during fermentation. When the control is not rigid, it is possible for the fermentation to reach temperatures high enough to kill the yeast and stop fermentation. These "stuck" fermentations usually occur, depending on strain of yeast, when the temperature reaches 32° or higher. The lethal effect cannot be attributed to heat alone; Jacob *et al.* (1964) found that a temperature of 55° for one minute was insufficient to kill wine yeasts. Inhibitors present in the fermenting must, for example ethanol, undoubtedly contribute to the heat-killing effect. Some oenologists believe that toxic substances are produced at high temperatures. It is true that it is difficult to referment "stuck" wines when the temperature is reduced to normal and a pure yeast culture is added.

Premium quality wines, especially white wines, are produced at low fermentation temperatures. Many reasons as to the need for this have been given (Amerine *et al.*, 1967, pp. 197–198; Ough and Amerine, 1966a). Part of the effect may be due to the amounts and kinds of byproducts formed by the yeast at various temperatures. For example, lower concentrations of volatile esters, acetaldehyde, and isoamyl and active amyl alcohols were found at lower fermentation temperatures by Ough *et al.* (1966), Ough and Amerine (1967b) and Rankine (1967b). Ferreira (1959) measured the products formed by six yeast strains at various temperatures, but he made intentional changes in temperatures during the fermentations.

V. Sparkling Wine

The problem of yeasts for sparkling wine production deserves special consideration. The secondary fermentation takes place at 10–12% ethanol at increased pressures and at relatively low temperatures (10°) in a wine already deprived, by primary fermentation, ageing and stabilization, of some of its natural growth factors. Yeast for the sparkling wine process must be selected not only for its ability to ferment under these conditions, but also for its ability to settle out rapidly and completely after the secondary fermentation. It is also desirable that the yeasts die or become inactive within a short period of time to prevent further fermentation after the *dosage* (the addition of sugar before final corking).

Ribéreau-Gayon and Peynaud (1960, p. 226) reported that Duclaux isolated a highly flocculent strain of yeast from wine of the Champagne region as early as 1887. Many strains of yeast, often called agglutinating strains, have been isolated and recommended for sparkling wine production. These are often strains of *Sacch. bayanus* or *Sacch. oviformis*. Rankine (1967b) also reported a "champagne" strain of *Sacch. fructuum*. According to Böhringer (1960), six pure yeast cultures are available for sparkling wine production, but apparently a number of other strains can be used. Local strains have been successfully used in Spain, Australia, the Soviet Union and elsewhere.

Most oenologists consider that long ageing on the yeast is necessary for proper development of the sparkling wine bouquet. For example, Datunashvili (1964a, b) reported improved flavour when the wine was left in contact with the yeast. The improvement was greatest when the yeast reached the starvation stage, at which time its alcohol dehydrogenase activity had almost disappeared. He suggested that this stage of starvation led to induction of an enzyme complex in the yeast cell which led to improved wine flavour. Bergner and Wagner (1965) noted that the storage of the wine in the presence of the yeast increased the amino acid content of the wine, particularly of lysine. Alternatively, Rodopulo (1961) reported that yeast autolysis led to enrichment of the sparkling wine in surface-active materials which improved flavour. This occurred to a greater extent in the complete absence of oxygen. In fact, in the Soviet Union, yeast autolysates have been added to sparkling wines to improve their flavour (see Smirnova, 1955, and Nilov and Datunashvili, 1961).

When a continuous system of sparkling wine production is desired, fermentation may have to commence under high pressure. A continuous system of this sort has been successfully operated (Brusilovskii, 1963), and the process has been granted a United States patent (Agabalianz et al., 1962). The system required the continuous addition of large amounts of yeast (the champagne strains Kakhouri-7 and Shampanskaya-7 were used) and fermentation at a relatively high sugar concentration. Kunkee and Ough (1966) found high initial carbon dioxide pressure to be inhibitory to yeast growth, especially at the sugar concentrations used (2·5% glucose). One of the advantages claimed (Agabal'yants and Avakyants, 1966) for the continuous sparkling wine process discussed above is that yeast autolysis occurs in the final tanks. This results in increased enzymic activity and in amounts of surface-active substances, and nitrogenous materials and flavours in the wine. It is interesting that Tyurina (1967) reported that Sacch. oviformis overran Sacch. cerevisiae var. ellipsoideus in the last tanks of a continuous system of sparkling wine production. The Sacch. oviformis increased the acetaldehyde and acetal content, and the resulting wine was considered undesirable.

Schanderl (1965) noted higher production of lipids in sparkling wines which had oxidative flora. If the bottled-fermented bottles were shaken too much (to loosen sediment sticking to the sides of the bottle) or if the wine in the tank process was stirred too much, the fats might separate from the yeast cells and float on the surface. Neither disgorging nor filtration will remove the fat.

VI. Spanish-type Sherry

For a century or more a process of making flor sherry, a type of high-aldehyde low-sugar wine, has developed in the south of Spain in Jerez de la Frontera. A wide variety of types are produced. A similar process is used in the Jura region of France, and was studied by Pasteur. The process involves a film stage of yeast growing on the surface of wines containing about 15% ethanol. The relation of yeast to the production of this type of wine includes: (i) the nature of species and strain of yeast involved; (ii) the conditions for the best growth of yeast and flavour production; (iii) the nature of the products of film-yeast growth; (iv) the substitution of submerged culture or other processes in place of the film culture of yeast.

The special nature of the film-yeast process was first noted by Prostoserdov and Afrikian (1933) and Prostoserdov (1934) in Armenia. Later reports by Schanderl (1936) and Marcilla Arrazola et al. (1936) already identified the yeast as a species of Saccharomyces. The history of these studies may be noted in the references above and in Niehaus

(1937), Cruess *et al.* (1938), Cruess (1948), Amerine *et al.* (1967), Fornachon (1953a), Joslyn and Amerine (1964), Saenko (1964) and elsewhere.

The early classification of the primary film of the "Jerez" yeast as
a species or variety of *Saccharomyces* has been generally substantiated.
As to which species is involved in each case is still not clear. The name
Sacch. beticus, proposed in Spain and often accepted in California, has
not been accepted elsewhere; *Sacch. fermentati* appears to be the present
favourite. Lodder and Kreger-van Rij (1952) give *Sacch. beticus* as an
obsolete synonym for *Sacch. fermentati.* Ohara *et al.* (1960) used two
strains of yeast for sherry production, one identified as *Sacch. beticus*
and the other as *Sacch. fermentati.* The former yielded the best quality
product. The yeast identification problem is complicated by the fact
that many species of *Saccharomyces*, as well as pellicle-forming species
of *Pichia*, *Candida* and *Hansenula*, are capable of forming films on
wines. The crucial difference between film yeasts is whether they produce sherry or not. The non-sherry-producing film yeasts grow well, but
generally utilize more ethanol and produce a less desirable product.
Nevertheless, there seems to be no doubt that, under practical plant
conditions, non-sherry-producing film yeasts are present in the film and
are responsible for some of the flavour formation. For example, Zhuravleva and Timuk (1965) found that 88% of the yeasts isolated from
films were species of *Saccharomyces* and that 12% were species of
Pichia. Of the former, 74% were *Sacch. oviformis* and the rest were
Sacch. cerevisiae var. *ellipsoideus.* Shakhsuvarian (1960) also reported
that most of the film-forming yeasts of Uzbekistan were *Sacch. oviformis.*
Strains which fermented galactose were reported to give higher quality
sherry, but further studies on this subject would be desirable. Tsyb
(1966) reported a strain of *Sacch. oviformis* which produced twice as
much aldehyde as one of *Sacch. cerevisiae* var. *ellipsoideus.* However,
one strain of the latter, with aeration, produced a high aldehyde content. According to Saenko and Sakharova (1963) the superiority of
Sacch. oviformis is due to its faster rate of oxidation of ethanol. Abramov
and Kotenko (1966) used specially selected strains of film yeasts,
apparently those of *Sacch. cerevisiae* var. *ellipsoideus.* In Spain, Iñigo
Leal *et al.* (1961) tested *Sacch. beticus*, *Sacch. cheresiensis*, *Sacch.
montuliensis*, *Sacch. rouxii*, *C. mycoderma*, *H. anomala* and *Zygosacch.*
[*Sacch.*] *acidifaciens*, all of which were suitable. In another study, Iñigo
Leal (1958) favoured *Sacch. oviformis.* Some competitive action on
film-formation of one yeast over another was exhibited when the yeasts
were added in stepwise fashion. *Saccharomyces veronae* followed by
Sacch. oviformis, for example, gave no film-formation for 120 days.
Cabezudo *et al.* (1968) used *Sacch. beticus*, *Sacch. cheresiensis*, *Sacch.
rouxii*, *Sacch. montuliensis*, *Sacch. mangini* and *Sacch. oviformis* as

film-formers. All were isolated from films in the sherry or Montilla districts of Spain. Kondo *et al.* (1965) reported that lactic acid bacteria inhibited aldehyde and acetal formation by film yeasts. This result needs further study.

Whatever the strain used it is necessary to acclimatize it to grow at 15–16% ethanol. At ethanol concentrations below 15% there is danger of contamination by acetic acid bacteria. Saenko (1961) found no difficulty in adapting two strains of yeast to form films at 16–17% ethanol by successive transfer to media of higher ethanol content. In addition to the advantage of suppressing *Acetobacter* growth, yeasts adapted to grow at high ethanol concentration produced more acetaldehyde and acetal, both of which are favourable to the quality of sherry-type wines.

Apparently a variety of strains of yeasts will form films and produce acetaldehyde under suitable conditions. However, the best conditions for each yeast must still be determined. Castor and Archer (1957) reported pantothenate to be an essential growth factor for film formation by *Sacch. beticus*, and a fastidious requirement for amino acids was indicated. Freiberg and Cruess (1955) made a similar study on *Sacch. cheresiensis.* Yeast extract, asparagine, aspartic acid, glutamic acid, glycine, peptone, urea, ammonium phosphate and ammonium sulphate had a stimulating effect on the weight of yeast film.

Fornachon (1953a) demonstrated that aeration of yeast cultures led to a rapid accumulation of acetaldehyde. The process was tested in Canada by Crowther and Truscott (1955, 1957) using agitation, in California by Amerine (1958), Ough and Amerine (1960), Ough (1961) and Farafontoff (1964) using agitation and slight or no pressure, and in New York by Lüthi *et al.* (1965) using a laboratory fermenter. A wide variety of similar processes have been tested in the Soviet Union by Preobrazhenskii (1964) and by Martakov *et al.* (1966a), and in Spain by Cabezudo *et al.* (1968). In all these processes, in contrast to the film yeast process used in Spain and other parts of the world, there is a very rapid accumulation of acetaldehyde, sometimes as much as 500–900 mg/litre in 10 days. According to Amerine and Ough (1964) the accumulation of acetaldehyde is dependent on the $NADH_2/NAD$ ratio which is most favourable during yeast growth.

The flavour of *flor* sherry is, of course, of great importance. Mere accumulation of acetaldehyde does not produce a high quality sherry-type wine. Besides the appearance of acetaldehyde and acetal, there is considerable utilization of acetic acid and to a lesser extent of ethanol. Garrido and Saavedra (1959) noted that 2,3-butylene glycol and acetoin were produced from ethanol, and through pyruvic acid some was converted to the carbohydrate, fat and protein of the yeast. They

(Saavedra and Garrido, 1962) also found formation of oxaloacetic and malic acids.

The volatile compounds have been widely studied. Galetto *et al.* (1966) reported a number of acetaldehyde acetals. Rodopulo *et al.* (1965) identified a variety of aldehydes and esters in film-yeast wines. Some esters and isopentanol and hexanol increased during ageing, which led them to suggest that these compounds were important in sherry flavour. Martakov *et al.* (1966) also considered that ester formation was important in sherry flavour.

Suomalainen and Nykänen (1966a) studied the relative amounts of neutral aroma components and fatty acids in three types of Spanish sherry, namely, amontillado, fino and extra dry fino. The differences were not marked enough to account for the differences in type. Benzyl alcohol was a noteworthy constituent. Fourteen fatty acids were identified. These results with film-yeasts may be compared with the studies of Webb and Kepner (1962) on Australian *flor* sherry and of Webb *et al.* (1964) comparing *flor*, submerged culture and baked sherries.

It is well known that the acetic acid content of wines under a film or in the submerged yeast culture procedure decreases. Cantarelli (1955) proposed a film stage on red wines; see also Cruess and Podgorny (1937) for an earlier similar proposal.

VII. Organic Acid Metabolism

A. ACID FORMATION

The production of organic acids by yeast during alcoholic fermentation is a minor but not a negligible consideration for the wine-maker. Peynaud (1939–40) reported the formation of acetic, citric, lactic and succinic acids. Probably the most important of these, because of its strong odour, is acetic acid. *Saccharomyces cerevisiae* produces large amounts of acetic acid under alkaline conditions, but Joslyn and Dunn (1939) showed that acetic acid formation also occurred at the pH value of wine. The amount formed depended on the redox potential, and greater amounts were found at the beginning of the fermentation. Spoilage yeasts can produce large amounts of acetic acid or ethyl acetate. These yeasts include *Sacch. acidifaciens* and species of *Kloeckera*, *Pichia* and *Brettanomyces*.

Succinic acid, as well as lactic acid, was early discovered as a by-product of alcoholic fermentation (Pasteur, 1872). Both these acids are essentially absent in grapes, but are found in wine. Peynaud (1947) and Thoukis *et al.* (1965) studied the formation of succinic acid, the concentration of which may be 0·1–0·4% in wine. Succinic acid is thought to be produced as part of the tricarboxylic acid cycle, possibly involving the glyoxylic acid shunt. Lactic acid is often found in higher concentrations than succinic acid; but less then 0·1% of the acid is due to lactic acid from alcoholic fermentation by yeast, the rest being formed by bacteria. Ţîrdea (1964) found only 0·005% lactic acid in wine which had been settled and treated with 250 mg sulphur dioxide per litre before inoculation with yeast. Lactic acid formed by yeast can be distinguished from that formed from malic acid by bacteria ("malo-lactic fermentation", Kunkee, 1967) on the basis of optical activity: D(−)-lactic acid is formed by yeast under anaerobic conditions, and L(+)-lactic acid is formed by malo-lactic bacteria from L-malic acid (Bréchot et al., 1966; Peynaud et al., 1966a, b). Small amounts of D-lactic acid may also be formed by certain species of lactic acid bacteria from residual hexoses and pentoses.

Mayer et al. (1964) reported that succinic acid, and to a lesser extent, lactic acid, were produced from malic acid by wine yeasts (four strains were tested). The production of malic and tartaric acids by strains of Sacch. cerevisiae var. ellipsoideus was demonstrated by Drawert et al. (1965a, b). The amount of malic acid formed was related to the nitrogenous composition of the medium, especially to proline and methionine. Bhattacharjee et al. (1968) also reported malic acid formation by Saccharomyces sp.

Growing yeasts release organic acids, particularly malic and citric, into the medium. These can be taken up again and metabolized. Citramalic and dimethylglyceric acid are thus produced (Carles et al., 1961; Sai, 1966). With aspartic acid as the nitrogen source, 7% and 8% of the glucose was converted to citramalic and dimethylglyceric acids, respectively. When ammonium sulphate was the nitrogen source, 2–3% was converted to citramalic acid and 20% to dimethylglyceric acid.

Iñigo Leal and Bravo Abad (1963) studied organic acid production by a number of yeasts. Saccharomyces beticus [fermentati] seemed to be the largest producer of acetic acid under the conditions used. Under film conditions it catabolizes acetic acid. Only C. pulcherrima produced fumaric acid. Ough and Kunkee (1967) reported inhibition of yeast growth and fermentation by citric acid, perhaps due to inhibition of the enzyme phosphofructokinase of the glycolytic pathway (Salas et al., 1965). Other organic acids are present in must and wine, e.g., formic, butyric, propionic and glyoxylic acids. These compounds may have substantial influences on flavour, on secondary bacterial fermentations and on the biochemical reactions of wine ageing. However, except as their concentrations was affected by yeast strain (see Section III.A, p. 10), it would seem that their involvement during alcoholic fermentation is minor.

We have already mentioned the production of pyruvic acid by various strains of yeast. Although it is produced in small amounts, pyruvic acid is important because of its capacity to bind sulphur dioxide (Peynaud and Lafon-Lafourcade, 1966). The effect of the sulphur dioxide formed on subsequent malo-lactic fermentation was studied by Fornachon (Rankine, 1965).

B. DECREASE IN ACID CONTENT

In many regions of the world the climate is not warm enough to ripen grapes properly for wine-making. The musts are, therefore, very high in total acidity, and the resulting wines are unbalanced in their ethanol:acid ratio. To decrease the high acidity, two traditional methods have been: (i) to induce a malo-lactic fermentation by various bacteria in which malic acid is decarboxylated to lactic acid; or (ii) by addition of calcium carbonate or other alkaline material. Neither of these procedures falls within the subject of this chapter.

Biological procedures involving yeasts have been suggested for decreasing the acid content. Kluyver (1914) first reported the conversion of malic acid to ethanol and carbon dioxide by yeast. This conversion would be an effective means of acid reduction. Şeptilici et al. (1966), for example, isolated four strains of Sacch. cerevisiae var. ellipsoideus which decreased the total acidity by about one-third. Peynaud (1938, 1948) reported species of Saccharomyces which reduced the malic acid by 10–24%. Rankine (1966a), using a wider range of yeasts, found that 3–45% of the malic acid disappeared in fermentations with various species of Saccharomyces. Drawert et al. (1967) reported the interesting observation that the malic acid metabolized by Sacch. cerevisiae was nearly completely converted to alcohols: 34% of the malic acid was converted to ethanol, 32% to isobutanol, 30% to amyl alcohol.

Biochemical studies have also been made on the utilization of malic acid by Schizosaccharomyces spp. Mayer and Temperli (1963) found complete conversions of malic acid to carbon dioxide and ethanol under aerobic conditions; Dittrich (1963, 1964) reported that 48–68% of the malic acid was converted to ethanol with very little residual acid. Dittrich (1963) felt that Schizosaccharomyces spp. were inferior to Saccharomyces spp. as fermenting agents for wine-making because of the inferior quality of the product. Gandini and Tarditi (1966), however, recommended Schizosacch. pombe plus a species of Saccharomyces. Their results showed that only 32–44% of the malic acid was removed. Rankine (1966a) reported that Schizosacch. malidevorans removed all the malic acid of musts. This strain, however, produced too much hydrogen sulphide for commercial use. Other workers (Ribéreau-Gayon and Peynaud, 1962; Mayer and Temperli, 1963; Peynaud, 1964; Peynaud and Sudraud, 1964; Benda and Schmitt, 1966) have tried Schizosacch.

pombe in wine-making. It is apparently quickly overrun by *Saccharomyces* spp. in mixed cultures. Bujak and Dabkowski (1961) indicated that the inhibition of *Schizosacch. pombe* by other yeasts depended mainly on the fermentation ability of the competing yeast. Jakubowska and Piątkiewicz (1965) found *Schizosacch.* [*pombe* var.] *acidodevoratus* utilized malic acid better than *Schizosacch. pombe* unless the latter was adapted on a malic acid medium.

Disadvantages of *Schizosaccharomyces* spp. are their high optimum fermentation temperatures and the poor flavour of the products (see, for example, Benda and Schmitt, 1966). The high production of acetoin may also be a disadvantage (Dittrich and Eschenbruch, 1965). Fermentation of sugars is about 50% slower with *Schizosaccharomyces* spp. than with *Saccharomyces* spp.

VIII. Flavour Formation

Although some flavour materials in wine come directly from the grape itself, most arise from the action of the yeast. The main kinds of materials found after fermentation of a simple synthetic medium, as compared to a natural juice, can be amazingly similar (see Suomalainen and Nykänen, 1966a). Other things being equal, one might suppose that flavour of the product could be influenced by the wine yeast used (see Section III, p. 10). Wahab *et al.* (1949) hoped to improve some wines by fermentation with yeast which might produce high concentrations of esters. In tests with many strains of yeast, they found a variation in the amounts of esters formed; however, the authors made no recommendations as to the use of these yeasts for wine production. Castor (1954) attempted to characterize yeast strains by their "flavour profiles" by comparing the amounts of major components formed. Crowther (1953) also reported a variety of flavour and odour differences with various yeasts.

It was not until the advent of gas-liquid chromatography that oenologists were able to detect and identify the aromatic components of wine which are present in minute amounts. Since the development of this technique, many papers on yeasts as flavour-producers have appeared. Using gas chromatography, Drawert and Rapp (1966) and Webb (1967a) found about 120 volatile components in wine, although not all of these need be the result of yeast metabolism. In addition to these compounds, Webb (1967b) listed 25 organic acids. Diacetyl (2,3-butanedione), acetoin, fusel oil, esters and hydrogen sulphide have received special attention, and they will be discussed separately at the end of this section.

According to Suomalainen and Nykänen (1966b) the aroma of

alcoholic beverages consists mainly of alcohols, esters, fatty acids and carbonyl compounds, the esters being the most numerous. A quantitative similarity in the composition of the odour fraction of these products led them to suggest that yeasts and the fermentation process play an important role in the formation of beverage aroma. They analysed the products of baker's yeast fermentation of synthetic media containing only ammonia as a nitrogen source. The distillate contained isoamyl alcohol, 2-phenethyl alcohol, and the ethyl esters of acetic, caproic, caprylic, capric, lauric, palmitic, palmitolic and other acids in various amounts. These acids, as well as isobutyric and isovaleric acids, were found in the free form, in most cases abundantly. Traces of stearic, n-propionic, n-butyric, n-valeric, oleic, linoleic, pelargonic and myristic acids were identified.

Ronkainen *et al.* (1967) reported the following carbonyl compounds produced by fermentation of glucose by baker's yeast: acetaldehyde, propionaldehyde, isobutyraldehyde, butyraldehyde and isovaleraldehyde (3-methylbutyraldehyde). Pyruvic and 2-ketoglutaric acids were predominant keto acids: others present were 2-ketobutyric, 2-ketoisovaleric, 2-ketoisocaproic and a trace of 2-keto-3-methyl valeric acids.

Van Zyl *et al.* (1963) studied 14 different yeasts in an attempt to determine the effect of their fermentation on the composition of the fermented product. With the use of gas chromatography they were not able to make any clear distinctions among the eight species of *Saccharomyces* tested. The most prominent difference among these species was the amount of ethyl octoate formed. The yeasts of the other genera did, however, produce notable amounts of some products; e.g. *H. anomala*, *Kl. apiculata* and *Brettanomyces* sp. produced, respectively, large amounts of ethyl acetate, ethyl butyrate and an unidentified compound. The amount of ethyl acetate formed by *Kl. apiculata* was lower than that produced by some of the *Saccharomyces* species.

Peynaud (1956) noted the generally low rate at which *Saccharomyces* sp., *Torulaspora* sp. and *Torulopsis* sp. produced ethyl acetate, and the high rate of formation by *Kl. apiculata, S. ludwigii,* and especially *Pichia* sp. and *Hansenula* sp. The flavour threshold given for ethyl acetate was 180–200 mg/litre, although experienced wine judges can detect and identify lower concentrations. Other aromatic and flavour materials besides those mentioned above are produced by yeast under oxidative conditions during Spanish-type sherry production. These compounds have been discussed in Section VI (p. 39).

A. DIACETYL AND ACETOIN

Diacetyl and acetoin are closely related compounds; both are formed from pyruvate. Acetoin is relatively odourless and tasteless, but its general sensory effect can be pleasant (Niven, 1952); however, diacetyl has a potent odour often associated with cheese or sauerkraut. Diacetyl has usually been measured in combination with acetoin, and this method of analysis has led to confusion in the understanding of the origin and the control of diacetyl and acetoin formation. Speckman and Collins (1968a) and Chuang and Collins (1968) have recently shown in yeast and bacteria that diacetyl is formed by a pathway different from that for acetoin, and that the amounts of the two formed are not necessarily related. Therefore, much of the literature information on diacetyl formation in wine would seem to be of dubious value. However, the procedure mentioned above (Speckman and Collins, 1968b) for the determination of diacetyl may not be as sensitive as oenologists would wish. It is clear that the bulk of the acetoin plus diacetyl found in wine is produced by bacteria, and not by yeast. Nevertheless, some is produced by yeast; Radler (1962), Kunkee et al. (1965) and Pilone et al. (1966) all found acetoin plus diacetyl in wines which had not undergone bacterial fermentations. Fornachon and Lloyd (1965) measured diacetyl and acetoin separately in wines with alcoholic fermentation by yeast only. They found a range of 0·1–2·7 mg diacetyl and 0·7–8·6 mg acetoin per litre. The relationship of yeast strain to formation of acetoin and 2,3-butylene glycol has been discussed in Section III.A (p. 10).

Guymon and Crowell (1965) found the maximum concentration of diacetyl and acetoin midway through alcoholic fermentation. They attributed the loss in the later stages of fermentation to the conversion of these compounds to 2,3-butylene glycol.

B. FUSEL OIL

Gas chromatography has allowed a detailed study of the formation, by yeasts, of higher alcohols, i.e., fusel oil. The main components of fusel oil are usually thought of as n-propanol, isobutanol and isoamyl and active amyl alcohols. They have an unpleasant flavour and odour, but are usually present in wines at concentrations so low that they are not necessarily unfavourable, and may even contribute to the quality of the wine. Their presence in distilled alcoholic beverages such as brandy, where they may become more concentrated, is of greater importance.

It has generally been accepted that the ability to produce fusel oil is about the same for all strains of wine yeast, but the ability in other yeasts varies greatly (Webb and Ingraham, 1963). For example, Guymon et al. (1961a) found a ten-fold variation in fusel oil formation among several genera of yeast, while Peynaud and Guimberteau (1962) found only a two-fold variation among wine species of Saccharomyces.

However, in the study by Van Zyl *et al.* (1963) the ranges in amounts of isobutyl and isoamyl alcohols found were about the same for the eight strains of *Saccharomyces* wine yeasts as the ranges for the other six genera tested. Rankine (1967b), in a detailed study of fusel oil formation with 12 strains of wine yeast and four varieties of grapes, found as much as a nine-fold difference, depending on yeast strain, in *n*-propanol formed, and nearly a three-fold difference in the amyl alcohols.

The mechanism of the formation of fusel oil by yeasts is understood and has been reviewed by Webb and Ingraham (1963) (see also Chapter 4, p. 147). The metabolic pathway of formation of these higher alcohols involves, in part, the pathway of formation of branched-chain amino acids. Thus, fusel oil is formed directly from the corresponding branched-chain amino acids by transamination, decarboxylation and reduction as well as from glucose. Drawert *et al.* (1967) have shown that fusel oil is also produced from malic acid. Ingraham and coworkers (Ingraham and Guymon, 1960; Ingraham *et al.*, 1961) obtained unusually low amounts of isobutanol, active and isoamyl alcohols when they used mutant strains of *Sacch. cerevisiae* which had nutritional requirements for valine, isoleucine and leucine, respectively. Their results confirmed the experiments of Thoukis (1958a, 1958b) which showed that fusel oil is formed from sugar. The mutants used by Ingraham and coworkers grew and fermented much more slowly than normal wine yeast, and would probably not be useful for wine-making.

While the biochemical pathways for fusel oil formation are known, the metabolic control mechanisms of the pathways have not been elucidated. The substrate itself has an influence on the amounts and kinds of higher alcohols formed (Rankine, 1967b), and so has the redox potential; aeration favours the formation of higher alcohols (Guymon *et al.*, 1961b). Kunkee *et al.* (1967) obtained inhibition of isobutanol formation by mixtures of leucine, isoleucine and valine, presumably by multivalent repression or concerted feed-back inhibition mechanisms.

Other factors which influence fusel oil formation are the amounts of ammonia and amino acids (Ehrlich, 1907; Peynaud and Guimberteau, 1962), temperature (Ough *et al.*, 1966a; Rankine, 1967b) and insoluble solids (Dietrich and Klammerth, 1941; Crowell and Guymon, 1963). Studies on the formation of higher alcohols have also been made with cell-free extracts of yeast by Yoshizawa (1963) and Kunkee *et al.* (1966).

2-Phenethyl alcohol has low volatility and is usually not found in distilled spirits, and therefore it is not classified as a component of fusel oil. However, Äyräpää (1963) has shown the mechanism of its formation to be analogous to that of the branched-chain higher alcohols. 2-Phenethyl alcohol, which has a strong rose-like odour, is found in wine at concentrations as high as 50 mg/litre (Äyräpää, 1962; Usseglio-Tomasset, 1967). More attention to this compound is expected in the future because of its intoxicating properties (Wallgren et al., 1963) and its effect on cell permeability (Silver and Wendt, 1967).

C. VOLATILE ESTERS

In a discussion of volatile compounds in wine, Webb (1967a) suggested that the importance of esters as aroma materials in wine has probably been exaggerated. Of course ethyl acetate is easily detected by sensory means, but there is little knowledge on the importance of other esters as flavour components of wine. The esters were once thought to be formed by direct esterification, but it has been pointed out (Nordström, 1963) that the non-enzymic reaction is much too slow. In a series of papers, Nordström presented results (Nordström, 1963, 1964a, b, c) implicating yeast as the major agent in ester formation. The composition of the volatile ester fraction depended on the growth of the yeast, the kind of yeast and the conditions of fermentation. He showed that the ester formation involved reaction of alcohols in the medium with various acyl groups activated by coenzyme A.

D. HYDROGEN SULPHIDE

Hydrogen sulphide has an obnoxious odour, and its presence in more than traces in wine is certainly objectionable. Macher (1952) showed that the amount of hydrogen sulphide formed by yeast during alcoholic fermentation depended on the strain of yeast and the temperature of fermentation. Zambonelli (1965) also found differences in yeast strains in this respect, and that the differences were genetic in character; mercury-resistant strains were capable of reducing sulphate to hydrogen sulphide but the other strains were not. Rankine (1964) demonstrated the differences in sulphide production between yeast cultures. In a definitive work, he (Rankine, 1963) also confirmed earlier studies by Rentschler (1951) which showed that yeasts could convert elemental sulphur and sulphur dioxide, but not sulphur-containing amino acids, to hydrogen sulphide. Rankine (1963) reported the effects of temperature, pH value, redox potential and the particle size of the sulphur on the formation of hydrogen sulphide. He concluded that the rate of formation depended on fermentation rate; at rapid stages of fermentation there was a greater use of sulphur or sulphur dioxide as hydrogen acceptors. However, R. W. Forman and R. E. Kunkee (unpublished results) compared strains of Sacch. cerevisiae var. ellipsoideus which had equally rapid fermentation rates and observed variation in hydrogen sulphide production. Rankine (1963) also showed that the variety of grapes had little effect, except in that they might be sources of elemental sulphur which has been applied for mildew inhibition. In the absence of elemental sulphur, the duration of storage of wine on the lees had no effect on hydrogen sulphide production. In the light of this information, yeast autolysis during storage on the lees as a cause of hydrogen sulphide formation should be re-examined. The mechanism of formation of hydrogen sulphide seems to involve pantothenic acid (Wainwright, 1961), but it is not related to the resistance of the yeast strain to sulphur dioxide (Rankine, 1963).

IX. Yeast Autolysis

Yeast autolysis influences the development of odours and flavours in wines, and thus is of importance to the oenologist. Joslyn and Vosti (1955) studied the conditions which influence autolysis of yeast, and Joslyn (1955) pointed out the importance of yeast autolysis to wine and beer stability. Nilov and Datunashvili (1961) obtained increases in autolysis by elevated temperatures and increases in ethanol concentrations. They also reported on the influence of pH value, aerobiosis and age.

The effects of autolysis on sensory characteristics are well appreciated. The increase in content of nutrients in wine resulting from yeast autolysis can have an important effect on wine quality by encouragement of secondary bacterial fermentations. Stimulation either of malo-lactic fermentation (Fornachon, 1957) or of bacterial spoilage by *Lactobacillus trichodes* (Fornachon *et al.*, 1949) occurs when wine is left too long on the lees.

Material released from the cell during autolysis has a direct influence on wine quality. Loza (1961) found an improvement in white wine with addition of yeast autolysate and storage at elevated temperature. Deibner and Benard (1963) also used yeast autolysates and heat for changes in sensory characteristics of sweet table wines. Manchev (1962) reported improvement by continual passage of wine through yeasts.

We have already mentioned in Section V (p. 38) the importance of autolysis for flavour of sparkling wine. During ageing of the latter, there is a release of some amino acids from the yeast into the wine. Bergner and Wagner (1965) reported differences in the concentrations of 17 amino acids during ageing of both bottle- and tank-fermented sparkling wines. The amount of amino acids found in the wine increased with increasing duration of storage of the wine in the presence of the yeast. Lysine was found to be released in greatest amounts. However, Popova and Puchkova (1960) found only a few instances where the quality of sparkling wine was improved by yeast autolysate addition and heat treatment.

X. Yeasts as Wine Spoilage Organisms

Wine spoilage can be caused by growth of yeasts in finished wine, by the formation of undesirable by-products during alcoholic fermentation by certain strains of yeasts (Sections III.C, p. 20; and VIII, p. 45), and by the associative growth of yeasts and acetic acid bacteria leading to the production of large amounts of acetic acid (Vaughn, 1938). Sudraud (1967) also reported high volatile acidity in high-sugar musts of the Sauternes region of Bordeaux during the 1966 vintage. Whether this was due to a symbiosis of acetic acid bacteria and yeasts, as in Vaughn's study, is not known.

There is abundant evidence that yeasts can survive alcoholic fermentation and remain viable for many years. Under favourable conditions, especially of temperature, they may later grow and cause turbidity. This is true even though the yeasts may not carry on an alcoholic fermentation. Yeast growth in bottled wines depends on the pH value, ethanol content, vitamins, amino acids and other nutrients, temperature, metals, amounts of sulphur dioxide and of other antiseptics, sugar and oxygen contents and the amount and nature of the yeasts present.

Dubourg (1897) found yeasts in cloudy Bordeaux Sauternes. Gayon and Dubourg (1912) also reported on yeasts as a cause of cloudiness. For early reports of yeasts as a cause of cloudiness in Germany, see Zimmermann (1938a, b). Cruess (1918) gives some very early data for California.

Castor (1952) has summarized the pre-1952 work on yeast cloudiness in wines. He reported that 10^5 yeast cells/ml would cause a haze and 10^6/ml a light cloud. He noted that yeast cells may multiply rapidly so that wines which are brilliantly clear when bottled may become cloudy in a period of two weeks (assuming only 60 viable cells/100 ml in the wine when bottled). The work of Kroemer and Krumbholz (1931, 1932), Krumbholz (1931a, 1931b) and Karamboloff and Krumbholz (1932) indicated that sugar-tolerant yeasts were responsible for spoilage. Baker (1936) isolated, but did not identify, four cultures of yeast from cloudy California white wines. Phaff and Douglas (1944) reported the presence of *Sacch. mellis* in a cloudy California dessert wine.

In California, Scheffer and Mrak (1951) reported *Pi. alcoholphila* and *C. rugosa* and four species of *Saccharomyces* (*cerevisiae*, *chevalieri*, *carlsbergensis* var. *monacensis* and *oviformis*) in yeast-cloudy wines. Several strains were resistant to sulphur dioxide and grew at relatively high concentrations of ethanol (over 18%). Their strain of *Sacch. chevalieri* resembled strains of film-forming yeasts used for Spanish sherry production (Section VI, p. 39). Barret *et al.* (1955) also noted possible dangers from alcohol-tolerant film yeasts, in their case *Sacch. oviformis.*

Schanderl and Draczynski (1952) discovered that *Brettanomyces* could grow and produce masking (adherence of sediment to the sides of the bottle), turbidity and volatile acidity in sparkling wines. In South Africa, Van der Walt and Van Kerken (1958, 1961), Van Zyl (1962) and Van Kerken (1963) reported *Brettanomyces* as a major cause of turbidity in bottled wines. Incidence of *Brettanomyces* contamination was 30 times as great as that reported by Peynaud and Domercq (1956) in Bordeaux. However, *Sacch. oviformis, Sacch. cerevisiae, Sacch. acidifaciens, Pi. membranaefaciens* and *Pi. fermentans* were also found. Mogilyanskï (1955) reported *Sacch. cerevisiae* var. *ellipsoideus, Sacch. pastorianus, S. ludwigii, Torulopsis* sp., *Monilia* sp., *Rh. glutinis, Schizosaccharomyces* sp., *Torulospora* sp., *Debaryomyces* sp., *Apiculatus* [*Kl. apiculatus*] sp. and *Willia* sp. in the precipitated material in bottled wines. *Pichia membranaefaciens* resulted in films and an off-flavour in Rankine's (1966b) study.

The source of yeasts in finished wine is incomplete clarification or secondary infection. Proper clarification, sterile filtration, pasteurization and the use of sulphur dioxide normally prevent problems of yeast cloudiness.

XI. Wine Preservation with Respect to Yeast

In order to prevent growth of undesirable yeasts in musts and wines, or to prevent refermentation of semi-dry wines by wine yeast, winemakers must add antiseptics or employ some physical process to promote biological stability.

A. CHEMICAL ANTISEPTICS

1. *Sulphur Dioxide*

Rarely are wines of any worth made without the use of sulphur dioxide. This antiseptic inhibits nearly all undesirable yeasts and bacteria. Wine yeasts as a class are more resistant to sulphur dioxide than wild yeasts, and they are capable of adapting (acclimatizing) to relatively high concentrations. The mechanics of adaptation has not received as much attention as it deserves. Scardovi (1951a, b, 1953) has shown that, although the enzyme complements of adapted and non-adapted cells seem to be similar, the adapted cells contain higher concentrations of glutathione.

Sulphur dioxide in wine is largely in the bisulphite form, the percentage being dependent on the pH value. The bisulphite form is inhibitory to fermentation because of its ability to combine with acetaldehyde, the normal hydrogen acceptor required for glycolysis. However, the acetaldehyde-bisulphite addition product is itself inhibitory to yeast growth and fermentation, and so are other bisulphite addition compounds (Rehm and Wittmann, 1962, 1963; Rehm et al., 1964, 1965; Wallnöfer and Rehm, 1965). Sulphur dioxide also reduces disulphide bonds of protein; this no doubt accounts for some of its inhibitory activity (Prillinger, 1963). Other information on sulphur dioxide and its use in wine can be found in articles by Kielhöfer (1963), Blouin (1966), Schroeter (1966) and Amerine and Joslyn (1959).

While the use of sulphur dioxide is of prime importance for the production of sound wine, at legal concentrations it in itself will not prevent further growth of some yeasts, especially wine yeast in semi-dry wine. However, in the presence of sulphur dioxide, other antiseptics may be used at relatively lower concentrations.

2. Diethyl Pyrocarbonate

Hennig (1959) showed diethyl pyrocarbonate [$(CH_3CH_2OCO)_2O$] (DEPC, PKE, Baycovin or Piref) to be a potent inhibitor of yeast (see also Mayer and Lüthi, 1960; Ough and Ingraham, 1961; Koch, 1962; Pauli, 1967). DEPC was found as a natural component of sparkling wine by Parfent'ev and Kovalenko (1951), but Kielhöfer and Würdig (1963) were unable to confirm this. DEPC is now widely used in semi-dry wine production to prevent regrowth and fermentation by yeast. It has the advantage of being nearly completely dissipated in wine within a few hours after its addition by its hydrolysis to ethanol and carbon dioxide. This advantage is, of course, coupled with the requirement that the treated wine be bottled rapidly before the activity of the DEPC has disappeared. DEPC also will react to some extent with ethanol in wine. This reaction results in the formation of low but sensory-detectable concentrations of diethyl carbonate [$(CH_3CH_2O)_2CO$] (DEC). The inhibitory activity of DEPC seems to be related to its ability to react with amino, hydroxyl and sulphydryl groups of proteins (Genevois, 1963; Duhm et al., 1966; Pauli, 1967). It has been shown to inhibit several enzymes, including alcohol dehydrogenase (Melamed, 1962; Rosén and Fedorčak, 1966; F. Herwatt, G. Thoukis and M. Ueda, personal communication, 1967). The lethal effect of DEPC is correlated with the concentration of yeast cells present (Van Zyl, 1962; Adams, 1965; Turtura, 1966; Pauli, 1967). A process patent for the use of DEPC and sulphur dioxide in wine production has been issued (Bouthilet, 1965). See Amerine and Kunkee (1968) for more information on the use of DEPC in wine preservation.

3. Other Antiseptics

Sorbic acid (2,4-hexadienoic acid) has the advantage over DEPC in the retention of its inhibitory activity in wine. It, too, is used commercially for the stabilization of semi-dry wine. The mechanism of activity of sorbic acid seems to involve its reaction with thiol compounds (York and Vaughn, 1964). Sorbic acid has no odour or taste; however, upon storage in wine it is transformed into an easily sensory-detectable compound (Peynaud, 1963). The presence of both ethanol and sulphur dioxide contributes to the inhibitory effect of sorbic acid. More information on sorbic acid can be found in Nomoto *et al.* (1955), Auerbach (1959), Bell *et al.* (1959), Ough and Ingraham (1960), Peynaud (1963) and Lück and Neu (1965).

The alkyl esters of *p*-hydroxybenzoic acids, especially the *n*-heptyl ester (WS-7, Staypro), are being successfully used by brewers for the control of yeast and bacteria (Kushida and Taki, 1955; Strandskov *et al.*, 1965). Some experiments by Rice (1968) with mixtures of the homologues of WS-7 in wine were promising. The legal restrictions prohibiting the use of these chemicals in wine may be changed as more information about them becomes available.

B. PHYSICAL STERILIZATION

Semi-dry wines can be rendered stable by the use of heat to kill the yeast (pasteurization) or by removal of the yeast by filtration. Ough and Amerine (1966a) discussed heat requirements for this purpose; however, pasteurization is being used less and less because of the deleterious effect of heat on the quality of the wine.

Membrane filters which allow a relatively high rate of flow and have uniformly small pores are now available for yeast stabilization of wine. The process has the disadvantage of requiring aseptic equipment and conditions after filtration and before bottling. Furthermore, this type of filter tends to clog relatively quickly; the wines must therefore be thoroughly prefiltered before the use of the membrane filters. Amerine and Kunkee (1968) pointed out that little data are available on the results obtained with this kind of stabilization, but there was no doubt that some large wineries, at least in California, have been using the method with confidence for the removal of yeast.

Appendix

YEASTS REPORTED ON GRAPES, IN MUSTS AND IN WINES

Yeasts marked [a] are species or varieties accepted by
Lodder and Kreger-van Rij (1952)

Genus and species	Notes, or where reported
Anthoblastomyces	
campinensis	
cryptococcoides	
saccharophileas	
Brettanomyces	
[a] *bruxellensis* Kufferath et Van Laer	In musts and wines
[a] *bruxellensis* var. *lentus*	Differs in rate of growth
[a] *bruxellensis* var. *non-membranaefaciens*	In a must
[a] *claussenii* Custers	
custersii	In musts and wines
intermedius Krumbholtz et Tauschanoff	In South African wines
italicus (see *Torulopsis bacillaris*)	
[a] *lambicus* Kufferath et Van Laer	In musts and wines
patavinus	In musts and wines
schanderlii Peynaud et Domercq	In South African wines
vini Peynaud et Domercq	In wines
Candida	
[a] *albicans*	
boidinii Ramirez	
[a] *brumptii* Langeron et Guerra	In wines, rare
[a] *catenulata*	From the sherry district of Spain
[a] *guilliermondii* (Castelli) Langeron et Guerra	In a dry table wine and in winery
[a] *guilliermondii* var. *membranaefaciens*	
intermedia var. *ethanophila* Verona et Toledo	
ingens Van der Walt et Van Kerken	
[a] *krusei* (Castelli) Berkhout	Common in musts and in wines
[a] *lipolytica*	
[a] *melinii* Diddens et Lodder	In musts and wines, but rare in Greek and South African musts
[a] *mycoderma* (Reess) Lodder et Kreger-van Rij	Very common
[a] *parapsilosis*	In a cloudy wine and in the winery, in South African and Czechoslovakian musts and wines
[a] *pulcherrima* (Lindner) Windisch	Common in musts and wines; on grapes
[a] *reukaufii*	
[a] *rugosa* (Anderson) Diddens et Lodder	In a California wine; rare

Genus and species	Notes, or where reported
Candida (cont.)	
[a] *scottii*	In Czechoslovakia; on green grapes from New Zealand
[a] *solani* Lodder et Kreger-van Rij	In Greek musts
sorbosa	
[a] *stellatoidea* (Jones et Martin) Lang et Guerra	In South African musts; on grape flowers
[a] *tropicalis* (Castelli) Berkhout	Isolated from a film on grape juice; in Greek musts
[a] *utilis*	In Spanish musts, in cork stopper
vanriji Peynaud (1964)	In grape juice
vinaria Ohara, Nonomura et Yunome	In Czechoslovakia
[a] *zeylanoides* (Castelli) Langeron et Guerra	In wines
Cryptococcus	
[a] *albidus* (Saito) Skinner	In wine, rare
[a] *diffluens*	On green grapes in New Zealand
[a] *laurentii*	In wine, including a fortified wine, but rare
[a] *luteolus*	In a table wine
Debaryomyces	
dekkeri Mrak, Phaff, Vaughn et Hansen	
globosus Klöcker	On grapes and in grape juice
[a] *hansenii* (Zopf) Lodder et Kreger-van Rij	In a must, rare
[a] *kloeckeri* Guilliermond et Péju	Very rare; reported once in California
kursanovi Kudryavtsev	In German musts
[a] *nicotianae*	In musts
[a] *vini* Zimmermann	In a spoiled wine
Endomycopsis	
[a] *lindneri* Saito	On grapes and in must in Brazil
vini Kreger-van Rij (1964)	Isolated in Brazil
Hanseniaspora	
apuliensis	In Italian musts; in wine (Galzy rejects the species)
guilliermondii	
uvarum	
[a] *valbyensis* Klöcker	In wine, especially in warm countries
vinae Van der Walt et Tscheuschner	From vineyard soil
Hansenula	
[a] *anomala* (Hansen) H. et P. Sydow	Common
[a] *saturnus* (Klöcker) H. et P. Sydow	In wine made from late-harvested grapes
[a] *schneggii*	In wine
[a] *suaveolens*	In wine

Genus and species	Notes, or where reported
Hansenula (cont.)	
[a] *subpelliculosa*	In Spanish musts
subpelliculosa var. *jerezana*	From the sherry district of Spain
Hyalodendron	In cellars
Issatchenkia	
orientalis Kudryavtsev	
Kloeckera	
[a] *africana* (Klöcker) Janke	Fairly common on grapes and in musts and wines
[a] *apiculata* (Reess) Janke	Very common
[a] *corticis* (Klöcker) Janke	In musts and wines; rare
[a] *jensenii* (Klöcker) Janke	In wines in Sardinia
[a] *magna* (De'Rossi) Janke	In musts in Italy
Pichia	
alcoholophila	(Possibly same as *Pichia membranaefaciens*)
etchellsii Kreger-van Rij (1964)	
[a] *farinosa* (Lindner) Hansen	In wine
[a] *fermentans* Lodder	On grapes, and in musts and wine
[a] *membranaefaciens* Hansen	Relatively common in musts at start of fermentation or in films or in wine
Rhodotorula	
[a] *aurantiaca* (Saito) Lodder	Very rare
[a] *glutinis* (Fresenius) Harrison	Rarely reported on grapes, musts or wines
[a] *minuta*	From green grapes in New Zealand
[a] *mucilaginosa* (Jörgensen) Harrison	In dry and sweet table wines; in Greek musts
[a] *pullida* Lodder	From southern Italy
[a] *rubra* (Demme) Lodder	From northern Italy
vini	From wine, rare
Saccharomyces	
aceti Santa Maria (1959)	In Spanish wine
[a] *acidifaciens* (Nickerson) Lodder et Kreger-van Rij	In spoiled wines and on grapes, resistant to sulphur dioxide (a fructophile)
[a] *bailii* (Lindner) Guilliermond	In South African and Italian dry table wine
[a] *bayanus* Saccardo	On grapes and in musts and wines, common but low frequency; in sparkling wine
beticus	Flor yeast, probably *Saccharomyces fermentati*
[a] *bisporus* (Naganishi) Lodder et Kreger-van Rij	On grapes; low alcohol yield

Genus and species	Notes, or where reported
Saccharomyces (cont.)	
capensis Van der Walt et Tscheuschner	
[a] *carlsbergensis* Hansen	On grapes and less often in wine (used for lager beer fermentation)
[a] *cerevisiae* Hansen	Probably includes the variety *ellipsoideus*
[a] *cerevisiae* var. *ellipsoideus* (Hansen) Dekker	The classical wine yeast, and possibly the most widely distributed
[a] *chevalieri* Guilliermond	In wines in Africa and Italy and on grapes in Czechoslovakia
coreanus Saito	Rare, from grapes
coreanus var. *armeniensis* Sarukhanyan, Sevoyan, Movesyan et Karpentyan (1965)	In Russian wine
[a] *delbrueckii* Lindner	In beer and wine
[a] *delbrueckii* var. *mongolicus* (Saito) Kreger-van Rij	In musts and wine
[a] *elegans* Lodder et Kreger-van Rij	On grapes and in wine (a fructophile)
elegans var. *intermedia* Verona et Zardetto de Toledo	On grapes in Brazil
eupagycus	
[a] *exiguus* Hansen	Originally found in pressed yeast, later in grape juice and wine
[a] *fermentati* (Saito) Lodder et Kreger-van Rij	(Believed the same as *Saccharomyces beticus*)
[a] *florentinus* (Castelli) Lodder et Kreger-van Rij	In musts and rarely in wines
[a] *fructuum* Lodder et Kreger-van Rij	In musts and in grape juice
globosus Osterwalder	On grapes and grape juice and wine, rare (same as *Saccharomyces delbrueckii*)
[a] *heterogenicus* Osterwalder	On grapes, in musts, and in grape juice but rarely
[a] *italicus* Castelli	On grapes and in grape juice from warm climates
kluyveri	In Greek musts and wines
[a] *lactis*	In the sherry district of Spain; in sparkling wine
[a] *marxianus*	The sporogenous form of *Candida macedoniensis*
[a] *mellis*	In Japanese wines and wineries; from the sherry district of Spain; originally from honey

Genus and species	Notes, or where reported
Saccharomyces (cont.)	
[a] *microellipsodes*	In Greek musts and wines
montuliensis	
[a] *oviformis*　Osterwalder	On grapes, in musts, grape juice and wines; common
oxidans　Santa Maria (1959)	In Spanish wines
[a] *pastorianus*　Hansen	In grapes, in musts and frequently in Loire fermentations
prostoserdovii　Kudryavtsev	
[a] *rosei*　(Guilliermond) Lodder et Kreger-van Rij	Common on grapes and in wines (formerly *Torulaspora*)
[a] *rouxii*　Boutroux	From over-ripe grapes
rouxii var. *jerezana*	From the sherry district of Spain
[a] *rouxii*　var.　*polymorphus*　(Kroemer et Krumbholz) Lodder et Kreger-van Rij	In musts
[a] *steineri*	In wine and on grapes (same as *Saccharomyces italicus*)
transvaalensis　Van der Walt	In Greek musts
unisporus	
[a] *uvarum*　Beijerinck	In musts and frequently in wines
vanudenii　Van der Walt et Nel	
[a] *veronae*　Lodder et Kreger-van Rij	In *Drosophila* and in wines
[a] *willianus*　Saccardo	Found more in beer than wine
Saccharomycodes	
bisporus　Castelli	In Italian wine, probably a variety of the following
[a] *ludwigii*　Hansen	From grape juice and wine; common, resistant to sulphur dioxide
Schizosaccharomyces	
acidodevoratus　Chalenko	
[a] *octosporus*　Beijerinck	Originally found in a must from sulphited currants
[a] *pombe*　Lindner	Utilizes malic acid
pombe var. *acidodevoratus*　Chalenko	
pombe var. *liquefaciens*　Osterwalder	
[a] *versatilis*	Originally isolated from home-canned grape juice
Sphaerulina	
intermixta	
Sporobolomyces　Minárik (1969)	In cork stopper; in film on ageing wine
Torulopsis	
[a] *anomala*	In Czechoslovakian musts and wines

Genus and species	Notes, or where reported
Torulopsis (cont.)	
ᵃ *bacillaris* (Kroemer et Krumbholz) Lodder	On grapes and in wine (a fructophile; synonymous with *Torulopsis stellata*)
behrendi Lodder et Kreger-van Rij	In Greek musts
burgeffiana Benda (1962b)	From grapes and in musts
ᵃ *candida*	In sweet table wine
cantarellii Van der Walt et Van Kerken	From grape musts
capsuligenus Van der Walt et Van Kerken	From a winery culture
ᵃ *colliculosa*	In wine
domercquii Van der Walt et Van Kerken	
ᵃ *famata* (Harrison) Lodder et Kreger-van Rij	In musts and wines
ᵃ *glabrata*	In musts and wines
ᵃ *globosa*	In a bottle storage room
ᵃ *inconspicua*	In winery; in musts and wines
pseudaeria Zsolt (1958)	From vineyard soil
pulcherrima	
stellata	Synonymous with *Torulopsis bacillaris*
vanzylii Van der Walt et Van Kerken	From a refrigerated cellar floor
ᵃ *versatilis*	In Czechoslovakian musts and wines
Trichosporon	
ᵃ *cutaneum*	In musts in Brazil
ᵃ *fermentans*	In musts or on diseased grapes
hellenicum Verona et Picci	In Greek musts
intermedium	In musts
ᵃ *pullulans* (Lindner) Diddens et Lodder	In musts
veronae	In musts

References

Abramov, Sh. A. and Kotenko, S. Ts. (1966). *Vinod. vinogr. SSSR* **26** (6), 18–20.
Abramovich, V. V. (1952). *Vinod. vinogr. SSSR* **12** (2), 14–16.
Adams, A. M. (1961). *Devs ind. Microbiol.* **3**, 341–346.
Adams, A. M. (1965). *Rep. hort. Exp. Stn Prod. Lab. Vineland* 133–145.
Agabal'yants, G. G. and Avakyants, S. P. (1966). *Vinod.vinogr.SSSR* **26** (1), 17–20.
Agabalianz, G. G., Merzhanian, A. A. and Broosilovski, S. A. (1962). U.S. Patent 3,062,656.
Amerine, M. A. (1958). *Appl. Microbiol.* **6**, 160–168.
Amerine, M. A. and Joslyn, M. A. (1951). "Table Wines: the Technology of their Production". University of California Press, Berkeley, California. 2nd Ed. (1969) in press.
Amerine, M. A. and Kunkee, R. E. (1965). *Vitis* **5**, 187–194.
Amerine, M. A. and Kunkee, R. E. (1968). *A. Rev. Microbiol.* **22**, 323–358.

Amerine, M. A. and Ough, C. S. (1957). *Am. J. Enol.* **8**, 18–30.

Amerine, M. A. and Ough, C. S. (1964). *Am. J. Enol. Vitic.* **15**, 23–33.

Amerine, M. A. and Winkler, A. J. (1942). *Proc. Am. Soc. hort. Sci.* **70**, 313–324.

Amerine, M. A., Martini, L. P. and De Mattei, W. (1942). *J. ind. Engng Chem.* **34**, 142–157.

Amerine, M. A., Berg, H. W. and Cruess, W. V. (1967). "The Technology of Wine Making", 2nd Ed. Avi Publ., Westport, Connecticut.

Anon. (1967). *Bull. Off. int. Vin.* **40**, 973–1001.

Antunes, A. A., Jr. (1916). "O Saccharomyces Ellipsoideus". E. E. d'O Debate, Santarem.

Ásvány, Á. (1952–57). *Szölész. Kut. Intéz. Évk.* **11** (2), 187–201.

Auerbach, R. C. (1959). *Wines Vines* **40** (8), 26–28.

Äyräpää, T. (1962). *Nature, Lond.* **194**, 472–473.

Äyräpää, T. (1963). *Proc. Congr. Eur. Brew. Conv.* **1963**, 276–287.

Babo, A., von and Mach, E. (1922–27). "Handbuch der Kellerwirtschaft". Paul Parey, Berlin.

Baker, E. E., Jr. (1936). *Wine Rev.* **4** (7), 16–18.

Barini-Banchi, G. (1956). *Riv. Vitic. Enol.* **9**, 173–181.

Barnett, J. (1965). *In* "Penguin Science Survey 1965 B" (S. A. Barnett and A. McLaren, eds.), pp. 163–181. Penguin Books, Baltimore, Maryland.

Barret, A., Bidan, P. and André, L. (1955). *C. r. hebd. Séanc. Acad. Agric. Fr.* **41**, 426–430.

Barton-Wright, E. C. and Thorne, R. S. W. (1949). *J. Inst. Brew.* **55**, 383–430.

Bashtannaya, I. I. (1966). *Trudȳ moldav. nauchlo-issed. Inst. Sadov.Vinogr.Vinod.* **12**, 83–97.

Bell, T. A., Etchells, J. L. and Borg, A. F. (1959). *J. Bact.* **77**, 573–580.

Benda, I. (1962a). *Bayer. landw. Jb.* **39**, 595–614.

Benda, I. (1962b). *Antonie van Leeuwenhoek* **28**, 208–214.

Benda, I. and Schmitt, A. (1966). *Weinberg Keller* **13**, 239–254.

Béraud, P. (1937). *Arch. Inst. Pasteur Tunis* **26**, 723–727.

Bergner, K. G. and Wagner, H. (1965). *Mitt. Klosterneuburg Ser. A Rebe Wein* **15**, 181–198.

Bersch, J. (1878). "Der Wein und sein Wesen". Alfred Hölder, Vienna.

Bertin, C. (1927). *Annls Falsif. Fraudes* **20**, 279–281.

Bhattacharjee, J. K., Maragoudakis, M. E. and Strassman, M. (1968). *J. Bact.* **95**, 495–497.

Bizeau, C. (1963). *Annls Technol. agric.* **12**, 247–276.

Blouin, J. (1966). *Annls Technol. agric.* **15**, 223–287, 359–401.

Böhringer, P. (1960). *In* "Die Hefen. II. Technologie der Hefen" (F. Reiff, R. Kautzmann, H. Lüers and M. Lindemann, eds.), pp. 157–270. Verlag Hans Carl, Nürnberg.

Bouthilet, R. J. (1965). U.S. Patent. 3,198,636.

Bréchot, P., Chauvet, J. and Girard, H. (1962). *Annls Technol. agric.* **11**, 235–244.

Bréchot, P., Chauvet, J., Croson, M. and Irrmann, R. (1966). *C. r. hebd. Séanc. Acad. Sci., Paris.* **C262**, 1605–1607.

Brusilovskii, S. A. (1963). *Bull. Off. int. Vin.* **36**, 206–217, 319–340.

Bujak, S. and Dabkowski, W. (1961). *Acta microbiol. pol.* **10**, 409–416.

Cabezudo, M. D., Llanguno, C. and Garrido, J. M. (1968). *Am. J. Enol. Vitic.* **19**, 63–69.

Cantarelli, C. (1955). *Biochim. appl.* **2**, 167–190. Also *Riv. Vitic. Enol.* **8**, 221–232 (1955).

Capriotti, A. (1966). *Studi sassar.* **13**, 287–322.

Carles, J., Lamazou-Betbeder, M. and Peck, R. (1961). *Annls Technol. agric.* **10**, 61–71.

Castan, P. (1927). *Landw. Jb. Schweiz* **41**, 311–319.

Castelli, T. (1952). *Bull. Res. Coun. Israel* **1** (4), 1–9.

Castelli, T. (1954). *Arch. Mikrobiol.* **20**, 323–342.

Castelli, T. (1960). "Lieviti e Fermentazione in Enologia". Luigi Scialpi Editore, Rome.

Castelli, T. (1965). *Annali Fac. Agr. Univ. Perugia* **20**, 1–33.

Castelli, T. (1967). *Vini Italia* **9**, 245–246.

Castelli, T. and Iñigo Leal, B. (1958). *Annali Fac. Agr. Univ. Perugia* **13**, 5–30, 186–203.

Castor, J. G. B. (1952). *Proc. Am. Soc. Enol.* **1952**, 139–159.

Castor, J. G. B. (1953a). *Wines Vines* **34** (8), 27; (9), 33.

Castor, J. G. B. (1953b). *Appl. Microbiol.* **1**, 97–102.

Castor, J. G. B. (1954). *Wines Vines* **35** (8), 29–31.

Castor, J. G. B. and Archer, T. E. (1957). *Appl. Microbiol.* **5**, 56–60.

Challinor, S. W. and Rose, A. H. (1954). *Nature, Lond.* **174**, 877–878.

Chuang, L. F. and Collins, E. B. (1968). *J. Bact.* **95**, 2083–2089.

Cook, A. H., Ed. (1958). "The Chemistry and Biology of Yeasts". Academic Press, New York.

Corrao, A. (1956). *Annali Sper. agr.* **11**, 495–504.

Crabtree, H. G. (1929). *Biochem. J.* **23**, 536–545.

Crowell, E. A. and Guymon, J. F. (1963). *Am. J. Enol. Vitic.* **14**, 214–222.

Crowther, R. F. (1953). *Rep. hort. Exp. Stn Vineland* **1951/52**, 80–83.

Crowther, R. F. and Truscott, J. H. L. (1955). *Can. J. agric. Sci.* **35**, 211–212.

Crowther, R. F. and Truscott, J. H. L. (1957). *Am. J. Enol.* **8**, 11–17.

Cruess, W. V. (1912). *J. ind. Engng Chem.* **4**, 581–585.

Cruess, W. V. (1918). *Univ. Calif. Publs agric. Sci.* **4** (1), 1–66.

Cruess, W. V. (1948). *Bull. Calif. agric. Exp. Stn.* **710**, 1–40.

Cruess, W. V. and Hohl, L. A. (1937). *Wine Rev.* **5** (11), 12, 24–25.

Cruess, W. V. and Podgorny, A. (1937). *Fruit Prod. J. Am. Vineg. Ind.* **17**, 4–6.

Cruess, W. V., Brown, E. M. and Flossfeder, F. C. (1916). *J. ind. Engng Chem.* **8**, 1124–1126.

Cruess, W. V., Weast, C. A. and Gilliland, R. (1938). *Fruit Prod. J. Am. Vineg. Ind.* **17**, 229–231, 251.

Datunashvili, E. N. (1964a). *Trudy̆ vses. nauchno-issled. Inst. Vinod. Vinogr. "Magarach"* **13**, 68–77.

Datunashvili, E. N. (1964b). *Vinod. Vinogr. SSSR* **24** (7), 15–18.

Deibner, L. and Benard, P. (1963). *Ind. aliment. agric.* **80** 511–518.

Delle, P. N. (1911). *Odessa. Otch.* "*Vinodiel'cheskoi Stantsii Russiakh*" *Vinogr. Vinod.*" **1908g i 1909g**, 118–160.

De'Rossi, G. (1920). *Staz. sper. agric. ital.* **53**, 233–297.

Descoffre, P. L. (1904). Thesis, Pharmacie, Université Bordeaux.

De Soto, R. T. (1955). *Am. J. Enol.* **6** (3), 26–29.

Dietrich, K. R. and Klammerth, O. (1941). *Z. Spiritusind.* **64**, 160.

Dittrich, H. H. (1963). *Wein-Wiss. Beil. Fachz. Dt. Weinbau* **18**, 392–405, 406–410.

Dittrich, H. H. (1964). *Zentbl. Bakt. ParasitKde Abt. II* **118**, 406–421.

Dittrich, H. H. and Eschenbruch, R. (1965). *Arch. Mikrobiol.* **52**, 345–352.

Dixon, K. C. (1937). *Biol. Rev.* **12**, 431–459.

Domercq, S. (1956a). *Annls Technol. agric.* **6**, 5–58, 139–183.
Domercq, S. (1956b). Doctorate Thesis: University of Bordeaux.
Drawert, F. and Rapp, A. (1966). *Vitis* **5**, 351–376.
Drawert, F., Rapp, A. and Ulrich, W. (1965a). *Náturwissenschaften* **52**, 306.
Drawert, F., Rapp, A. and Ulrich, W. (1965b). *Vitis* **5**, 20–23.
Drawert, F., Rapp, A. and Ullemeyer, H. (1967). *Vitis* **6**, 177–197.
Dubos, R. (1960). "Pasteur and Modern Science", pp. 56–58. Anchor Books, Garden City, New York.
Dubourg, E. (1897). *Revue Vitic.* **8**, 467–472.
Duclaux, E. (1896). "Pasteur, Histoire d'un Esprit", pp. 274–277. Imprimerie Charaire, Sceaux.
Duhm, B., Maul, W., Medenwald, H., Patzschke, K. and Wegner, L. A. (1966). *Z. Lebensmittelunters. u. -Forsch.* **132**, 200–216.
Egamberdiev, N. B. (1967). *Prikl. Biokhim. Mikrobiol.* **3**, 458–463.
Ehrlich, F. (1907). *Ber. dt. chem. Ges.* **40**, 1027–1047.
Engel, F. (1950). *Mitt. VersAnst. GärGew., Wien* **7**, 63–68, 98–100.
Farafontoff, A. (1964). *Am. J. Enol. Vitic.* **15**, 130–133.
Ferenczi, S. (1961–62). *Acta agron. hung.* **11**, 239–264.
Ferreira, J. D. (1959). *Am. J. Enol. Vitic.* **10**, 1–7.
Fischler, F. (1930). *Wein Rebe* **11**, 403.
Flanzy, M. and Ournac, A. (1963). *Annls Technol. agric.* **12**, 65–84.
Florenzano, G. (1961). *Italia vinic. agr.* **51**, 189–192.
Fornachon, J. C. M. (1953a). "Studies on the Sherry Flor". Australian Wine Board, Adelaide, South Australia.
Fornachon, J. C. M. (1953b). *Aust. J. biol. Sci.* **6**, 222–223.
Fornachon, J. C. M. (1957). *Aust. J. appl. Sci.* **8**, 120–129.
Fornachon, J. C. M. and Lloyd, B. (1965). *J. Sci. Fd Agric.* **16**, 710–716.
Fornachon, J. C. M., Douglas, H. C. and Vaughn, R. H. (1949). *Hilgardia* **19**, 129–132.
Franz, B. (1961). *Nahrung* **5**, 457–481.
Freiberg, K. J. and Cruess, W. V. (1955). *Appl. Microbiol.* **3**, 208–212.
Galetto, W. G., Webb, A. D. and Kepner, R. E. (1966). *Am. J. Enol. Vitic.* **17**, 11–19.
Gandini, A. and Tarditi, A. (1966). *Ind. Agric. Florence* **4**, 411–420.
Garrido, J. M. and Saavedra, I. J. (1959). *Revta esp. Fisiol.* **15**, 189–192.
Gayon, U. and Dubourg, E. (1890). *C.r.hebd.Séanc. Acad. Sci., Paris* **110**, 865–868.
Gayon, U. and Dubourg, E. (1912). *Revue Vitic.* **38**, 5–7.
Geiss, W. (1952). "Gezügelte Gärung". Sigurd Horn Verlag, Frankfurt.
Genevois, L. (1936). *Bull. Soc. Chim. biol.* **18**, 295–300.
Genevois, L. (1938). *Revue Vitic.* **88**, 103–125, 382–386.
Genevois, L. (1963). *Annls Technol. agric.* **12** (*numéro hors sér.* 1), 127.
Gilliland, R. B. and Lacey, J. P. (1966). *J. Inst. Brew.* **72**, 291–393.
Gottschalk, A. (1946). *Biochem. J.* **40**, 621–626.
Gray, W. D. (1941). *J. Bact.* **42**, 561–574
Gray, W. D. (1945). *J. Bact.* **49**, 445–452.
Guymon, J. F. (1966). *Devs ind. Microbiol.* **7**, 88–96.
Guymon, J. F. and Crowell, E. A. (1965). *Am. J. Enol. Vitic.* **16**, 85–91.
Guymon, J. F., Ingraham, J. L. and Crowell, E. A. (1961a). *Am. J. Enol. Vitic.* **12**, 60–66.
Guymon, J. F., Ingraham, J. L. and Crowell, E. A. (1961b). *Archs Biochem. Biophys.* **95**, 163–168.

64   RALPH E. KUNKEE AND MAYNARD A. AMERINE

Hansen, E. C. (1881). *Meddr Carlsberg Lab.* **1**, 293–327.
Hansen, E. C. (1886). *Meddr Carlsberg Lab.* **2**, 157–167.
Hansen, E. C. (1888). *Meddr Carlsberg Lab.* **2**, 257–322.
Hennig, K. (1959). *Dt. LebensmittRdsch.* **55**, 297–298.
Holm, H. C. (1908). *Bull. Calif. agric. Exp. Stn* **197**, 169–175.
Holzer, H. (1961). *Cold Spring Harb. Sym. quant. Biol.* **26**, 277–288.
Hulač, V. (1955). *Kvasný Prům.* **1**, 135–136.
Ingraham, J. L. and Guymon, J. F. (1960). *Archs Biochem. Biophys.* **88**, 157–166.
Ingraham, J. L., Guymon, J. F. and Crowell, E. A. (1961). *Archs Biochem. Biophys.* **95**, 169–175.
Iñigo Leal, B. (1958). *Revta Ciencia apl.* **12**, 318–324.
Iñigo Leal, B. and Bravo Abad, F. (1963). *Revta Ciencia apl.* **17**, 317–319, 406–409.
Iñigo Leal, B., Arroyo Varcia, V., Bravo Abad, F. and Llaguno, C. (1961). *Revta Agroquim. Tecnol. Alimentos* **1** (2), 11–17.
Jacob, F. C., Archer, T. E. and Castor, J. G. B. (1964). *Am. J. Enol. Vitic.* **15**, 69–74.
Jacquemin, G. (1893). "Étude des Perfectionnements Apportés dans la Culture et l'Emploi des Levures Destinées à la Production des Boissons Alcooliques". Imprimerie Nancéienne, Nancy.
Jacquemin, G. and Alliot, H. (1903). "La Vinification Moderne" 2 Vol. J.-B. Baillière et Fils, Paris.
Jakubowska, J. and Piątkiewicz, A. (1965). *Acta microbiol. pol.* **14**, 67–71.
Jörgensen, A. (1900). "Micro-organisms and Fermentation". MacMillan, London.
Jörgensen, A. P. C. and Hansen, A. (1956). "Mikroorganismen der Gärungsindustrie". H. Carl, Nürnberg. "Micro-organisms and Fermentation". Charles Griffin and Co., London (1948).
Joslyn, M. A. (1951). *Mycopath. Mycol. appl.* **5**, 260–276.
Joslyn, M. A. (1955). *Wallerstein Labs Commun.* **18**, 107–121.
Joslyn, M. A. and Amerine, M. A. (1964). "Dessert, Appetizer and Related Flavored Wines". University of California, Division of Agricultural Sciences, Berkeley, California.
Joslyn, M. A. and Dunn, R. (1939). *J. Am. chem. Soc.* **60**, 1137–1141.
Joslyn, M. A. and Vosti, D. C. (1955). *Wallerstein Labs Commun.* **18**, 191–201.
Karamboloff, N. and Krumbholz, G. (1932). *Arch. Mikrobiol.* **3**, 113–121.
Kielhöfer, E. (1963). *Annls Technol. agric.* **12** (*numéro hors sér.* 1), 77–92.
Kielhöfer, E. and Würdig, G. (1963). *Dt. LebensmittRdsch.* **59**, 197–200.
Kir'yalova, E. N. (1955). *Trudy vses. nauchno-issled. Inst. Sel'skokhoz. Mikrobiol.* **12**, 123–129.
Kir'yalova, E. N. (1958). *Trudy vses. nauchno-issled. Inst. Sel'skokhoz. Mikrobiol.* **15**, 238–259.
Kir'yalova, E. N. (1959). *Mikrobiol. na Sluzhbe Sel'skn. vses. Akad. Sel'skokhoz. Nauk.* **1959**, 224–229.
Kir'yalova, E. N. and Shklyar, M. Z. (1953). *Trudy vses. nauchno-issled. Inst. Sel'skokhoz. Mikrobiol.* **12**, 141–142.
Kliewer, W. M., Howarth, L. and Omori, M. (1967). *Am. J. Enol. Vitic.* **18**, 42–54.
Kluyver, A. J. (1914). Doctorate Thesis, University of Delft.
Koch, J. (1962). *Weinberg Keller* **9**, 18–25.
Kolek, J. (1961). *Kvasný Prům.* **7**, 37–38.
Kondo, G. F., Belova, V. K. and Fadenko, P. S. (1965). *Sadov. Vinogr. Vinod. Mold.* **1965** (8), 30–34.

Kreger-van Rij, N. J. W. (1964). *Antonie van Leeuwenhoek* **30**, 428–432.

Kroemer, K. (1912a). *Ber. K. Lehranst. Wein-Obst- u. Gartenb. Geisenheim* **1910**, 137–141.

Kroemer, K. (1912b). *Landw. Jbr* **43**, *Ergänzungbd.* 1, 170–172.

Kroemer, K. and Krumbholz, G. (1931). *Arch. Mikrobiol.* **2**, 352–410.

Kroemer, K. and Krumbholz, G. (1932). *Arch. Mikrobiol.* **3**, 384–396.

Krumbholz, G. (1931a). *Arch. Mikrobiol.* **2**, 411–492.

Krumbholz, G. (1931b). *Arch. Mikrobiol.* **2**, 601–619.

Kunkee, R. E. (1967). *Adv. appl. Microbiol.* **9**, 235–279.

Kunkee, R. E. and Amerine, M. A. (1968). *Appl. Microbiol.* **16**, 1067–1075.

Kunkee, R. E. and Ough, C. S. (1966). *Appl. Microbiol.* **14**, 643–648.

Kunkee, R. E., Pilone, G. J. and Combs, R. E. (1965). *Am. J. Enol. Vitic.* **16**, 219–223.

Kunkee, R. E., Guymon, J. F. and Crowell, E. A. (1966). *J. Inst. Brew.* **72**, 530–536.

Kunkee, R. E., Guymon, J. F. and Crowell, E. A. (1967). *Bact. Proc.* **1967**, 19.

Kusewicz, D. (1965). *Acta microbiol. pol.* **14**, 155–160.

Kushida, T. and Taki, C. (1955). *Nippon Jozo Kyokai Zasshi* **50**, 530–526.

Lafon, M. (1956). *Annls Inst. Pasteur, Paris* **91**, 91–99.

Lafon-Lafourcade, S. and Peynaud, E. (1965). *C. r. hebd. Séanc. Acad. Sci., Paris* **261**, 1778–1780.

Leeuwenhoek, A. van (1687). *Phil. Trans. R. Soc.* **15** (170), 963–979; plate 170 (facing p. 947).

Lescure, L. A., Jr. (1956). Master's Thesis, University of California, Davis, California.

Llinca, P. (1961). *Lucr. Inst. Cerc. aliment.* **5**, 145–156.

Lodder, J. and Kreger-van Rij, N. J. W. (1952). "The Yeasts: A Taxonomic Study". Interscience Publishers, New York.

Loza, V. M. (1961). *Trudy krasnodar. Inst. Pishch. Prom.* **1961** (22), 180–187.

Lück, E. and Neu, H. (1965). *Z. Lebensmittelunters.-Forsch.* **126**, 325–335.

Lüthi, H. R., Stoyla, B. and Mayer, J. C. (1965). *Appl. Microbiol.* **13**, 511–514.

Macher, L. (1952). *Dt. LebensmittRdsch.* **48**, 183–189.

Malan, C. E. (1956). *Riv. Vitic. Enol.* **9**, 11–22.

Manchev, S. (1962). *Nauchni Trud. vissh. Inst. khranit. Prom. Plovidiv* **8** (2), 75–99.

Marcilla Arrazola, J., Alas, G. and Feduchy Marino, E. (1936). *An. Cent. Invest. Vinicola* **1**, 1–230.

Marques Gomes, J. V. and Vaz de Oliviera, M. M. F. (1963–64). *Anais Inst. Vinho Porto* **20**, 51–107.

Marques Gomes, J. V., Silva Babo, M. F. da and Guimaraes, A. F. (1961). *Bull. Off. intern. Vin.* **34** (368), 39–49.

Martakov, A. A., Levchenko, T. N. and Kolesnikov, V. A. (1966a). *Prikl. Biokhim. Mikrobiol.* **2**, 584–588.

Martakov, A. A., Kolesnikov, V. A. and Ignatov, M. P. (1966b). *Vinod. Vinogr. SSSR* **26** (1), 10–17.

Martini, A. (1960). *Riv. Vitic. Enol.* **13**, 263–273.

Matinand, V. (1909). *Revue Vitic.* **32**, 174–178, 206–210.

Mayer, K. and Lüthi, H. (1960). *Mitt. Geb. Lebensmittelunters. u. Hyg.* **51**, 132–137.

Mayer, K. and Temperli, A. (1963). *Arch. Mikrobiol.* **46**, 321–328.

Mayer, K., Busch, I. and Pause, G. (1964). Z. Lebensmittelunters. u. -Forsch. 125, 375–381.

Meissner, R. (1920). "Technische Betriebskontrolle im Weinfach". Verlag E. Ulmer, Stuttgart.

Melamed, N. (1962). Annls Technol. agric. 11, 5–31.

Mensio, C. (1910). Staz. sper. agrar. ital. 43, 797–931.

Mensio, C. and Forti, C. (1928). "Enologia". Unione Tip.-Editrice Torinese, Turin.

Michel, M. (1958). Vignes Vins No. 70, 6–8.

Minárik, E. (1963). Kvasný Prům. 9, 242–245.

Minárik, E. (1966). Biol. Práce 12, 1–107.

Minárik, E. (1967a). Vitis 6, 82–88.

Minárik, E. (1967b). Vitis 6, 89–98.

Minárik, E. (1967c). Wein.-Wiss. Beil. Fachz. Deut. Weinbau 22, 67–74.

Minárik, E. (1969). Mitt. Klosterneuburg Ser. A. Rebe Wein 19, 40–45.

Minárik, E., Laho, L. and Navara, A. (1960). Mitt. Klosterneuburg Ser. A. Rebe Wein 10, 218–223.

Minárik, E., Kochová-Kratochvílová, A. and Laho, L. (1965). Wein.-Wiss. Beil. Fachz. Dt. Weinbau 20, 193–205.

Mogilyanskiĭ, N. K. (1955). Sadov. Vinogr. Vinod. Mold. 10 (4), 48–50.

Mosiashvili, G. I. (1955). Sadov. Vinogr. Vinod. Mold. 10, (6), 48–59.

Mosiashvili, G. I. (1958). Vinod. Vinogr. SSSR 18 (2), 6–9.

Mrak, E. M. and McClung, L. S. (1940). J. Bact. 40, 395–407.

Müller-Thurgau, H. (1889). Weinb. Weinhandel 32, 135–154.

Murolo, G. (1966). Industria Agric. Florence 4, 585–588.

Musso, L. (1932). "Les Levures Selectionnées en Vinification". Institute Pasteur d'Algerie, Algiers.

Niehaus, C. J. G. (1937). Fg S. Afr. 12, 82, 85.

Nilov, V. I. and Datunashvili, E. N. (1961). Sadov. Vinogr. Vinod. Mold. 16 (10), 29–30.

Nilov, V. I. and Valuĭko, G. G. (1958). Vinod. Vinogr. SSSR 18 (8), 4–7.

Nitescu, M. A. (1915). Doctorat ès Sciences Thesis: Université Paris.

Niven, C. F., Jr. (1952). Bact. Rev. 16, 247–254.

Nomoto, N., Narahaski, Y. and Niikawa, Y. (1955). Nippon Nogeihagaku Kaishi 28, 805–809.

Nord, F. F. and Weiss, S. (1958). In "The Chemistry and Biology of Yeasts" (A. H. Cook, ed.), pp. 323–368. Academic Press, New York.

Nordström, K. (1963). J. Inst. Brew. 69, 310–322.

Nordström, K. (1964a). J. Inst. Brew. 70, 42–55.

Nordström, K. (1964b). J. Inst. Brew. 70, 226–233.

Nordström, K. (1964c). J. Inst. Brew. 70, 328–336.

Nyerges, P. (1952–57). Szölész. Kut. Inté. Évk. 11 (2), 203–222.

Nyerges, P. (1963). Szölész. Kut. Inté. Évk. 12 (7), 273–287.

Ohara, Y., Kagami, M. and Nonomura, H. (1960). Yamanashi Daigaku Hakko Kenkyusho Kenkyu Hokoku No. 7, 19–25.

Osterwalder, A. (1934). Zentbl. Bakt. ParasitKde Abt. II 90, 226–249.

Osterwalder, A. (1942a). Landw. Jb. Schweiz 56, 169–201.

Osterwalder, A. (1942b). Landw. Jb. Schweiz 56, 131–132.

Ostroukhova, Z. A. (1961). Mikrobiologiya 30, 297–300.

Ough, C. S. (1961). Appl. Microbiol. 9, 316–319.

Ough, C. S. (1964). Am. J. Enol. Vitic. 15, 167–177.

Ough, C. S. (1966a). *Am. J. Enol. Vitic.* **17**, 20–26.
Ough, C. S. (1966b). *Am. J. Enol. Vitic.* **17**, 74–81.
Ough, C. S. and Amerine, M. A. (1960). *Fd Technol., Champaign* **14**, 155–159.
Ough, C. S. and Amerine, M. A. (1966a). *Bull. Calif. agric. Exp. Stn.* **827**, 1–36.
Ough, C. S. and Amerine, M. A. (1966b). *Am. J. Enol. Vitic.* **17**, 163–173.
Ough, C. S. and Amerine, M. A. (1967a). *Wines Vines* **48** (5), 23–27.
Ough, C. S. and Amerine, M. A. (1967b). *Am. J. Enol. Vitic.* **18**, 157–164.
Ough, C. S. and Ingraham, J. L. (1960). *Am. J. Enol. Vitic.* **11**, 117–122.
Ough, C. S. and Ingraham, J. L. (1961). *Am. J. Enol. Vitic.* **12**, 149–151.
Ough, C. S. and Kunkee, R. E. (1967). *Am. J. Enol. Vitic.* **18**, 11–17.
Ough, C. S. and Kunkee, R. E. (1968). *Appl. Microbiol.* **16**, 572–576.
Ough, C. S., Guymon, J. F. and Crowell, E. A. (1966). *J. Fd Sci.* **31**, 620–625.
Ournac, A. and Flanzy, M. (1967). *Annls Technol. agric.* **16**, 41–54.
Parfent'ev, L. N. and Kovalenko, V.I. (1951). *Vinod. Vinogr. SSSR* **11** (3), 16–19.
Parle, J. N. and Di Menna, M. E. (1966). *N. Z. Jl agric. Res.* **9**, 98–107.
Paronetto, L. (1955). *Riv. Vitic. Enol.* **8**, 277–282.
Pasteur, L. (1858). *Annls Chim. Phys. Sér. 3* **52**, 404–418.
Pasteur, L. (1861). *C. r. hebd. Séanc. Acad. Sci., Paris* **52**, 1260–1264.
Pasteur, L. (1866). "Études sur le Vin". Masson, Paris. Also "Oeuvres de Pasteur" Vol. 3, pp. 111–385. Masson, Paris (1924).
Pasteur, L. (1872). *Annls Chim. Phys. Sér. 4* **25**, 145–151.
Pasteur, L. (1873). "Études sur le Vin", 2nd. Ed. Savy, Paris.
Pauli, O. (1967). *Bull. Off. int. Vin.* **40**, 764–772.
Peynaud, E. (1938). *Annls Falsif. Fraudes* **31**, 332–347.
Peynaud, E. (1939–40). *Annls Ferment.* **5**, 321–337, 386–401.
Peynaud, E. (1947). *Inds aliment. agric.* **64**, 87–95, 167–188, 301–317.
Peynaud, E. (1948). "Contribution à l'Étude Biochimique de la Maturation du Raisin et de la Composition des Vins". Imprimerie G. Santai et Fils, Lille.
Peynaud, E. (1956). *Inds aliment. agric.* **73**, 253–257.
Peynaud, E. (1963). *Annls Technol. agric.* **12** (numéro hors sér. 1), 99–114.
Peynaud, E. (1964). *Arch. Mikrobiol.* **47**, 219–224.
Peynaud, E. and Domercq, S. (1955). *Annls Inst. Pasteur, Paris* **89**, 346–351.
Peynaud, E. and Domercq, S. (1956). *Arch. Mikrobiol.* **24**, 266–280.
Peynaud, E. and Domercq, S. (1959). *Am. J. Enol. Vitic.* **10**, 69–77.
Peynaud, E. and Guimberteau, G. (1962). *Annls Technol. agric.* **11**, 85–105.
Peynaud, E. and Lafon-Lafourcade, S. (1962). *Revue Ferment. Ind. aliment.* **17**, 11–21.
Peynaud, E. and Lafon-Lafourcade, S. (1963). *Qualitas Pl. Mater. veg.* **9**, 365–380.
Peynaud, E. and Lafon-Lafourcade, S. (1966). *Inds aliment. agric.* **83**, 119–126.
Peynaud, E. and Sudraud, P. (1964). *Annls Technol. agric.* **13**, 309–328.
Peynaud, E., Lafon-Lafourcade, S. and Guimberteau, G. (1966a). *C. r. hebd. Séanc. Acad. Sci., Paris* **D263**, 634–636.
Peynaud, E., Lafon-Lafourcade, S. and Guimberteau, G. (1966b). *Am. J. Enol. Vitic.* **17**, 302–307.
Peynaud, E., Lafon-Lafourcade, S. and Guimberteau, G. (1967). *Revue fr. Oenol.* **7** (27), 17–20.
Phaff, H. J. and Douglas, H. C. (1944). *Fruit Prod. J. Am. Fd Mfr* **23**, 332–334.
Phaff, H. J., Miller, M. W. and Mrak, E. M. (1966). "The Life of Yeasts". Harvard University Press, Cambridge, Massachusetts.
Picci, G. (1955). *Agricoltore Ital.* **10**, 310–313.
Pieri, G. and Contini, L. (1955). *Riv. Vitic. Enol.* **8**, 347–351.

Pilone, G. J., Kunkee, R. E. and Webb, A. D. (1966). *Appl. Microbiol.* **14**, 608–615.
Popova, E. M. and Puchkova, M. G. (1960). *Biokhim. Vinodel.* **6**, 53–59.
Porchet, B. (1931). *Annu. agric. Suisse* **32** (2), 135–154.
Porchet, B. (1938). *Annls Ferment.* **4**, 578–600.
Preobrazhenskii, A. A. (1964). *Vinod. Vinogr. SSSR* **24** (2), 21–26.
Prillinger, F. (1963). *Annls Technol. agric.* **12** (*numéro hors sér.* 1), 159–169.
Prostoserdov, N. N. (1934). *Weinland* **6**, 72–73.
Prostoserdov, N. N. and Afrikian, R. (1933). *Weinland* **5**, 389–391.
Radler, F. (1962). *Vitis* **3**, 136–143.
Rankine, B. C. (1955a). *Austr. J. appl. Sci.* **6**, 408–413, 414–420, 421–425.
Rankine, B. C. (1955b). *Am. J. Enol.* **6** (3), 11–15.
Rankine, B. C. (1963). *J. Sci. Fd Agric.* **14**, 79–91.
Rankine, B. C. (1964). *J. Sci. Fd Agric.* **15**, 872–877.
Rankine, B. C. (1965). *J. Sci. Fd Agric.* **18**, 41–44.
Rankine, B. C. (1966a). *J. Sci. Fd Agric.* **17**, 312–316.
Rankine, B. C. (1966b). *Am. J. Enol. Vitic.* **17**, 82–86.
Rankine, B. C. (1967a). *J. Sci. Fd Agric.* **18**, 41–44.
Rankine, B. C. (1967b). *J. Sci. Fd Agric.* **18**, 583–589.
Rebelein, H. (1957). *Z. Lebensmittelunters. u. -Forsch.* **105**, 403–420.
Rehm, H.-J. and Wittmann, H. (1962). *Z. Lebensmittelunters. u. -Forsch.* **118**, 413–429.
Rehm, H.-J. and Wittmann, H. (1963). *Z. Lebensmittelunters. u. -Forsch.* **120**, 465–478.
Rehm, H.-J., Sening, E., Wittmann, H. and Wallnöfer, P. (1964). *Z. Lebensmittelunters. u. -Forsch.* **123**, 425–432.
Rehm, H. J., Wallnöfer, P. and Keskin, H. (1965). *Z. Lebensmittelunters. u. -Forsch.* **127**, 72–85.
Renaud, J. (1941). "Les Levures des Vins du Val de Loire, Recherches Morphologiques, Biologiques et Cytologiques". Librairie Génerale de l'Enseignement, Paris.
Rentschler, H. (1951). *Schweiz. Z. Obst- u. Weinb.* **60**, 197–200.
Requinyi, G. and Soós, I. (1935). *Ampelológiai Int. Evk.* **9**, 80.
Ribéreau-Gayon, J. and Peynaud, E. (1960). "Traité d'Oenologie" Vol. 1. Librairie Polytech. Béranger, Paris.
Ribéreau-Gayon, J. and Peynaud, E. (1961). "Traité d'Oenolgie" Vol. 2. Librairie Polytech. Béranger, Paris.
Ribéreau-Gayon, J. and Peynaud, E. (1962). *C. r. hebd. Séanc. Acad. Agric. Fr.* **48**, 558–560.
Rice, A. C. (1968). *Am. J. Enol. Vitic.* **19**, 101–107.
Rodopulo, A. K. (1961). *Vinod. Vinogr. SSSR* **21** (3), 12–17.
Rodopulo, A. K., Egorov, I. A. and Yashina, V. E. (1965). *Prikl. Biokhim. Mikrobiol.* **1**, 95–101.
Ronkainen, P., Brummer, S. and Suomalainen, S. (1967). *J. Chromat.* **27**, 443–445.
Rosell, R. F., Ofria, H. V. and Palleroni, N. J. (1968). *Am. J. Enol. Vitic.* **19**, 13–16.
Rosén, C.-G. and Fedorćak, I. (1966). *Archs Biochem. Biophys.* **130**, 401–405.
Rzędowska, H. and Rzędowski, W. (1955). *Pr. Instw. Lab. badaw. Przem. roln. Spożyw.* **5** (2), 18–25.
Saavedra, I. J. and Garrido, J. M. (1962). *Revta Agroquim. Tecnol. Alimentos* **2**, 150–158.
Saenko, N. F. (1961). *Trudȳ. Inst. Mikrobiol. Mosk.* **10**, 96–102.

Saenko, N. F. (1964). "Kheres". Izdatel'stvo Pischevaya Promyshlennost', Moscow.

Saenko, N. F. and Sakharova, T. A. (1963). *Vinod. Vinogr. SSSR* **23** (2), 4–6.

Sai, T. (1966). *Jozo Kagaku Kenkyu Hokoku* **12** (1), 35–42.

Salas, M. L., Viñuela, E., Salas, M. and Sols, A. (1965). *Biochem. biophys. Res. Commun.* **19**, 371–376.

Saller, W. (1954). *Dt. Weinztg.* **90**, 540, 562, 564.

Saller, W. and Stefani, C. de (1962). *Mitt. Klosterneuburg Ser. A Rebe Wein* **12**, 11–18.

Santa Maria, J. (1959). *An. Inst. nac. Invest. agron.* **8**, 713–736.

Sarishvili, N. G. (1963). *Vinod. Vinogr. SSSR* **23** (2), 6–10.

Sarukhanyan, F. G., Sevoyan, A. G., Movesyan, G. P. and Karpetyan, I. O. (1965). *Izv. Akad. Nauk Armyan SSR, Biol. Nauk.* **18** (3), 3–10.

Scardovi, V. (1951a). *Annali Microbiol.* **4**, 131–172.

Scardovi, V. (1951b). *Annali Microbiol.* **5**, 5–16.

Scardovi, V. (1953). *Annali Microbiol.* **5**, 140–161.

Schanderl, H. (1936). *Wein Rebe* **18**, 16–25.

Schanderl, H. (1938). *Wein Rebe* **20**, 10–17.

Schanderl, H. (1957). *In* "Yeasts" (W. Roman, ed.), pp. 127–139. Academic Press, New York.

Schanderl, H. (1959). "Mikrobiologie des Mostes und Weines", 2nd Ed. Verlag E. Ulmer, Stuttgart.

Schanderl, H. (1965). *Mitt. Klosterneuburg Ser. A Rebe Wein* **15**, 13–20.

Schanderl, H. and Draczynski, M. (1952). *Dt. Weinztg.* **88**, 462–464.

Scheffer, W. R. and Mrak, E. M. (1951). *Mycopathol. Mycol. appl.* **5**, 236–249.

Schmitthenner, F. (1937). *Wein Rebe* **19**, 65–82, 93–106.

Schmitthenner, F. (1950). "Die Wirkung der Kohlensäure auf Hefen und Bakterien". Seitz-Werke, Bad Kreuznach.

Schroeter, L. C. (1966). "Sulfur Dioxide. Applications in Foods, Beverages, and Pharmaceuticals". Pergamon Press, London.

Schulle, H. (1953a). *Arch. Microbiol.* **18**, 133–148.

Schulle, H. (1953b). *Arch. Microbiol.* **18**, 342–348.

Schulle, H. (1954). *Dt. Weinztg.* **90**, 736–738.

Şeptilici, G., Cîmpeanu, H., Gherman, M., Sandu-Ville, S., Tîrdea, C. and Giosanu, T. (1966). *Inst. Cercet. Horti-Viticole Lucrari Ştinţifice* **9**, 439–449.

Shakhsuvarian, A. V. (1960). *Biokhim. Vinodel.* **6**, 79–87.

Shimatani, Y. and Nagata, Y. (1967). *J. Ferment. Technol., Osaka* **45**, 179–190.

Siegel, A., Rotter, R. G. and Schmid, L. (1965). *Z. Lebensmittelunters. u. -Forsch.* **126**, 321–324.

Šikovec, S. (1966). *Mitt. Klosterneuburg Ser. A Rebe Wein* **16**, 127–138.

Silva, H. D., da (1953). *Anais Junta Nac. Vinho* **5**, 69–76.

Silver, S. and Wendt, L. (1967). *J. Bact.* **93**, 560–566.

Smirnova, A. P. (1955). *Vinod. Vinogr. SSSR* **15** (7), 41–44.

Soós, I. and Ásvány, Á. (1952). *Ampelológiai Int. Évk.* **10**, 255–290.

Speckman, R. A. and Collins, E. B. (1968a). *J. Bact.* **95**, 175–180.

Speckman, R. A. and Collins, E. B. (1968b). *Analyt. Biochem.* **22**, 154–160.

Strandskov, F. B., Ziliotto, H. L., Brescia, J. A. and Bockelmann, J. B. (1965). *Proc. Am. Soc. Brew. Chem.* **1965**, 129–134.

Sudraud, P. (1967). *C. r. hebd. Séanc. Acad. Agric. Fr.* **53**, 339–342.

Sulkhanishvili, N. V. (1963). *Trudy Inst. Sadov.Vinogr. Vinod. Gruz. SSR* **15**, 330–336.

Suomalainen, H. and Nykänen, L. (1966a). *J. Inst. Brew.* **72**, 469–474.
Suomalainen, H. and Nykänen, L. (1966b). *Suom. Kemistilehti B* **39**, 252–256.
Szabó, J. and Rakcsányi, L. (1937). *5th Congr. Intern. Tech. Chim. Agr.* **1**, 936–949. Also *Magy. Ampelól. Évk.* **9**, 346–361 (1935).
Tarantola, C. (1945–46). *Annali Accad. Agric. Torino* **88**, 115–133.
Thorne, R. S. W. (1949). *J. Inst. Brew.* **55**, 201–222.
Thorne, R. S. W. (1950). *Wallerstein Labs Commun.* **13**, 319–338.
Thoukis, G. (1958a). Ph.D. Thesis. University of California, Davis, California.
Thoukis, G. (1958b). *Am. J. Enol. Vitic.* **9**, 161–167.
Thoukis, G., Reed, G. and Bouthilet, R. J. (1963). *Am. J. Enol. Vitic.* **14**, 148–154; also see *Wines Vines* (1963) **44** (1), 25–26.
Thoukis, G., Ueda, M. and Wright, D. (1965). *Am. J. Enol. Vitic.* **16**, 1–8.
Ţîrdea, C. (1964). *Anal. Ştiinţ. Univ. A. I. Cuza Sect. II* **10**, 193–196.
Tsyb, T. S. (1966). *Vinod. Vinogr. SSSR* **26** (6), 15–17.
Tyurina, L. V. (1960). *Trudȳ vses. nauchno-issled. Inst. Vinod. Vinogr. "Magarach"* **9**, 96–106.
Tyurina, L. V. (1967). *Trudȳ vses. nauchno-issled. Inst. Vinod. Vinogr. "Magarach"* **15**, 67–80.
Turtura, G. C. (1966). *Ricerca scient.* **36**, 638–645.
Usseglio-Tomasset, L. (1967). *Rev. Vitic. Enol.* **20**, 10–34.
Van der Walt, J. P. and Van Kerken, A. E. (1958). *Antonie van Leeuwenhoek* **24**, 239–252.
Van der Walt, J. P. and Van Kerken, A. E. (1961). *Antonie van Leeuwenhoek* **27**, 81–90.
Van Kerken, A. E. (1963). "Contribution to the Ecology of Yeasts Occurring in Wine". University of the Orange Free State, Pretoria.
Van Zyl, J. A. (1962). *Bull. S. Afr. Dept. Agric. Tech. Serv. Sci.* **381**, 1–42.
Van Zyl, J. A. and Du Plessis, L. de W. (1961). *S. Afr. J. agric. Sci.* **4**, 393–401.
Van Zyl, J. A., De Vries, M. J. and Zeeman, A. S. (1963). *S. Afr. J. agric. Sci.* **6**, 165–179.
Vaughn, R. H. (1938). *J. Bact.* **36**, 357–367.
Ventre, J. (1913). "Influence de Quelques Levures Elliptiques sur la Constitution des Vins". Coulet et Fils, Montpellier.
Ventre, J. (1930). "Traité de Vinification" Vol. 1. Librairie Coulet, Montpellier.
Ventre, J. (1934). *Revue Vitic.* **80**, 181–346.
Verona, O. (1933). *Boll. R. Ist. sup. agr. Pisa* **9**, 231–234.
Verona, O. (1952). *Annls Inst. Pasteur, Paris* **82**, 245–247.
Verona, O. and Florenzano, G. (1956). "Microbiologia Applicata all'Industria Enologica". Officine Grafiche Calderini, Bologna.
Verona, O. and Zardetto de Toledo, O. (1954). *Annali Fac. Agr. Univ. Pisa* **15**, 163–191.
Wahab, A., Witzke, W. and Cruess, W. V. (1949). *Fruit Prod. J. Am. Fd Mfr* **28**, 198–200, 202, 219.
Wainwright, T. (1961). *Biochem. J.* **80**, 27P–28P.
Wallgren, H., Sammalisto, L. and Suomalainen, H. (1963). *J. Inst. Brew.* **69**, 418–420.
Wallnöfer, P. and Rehm, P. (1965). *Z. Lebensmittelunters. u. -Forsch.* **127**, 195–206.
Walters, L. S. and Thiselton, M. R. (1953). *J. Inst. Brew.* **59**, 401–404.
Webb, A. D. (1967a). *In* "Chemistry and Physiology of Flavors" (H. W. Schultz, E. A. Day and L. M. Libbey, eds.), pp. 204–227. Avi Publ., Westport, Connecticut.

Webb, A. D. (1967b). *Biotechnol. Bioengng* **9**, 305–319.

Webb, A. D. and Ingraham, J. L. (1963). *Adv. appl. Microbiol.* **5**, 317–353.

Webb, A. D. and Kepner, R. E. (1962). *Am. J. Enol. Vitic.* **13**, 1–14.

Webb, A. D., Kepner, R. E. and Galetto, W. G. (1964). *Am. J. Enol. Vitic.* **15**, 1–10.

Wikén, T. and Richard, O. (1951). *Antonie van Leeuwenhoek* **17**, 209–236.

Wikén, T. and Richard, O. (1952). *Antonie van Leeuwenhoek* **18**, 31–44.

Wikén, T. and Richard, O. (1955). *Antonie van Leeuwenhoek* **21**, 337–361.

Wood, W. A. (1961). *In* "The Bacteria" (I. C. Gunsalus and R. Y. Stanier, eds.), Vol. 2, pp. 59–149. Academic Press, New York.

Wortmann, J. (1892). *Landw. Jbr* **21**, 901–936.

Wortmann, J. (1895). *Mitt. Weinb Kellerw.* **7**, 65–71.

Wortmann, J. (1905). "Die Wissenschaftlichen Grundlagen der Weinbereitung und der Kellerwirtschaft". Berlin.

York, G. K. and Vaughn, R. H. (1964). *J. Bact.* **88**, 411–417.

Yoshizawa, K. (1963). *Agric. Biol. Chem., Japan* **27**, 162–164.

Zájara Jiménez, J. (1958). *Microbiologia esp.* **11**, 313–322.

Zambonelli, C. (1965). *Annali Microbiol.* **15**, 89–97, 99–106, 181–195.

Zardetto de Toledo, O. and Gonzalves Teixeira, C. (1955). *Agricoltore Ital.* **55**, 155–164.

Zardetto de Toledo, O., Gonzalves Teixeira, C. and Verona, O. (1959). *Annali microbiol. Enzimol.* **9**, 22–34.

Zhuravleva, V. P. (1963). *Izv. Akad. Nauk turkmen. SSR Ser. Biol. Nauk.* **1963** (2), 19–24.

Zhuravleva, V. P. and Timuk, O. E. (1965). *Izv. Akad. Nauk turkmen. SSR Ser. Biol. Nauk.* **1965** (1), 36–40.

Zimmermann, J. (1938a). *Zentbl. Bakt. ParasitKde Abt. II* **98**, 36–65.

Zimmermann, J. (1938b). *Revue Vitic.* **90**, 472–473.

Zsolt, J. (1958). *Antonie van Leeuwenhoek* **24**, 210–214.

Zsolt, J. (1959). *Acta bot. hung.* **5**, 233–257.

Chapter 3

The Role of Yeasts in Cider-making

F. W. Beech and R. R. Davenport

University of Bristol, Research Station, Long Ashton, Bristol, England

I. Introduction

The term cider is used in this chapter to describe the fermented juice of the apple, as it is understood in the United Kingdom and in Europe. It is necessary to make this distinction as unfermented juice is called cider in North America, the fermented product being described there as "hard" cider. References to perry, a similar product made from pears, will be included with cider, since the two products are closely related (Pollard and Beech, 1963).

The composition of the apples used for cider-making varies from one country to another. In the United Kingdom, Northern France and

Spain, special varieties are grown containing a proportion of tannins and phenolics that impart a characteristic bitterness and astringency to the drink (Burroughs, 1962; Williams and Child, 1966). Culinary and dessert apples, unsuitable for the fresh fruit market, by virtue of unsuitable size or skin blemishes, are used in England to dilute any excessive bitterness but are the main source of fruit in all other cider-producing countries except France and Spain. Cider normally contains from three to seven per cent alcohol by volume; it may be sold free from sugar (dry) or with varying amounts of natural juice or added sugars (sweet). The process of cider-making has been known for centuries. Basically it consists of grinding the fruit to pulp and expressing the juice under pressure; preliminary enzyme depectinization, as required for soft fruits, is not necessary. The freshly pressed juice may be allowed to ferment naturally with the organisms it has acquired from the fruit or pressing equipment, but in the United Kingdom and Switzerland, sulphur dioxide is added in amounts up to 150 p.p.m. to suppress all but the fermenting yeasts, *Saccharomyces* spp. With higher standards of factory hygiene, naturally-occurring strains of fermenting yeasts can be rare in the juice, and a suitable culture yeast is added once the action of the sulphite is complete. The process of fermentation can take from two weeks to several months, according to the nutrient status of the juice and the fermentation conditions. In France the rate of fermentation is deliberately reduced by chilling to assist in the production of ciders that are still sweet, by virtue of some juice remaining unfermented (Beech and Pollard, 1966). Cider is normally clarified and blended after fermentation and stored in airtight vessels. If the bacterial conversion of malic acid to lactic acid and carbon dioxide has not taken place during the yeast fermentation, it normally does so during storage, thereby imparting a more mature flavour to the cider. Finally, the beverage is sold in glass, plastic or wooden containers after any necessary degree of sweetening and sterilizing treatment that may be required. Detailed accounts of the process have been given by Pollard and Beech (1957), Carr (1964b), Buckle (1967) and Pollard *et al.* (*in press*).

In spite of the antiquity of the process, relatively little was known until recently of the changes that take place in the yeast flora from the burgeoning flower buds in the orchard to the final product. The earlier work has been detailed by Kayser (1890), Müller-Thurgau (1899), Guilliermond (1920), Osterwalder (1912, 1924), Pearce and Barker (1908) and Barker (1950b, 1951). The greater part of this chapter describes the results of research work carried out at Long Ashton in the last fifteen years which has demonstrated the succession of yeast species that occurs in cider-making. As far as possible only references related directly to this succession will be quoted. Methods for isolating the

(a) (b) (c)

(d) (e)

FIG. 1. Stages in apple fruit bud development as used by Davenport (1968) for
east ecological survey of a cider apple orchard.

(a) Dormant fruit bud (b) Bud burst (c) Mouse ear
(d) Green cluster (e) Pink bud (f) Petal fall
(g) Early fruitlet

Pictures—Bayer, Germany

(f)

(g)

FIG. 1 (f), (g). For legend see over.

yeasts have been detailed elsewhere by Beech and Davenport (1969a, b). Methods used for identifying the cultures have been those described by Lodder and Kreger-van Rij (1952), Wickerham (1951) and Beech et al. (1968). In any survey of a yeast flora made over a period of months or years, a species in the Endomycetaceae may be isolated on numerous occasions. If some of the isolates form ascospores subsequently, but the remainder, identical in all respects, cannot be induced to form spores, are they then to be classed as a totally different species? Such a rule blurs the exactness of an ecological survey. Hence sporing and non-sporing partners have been coupled together, the first name being indicative of whether spores have beeen observed or not. Further, because *Aureobasidium pullulans*, although classed as a mould, occurs so regularly in association with yeasts in the orchard, its presence has been recorded; some workers consider that this organism and other so-called "black yeasts" should be grouped with the yeasts (Mrak and Phaff, 1948; Cooke, 1959). Some mention will be made of changes in the bacterial flora, but only where these have any effect on the yeasts: detailed accounts of cider bacteria have been given by Millis (1951), Lüthi (1957a), Carr (1953, 1957, 1958, 1959a, b, 1960, 1964a), Beech and Carr (1953), Carr and Whiting (1956).

II. The Orchard

Unless stated to the contrary, the results quoted in Section II are largely a summary of a five-year survey made by Davenport (1968) on the yeast flora of two cider apple cultivars, Yarlington Mill and Court Royal, growing in Orchard 28 at the Research Station, Long Ashton.

Samples were taken from trees of known physiological age and, as this is influenced by weather conditions, a biological rather than a lunar calendar was chosen. Thus, the sampling times are based on the flowering and fruiting states of the apple, described in the Ministry of Agriculture Bulletin No. 137 (Fig. 1, facing).

The yeast flora of the apple as it hangs on the tree is the sum total of those derived from vectors within the orchard and modified by such environmental conditions as exudates, dehydration, ultraviolet radiation and ambient temperature. The effect of these external conditions will not be detailed in this chapter, but they have been investigated by Lund (1954) and di Menna (1962).

A. TERRESTRIAL VECTORS

1. *Soil*

The soil is a reservoir for many yeasts, the species present in any sample being influenced by factors such as the cover crop and the amount of cultivation. The yeast flora of soils has been examined by many workers but, wherever possible, references will be selected from those relating to orchard soils. Lund (1954) found the greatest number of yeasts in soils under *Ribes* sp. The maximum count, $2 \cdot 45 \times 10^5$/g of dry soil, was obtained at the surface; the values decreased with increasing depth until, at 20–30 cm below the surface, yeasts were virtually absent. The number of yeasts in the surface layer appeared to vary seasonally, rising to a maximum in March; in contrast, the counts in the lower layers were more constant. Lund isolated *Candida mycoderma*, *C. pulcherrima*, *Kloeckera apiculata*, *Torulopsis molischiana*, *Hanseniaspora valbyensis*, *Hansenula suaveolens*, *H. angusta*, *H. californica* and *Rhodotorula glutinis*. There was no consistent evidence for certain species being found at particular depths of soil.

Again, Miller and Webb (1954) found their highest yeast count, $5 \cdot 5 \times 10^4$/g, in soil from an old apple orchard; unfortunately the yeasts were not identified. Adams (1961) also examined soil from Canadian orchards, taking his samples from the top 5–8 cm throughout the growing season. The greatest counts were obtained from grass-covered cherry and peach orchards but, strictly speaking, these were not comparable with the grape, pear and apple orchards that had been cultivated prior to sampling. Numbers were at a maximum during September and showed wide variations from site to site. No replication of samples was reported by these authors, so that it is difficult to decide on the significance of the variations. Adams did not detail the source of his isolates but, of of the 160 from all soil samples, 32 were ascosporogenous, 19 being *Saccharomyces* sp., the remainder being *Pichia* and *Hansenula*. Of the anascosporogenous yeasts, the largest group were *Torulopsis* spp. (42), followed by *Kloeckera* (29), *Candida* and *Cryptococcus* (23 each), the remainder being mainly *Rhodotorula* sp. Thus, even though the soil was sampled when rotten fruit was lying on the ground, very few fermenting yeasts were observed, only a small percentage of which were capable of producing sufficient alcohol to be of commercial interest. Results for rotten fruit are reported in Section II.D.2, p. 89. In contrast to these results, di Menna (1957) found *Cr. terreus* was the most frequently occurring yeast in the pasture soils of New Zealand, with *Cr. albidus*, *C. curvata* and *Schizoblastosporion starkeyi-henricii* dominant on occasion. No *Saccharomyces* sp. was isolated.

The soil under both cider-apple cultivars contained twice as many yeast species (10) in the autumn as in the spring (5). Again, in the autumn there were quantitatively up to 100 times more yeasts under Court Royal (the highest count being 2×10^5/g) than under Yarlington Mill. However, the soil counts in the spring were very similar, up to 6×10^3/g. The species of yeasts found at the different sampling periods are given in Table I. *Pichia polymorpha*, originally isolated by Klöcker (1912, 1913), seemed to be the only yeast characteristic of one cultivar, Court Royal. The soil could be the ultimate source of the *Trichosporon* sp. found on mature and mummified fruit. Again, the film yeast, *Pi. membranaefaciens/C. mycoderma*, was confined entirely to the soil and associated vectors such as grass, slugs and earthworms, and was never found on any aerial or tree vector or flower or fruit bud state. The genera isolated resembled those described by Adams (1961), except that *Kloeckera* spp. were absent, as also noted by Lund (1954).

TABLE I. *Seasonal Variation of the Yeast Flora of the Soil in Orchard 28, Research Station, Long Ashton (Reproduced with permission from Davenport, 1968)*

[*Apple cultivars: Court Royal and Yarlington Mill. + Indicates presence of organism*]

Yeast	Spring: Swelling bud		Autumn: Mature fruit	
	Court Royal	Yarlington Mill	Court Royal	Yarlington Mill
Aureobasidium pullulans	+	+	+	+
Saccharomyces cerevisiae	+			
Hansenula silvicola		+	+	+
Hansenula/Torulopsis sp. B.			+	+
Hansenula beckii	+			
Pichia polymorpha	+		+	
Pichia membranaefaciens/ *Candida mycoderma*			+	+
Candida pulcherrima			+	
Candida reukaufii			+	+
Torulopsis famata/ *Debaryomyces hansenii*	+		+	+
Rhodotorula glutinis/ *Sporobolomyces roseus*	+	+	+	+
Rhodotorula aurantiaca/ *Sporobolomyces salmonicolor*				+
Trichosporon sp.		+	+	+

2. Vegetation

In the context of this section, vegetation means the grasses and plants, such as clover, found on the orchard floor. Di Menna (1959), studying the yeast flora of leaves of New Zealand pasture plants, found that the numbers of yeasts varied seasonally from 3×10^4 to $3\cdot5 \times 10^6$/g leaves, rising to peaks of 10^8/g in the February–March period when the flora consisted of carotenoid-producing species. Similar conclusions were reached by Last and Deighton (1965).

As far as is known the yeast flora of orchard grasses has not been examined recently, so that it is difficult to draw comparisons with the results obtained at Long Ashton (Table II). The highest counts were obtained under Court Royal in the autumn, when values were up to $10^6/g$, while in the spring the average was $1.5 \times 10^3/g$. The flora in the spring was very sparse, sporulating yeasts being found under Yarlington Mill but not under Court Royal. As with the soils (Section II.A.1,

TABLE II. *Seasonal Variation of the Yeast Flora of the Surface Vegetation in Orchard 28, Research Station, Long Ashton (Reproduced with permission from Davenport, 1968)*

[*Apple cultivars: Court Royal and Yarlington Mill.* + *Indicates presence of organism*]

Yeast	Spring: Dormant → Swelling bud		Autumn: Mature fruit	
	Court Royal	Yarlington Mill	Court Royal	Yarlington Mill
Aureobasidium pullulans	+			+
Saccharomyces florentinus			+	+
Hansenula mrakii			+	
Hansenula beckii		+	+	
Pichia membranaefaciens/ Candida mycoderma		+	+	+
Candida pulcherrima			+	
Candida reukaufii		+	+	
Kloeckera apiculata/ Hanseniaspora valbyensis	+		+	
Torulopsis famata/ Debaryomyces hansenii			+	+
Torulopsis famata/Cryptococcus luteolus var. A.			+	+
Torulopsis nitratophila			+	+
Rhodotorula glutinis/ Sporobolomyces roseus	+		+	+
Rhodotorula aurantiaca/ Sporobolomyces salmonicolor	+	+	+	+
Trichosporon sp.			+	+

p. 75) the autumn flora was much richer, and again the greatest variety of species was found under Court Royal. The species isolated from the orchard vegetation, whilst they included some carotenoid-forming yeasts, were totally different from those of pasture grasses. The rhizosphere was not examined (Bab'eva and Savel'eva, 1963).

Flowers sampled under the trees at 80% petal fall had a simple yeast flora consisting of the yeasts commonly found in this cider apple orchard. Thus, buttercups (*Ranunculus bulbosus*) and the seed heads of dandelions (*Taraxacum officinale*) carried *A. pullulans*, *T. famata/D. hansenii*, *Rh. glutinis/Sp. roseus*, *C. reukaufii*, *C. pulcherrima*, *Rh. aurantiaca/Sp. salmonicolor*. The dandelion flowers held all but the last two yeasts, but included *H. anomala*, the only one of this group that was not also isolated from the soil.

Fallen leaves carried a quantitatively smaller flora than those on the tree (750–1,900 yeasts/cm^2) but qualitatively there were more species. Thus, at the dormant-bud stage, the fallen leaves under both Yarlington Mill and Court Royal carried *A. pullulans*, *Rh. glutinis/Sp. roseus* and *C. pulcherrima*; as did the leaves on the tree, together with *T. famata/D. hansenii* and *C. reukaufii*. Fallen leaves of Court Royal also had *Kl. apiculata/Ha. valbyensis* and *C. melinii* in addition, these and small slugs (Section II.A.3) being the only sources of the latter yeast in the orchard.

3. *Invertebrates*

It might be expected that earthworms could act as agents in transmitting yeasts and other micro-organisms through the soil. Parle (1963a) found that the species of earthworm was important: the gut of *Allolobophora caliginosa* contained more yeasts than *Lumbricus terrestris*, while *A. longa* contained more kinds of yeast than the soil. None of the counts were as important as those of actinomycetes and bacteria, which reproduced in the gut, whereas yeasts and bacteria did not. The reverse effect was found in earthworm casts (Parle, 1963b) where the number of filamentous fungi and yeasts increased rapidly after the casts were produced, but not the bacteria and actinomycetes. The commonest yeasts in the casts were *Cr. diffluens*, *C. humicola*, *Trichosporon pullulans* and, less frequently, *C. curvata*, *D. hansenii*, *Trich. cutaneum* and *Rh. glutinis*. Unfortunately no details were given of the yeast flora of the soil, so it must be assumed they had been there originally and were merely increasing in numbers as the casts aged. Certainly the flora is more reminiscent of pasture land, where fermenting yeasts are rare (Parle, 1963b), than of orchard soils (Section II.A.1, p. 75).

Davenport examined the total yeast flora of earthworms collected in the soil under the two cider apple cultivars (Table III). Comparing these results with those given for the soil sampled at the same time (Table I), earthworms under Court Royal had a higher count than those from under Yarlington Mill, reflecting the counts found in the soil. Nine out of the 11 species in the soil were also isolated from earthworms. *Saccharomyces* sp., *Hansenula/Torulopsis* spp. B and *C. reukaufii* were not obtained, while *Kl. apiculata/Ha. valbyensis* and *Pi. guiller-*

mondii were absent from the soil; the occurrence of the other two *Pichia* spp. has been discussed already.

Large and small slugs (*Lomax* sp.) were also examined; the small slugs in particular had a greater population than worms, approximately 1,000 times as great, the numbers in large slugs being ten times greater than in worms. The microflora of these three invertebrates is given in

TABLE III. *Complete Microflora of Earthworms Collected under two Cider-apple Cultivars in Orchard 28 of the Research Station, Long Ashton (Reproduced with permission from Davenport, 1968)*

[*Apple cultivars*: Court Royal and Yarlington Mill. + Indicates presence of organism]

Yeast	Spring: Swelling bud		Autumn: Mature fruit	
	Court Royal	Yarlington Mill	Court Royal	Yarlington Mill
Aureobasidium pullulans	+			+
Hansenula silvicola	+	+		
Hansenula beckii	+	+		
Pichia membranaefaciens/				
Candida mycoderma	+			
Pichia guilliermondii	+			
Candida pulcherrima			+	
Kloeckera apiculata/				
Hanseniaspora valbyensis			+	
Torulopsis famata/				
Debaryomyces hansenii				+
Rhodotorula glutinis/				
Sporobolomyces roseus	+	+	+	+
Rhodotorula aurantiaca/				
Sporobolomyces salmonicolor				+
Trichosporon sp.				+
Total count/Specimen	250	10	1,580	150

Table IV for the autumn sampling when the fruit was mature and part had fallen on the ground. Qualitatively, the slug flora was also more varied than that of the worm, the small slug flora being more varied than that of the large slug. Several yeasts were isolated from the two types of slugs that were absent both from the Yarlington Mill fruit or any other vector associated with this sampling period. The species concerned were *Kl. antillarum*, *Rh. macerans/Sporobolomyces* sp. A, *Hansenula* sp. X (Bowen, 1962), *C. solani* var. N and *T. candida*; they were not isolated from any other source, while *H. californica* was only found otherwise on rotten fruit and in fresh bird-droppings in the grass in the spring.

TABLE IV. *Internal and External Yeast Flora of Earthworms and Slugs (Lomax sp) Found in the Soil under Cultivar Yarlington Mill from Orchard 28, Research Station, Long Ashton (Reproduced with permission from Davenport, 1968)*
[+ *Indicates presence of organism*]

Yeast	Autumn: Mature fruit					
	External			Internal		
	Earth-worm	Small slug	Large slug	Earth-worm	Small slug	Large slug
Aureobasidium pullulans		+		+		
Hansenula californica		+	+			+
Pichia membranaefaciens/Candida mycoderma		+	+			
Hansenula sp. X		+				
Candida pulcherrima		+	+		+	+
Candida reukaufii		+	+	+	+	+
Candida solani var. N		+				
Kloeckera apiculata/Hanseniaspora valbyensis			+		+	+
Kloeckera antillarum		+			+	
Torulopsis famata/Debaryomyces hansenii	+	+			+	
Torulopsis candida/Debaryomyces subglobosus		+				
Rhodotorula glutinis/Sporobolomyces roseus		+	+	+		
Rhodotorula aurantiaca/Sporobolomyces salmonicolor	+	+			+	
Rhodotorula macerans/Sporobolomyces sp. A		+	+		+	+
Trichosporon sp.		+		+	+	

B. AERIAL VECTORS

The air and aerial vectors in the orchard are the means of transferring yeasts from the orchard floor to the fruit.

1. *Air*

Adams (1964) found that only 5% of the aerial flora over fruit orchards at Vineland Ontario Experiment Station were yeasts, the rest being moulds. Initial trials with agar plates exposed at ground level showed the development of yeasts, moulds and bacteria; 739 yeast-like colonies were observed, and 180 isolates from them gave the following genera in decreasing order of importance: *Kloeckera, Cryptococcus, Torulopsis, Rhodotorula, Candida* and *Brettanomyces*. No *Saccharomyces* sp. was isolated, but colonies may have been missed because of heavy mould growth on the plates.

Fewer organisms were trapped in a subsequent trial with plates 18 in. from the ground. Of the 445 isolates made in 1958/9, 178, or nearly half, came from the apple-growing site, fewest (37) came from the cherry orchard, while the remaining sites yielded between 70 and 80. The number of yeasts collected per month rose to a maximum in September, decreasing again thereafter: no yeasts were found during the winter months, due to the snow cover. Adams did not identify the isolates in this experiment, his object being to select "fermentative" yeasts (i.e. those capable of producing more than 12% alcohol by volume). From May to August 1958, 1% of the isolates were of this type, whereas in September the proportion had risen to 16%, nearly half of which came from the apple site: 13·5% of the isolates made during the remainder of the year were also fermenting yeasts, all of which were strains of *Sacch. cerevisiae*. Thus, of his 445 isolates, 13·2% were fermenting yeasts, most of which came from the apple site. In the following year only 2·5% were fermentative, and most of these came from the vineyards. Hence, isolations would need to be taken over several years in order to determine whether there was any consistent pattern in the aerial flora at these sites. It is interesting to compare the aerial flora of these horticultural sites, rich in fermenting yeasts (14% of the 1,000 yeasts isolated), with the very low yeast counts obtained away from fruit-growing areas. Di Menna (1955) found that, of 431 yeasts obtained from air samples taken in and around a building in the New Zealand town of Dunedin, 42% were *Cryptococcus* sp., 26% *Debaryomyces*, 18% *Sporobolomyces/ Rhodotorula* with trace amounts of *Torulopsis, Candida* and *Trichosporon*. A two-year survey of the air at Manhatten, Kansas gave the following results:

Organism	Rogerson (1958)	Kramer et al. (1961)
Pullularia (Aureobasidium)	15	819
Sporobolomyces	273	469
Candida	1	1
Rhodotorula	1	1
Yeasts	139	9,563
Total colonies	13,685	113,667

Davenport demonstrated yeasts in the air of apple orchards using the Rotorod technique (Asai, 1960; Carter, 1961). In four separate trials he obtained A. pullulans and Sp. pararoseus/Rh. mucilaginosa from the air in Orchard 28, as well as mould spores. However, in another orchard in which cider-apple press cloths had been hung to dry for generations, he obtained Sacch. uvarum and acetic acid bacteria on two samplings twelve months apart. Further, the air in the Long Ashton Research Station cider factory during the milling and pressing season contained Sacch. uvarum, C. mycoderma and acetic acid bacteria. The first yeast has been isolated repeatedly from Long Ashton ciders.

2. Insects

Insects are probably the most important aerial vectors for yeasts in nature; while crawling insects are not strictly speaking aerial vectors, those that inhabit the trees have been considered to be so for this purpose.

Insects collected by Davenport from the leaves were either microbiologically sterile, or carried A. pullulans or moulds; the "insects" included caterpillars of the summer fruit tortrix, Adoseophyes orana and mites. Grubs of the apple blossom weevil (Anthonomes pomorum) collected at the blossom stage also carried moulds only. The yeast flora of the remaining insects collected in the tree is given in Table V. There was very wide variation in the numbers of yeasts per specimen, some may have been lost in the digestive tract between sampling and processing prior to plating, but even specimens with low counts carried at least three yeast species. One of these, the froghopper, was important for being a source of Sacch. carlsbergensis, a yeast closely related to Sacch. uvarum. The frequent occurrence of C. pulcherrima, C. reukaufii and Rh. glutinis/Sp. roseus is striking. Except for Kl. apiculata, none of the yeasts described by Phaff et al. (1966) was found in the samples of Drosophila, but firm conclusions should not be drawn from a single sample. The flora of the bees resembled that found by Lund (1954), but the wasp flora was more extensive than he reported.

TABLE V. *Yeast Flora of Insects Collected in Orchard 28, Research Station, Long Ashton (Reproduced with permission from Davenport, 1968)*

[Apple cultivars: Court Royal (CR) and Yarlington Mill (YM). + Indicates presence of organism]

Yeast	Spiders[a]	St. Mark fever flies[b]		Bees	Crane flies[c]	Aphids	Manure flies[d]	Red and black frog-hoppers[e]	Ants	Wasps	Fruit flies[f]
	Dormant bud	Blossom		Green cluster →80% petal fall	80% Petal fall	Fruit-let	Mature fruit				
	CR & YM	YM	CR	CR & YM	CR	+	CR & YM	CR & YM	CR & YM	CR & YM	CR
Aureobasidium pullulans				+	+	+					
Saccharomyces carlsbergensis	+									+	
Hansenula anomala	+							++			
Kloeckera apiculata/			+								
Hanseniaspora valbyensis			+	+++			+		+		+
Torulopsis famata/											
Debaryomyces hansenii		++		+++							
Candida pulcherrima	++				+		++	++	++	+	++
Candida reukaufii							++				
Candida parapsilosis								+			
Cryptococcus/Torulopsis famata var. A											
Cryptococcus sp.						+					
Rhodotorula glutinis/	+										
Sporobolomyces roseus		+		+			+	+		+	+
Rhodotorula aurantiaca/										+	
Sporobolomyces salmonicolor											
Rhodotorula/Sporobolomyces sp. C					+						
Trichosporon sp.	+					+					
Total count/specimen	1.6×10^4	Very few	1.25×10^6	1.4×10^4 →1.4×10^6	2.16×10^5		1.5×10^6	3.55×10^3	2.1×10^3	1.65×10^5	440

[a] *Arachnida* sp.; [b] *Bilba* sp.; [c] *Ctenophora ornata*; [d] *Scotophaga sterioranum*; [e] *Cercopsis vulnerata*; [f] *Drosophila* sp.

3. *Birds*

Kawakita and van Uden (1965) examined the yeasts in the digestive tract of gulls and terns, but no studies can be found of the flora of birds in apple orchards, except possibly of searches for *Cr. neoformans* in nests and droppings. Perch sites in the rows of Yarlington Mill and Court Royal, sampled from swelling bud to green cluster, disclosed only *A. pullulans*, an organism previously found on apple bark (Section II.C.1). Leaves covered with bird droppings sampled at 80% petal fall, had an average count of 3×10^5 yeasts/leaf (approximately $10^4/cm^2$) that was greater than found for unmarked leaves. The flora was made up exclusively of carotenoid-producing yeasts, *Rh. glutinis/Sp. roseus*, *Rh. aurantiaca/Sp. salmonicolor*, *Rhodotorula/Sporobolomyces* sp. C; only the first had been present on unmarked leaves (Section II.C.3).

Fresh bird droppings, collected from the grass beneath the trees at dormant bud and 80% petal fall, showed some differences according to the cultivar under which they were collected. Thus, *C. albicans*, a yeast common to birds, was the dominant yeast under Court Royal, followed by *C. pulcherrima*, *Rh. glutinis/Sp. roseus*, *Rhodotorula/Sporobolomyces* sp. C, *Rh. aurantiaca/Sp. salmonicolor*. By contrast, the ubiquitous *Rh. glutinis/Sp. roseus* was common under Yarlington Mill, followed by *Rhodotorula/Sporobolomyces* sp. C, *H. californica* and *Sacch. florentinus*. Both these floras should be compared with those given in the "Spring" columns of Table II (p. 78); apart from some of the carotenoid-producing yeasts, they are very different. It is interesting to note that *Sacch. florentinus* was isolated from the "Autumn" grass flora.

C. TREE FACTORS

1. *Bark*

The bark of the two cider-apple cultivars had less than one yeast per cm^2, the species identified being *A. pullulans*, *C. reukaufii* and *Rh. glutinis/Sp. roseus*, the first being most rare. Similarly, Lund (1954) found only a single yeast species, *Sacch. fermentati*, on the bark of forest trees.

2. *Buds*

Keener (1950, 1951) showed that buds of some woody plants have both an internal and external microflora. Apple buds, free from obvious fungal growth, were examined at different bud states (Davenport, 1968). The internal flora of the buds varied from 240 to $5 \times 10^5/g$. From the frequency with which the different yeasts occurred, it could be said that moulds, *A. pullulans*, *T. famata/D. hansenii*, *Rh. aurantiaca Sp. salmonicolor*, *Rh. glutinis* var. *rubescens* and *Rh. glutinis/Sp. roseus* made up the resident microflora. On the other hand, bacteria, *C. pulcherrima*, *C. reukaufii* and *Sporobolomyces* sp. X, occurred much less frequently; the presence of the first two was probably due to insect vectors, the last was not isolated anywhere else in the orchard. Examination of the complete bud microflora gave the same results for the resident population, but now it included bacteria and *C. reukaufii*, whereas the temporary population included *C. pulcherrima*, *Kl. apiculata/Ha. valbyensis*, *Sp. pararoseus/Rh. mucilaginosa*, *H.anomala*, *Hansenula* sp. II. Thus, bacteria and *C. reukaufii* were probably part of the resident external bud flora, while being only rare internally. As well as obtaining the above results by the maceration technique, Davenport (1967) demonstrated the presence of living yeast cells *in situ* in apple buds by using Sellotape impressions overlaid with apple juice-yeast extract agar (Beech and Davenport, 1969a, b). Further, the lumen of the hairs within the buds contained living cells resembling *A. pullulans* and similar saprophytic fungi.

3. *Leaves*

The two cultivars were found to have the same population of epiphytic yeasts, namely, *A. pullulans*, *C. pulcherrima*, *Rh. glutinis/Sp. roseus* and moulds. The numbers varied widely, being greater on Court Royal. Comparing leaves at the fruitlet stage, those on short spurs had waxy upper and dull lower surfaces, whereas leaves on the current year's extension growth tended to be pubescent. The spur leaves had more yeasts per cm² on their surfaces and petioles than the extension leaves, being of the order of 25–30/cm². Microcolonies were found at the base of the petioles of some of the spur leaves but not of the extension leaves, presumably because the latter were pubescent and prevented the yeasts being moved by rain.

4. *Apple Flowers*

There are virtually no references in the literature to the yeast flora of apple flowers. Lund (1954) found that a flower of *Pyrus communis* contained 4.5×10^3 yeasts, made up of *T. famata*, *C. reukaufii* and *Rh. glutinis*. In contrast, a flower of *Pyrus malus* contained only 20 cells of *Cr. albidus*. The flowers of the two cider-apple cultivars contained fewer species of yeasts than the closed fruit buds, but much greater total numbers. Hence, the flowers of Yarlington Mill contained 9.3×10^4 yeasts/g that consisted of *A. pullulans*, *T. famata/D. hansenii* and *C. reukaufii*. The yeasts in Court Royal flowers were again more numerous, both qualitatively and quantitatively, *Rh. glutinis/Sp. roseus*, *Rh. aurantiaca/Sp. salmonicolor*, but not *C. reukaufii* being present, in addition to those found on Yarlington Mill: the count increased to

$2 \cdot 5 \times 10^5/\text{g}$. Phaff *et al.* (1966) have commented that yeasts from other flowers form a well defined group, *A. pullulans* being most common, followed by species of *Cryptococcus*, *Rhodotorula* and *Sporobolomyces*, with *Candida* sp. and *Torulopsis* sp. being next in importance. They commented particularly on the presence of *C. reukaufii* and the closely related species *C. pulcherrima*. With the exception of the latter, which is very common on the fruit, the yeast flora of apple flowers is similar to that of other flowers.

D. FRUIT STATES

1. *Yeast Distribution Throughout the Apple*

It is not always appreciated that apples, like their fruit buds, have both an internal and an external yeast flora. Romwalter and von Király (1939) isolated *Sacch. cerevisiae* var. *ellipsoideus* and some *Torulopsis* spp. from fruit tissue. Marcus (1942) and Niethammer (1942) found several asporogenous yeasts in the flesh of apples, pears and other fruits (see also Tanner, 1944). Lüthi (1959b), in his review of the microflora of non-citrus juices, considered that internal infections probably occurred at the flower stage. Marshall and Walkley (1951a) stated that the cores of sound Bramley Seedling apples contained 790 yeasts/g, but failed to find any in the flesh. Where apples had developed internal rots round the core, the obviously rotted flesh contained $2 \cdot 3 \times 10^4$ to $2 \cdot 9 \times 10^4$ yeasts/g, but in the areas apparently free from rot the counts lay between 100 and 300/g. The cores, eye and stalk cavities of aseptically dissected samples of the cultivar Kingston Black, taken from a washer, were all found (Beech, 1957, 1959a) to contain yeast populations that were qualitatively and quantitatively different from one another, as shown in Table VI. Sections dissected aseptically from an unwashed sample of the same bulk of fruit had a more varied yeast

TABLE VI. *Yeast Flora of Sections of the Cider-apple Cultivar Kingston Black (Reproduced with permission from Beech, 1959a)*

Apple section	Major components	Minor components
Skin	*Candida catenulta*, *Hansenula* sp. B	*Candida parapsilosis* var. *intermedia*
Stalk cavity	*Torulopsis* sp. A	*Candida parapsilosis* var. *intermedia* *Candida pulcherrima*
Core	*Candida* sp. F, *Hansenula* sp. B	*Candida pulcherrima*
Eye cavity	*Candida* sp. F, *Torulopsis* sp. C	*Candida pulcherrima*, *Hansenula* sp. B

TABLE VII. *Distribution of Yeasts within Apples of the Cultivar Court Royal (Reproduced with permission from Davenport, 1968)*

[+ *Indicates presence of organism*]

Yeast	Skin	Eye		Core	Pips	Stalk		Flesh plugs
		External	Internal		Internal	External	Internal	
Aureobasidium pullulans	+		+	+		+		
Saccharomyces florentinus			+		+	+		
Hanseniaspora sp. A	+					+	+	+
Candida pulcherrima	+	+	+	+	+	+		
Candida reukaufii		+	+				+	
Kloeckera apiculata/Hanseniaspora valbyensis	+	+	+	+				
Torulopsis famata/Debaryomyces hansenii	+	+	+	+		+	+	+
Torulopsis famata/Cryptococcus luteolus var. A								
Torulopsis nitratophila		+						
Rhodotorula glutinis/Sporobolomyces roseus		+						
Trichosporon sp.			+					
Total viable count	10^6/cm^2	3.3×10^3/cm^2	330/g	1.4×10^3/g	1/g	10^3/cm^2	2.5×10^4/g	750/g

flora that included *Pi. fermentans*, two *Debaryomyces* spp., and three *Candida* spp. not present on the washed fruit.

Because yeasts had been found inside closed fruit buds, Davenport extended the above work by comparing macerates of fresh plugs taken aseptically from defined areas within sound apples as well as from sections examined by Beech. From the results given in Table VII it is obvious that the greatest numbers of yeasts were on the skin and in the flesh under the stalk (even though one cannot compare numbers of yeasts measured per unit area and per unit weight). Only the external surfaces of the pips were sterile, but even these contained two yeast species internally. All the fleshy parts of the fruit contained yeasts, yet the distribution was not uniform. The same effect was noted by Samish *et al.* (1963) for bacteria in sound tomatoes and cucumbers. The greatest number of species were in the flesh under the eye, while the flesh not underneath specialized structures contained two species, *C. pulcherrima* and *Kl. apiculata/Ha. valbyensis*. There was a simpler pattern of yeasts in Yarlington Mill, but again all the flesh plugs contained *C. pulcherrima* It is interesting to speculate how the apple, or even the dormant bud, acquires this internal flora.

2. *Changes in the Yeast Flora with Fruit Maturity*

In contrast to the paucity of published information on the yeast flora of apple flowers, there is a wealth of detail on the flora of the fruit. However, this usually relates to the mature fruit on the tree or even after harvesting whereas, as shown in Table VIII, the yeast flora changes from the fruitlet to the mummified fruit. The simple yeast flora of the immature fruit became much more complex with increasing maturity. Several species were then found that had not been present in the bud stages, including *Sacch. florentinus*, *Hanseniaspora* var. A, *T. famata/Cr. luteolus* var. A, *T. nitratophila* and, for the first time on apples, a species of *Trichosporon*.

If the comparison is extended to the flora of the bulk sample of mature fruit, as distinct from the sum total found on separate parts of individual apples, then *H. suaveoleus*, *Kloeckera* sp. A and *Saccharomycodes ludwigii* were also found on mature fruit only. The flora changed once the fruit began to mummify; the population increased considerably in numbers, probably due to fermentation of the juice sugars, since *Saccharomyces* sp. increased at the expense of those characteristic of buds and flowers. No species of *Citeromyces* has been isolated from apples prior to this survey. As the mummified fruit became increasingly desiccated in the following year, *Saccharomyces* sp. were replaced by species of *Hansenula* and other genera more characteristic of the mature fruit. Finally, the flora of the fully mummified fruit was very similar to

TABLE VIII. Yeast Flora of Whole Apples (Cultivar Court Royal) at Different Stages of Development (Reproduced with permission from Davenport, 1968)

[+ Indicates presence of organism]

Yeast	Immature fruit	Mature fruit		Mummified fruit		
		Individual	Bulk	Dormant bud	Swelling/breaking	Mouse ear/green cluster
Aureobasidium pullulans	+	+	+			+
Saccharomyces florentinus		+				
Saccharomyces fructuum var. A						
Citeromyces sp. A				+	+	+
Hansenula anomala				+	+	
Hansenula mrakii				+	+	
Hansenula californica						
Hansenula suaveolens			+			
Pichia fermentans					+	
Pichia polymorpha					+	
Pichia farinosa						
Hanseniaspora var. A				+		
Candida pulcherrima	+	+	+		+	+
Candida reukaufii	+	+	+	+		+
Kloeckera apiculata/Hansenula valbyensis		+				
Kloeckera sp. A						
Kloeckera sp. Y						
Torulopsis famata/Debaryomyces hansenii		+		+	+	
Torulopsis famata/Cryptococcus luteolus var. A		+			+	+
Torulopsis nitratophila		+	+			
Rhodotorula glutinis/Sporobolomyces roseus						
Rhodotorula aurantiaca/Sporobolomyces salmonicolor	+				+	+
Saccharomycodes ludwigii				+		
Trichosporon sp.		+				

that of the immature fruit. It cannot, however, be assumed that all cider apples will pass through a similar cycle of yeasts. Yarlington Mill, in contrast, showed only slight variations between the incidence of *A. pullulans*, *C. pulcherrima*, *Kl. apiculata/Ha. valbyensis*, *T. famata/D. hansenii* and *Rh. glutinis/Sp. roseus*, yeasts characteristic of apple buds and leaves. If the phenolics and tannins of cider apples are inhibitory to some yeasts, then a simpler microflora might be expected in the bitter-sweet Yarlington Mill than in the neutral cultivar Court Royal. Alternatively, the results could be construed as evidence that the microflora of an apple is a cultivar characteristic, irrespective of tannin content. The case could be argued either way when one considers the difference in the microflora of soil, grass, vectors, etc., associated with them. The results for Yarlington Mill agree with those obtained by Bowen (1962) for mature fruit from this same orchard (Bowen and Beech, 1964). Unfortunately he did not examine any cultivars with such low tannin contents as Court Royal.

Some of the fruit lay on the ground under the trees for three months after harvesting, and was then sampled when rotten. The count had risen considerably compared with mature fruit, being greater than 27×10^6/g for Court Royal and 4×10^6/g for Yarlington Mill. Again the yeast flora of the Yarlington Mill fruit was very simple, *H. californica*, *Candida/Hansenula* sp. A, *C. reukaufii*, *Rh. aurantiaca/Sp. salmonicolor*, *Rh. glutininis/Sp. roseus*, *Rh. glutinis* var. *rubescens* and *Trichosporon* sp. It was more complex than the fruit on the tree, whether the latter were mature or mummified. *Hansenula californica* could have come from bird droppings or small slugs, and some of the rhodotorulae from the soil or grasses.

E. GEOGRAPHICAL EFFECTS

The yeast flora of an orchard could be affected by the aspect and terrain and by the position of the orchard within a country. It is difficult to assess the importance of country to country variation, except possibly with a dessert apple such as Golden Delicious which is grown over large areas of the globe.

1. *Yeast Variation within an Orchard*

a. Section of an orchard. In the season 1963/4 all the bud, flower, fruit states and vectors in Rows 5 and 6 of Orchard 28 at Long Ashton Research Station were sampled as often as possible. This involved a minimum of 97 samplings and the isolation and identification of nearly 500 yeasts. Some of the results have been given in Sections II.A–II.D (pp. 75–87). By combining all these results it should be possible to estimate the total yeast population of a swathe cut through the orchard

TABLE IX. *Yeasts Found in Rows 5 and 6 of Orchard 28, Season 1963/4 at Long Ashton Research Station (Reproduced with permission from Davenport, 1968)*

Yeast	No. of times isolated as a percentage of total no. of samplings (97)	No. of times isolated from bud, flower or mature fruit as a percentage of the total possible (14)
(a) Yeasts associated with apple buds, flowers and/or mature fruit		
Aureobasidium pullulans	65	100
Rhodotorula glutinis/Sporobolomyces roseus	70	71
Torulopsis famata/Debaryomyces hansenii	47	71
Candida reukaufii	51	64
Rhodotorula aurantiaca/Sporobolomyces salmonicolor	40	64
Candida pulcherrima	48	43
Kloeckera apiculata/Hanseniaspora valbyensis	26	22
Trichosporon sp.	17	7
Saccharomyces florentinus	8	7
Torulopsis famata/Cryptococcus luteolus var. A	4	7
Torulopsis nitratophila	4	7
Hanseniaspora sp. A	1	7
Hansenula suaveolens	1	7
Kloeckera sp. A	1	7
Saccharomycodes ludwigii	1	7
(b) Yeasts associated with apple buds and flowers but NOT with mature fruit		
Rhodotorula glutinis var. *rubescens*	16	43
Hansenula anomala/Candida pelliculosa	10	7
Rhodotorula mucilaginosa/Sporobolomyces pararoseus	3	7
Saccharomyces cerevisiae/Candida robusta	2	7
Hansenula sp. II	1	7
(c) Yeasts associated with mummified or rotten fruit but NOT with mature fruit		
Torulopsis etchellsii	6	
Candida parapsilosis	3	
Pichia farinosa	2	
Citeromyces sp. A	1	
Kloeckera sp. Y	1	
Pichia fermentans	1	
Saccharomyces fructuum var. A	1	

TABLE IX (*cont.*)

Yeast	No. of times isolated as a percentage of Total No. of samplings (97)
(d) Yeasts associated with vectors only	
Hansenula californica	14
Pichia membranaefaciens	14
Hansenula silvicola	10
Hansenula beckii	7
Hanseniaspora uvarum	6
Rhodotorula/Sporobolomyces sp. C	6
Candida melinii	4
Hansenula mrakii	4
Pichia polymorpha	4
Candida solani var. N	3
Hansenula sp. X	3
Kloeckera antillarum	3
Rhodotorula macerans/Sporobolomyces sp. A	3
Torulopsis candida/Debaryomyces subglobosus	3
Candida albicans	2
Cryptococcus/Torulopsis famata var. A	2
Cryptococcus sp.	2
Hansenula/Torulopsis sp. B	2
Pichia guilliermondii	2
Saccharomyces carlsbergensis	2

from north to south and occupying roughly one sixth of its area. Not every vector was examined at each of the 14 samplings, i.e. two for each stage of bud and fruit development (see Fig. 1, facing p. 74). With this proviso, it can be seen from Table IXa that, of the 47 species or varieties of yeasts, the following were associated with the mature fruit: *A. pullulans, Rh. glutinis/Sp. roseus, T. famata/D. hansenii, C. reukaufii, Rh. aurantiaca/Sp. salmonicolor, C. pulcherrima, Kl. apiculata/Ha. valbyensis, Trichosporon* sp., *Sacch. florentinus, T. famata/Cr. luteolus* var. A, *T. nitratophila, Hanseniaspora* sp. A, *H. suaveolens, Kloeckera* sp. A and *S. ludwigii*. One yeast, *Rh. glutinis* var. *rubescens*, was found at many of the bud and flower stages but not on the mature fruit (Table IXb). It probably occupied an intermediate position between bud and fruit flora in the season 1963/4. On the other hand, *Rh. mucilaginosa/Sp. pararoseus, H. anomala, Sacch. cerevisiae* and

Hansenula sp. II, were isolated only occasionally from the bud stages. The number of times a yeast was associated with a particular stage does not seem significant, since seven yeasts were isolated, but only rarely, during the whole investigation, yet were found on the mature fruit. The sources of four of these, *Hanseniaspora* sp. A, *H. suaveolens, Kloeckera* sp. A, *S. ludwigii*, are unknown since they were not found anywhere else. Similarly, four of the seven yeasts found on mummified and rotten fruit, but not on mature fruit (Table IXc), i.e. *Citeromyces* sp. A, *Kloeckera* sp. Y, *Pi. fermentans* and *Sacch. fructuum* var. A, came from no other source. While some of the 47 yeasts came from vectors as well as bud, flower or fruit stages, there were another 25 yeasts species that were found on vectors only (Table IXd). Yet one cannot say that they were unimportant since, in another season possibly due to a change in climate or in management conditions, some of these yeasts could be found on the fruit. For example, in 1962, Bowen isolated *T. candida* from Row 6 Yarlington Mill in this same orchard.

b. Complete orchard. Davenport's analysis of the yeast flora of the whole of Orchard 28 in the season 1962/3 (Table X) showed that the major yeasts for Rows 5 and 6 were, with the exception of *C. reukaufii*, the same major yeasts found in 1963/4. However, *C. reukaufii* and *Kl. apiculata* were found on other cultivars. *Candida reukaufii* was less important than in 1963/4, while the *Trichosporon* sp. occurred more frequently, although we are now comparing the flora of a complete orchard with that of a section. Probably for the same reason, *H. californica, H. silvicola, Pi. guilliermondii* and *T. candida* were found on vectors only in 1963/4. *Sporobolomyces/Rhodotorula* sp. B, *T. anomala, Trich. pullulans* and *Trich. cutaneum* were present in a limited number of cultivars in 1962/3, probably the reason why they were not found in 1963/4. The remaining fruit yeasts of 1963/4, may well have been present in the previous year's vectors. It would need complete analyses of all sources within the orchard for both seasons to decide whether the season had an effect on the yeast flora of the orchard. One could say from the present evidence that the major components of the yeast flora would always be present on the fruit in any season, but that there would be seasonal variations in the minor components.

In 1962/3 (Table X) there was a distinct decline in the number of yeast species on the fruit going from the windward side of the orchard (YM 12) to the leeward (SC 1), irrespective of there being six apple cultivars in the survey. This change in species could be due to dispersion both of air-borne yeasts and aerial vectors. The range of counts did not show this clear-cut pattern, perhaps justifying the contention of Bowen and Beech (1964) that the proportional count was more informative than a total count.

TABLE X. *Yeast Flora of Mature Fruit, Orchard 28, Research Station, Long Ashton: Season 1962/3 (Reproduced with permission from Davenport, 1968)*

[+ Indicates presence of organism]

Cultivar	Sweet Coppin	Court Royal	Yarlington Mill	Stoke Red	Sweet Coppin	Foxwhelp	Tremlett's Bitter	Court Royal	Yarlington Mill	Frequency (%)
Row number in orchard	1	5	6	7	8	9	10	11	12	
Number of yeast species	5	7	6	7	7	9	9	9	12	
Count range (10^3/specimen)	3·6–8·4	3·6–57·2	0·4–17·2	2·0–36·4	2·0–35·6	7·2–37·6	0·4–13·6	0·8–80	7·2–25·2	
Yeast										
Aureobasidium pullulans	+	+	+	+	+	+	+	+	+	100
Torulopsis famata/Debaryomyces hansenii	+	+	+	+	+	+	+	+	+	100
Rhodotorula glutinis/Sporobolomyces roseus		+	+	+	+	+	+	+	+	89
Rhodotorula aurantiaca/Sporobolomyces salmonicolor	+		+	+	+	+	+	+	+	89
Candida pulcherrima		+	++	++	++	++	++	++	++	89
Kloeckera apiculata/Hanseniaspora valbyensis		+			++	+	++	+	++	67
Pichia guilliermondii	+				+	+	++	+		45
Trichosporon sp.				+		+	+		+	33
Torolopsis candida/Debaryomyces subglobosus			+	+			+		+	33
Trichosporon pullulans	+								++	22
Candida reukaufii								++	++	22
Trichosporon cutaneum						+				11
Hansenula californica								+		11
Hansenula silvicola		+								11
Torulopsis anomala var. A		+								11
Saccharomyces cerevisiae									++	11
Sporobolomyces/Rhodotorula sp. B									++	11

2. Comparison of Orchards within a Country

Bowen (1962) investigated the variations in the yeast flora of cider apples growing in four orchards at Long Ashton and in orchards in four cider apple-growing counties; as far as possible the same cultivars were chosen at each site. It was concluded (Bowen and Beech, 1964) that there was a restricted type of yeast flora on the Long Ashton estate as a whole, rather than on a particular site or cultivar. *Candida pulcherrima* and *Rhodotorula* sp. were present in all areas, one or the other being the predominant yeast; *C. krusei* was present universally as a trace component. *Debaryomyces kloeckeri* (now *D. hansenii*, Kreger-van Rij, 1964) was present in all except the Somerset orchards, but its imperfect partner, *T. famata*, was present in one: *D. kloeckeri* was the major yeast in one Herefordshire orchard. The authors suggested there was some limited association of certain yeasts with particular areas, and concluded that there was more evidence for a geographical influence rather than a distribution of yeast species based on apple cultivars. Castelli (1957) also found a geographical distribution of some wine yeasts in Italy, but the greater climatic differences there probably made the effect more obvious. Barker (1950b) found no evidence for any particular yeast or group of yeasts associated with the bittersweet cultivar, Kingston Black.

3. Comparison of English and Foreign Orchards

Clark *et al.* (1954) found that the yeast flora of hand-picked apples (McIntosh) from Quebec consisted of *C. malicola* (Clark and Wallace, 1954), *T. famata*, *Rh. glutinis* var. *rubescens*, *Rh. mucilaginosa*, *Cr. albidus* and *Cr. neoformans*. *Candida malicola* has since been reclassified as *A. pullulans* (Cooke, 1962). Previous isolates of the animal pathogen, *Cr. neoformans*, from vegetable sources have been very rare; Beech (1959a) isolated a yeast with similar biochemical properties from cider apples but did not test its pathogenicity. Wild apples gathered from various sites in Quebec supported *C. malicola* (*A. pullulans*), *Cr. albidus*, *Cr. laurentii*, *Cr. neoformans* and *C. scottii*. Such large proportion of *Cryptococcus* sp. would be unusual on English cider apples, and possibly they are more indigenous to Quebec apples than *A. pullulans*, as postulated by the authors. Williams *et al.* (1956) also examined the yeast flora of Quebec apples (McIntosh) during the ripening period. In three orchards the numbers decreased steadily from about 1.8×10^4/g at the beginning of August, to 300/g in September, rising to 8×10^3–1.2×10^4/g in October. Marshall and Walkely (1951a) reported the reverse trend in England for the skin flora of the cultivar Grenadier; numbers rose from 2/cm^2 in June to 5×10^3/cm^2 in September, then decreased steadily to 500/cm^2 in October. Possibly this difference can be explained by the changing surface-area to weight ratio as the fruit ripens. Certainly the number of species per unit weight changes for English cider apples, as described by Marshall and Walkley, but qualitative and quantitative changes do not necessarily parallel one another. The yeasts isolated by Williams *et al.* (1956), in order of importance, were *C. malicola* (*A. pullulans*), *Rh. glutinis* var. *rubescens*, *Cr. laurentii*, *T. famata*, *Rh. mucilaginosa*, *T. candida*, *C. krusei*, *Cr. albidus*, *Kl. apiculata* and *Sacch. bisporus* (one isolate out of 222). The preponderance of *A. pullulans*, *Rhodotorula* and *Torulopsis* spp., and the small numbers of *Saccharomyces* sp., is similar to results obtained with English apples. The results differ by the absence of *C. pulcherrima*, the few isolates of *Kloeckera* sp. and, again, the presence of *Cryptococcus* sp. Phaff *et al.* (1966) state that: "*Candida pulcherrima* can be isolated often from flowers as well as fruits, and on the latter they may originate from the blossoms." This is not true of the flowers of the English cider apples (Section II.C.4, p. 86). Further, although *C. pulcherrima* is commonly found on English and European apples (Beech, 1959a), we can find only one reference in the literature to this species having been found in North America, and then without the source being stated definitely (Spencer and Phaff, 1963).

Candida pulcherrima was found by Legarkis (1961) on cider apples growing in northern and central Normandy (France). This yeast was accompanied by *C. tropicalis*, *C. mycoderma*, *C. parapsilosis* var. *intermedia*, *Kl. apiculata*, *T. bacillaris*, *Pi. fermentans*, *C. tropicalis*, *Sacch. rosei*, *Sacch. cerevisiae* and *Sacch. carlsbergensis*. *Saccharomyces rosei* has not so far been isolated from English apples, juices or ciders.

Sasaki and Yoshida (1959) surveyed the superficial yeast flora of Japanese apples by rubbing their skins on the surfaces of agar media and incubating at 27°, a temperature that would inhibit the growth of many psychrophiles found on apples (Beech and Davenport, 1969a); an abstract of their results is given in Table XI. Insufficient evidence was given to decide whether the differences were seasonal, varietal or geographical. Again, the absence of *C. pulcherrima* is interesting; even though iron salts were not added to the medium, one would still expect the isolation of a yeast classed as a strain of *C. reukaufii* had *C. pulcherrima* been present. In fact, *C. pulcherrima* was absent from the wide range of fruits examined by Sasaki and Yoshida. For the remaining yeasts, *Candida*, *Rhodotorula* and *Sporobolomyces* spp. are common on English apples; Beech (1959a) and Bowen and Beech (1964) isolated *D. kloeckeri* from cider apples, but not from those grown at Long Ashton or any other site in Somerset. Similarly, Davenport (1968) isolated *T.*

famata from Long Ashton fruit, but could only induce the formation of ascopores in one isolate during the identification tests.

TABLE XI. *External Yeast Flora of Japanese Apples* (*After Sasaki and Yoshida, 1959*)

[+Indicates presence of organism]

Yeast	Early variety	Late variety
Dermatium (*Aureobasidium*) *pullulans*	+	+
Candida tropicalis	+	
Candida tenuis		+
Torulopsis candida	+	+
Torulopsis aeria	+	+
Debaromyces kloeckeri	+	+
Debaryomyces nicotianae		+
Sporobolomyces roseus	+	+
Rhodotorula glutinis	+	+
Rhodotorula mucilaginosa	+	+
Rhodotorula flava		+

Thus, in spite of the difficulties of obtaining a true assessment of all the yeasts on a sample (Beech and Davenport, 1969a, b), it is possible to make valid comparisons between the important components of the yeast flora of apples grown at different sites. It would need a concerted survey carried out in the same season, on the same apple cultivars in different orchards of the world, in order to determine whether the distribution of apple yeasts was determined varietally or geographically. Techniques for preparing the samples, the media and incubation conditions would need rigorous standardization beforehand to ensure that the results were valid. Some previous surveys have employed isolation techniques that would prevent the growth of some of the more exacting species; the complete absence of *Schizosaccharomyces* and *Brettanomyces* spp. from any orchard is noteworthy. Bibliographic references to the yeast species isolated from apples, juice and cider are given in the Appendix.

III. Juice Processing

Cider apples are usually allowed to fall as they ripen, and the crop is harvested by hand or mechanically. It is then transported to the factory in sacks or, more frequently nowadays, in bulk tipping trucks. Fruit may be stored for some time before pressing, formerly in sacks, but now almost universally in silos. Dessert apples, hand picked from the tree, are sorted and any misshapen fruit stored in bins until sufficient are present to send to the cider or fruit juice factory. The yeast flora can be modified to a greater or lesser extent by any one of the processes employed.

A. FRUIT HANDLING

1. *Harvesting and Storage*

Davenport (1968) compared the yeast flora of mature Court Royal apples as they hung on the tree (Table VIII) and as they lay on the floor of the orchard. The flora of the latter was modified, *A. pullulans*, *Rh. glutinis*/*Sp. roseus*, *Hanseniaspora* var. A, *Trichosporon* sp., *Sacch. florentinus* and *T. nitratophila* no longer being recovered, while four new species had been acquired, *Rh. aurantiaca*/*Sp. salmonicolor*, *Rh. rubra*, *Hansenula*/*Candida* sp. III and *Pi. polymorpha*. The first could have come from the grass (Table II) and the fourth from the soil (Table I), but the second and third had not been isolated previously.

Bowen and Beech (1967) examined this aspect in more detail. The yeast flora of the cider apple cultivar Dabinett hanging on the tree in Orchard 1 of the Long Ashton Research Station was exclusively *C. pulcherrima*, with a viable count of 155 yeasts/g. Fruit from the same crop, but gathered aseptically from the orchard floor, had acquired a much greater count, $1·25 \times 10^4$/g, the increase being mainly in *Torulopsis* sp. A and *T. candida*; the minor components of the flora, in order of importance, were *Rh. glutinis* (8%), *C. pulcherrima*, *C. krusei*, *Kl. apiculata* and a *Saccharomyces* sp. (0·2%). A similar effect was obtained in Orchard 20 with the same cultivar. The tree-hung fruit had 10^3 yeasts/g, exclusively *C. pulcherrima*, while the ground fruit had $8·05 \times 10^4$/g, mostly *Rhodotorula* sp. Y (93%), with smaller amounts of *Rh. glutinis*, *Torulopsis* sp. A, *C. pulcherrima* and *Torulopsis* sp. X. In both cases, the count for *C. pulcherrima* had increased from six to eight times, but this species had been overgrown by another yeast.

As shown by Bowen and Beech (1967), the increase in numbers continued during storage in the sack (Table XII). Thus, for the cultivar Sweet Alford, the count increased progressively from 480/g on the tree to $1·25 \times 10^3$/g on the ground, with two further ten-fold increases after storage for one and two weeks respectively in sacks at ambient temperature. The main increase occurred with *Kl. apiculata*, whose count changed from 50/g on the tree to over $4·5 \times 10^4$/g at the end of the storage period. Four species present on the tree fruit were still present after storage; of the remaining three, *Torulopsis* sp. X came from the ground, but the other two first appeared during storage.

Mechanical harvesting of cider apples is still in its infancy; there has

TABLE XII. *Yeasts of Fruit (Cultivar Sweet Alford) and of Pilot Plant Juice, Long Ashton* (*Reproduced with permission from Bowen and Beech, 1967*)

Yeast	Proportion (%) of total yeast population on					
	Fruit from tree	Fruit from ground	Fruit stored 1 week		Fruit stored 2 weeks	
			Fruit flora	Juice flora	Fruit flora	Juice flora
Candida krusei	21			22	7	7
Candida mycoderma	1	57				
Candida parapsilosis			6	24	9	19
Candida pulcherrima	60		45	5	5	3
Candida sp. F				9		25
Candida sp. X						5
Kloeckera apiculata	10	24	30	23	39	23
Kloeckera sp. A				4		
Rhodotorula sp. Y		1	< 1			
Torulopsis famata			1	2	12	
Torulopsis inconspicua						14
Torulopsis sp. A				2		3
Torulopsis sp. R	8	16			8	
Torulopsis sp. X		2	19	9	20	
Total viable count/g or ml	482	$1\cdot23 \times 10^3$	$1\cdot07 \times 10^4$	$8\cdot6 \times 10^4$	$1\cdot162 \times 10^5$	$7\cdot3 \times 10^4$

been only one publication (Beech, 1963a) on the effect of a rather crude spike-type harvesting machine on the yeast flora of apples. Mechanically-harvested Court Royal apples showed an 80-fold increased yeast count after storage for three weeks in the sack. This high yeast count accompanied the development of rotten fruit, which in turn caused a great increase in the sulphite-binding capacity of the juice (Burroughs and Sparks, 1963) and, therefore, rendered ineffective the control of the natural flora at a later stage of processing (Section IV.A.2, p. 109). Thus, when juices of three cultivars were treated with 150 p.p.m. sulphur dioxide, the residual sulphur dioxide next morning was as follows (Beech, 1963a):

Cultivar	Residual free sulphur dioxide content (p.p.m.)	
	Hand-harvested	Mechanically-harvested
Bulmer's Norman	77	14
Court Royal	63	16
Tremlett's Bitter	60	31

The juices of all the hand-harvested samples would be satisfactory in this respect, but only the juice of the mechanically-harvested Tremlett's Bitter. This harvesting method should be restricted to fruit to be pressed immediately or to late maturing cultivars high in tannins, whose wounds seal rapidly. Fruits containing more than 10% rots, however harvested, would have too high a concentration of sulphite-binding compounds. In any case, rotten fruit support an abnormally high count of moulds and yeasts that are difficult to control at a later stage (Marshall and Walkley, 1951a).

2. *Washing*

Following storage, and prior to milling, the apples are washed. Beech (1959b) found that the unwashed fruit had a richer flora (11 species) than washed (8 species). The fruit could not be rendered sterile by washing because of its internal flora. A similar conclusion had been reached by Marshall and Walkley (1951b). They showed that the surface counts of unwashed fruit samples ranging from 7×10^3 to 1.4×10^5 yeasts per apple were reduced to 100–200/apple. Needless to say, the very use of wash water creates its own problems. If used only once the expense is excessive, whereas extensive recirculation of the wash water can cause recontamination of the fruit, and produce a high biochemical oxygen demand (B.O.D.) in the effluent. Limited recirculation, some chlorination, and refreshment with clean water just prior to milling, seem to provide the best compromise.

B. JUICE HANDLING

Juice can be produced for direct fermentation or for intermediate storage as juice or concentrate. The microbiology of juice extraction is common to all three products. The pretreatment required for juice and concentrate storage is detailed in this Section: juice pretreatments prior to fermentation are given in Section IV.A (p. 107).

1. *Milling and Pressing*

Apples hanging on the tree support a small yeast population, but poor conditions of harvesting and storage cause great increases in numbers, although rarely of *Saccharomyces* sp. But, once the juice is expressed, it soon starts to ferment due to the latter yeasts. Their source in grape-juice factories was pin-pointed by Fabian (1933) as the frames and cloths of the pressing equipment. Lüthi (1950) demonstrated that the tubes and juice-holding vessels were further subsidiary sources of infection. Lüthi and Hochstrasser (1952) and Lüthi and Vetsch (1955) showed that the heat-resistant moulds *Paecilomyces varioti*, *Byssochlamys nivea* and *Phialophora mustea* could become established in fruit presses on farms and in juice factories, from which they were very difficult to remove.

Marshall and Walkley (1951b) investigated the sources of infection in an apple-juice factory. At the beginning of the day the pressed-out residue or pomace trapped some of the juice yeasts, but the cloths became contaminated with yeasts very rapidly. At the end of the day the cloths contained from 10^3–8×10^3 yeasts/cm^2; this alone would contribute up to 800 yeasts/ml of juice. Beech (1959a) demonstrated the dramatic change in yeast species from fruit to juice flora by comparing the flora of the cultivar Kingston Black, pressed out on a clean pilot scale and on a large press maintained under conditions common in commercial practice at the time (Table XIII). The flora of the juice from the clean press was very similar to that of the fruit, whereas the second was more like a juice undergoing incipient fermentation. Similarly, in the 1961 pressing season at the Long Ashton Research Station, *Kl. apiculata*, *C. pulcherrima* and *Saccharomyces* sp. occurred in over 70% of the juices pressed from different apple cultivars on the large press. Subsequently the standard of hygiene was improved, so that in the 1963 season *Saccharomyces* sp. had been eliminated and the species and counts in juices from both the large and small presses were more uniform (Table XIV). Even so the counts were much greater than on apples on the tree, although part of this increase was, of course, due to harvesting and storage effects. In the next year each batch of juice was sampled throughout the season. The juices from both presses contained species of seven yeast genera, *Kloeckera*, *Candida*, *Cryptococcus*, *Torulopsis*, *Hansenula*, *Pichia* and *Rhodotorula* in order of diminishing importance. No species of *Saccharomyces* was isolated from any juice, some samples of which did not ferment when incubated in the laboratory, producing instead surface growth of moulds or acetic acid bacteria. Further, there was no general increase in yeast count throughout the season, the numbers varying widely from one sample to the next. Marshall and Walkley (1951a) obtained similar variations with counts on different batches of sound fruit.

These results show that it is possible to keep a large-scale press free from serious yeast infection. This was also shown for the actual cider factories by Bowen and Beech (1967). The qualitative and quantitative features of the yeast flora of fruit going into an English factory in Norfolk, and of the juice produced, were very similar: in this particular factory the milling and pressing equipment had been cleaned just before the fruit was pressed. Similar results were obtained from a Herefordshire factory that maintains a rigorous sanitation programme. There were no *Saccharomyces* yeasts present in the juice, 83% of the yeasts present being *Rh. mucilaginosa* with small amounts of *Kl. apiculata*,

TABLE XIII. *Comparison of the Yeast Flora of Juice of the Cultivar Kingston Black obtained from Pilot- and Commercial-presses (Reproduced with permission from Beech, 1959a)*

Yeast	Juice from	
	Pilot press (%)	Commercial press (%)
Candida pulcherrima	53	2·25
Candida sp. F	10	12·0
Cryptococcus neoformans	3	
Unknown 2	3	
Rhodotorula minuta	3	
Torulopsis sp. A	12	0·5
Apiculate yeasts	3	
Unidentified yeasts	13	17·0
Pichia fermentans		0·25
Hanseniaspora osmophila		7·0
Saccharomyces uvarum		1·5
Saccharomyces cerevisiae var. II		7·0
Saccharomyces microellipsodes		52·5

TABLE XIV. *Yeast Flora of Apples and their Juices from the Pilot- and Commercial-scale Presses; Season 1963*

Yeast	Frequency (%) of yeasts in the season's fruits and juices		
	Macerated fruit	Juice from	
		Pilot press	Commercial press
Candida pulcherrima	75	73	86
Kloeckera apiculata	50	83	100
Carotenoid-forming yeasts	50	17	29
Torulopsis famata/Debaryomyces hansenii	50	50	71
Candida reukaufii	25	23	14
Torulopsis famata var. A		54	43
Hanseniaspora sp.		20	
Pichia membranaefaciens		17	57
Hansenula anomala			29
Average count (cells/g or ml.)[a]	$2·95 \times 10^3$	$2·49 \times 10^5$	$5·01 \times 10^5$

[a] Count/g $\equiv$ 1·4175 (count/ml).

Torulopsis and *Candida* spp. The juice counts in the remaining factories reflected the degree of hygiene maintained in the pressing equipment. The counts were between two and eight times those of the previous two factories. *Saccharomyces* sp. were present in all their juices and the troublesome contaminant of bottled ciders, *S. ludwigii,* was present in most of them.

Conventional rack-and-frame presses are slowly being displaced by continuous presses (Beech, 1962, 1963b; Swindells and Robbins, 1966). It is to be hoped that designers will bear in mind the problem of keeping them free of yeast infection during the height of the pressing season.

2. *Control of Juice Microflora*

Treatment of juice prior to storage or concentration is complicated by the presence of pectin and soluble protein. The first renders filtration uneconomical and both cause permanent hazes when the juice is heated sufficiently to render it sterile (Monties and Barret, 1965). Normally the pectin would be demethylated by the juice enzyme, pectin methyl-esterase, and the pectic acid so formed would be broken down into galacturonic acid by the enzyme, polygalacturonase, secreted by most *Saccharomyces* yeasts found in cider fermentations (Pollard and Kieser, 1959). If the first enzyme is destroyed by heat, the second cannot act and, while this is desirable in cloudy juice intended for consumption as such, it would render the appearance of a cider prepared from it un-appetizing. Juice intended for storage or concentration is normally depectinized with a commercial preparation made from extracts of a suitable mould (Neubeck, 1959). Two hours at 40° or overnight at 20° are usually sufficient; any tendency for the count to increase is often counteracted by the entrainment of cells in the deposit that forms. Clarification by centrifugation causes a reduction in the count to about 3–4% of the original value (Millis, 1951; Marshall and Walkley, 1951b; Beech *et al.,* 1952). Fining with gelatin and tannin is used sometimes but mechanical methods of clarification, especially automatically-fed kieselguhr filters and self-desludging centrifuges, are being used increasingly in large factories.

3. *Juice Storage*

There are several methods for storing juice for later consumption as such, ranging from the New England (U.S.A.) farm-gate sales of chilled preserved apple juice (Smith *et al.,* 1962; Do and Salunkhe, 1964) in the making season, to the modified Boehi (1912) method of storage under 3·5 atm. of carbon dioxide at 2°, or 8 g CO_2/litre (Jenny, 1940; Lüthi, 1959a). Probably only in France is juice stored, not for making cider, but for blending with dry cider prior to bottling. This is to conform with the French laws that prevent cider being sweetened with sugar or concentrated apple juice.

In France, juice is depectinized, clarified, flash-pasteurized and stored at 0°. Some 20% of the stocks usually ferment slowly to dryness, since absolute sterility is not attempted (Pollard and Beech, 1966b). Most French factories store sterile juice at ambient temperature under a blanket of nitrogen ((Ménoret and Gautheret, 1962), a method that needs very strict technical control. The juice is depectinized, centrifuged, filtered and flash-pasteurized at 110° for five seconds into 5,000 to 8,000-gal tanks, previously steam-sterilized at 18 lb/in^2 and filled with sterile nitrogen whilst cooling. A blanket of nitrogen is left over the juice, and all observation ports and valves are kept submerged in a solution of a quaternary ammonium compound. The tanks are fitted with bursting discs or pressure indicators. Frequent inspections are needed to detect incipient mould growth or fermentation when the contents are flash-pasteurized into a freshly sterilized tank. Even better control would be obtained if the gas layer were sampled automatically at intervals by a remotely-controlled gas–liquid chromatographic analyser. The success of the process depends on the rigid sterilization programme for tanks and pipe lines. According to Ménoret and Gautheret (1962), mould growth occasionally develops in juices stored under nitrogen. Some extra protection of this nature is required since *Penicillium brefeldianum* can survive 105° for 30 min (Anon., 1967), and several other moulds can stand 90° for several minutes (Lüthi and Vetsch, 1955). The growth of moulds in tank-stored juices has been attributed to the presence of traces of oxygen. However, Ruyle *et al.* (1946) showed that carbon dioxide had an inhibitory effect, since a mixture of 45% each of carbon dioxide and nitrogen prevented mould growth when 10% of oxygen was present (see also Marshall and Walkley, 1952). Beech *et al.* (1964) showed that some moulds such as *Penicillium expansum* could only be prevented from growing by 100% carbon dioxide; for the remaining moulds they tested, *Aspergillus niger*, *Rhizopus nigricans* and *Monascus ruber*, growth appeared within two weeks when the concentration of carbon dioxide was below 30% (nitrogen being 70%). Ascorbic acid, added at the rate of 200 p.p.m., prevented this growth, presumably by counteracting the effect of traces of oxygen (Peter, 1964). Unless ascorbic acid is added, it is extremely difficult to remove all traces of oxygen from juices. Daepp and Lüthi (1966) found oxygen in the head space of juices filled with carbon dioxide under pressure. Yates *et al.* (1967) suggested that the growth inhibition observed under high carbon dioxide concentrations was primarily due to displacement of any oxygen, and not to the direct toxicity of carbon dioxide. *Byssochlamys rivea*, a heat-resistant infection in many fruit products, had an absolute requirement for carbon dioxide under anaerobic conditions.

C. JUICE CONCENTRATION

1. *Production*

In the early days of vacuum concentration and aroma recovery, the raw juice was not sterile, and yeast growth often occurred in those parts of the evaporator where the temperature was below 30° (Charley *et al.*, 1941). Pipes could become blocked with incrustations of yeasts and juice deposits while the sugar content of the product was reduced by fermentation. This trouble disappeared with the advent of aroma recovery at higher temperatures (Milleville and Eskew, 1944), where 10% of the juice is flashed off at 95°, micro-organisms and enzymes being destroyed at this temperature.

Two types of concentrated juice are produced for use ultimately in cider-making, "half-concentrated" cloudy juice and "fully-concentrated" clear juice (Brun and Tarral, 1963; Pollard and Beech, 1966a, b). The production of half-concentrate is confined almost entirely to Switzerland, as a technical aid to processing within the factory. The fresh juice, fortified if necessary with ascorbic acid, is passed through the aroma recovery plant, then evaporated under vacuum at 45° to 55° to three or four times its original sugar concentration. The half-concentrate is chilled to 0°, injected with four volumes of carbon dioxide, and stored in clean tanks at 0° until it is diluted for juice products or fermented into cider. This type of concentrate cannot be sold in commerce since it is non-sterile and unstable at ambient temperatures; its sugar content, and therefore osmotic pressure, being insufficient to inhibit yeast growth. Concentrate for export is made by depectinizing and clarifying the aroma-stripped juice, evaporating to six or eight times its original sugar concentration, and storing at 6° under a blanket of carbon dioxide. Half-concentrate does not require removal of pectin as the concentration of pectin, sugar and acid are never high enough to form a gel.

2. *Storage Problems*

It is not easy to sterilize half-concentrated juices because of their viscous nature and susceptibility to flavour damage if over-heated. The storage tanks cannot be heat-sterilized since they are held uninsulated in cold rooms. The product is stabilized, therefore, by the low temperature and high carbon dioxide concentration. Beech *et al.* (1964) have shown that the behaviour of juice yeasts in stored concentrates depends, as might be expected, on their sensitivity or otherwise to osmotic pressure (or to the amount of "available" water). Osmophobic yeasts, such as *Sacch. uvarum, Ha. valbyensis, S. ludwigii* and *Br. claussenii*, did not survive storage at a specific gravity of 1·200 for more than a week at 15°. Osmo-tolerant yeasts from apples, such as *D. kloeckeri, C. pulcherrima* and *Kl. apiculata,* survived for three months at 5° and 15°, albeit in reduced numbers, in concentrates of sp. gr. 1·300 made from unoxidized juice. Osmophilic yeasts grew in and fermented this strength of concentrate. One culture, isolated from a cider factory, vigorously fermented the concentrate at 5° and 15°, whether made from oxidized or unoxidized juice. Two strains of *Sacch. rouxii* from culture collections were less vigorous, surviving better at 5° than at 15° in oxidized concentrates, and fermenting unoxidized juice concentrates at both temperatures. These results show that concentrates under sp. gr. 1·300 must be stored below 5°, and processing conditions should be changed to allow sterile storage. The latter technique is now being taught in Switzerland (H. Lüthi, private communication). From the results given in Section III.B.3 (p. 104), de-aeration plus high carbon dioxide concentrations would be necessary to inhibit surface mould growths. Even high gravity concentrates can succumb to the development of osmophilic yeasts, since these organisms will float to the top and start to ferment if condensation from the vapour space dilutes the surface layer (Ingram, 1958).

The importance of the state of oxidation of the concentrate, and its effect on the yeast, was shown by Lüthi (1958). Amino acids rapidly disappeared from apple concentrates stored at high temperatures; the first to go was glutamic acid, followed by asparagine and serine, and then the remaining amino acids. The vitamins calcium pantothenate, thiamine and pyridoxine were also attacked. Furthermore, the concentrate browned rapidly with the accompanying production of fermentation inhibitors, one of which was identified as hydroxymethylfurfural. Not unnaturally, diluted concentrates of this type are often difficult to ferment and need to be pretreated with charcoal to remove the inhibitors, followed by the addition of vitamins and nitrogenous compounds such as ammonium salts and amino acids. Pribela and Betusova (1964) have made more detailed examinations of the changes in the amino acid composition of concentrates heated to temperatures lower than those used by Lüthi, while Vasatko and Pribela (1965) have correlated the disappearance of sugars under the same conditions with the appearance of melanoidins and decomposition products.

IV. Fermentation

A. PRE-FERMENTATION TREATMENTS

1. *Pulp*

During the heyday of the English farm cider-maker, naturally sweet ciders were commonly produced by reducing the nutrient status of the juice. In the U.K. the process was called *keeving* (Beech and Challinor, (1951); while in France, where it is still used to some extent, it is known as *maceration* and *défécation*. Briefly, the milled pulp was held for a few hours before pressing to increase the juice pectin and enzyme content, then small quantities of calcium chloride and sodium chloride were added to the chilled ($<10°$) juice (Tavernier and Jacquin, 1946). The pectin was converted by the juice enzyme, pectin methylesterase, to pectic acid which reacted with the calcium salts to form a calcium pectate jelly (Kieser *et al.*, 1949). Subsequently the yeasts in the juice started to grow, utilizing a proportion of the nitrogen and vitamin content. Some of the newly-formed cells were trapped in the jelly particles as they coalesced, and these were carried to the surface by the rising gas bubbles to form a heavy jelly clot. The remaining crop of yeast formed a deposit in the container below a clear juice of reduced nutrient status. This juice, when syphoned off from between head and lees, tended to ferment more slowly; fermentation often ceased prematurely, leaving a stable naturally sweet cider. The microbiology of the process was described by Beech (1958), who showed that the yeasts in suspension tended to decrease in the chilled juice, but increased later with the growth of the main fermenting yeasts, *Saccharomyces* sp. As the jelly floccules coalesced, the suspended yeast count decreased dramatically until finally only 5×10^3 yeast cells/ml were present in the juice. At this stage the count was the same as in the control juice after pulp filtration. The effect on the subsequent rate of fermentation for juice of the cultivar Sweet Alford was as follows:

Juice	Time to ferment to sp. gr. 1·025 (days)
Keeved, from macerated pulp	263
Filtered, from macerated pulp	> 260

In contrast, unkeeved juices from untreated and macerated pulp took 68 and 86 days respectively to reach the same gravity. The flavour of ciders made successfully by this process was better than that of ciders made from untreated pulp, being full and fruity. However, the process needed considerable supervision in the early stages, and was not successful with juices of high nitrogen content. Further, sweet ciders with pH values above 3·7 were very susceptible to the disorder called "sickness", due to a motile gram-negative non-sporing bacillus, *Zymomonas anaerobia*, that produces alcohol, carbon dioxide and relatively high concentrations of acetaldehyde (Millis, 1951, 1956; Beech and Carr, 1953). For these reasons the process was not adopted for factory use in the U.K. Even holding the sweet cider chilled, as is still done in some French factories, does not prevent large volumes from being attacked by this organism. Masuda et al. (1964a) advocated the addition of apple pulp to the fermenting juice to obtain the improvement in flavour without having to retain natural juice sugars in the cider.

2. Juice

For many years it was standard practice to run freshly pressed juice into vats and to allow fermentation to occur naturally. The source of the yeasts responsible, Saccharomyces sp., was the processing equipment (Section III.B.1, p. 101). The ciders produced varied widely in quality, and bacterial disorders such as acetification, oiliness, etc. were more common than at the present time.

In 1952 Beech et al. demonstrated that the natural yeast flora of the juice could be controlled advantageously by the addition of sulphur dioxide. A concentration of 150 p.p.m. in the juice gave ciders that were superior in flavour to those produced by any other method (Beech, 1953). John Evelyn in 1664 had described the burning of sulphur in casks and the filling of these with apple juice while they still held the sulphur dioxide gas. A similar technique was practised by the better farm cider-makers in the succeeding centuries, but the process gradually ceased, presumably because of the uncertain amount of sulphur dioxide imparted to the juice and the unpleasant nature of the gas. Instead, the untreated juice was allowed to ferment naturally, and an appropriate amount of potassium metabisulphite was added as a preservative between clarification and storage or before sale. The beneficial effects of the sulphur dioxide on the flora and, therefore, on the quality of the cider, was sacrificed in a search for microbial stability in the final product. The wheel has now turned full circle; calculated amounts of sulphur dioxide or its derivatives are injected into the juice, and every effort is made to keep the concentration of sulphite-binding compounds to a minimum. Later, a small amount of sulphur dioxide is added to the sterile cider as a protection against accidental contamination during bottling, and as an anti-oxidant. The effect of these treatments on the yeast flora in the juice is to reduce the total viable yeast and bacteria count to 1·25% of the initial value (Beech, 1953). Sulphur dioxide does not act immediately on yeasts (Table XV). The glucose concentration (1·8%) was the average found in most apple juices (Warcollier and Le Moal, 1935); the principal sugar (fructose) does not combine with sulphur dioxide; solutions were allowed to equilibrate for 24 h before inoculation. The yeast typical of English apples, C. pulcherrima, was destroyed within three hours, whereas the fermenting yeast, Sacch. uvarum, showed only slight reduction in numbers, even after 24 h.

Obviously the pH value, as well as the time of contact, has a considerable effect, since the effectiveness of sulphur dioxide is said to be related to the concentration of undissociated sulphurous acid (Rehm and Wittmann, 1963), although Falk and Giguère (1958) state that molecular H_2SO_3 does not exist in solution. Thus, 15 p.p.m. at pH 3·0 is equivalent to 150 p.p.m. at 4·0 (Vas and Ingram, 1949): Table XVI shows the interaction of these two factors (Beech, 1959a). At the lowest pH level the numbers decreased in the presence of the relatively high concentration of sulphur dioxide, but when sampled at 168 h the juice was in active fermentation, with a count of 10^7 yeast cells/ml.

TABLE XV. *Effect of Sulphur Dioxide on the Viability of Yeasts Suspended in a Glucose/Buffer Mixture, pH 3·0*

Time of contact (h)	Viable count (cells/ml)			
	Candida pulcherrima		*Saccharomyces uvarum*	
	15 p.p.m. SO_2	25 p.p.m. SO_2	15 p.p.m. SO_2	25 p.p.m. SO_2
0	2,180	1,900	5,800	3,800
1	50	25	3,400	3,200
5	0	0	3,000	1,750
24	0	0	2,200	1,400

TABLE XVI. *Effect of Sulphur Dioxide (150 p.p.m.) on Saccharomyces uvarum in Sterile Apple Juice Adjusted to Different pH Values (Reproduced with permission from Beech, 1959a)*

Time of contact (h)	Viable count (cells/ml)		
	pH 3·4	pH 3·7	pH 4·0
0	3,500	850	5,300
2	1,900	800	5,200
5	650	900	6,500
24	Not sampled	35,000	> 10,000

Higher concentrations of sulphur dioxide are necessary to control the natural flora of fresh juice because of the presence of oxygen and sulphite-binding compounds (Whiting and Coggins, 1960a; Burroughs, 1964; Burroughs and Sparks, 1964). Thus, Burroughs (1959) found that nearly half the free sulphur dioxide disappeared within 20 min of addition to the juice, and was not detectable once fermentation started.

Hence, it is possible to sulphite a freshly-pressed juice inefficiently. Of the two examples shown in Table XVII, the juice from mixed low acid fruit did not have a large enough sulphite addition, since no free sulphur dioxide was present next morning. The high concentration of *Kl. apiculata*, *C. pulcherrima* and *Pichia* sp. after treatment was also typical of inefficient sulphiting. In contrast, juice from the cultivar Sweet Coppin was treated correctly, as evidenced by the much reduced yeast and bacterial count, and the absence of the above three species. Yeasts classed as *Torulopsis* could be *Saccharomyces* sp. that did not form spores during their identification tests. The presence of *S. ludwigii* after treatment is noteworthy; it resists sulphur dioxide to persist, not only in ciders, but also in juices preserved with high concentrations of sulphur dioxide (muté) where, with *Schizosaccharomyces* sp., it can cause serious economic losses (Chalenko and Korsakova, 1958; Jakubowska, 1963; Rankine, 1966).

TABLE XVII. *Effect of Sulphur Dioxide on the Microflora of Cider Apple Juices*

Organism	Viable count (cells/ml)			
	Mixed low acid juice		Sweet Coppin juice	
	Untreated	50 p.p.m. SO_2	Untreated	150 p.p.m. SO_2
Kloeckera apiculata	$1 \cdot 68 \times 10^6$	$1 \cdot 7 \times 10^5$	6×10^5	
Saccharomyces sp.	$1 \cdot 3 \times 10^5$	$2 \cdot 5 \times 10^5$	4×10^4	$7 \cdot 1 \times 10^3$
Pichia sp.	5×10^3	10^4	10^4	120
Candida pulcherrima	6×10^3	5×10^3	4×10^4	
Torulopsis sp.			7×10^4	$1 \cdot 4 \times 10^3$
Saccharomycodes ludwigii			4×10^4	$2 \cdot 7 \times 10^3$
Total yeasts	$1 \cdot 82 \times 10^6$	$4 \cdot 35 \times 10^5$	8×10^5	$1 \cdot 17 \times 10^4$
Acetic acid bacteria	+		7×10^6	10^4

Very recent research at Long Ashton has shown the importance of having 30 p.p.m. free sulphur dioxide still present in the juice at least six hours after sulphiting. In this connection it is vital that the percentage of rots in the fruit should be kept below 10% or it will be impossible to retain this amount of free sulphur dioxide (Section III.A.1, p. 99). In practice 100 p.p.m. sulphur dioxide is added to juices with a pH value less than 3·3, and 150 p.p.m. to juices with a higher pH. The pH level in juice from the blended fruit pressed out in cider factories usually lies between 3·5 and 3·7. Juices of single cultivars having pH values above 4·0 can rarely be sulphited satisfactorily without exceeding the legal limit of 200 p.p.m.

B. FERMENTATION

1. *Natural*

Under this heading will be considered fermentations carried out by yeasts naturally present on the fruit, acquired from the pressing equipment or remaining after some form of juice treatment.

a. Yeast succession. The classical picture of a cider fermentation is that it is begun by apiculate yeasts which are soon accompanied and then overtaken by *Saccharomyces* sp.: *S. ludwigii* is often present during fermentation, while *Zygosaccharomyces* (*Saccharomyces* sp.) and *Torula* forms are common (Barker, 1950b, 1951; Challinor, 1955). A similar succession of yeast species occurs in fermenting grape juice (Peynaud and Domercq, 1954, 1955; Domercq, 1956; Mosiashvili, 1956; Melas-Joannidis *et al.*, 1958). The minor microflora of the yeast lees found at the end of cider fermentations has been described by Sainclivier (1951). Clark *et al.* (1954) found that one-year old ciders made from Canadian wild apples contained *Sacch. oviformis, Sacch. cerevisiae, Sacch. steineri, D. kloeckeri, T. candida, C. mesenterica, Pi. polymorpha* and *Pi. membranaefaciens*: it is difficult to exclude film yeasts from cider stored in cask for any length of time.

The succession of yeasts in untreated and sulphited juice of the cider apple cultivar Kingston Black was examined by Beech (1957, 1958, 1959a); the results are summarized in Table XVIII. The effect of efficient sulphiting is shown in the diminution of the natural flora to 1·3% of its original value; *Torulopsis* sp. formed the major part of the flora, and *Saccharomyces* sp. must have been present in such low concentrations as to be undetectable, except possibly by enrichment methods. Excessive development of aerobic organisms during subsequent sampling was prevented by covering the surfaces of the two juices with liquid paraffin.

The yeast succession in the non-sulphited juice was characterized firstly by the complete disappearance of the juice flora with the onset of fermentation, and secondly by the persistence of apiculate yeasts during the height of the main yeast fermentation. A small fermenting yeast, unclassified but similar morphologically to the description of Barker's *Zygosaccharomyces* sp., appeared before half the sugar had disappeared. Bacteria were isolated for the first time at this stage and continued to increase in numbers until the cider was dry. Bacteria were absent at all stages in the fermentation of the sulphited juice; *Sacch. uvarum* grew more prolifically, and therefore fermented the juice more rapidly, than in the non-sulphited control. Apiculate yeasts were absent, and the small unclassified yeast was less important. Thus, sulphiting virtually produced a pure culture fermentation that was more

TABLE XVIII. *Succession of Yeasts Occurring during the Fermentation of Juice of the Cider Apple Cultivar Kingston Black*
(Reproduced with permission from Beech, 1958, 1959a)

Juice

Yeast	Viable count (cells/ml) Untreated Sp. gr. 1·060	Viable count (cells/ml) Sulphited (150 p.p.m. SO₂) Sp. gr. 1·060
Candida pulcherrima	$8·5 \times 10^4$	650
Candida parapsilosis	$1·5 \times 10^4$	
Candida solani	5×10^3	50
Cryptococcus sp. A	5×10^3	
Cryptococcus luteolus	5×10^3	
Torulopsis sp. A	2×10^4	800
Torulopsis sp. B	$1·5 \times 10^4$	550
Rhodotorula sp.	5×10^3	
Kloeckera sp.	5×10^3	
Total	$1·6 \times 10^5$	$2·05 \times 10^3$

Cider

Yeast	Untreated Sp. gr. 1·055	Untreated Sp. gr. 1·045	Untreated Sp. gr. 1·035	Untreated Sp. gr. 1·025	Untreated Sp. gr. 1·015	Untreated Sp. gr. 1·005	Sulphited Sp. gr. 1·055	Sulphited Sp. gr. 1·045	Sulphited Sp. gr. 1·035	Sulphited Sp. gr. 1·025	Sulphited Sp. gr. 1·015	Sulphited Sp. gr. 1·005
Saccharomyces uvarum	$1·165 \times 10^7$	$2·24 \times 10^7$	$1·235 \times 10^7$				$1·76 \times 10^7$	$3·75 \times 10^7$	$2·51 \times 10^7$			
Apiculate yeasts	$8·8 \times 10^5$	$2·5 \times 10^5$	$1·8 \times 10^5$									
Unknown yeast I			$1·15 \times 10^6$									
Bacteria			5×10^4									
Saccharomyces uvarum				$7·28 \times 10^6$	$5·8 \times 10^6$	5×10^6				$6·25 \times 10^6$	$3·85 \times 10^6$	$2·64 \times 10^6$
Apiculate yeasts				10^5	$1·5 \times 10^6$	9×10^5						
Unknown yeast I				$2·75 \times 10^6$	$1·8 \times 10^5$	3×10^5					$< 5 \times 10^4$	$3·6 \times 10^5$
Bacteria				$1·5 \times 10^5$								

efficient than the untreated control. This effect is characteristic of many of the cider fermentations that we have studied. The exceptions occur when *S. ludwigii* is present in the juice. This yeast persists throughout the fermentation of both sulphited and untreated juices, as shown in Table XIX for juice pressed from mixed fruit in a cider factory. Sulphiting reduced the initial count and delayed the growth of this yeast, but even halfway through the fermentation there was virtually only the difference of one generation between them.

TABLE XIX. *Persistence of* Saccharomycodes ludwigii *during a Cider Fermentation*

Sampling period	Saccharomycodes ludwigii (10^3/ml)	
	Untreated juice	Sulphited juice (150 p.p.m. SO_2)
Fresh juice	40	2·6
After 5 degrees loss in sp. gr. by fermentation	42	48
After 25 degrees loss	2,040	900

The influence of the initial sulphiting treatment persists, not only throughout the fermentation, but even to the fully-fermented ciders (Table XX). The low pH value of the Bramley Seedling juice normally restricts bacterial growth during fermentation; the high yeast count included many apiculate yeasts, whereas the cider from sulphited juice contained *Saccharomyces* sp. only. In the other ciders, bacterial activity had been prevented during fermentation by sulphiting the juice, con-

TABLE XX. *Influence of Sulphiting on the Final Microbial Count of Dry Ciders* (*Reproduced with permission from Beech, 1959a*)

Cultivar	Sulphite concentration (p.p.m.)	Original juice pH	Viable organisms after fermentation (10^6/r			
			Yeasts		Bacteria	
			No SO_2 in juice	SO_2 in juice	No SO_2 in juice	SO_2 in jui
Bramley Seedling	150	3·2	50	53·5	—	—
Kingston Black	150	3·5	8·5	60	57	—
Sweet Coppin	150	3·9	10	60	3	—
Mixed fruit	200	4·2	0·1	0·68	4·3	—

sequently the fermentations were more efficient and the flavour of the ciders fresh and free from taints. It is not surprising, therefore, that all English cider-makers now sulphite their juices before fermentation.

b. *Factory flora.* In Section III.B.1 (p. 101) it was shown that the yeast flora of the freshly pressed juice could vary with the standard of sanitation maintained in the factory. Further, the yeast flora of the fruit itself appeared to vary from county to county (Section II.E.2, p. 96). It would seem, therefore, that the yeast responsible for the natural fermentation of the sulphited juice should be different in each factory. Bowen (1962), in the U.K., isolated cultures of *Sacch. cerevisiae* from cider factories in Somerset and Devon, and found (Bowen and Beech, 1967) that different strains of the species were present in each factory (Table XXI). Similarly, differences can be demonstrated in physical properties, such as the ability to flocculate, and these have important economic advantages in filtration costs, etc. Some factories commonly have yeasts other than *Saccharomyces* sp. present in their juices or stored ciders. Thus, the ciders of one Somerset factory commonly contain a *Hansenula* sp., while those of another in the same county support populations of *Brettanomyces* sp. *Saccharomycodes ludwigii* and its variety *bisporus* can be isolated regularly from a number of factories, and only rigid microbiological control of the bottling plant prevents these last two yeasts from infecting the finished product.

TABLE XXI. *Carbon Assimilation Tests with Strains of* Saccharomyces cerevisiae *isolated from some English Cider Factories (Reproduced with permission from Bowen and Beech, 1967)*

[+ *Indicates assimilation*]

Carbon source	Source of yeast			
	Somerset factory			Devon factory
	No. 1	No. 2	No. 3	
Sorbose			+	
Mannitol				+
Sorbitol	+			
Erythritol		+		
Salicin		+		
Succinic acid	+			+
Cellobiose	+	+		

2. *Yeast Addition*

a. *Dominant fermentation.* Spoilage yeasts can still cause problems, even in factories that normally have a desirable fermenting yeast naturally present in their ciders. For this reason investigations into the succession of yeasts occurring in factory cider fermentations have given way to studies on the use of pure yeast cultures. Originally it was thought that a heavy inoculum added to untreated juice would allow the culture to dominate the fermentation. At the time, the techniques for identifying strains of yeasts were in their infancy, and organoleptic assessment of the ciders was virtually the only means of determining whether the added culture had carried out the fermentation (Barker, 1950a). Subtle differences in flavour were difficult or impossible to assess consistently and, in fact, this has become possible only with the advent of gas chromatography. Other workers have attempted a quantitative assessment of dominance, by statistical analysis of such parameters as alcohol production, etc. in replicated samples of control and yeasted ciders and wines (Graff, 1959; Rankine, 1955; Rankine and Lloyd, 1963). The direct assessment of dominance, by typing yeasts isolated from the fermenting ciders, was undertaken by Beech et al. (1952) and Beech (1953). They investigated the effect of different juice treatments on the initial flora, and the subsequent rate of yeast growth under factory conditions. Single clarification treatments, such as centrifugation and coarse filtration, were relatively ineffective. However, adequate sulphiting, either alone or in combination with a clarification treatment, reduced the natural flora very satisfactorily. Complete destruction or removal of the juice yeasts by flash pasteurization or sterile filtration required preliminary enzymic depectinization. Unlike sulphiting, such treatments gave no protection against subsequent infection of the juices and ciders. The yeasts isolated during and after fermentation were examined for colonial appearance, alcohol tolerance, fermentation of raffinose, growth in the presence of actidione, and vitamin requirements. This information was supplemented by measurement of the yeast counts and nitrogen contents of the supernatant ciders, together with rates of fermentation and the production of organic acids. The added yeasts could be recovered only from ciders made from juice in which the natural flora had been adequately suppressed, i.e. by the use of sulphur dioxide, flash pasteurization or sterile filtration. Subsequently the fusel oil content of ciders has proved to be invaluable for determining whether an added yeast has dominated the fermentation (Pollard et al., 1966). This is shown in Table XXII for a number of ciders made from sulphited juices with and without the addition of yeast Port 350R which, in apple juice at least, forms only small amounts of all fusel oil components except n-propanol. With each pair of ciders the added yeast appeared to have dominated the fermentation, since the amounts of both aliphatic and aromatic alcohols were much less than those produced by the Saccharomyces sp. yeast left after sulphiting.

TABLE XXII. *Fusel Oil Components of Ciders made from Sulphited Apple Juice with and without the Addition of Yeast Port 350R (Reproduced with permission from Pollard, Kieser and Beech, 1966)*

Cider number	Amount of fusel oil component (p.p.m.)			
	Higher aliphatic alcohols		2-Phenylethanol	
	Port 350R	Natural yeast	Port 350R	Natural yeast
1	150	237	33	254
2	162	223	54	209
3	167	335	37	241
4	151	197	65	127

The amount of inoculum added is important if the fermentation is to be dominated by the culture yeast. A minimum concentration of 10^6 cells of culture yeast/ml of treated apple juice has been found necessary; Rankine and Lloyd (1963) have given the same value for sulphited grape juice.

b. *Pure culture fermentations.* Until the early 1950s it was widely held that the pure culture fermentation of apple juice would give an apple wine, rather than a typical English cider (Barker, 1951). Some workers have suggested that a true cider flavour could be produced by the sequential addition of pure yeast cultures, derived originally from the juice. Thus Schanderl (1957) considered that *Kl. apiculata* formed esters that were responsible for the characteristic flavour of Normandy ciders and German apple wines. Van Zyl *et al.* (1963) found that *Kl. apiculata* formed large quantities of esters in wine; these consisted of small quantities of ethyl acetate and very large amounts of ethyl butyrate. *Hansenula anomala* was characterized by production of ethyl acetate in large quantities. Peynaud (1956) stated that both these yeasts, as well as *S. ludwigii* and *Pichia* sp., were major producers of ethyl acetate, but he did not use gas-liquid chromatography for his estimations. English ciders containing large amounts of ethyl acetate are considered to be of poor quality, having the objectionable aroma of an acetified cider, although without the increased volatile acidity of the latter. Masuda *et al.* (1964b) found that *Sacch. cerevisiae* produced the largest amount of ethanol and the smallest amount of volatile acid; *Sacch. rosei* gave a considerable amount of aldehydes and volatile esters, while *T. bacillaris* formed high concentrations of volatile acid. According to van Zyl *et al.* (1963), *Sacch. rosei* produced large amounts of ethyl acetate, and *T. bacillaris* much acetaldehyde. Fermentation with *Sacch. cerevisiae* alone was preferred to a combination of this yeast with *Kl. magna*, a grouping commonly found in the natural fermentation of untreated juice. Kir'yalova (1960) found that under both pilot-plant and industrial conditions, *Sacch. ellipsoideus* would suppress the growth of *Ha. apiculata* and *Torulopsis* sp., when all three were inoculated together originally. At low temperatures and with minimum concentrations of *Sacch. ellipsoideus* the juice was fermented by the last two yeasts. In grape juice fermentations Leal (1958) showed that *C. pulcherrima* could inhibit the activity of *Sacch. ellipsoideus* when the latter was added three to six days after the first. Leal *et al.* (1961) divided the yeasts of the natural flora of Spanish wines into four groups: (i) initial-phase yeasts, *K. apiculata*, *Ha. guilliermondii* (*Ha. valbyensis*) and *H. subpelliculosa*, that gave a high volatile acidity and a low concentration of acetaldehyde; (ii) intermediate-phase *Zygosacch.* (*Sacch.*) *veronae* and *Torulaspora* (*Sacch.*) *rosei* with intermediate sugar metabolism and producing low volatile acidity; (iii) final-phase yeasts, *Sacch. mangini*, *Sacch. oviformis*, *Sacch. pastorianus* and *Sacch. italicus*, with high alcohol yield per unit weight of sugar consumed and low volatile acid production; (iv) veil-forming yeasts, *Sacch. beticus* and *Sacch. cheresiensis* that produce acetaldehyde but are not found in cider-making. Pollard *et al.* (1966) showed that, while some of the juice or initial-phase yeasts such as *Hansenula* sp., *Kl. apiculata* and *C. pulcherrima* produced amounts of fusel oils similar to the main fermenting yeasts, others such as *Kl. magna* and two *Torulopsis* sp. produced very small quantities indeed. Combinations of such yeasts with *Sacch. uvarum* could produce some augmentation of the fusel oil content but, more frequently, showed no change (*Kl. magna*) or considerable reduction (*T. dattila, C. parapsilosis*). From the evidence at present available, it would seem that little would be gained from inoculating apple juice with yeasts other than *Saccharomyces* sp. The role of yeasts such as *Brettanomyces* sp., especially in maturing cider, has yet to be determined: van Zyl (1962) considered that they caused turbidity and flavour spoilage in finished wines. Mosiashvili *et al.* (1961) have advocated the addition of *Zygosacch. bailii* (*Sacch. bailii*), *Ha. apiculata* (*Ha. valbyensis*) and *C. pulcherrima* to stored wines in order to shorten the maturing period. Other Russian authors have advocated the addition of autolysed preparations of *Saccharomyces* sp. for the same purpose. It has yet to be shown whether the same changes could be induced in cider, which requires a much shorter maturation period than wine.

c. Selection of cultures. There are many references in the literature to possible tests for selecting fermenting yeasts suitable for wine-making and brewing. With few exceptions (Hough, 1957; Rankine, 1964, 1966, 1967a, b; Hudson, 1967) most papers describe only one characteristic of a group of yeasts, whereas the user requires the complete industrial characteristics of a particular yeast. Obviously a yeast suitable for cider, although having properties similar to those needed for wine, would not necessarily be suitable for beer, *saké* and other beverages. As will be shown later, the strain of yeast used in cider-making must be varied according to the method of fermentation, flavour required and method of packaging. However, the following properties would be desirable in any cider yeast.

It is imperative that the culture should not produce a taint in the cider under any circumstances. Thus, even a trace of hydrogen sulphide is objectionable; in wines it can be formed from residues of vineyard sulphur dusts and sprays applied shortly before harvest (Rankine, 1963, 1964): but, the correct choice of a suitable strain of *Saccharomyces* can prevent this taint, even in the presence of sulphur particles. In cider, hydrogen sulphide is more likely to be formed from other sources. Wainwright (1962) has shown that there is a complex of six enzymes in yeast extract capable of reducing sulphite to sulphide; three of these have been described by other workers. According to Zambonelli (1964), the amount of hydrogen sulphide produced from sulphite is highly reproducible under the same culture conditions, and details were given of a suitable test medium. However, hydrogen sulphide is normally formed from sulphate in pantothenate-deficient substrates. Under these circumstances the greatest amounts appear during the logarithmic phase of fermentation (Kodaira *et al.*, 1958), and production of hydrogen sulphide is stimulated by traces of glutamic acid, aspartic acid or alanine, but is completely suppressed by methionine or leucine (Kodaira and Uemura, 1961). A suitable testing medium was devised by Stewart *et al.* (1962); they found that ten out of 90 brewer's yeasts produced hydrogen sulphide consistently in the absence of pantothenate, 14 produced small amounts, while the remainder were incapable of producing any. Thus, hydrogen sulphide production, irrespective of the causal condition, is a strain-specific characteristic of yeasts. Its presence in ciders is usually caused by infection in the yeast culture vessel or by the culture yeast being overgrown by a contaminant during fermentation.

Similarly, some yeasts cause taints in ciders by the production of acetoin and diacetyl. Normal cells produce more diacetyl the less the strain is able to assimilate valine (Anh, 1966). High inoculation rates, increased fermentation temperatures (Portno, 1966), aeration (Kringstad and Rasch, 1966) and deficiencies of *myo*-inositol (Lewin and Smith, 1964) all increased diacetyl concentrations. Czarnecki and Engel (1959) found that respiration-deficient cells produced more diacetyl than normal cells. Although some yeasts can form up to 2 p.p.m. diacetyl, which is detectable by taste, the major culprits are found among the lactic acid bacteria, such as *Leuconostoc mesenteroides* (Fornachon and Lloyd, 1965), which uses pyruvic and citric acids as substrates. This organism is normally controlled by efficient juice sulphiting.

Even though a yeast does not cause flavour taints, it can still be unsuitable for cider-making if it produces sulphite-binding compounds (Burroughs and Whiting, 1961) in any quantity. The most important compound in this respect is acetaldehyde, which is produced by such yeasts as *H. anomala* and *Kl. apiculata*. However, *Sacch. fermentati* (van Zyl *et al.*, 1963) and *Saccharomyces* sp. used for sherry-making, i.e. *Sacch. beticus*, *Sacch. cheresiensis* (*Sacch. oviformis*), *Sacch. rouxii*, *Sacch. acidifaciens* as well as *C. mycoderma* (Leal *et al.*, 1961), can produce several hundred p.p.m. in their veil phase. (It is by virtue of its acetaldehyde production that the "sickness" bacterium, *Zymomonas anaerobia*, can grow in the presence of 500 p.p.m. sulphur dioxide.) The addition of sulphur dioxide by itself increases the total aldehyde content of a cider fermented by any *Saccharomyces* sp. Thus, a cider made from sulphited juice contained 69 p.p.m. total aldehyde, while its non-sulphited partner had only 29 p.p.m. (Burroughs, 1959). This aldehyde/bisulphite complex is tasteless, unlike excess free aldehyde, which is objectionable in ciders. Ohara *et al.* (1957), Veselov *et al.* (1963) and Schanderl and Staudenmayer (1964) have made similar observations on wines. The next important sulphite-binding compound is pyruvic acid; again the amount produced by *Saccharomyces* sp. is strain-specific (Rankine, 1967a). The same applies to α-ketoglutaric acid, another important sulphite-binding compound (Lewis and Rainbow, 1963; Rankine, 1968). Galacturonic acid, binding sulphur dioxide less strongly, is produced from the degradation of pectin by a number of cider and wine yeasts possessing the enzyme polygalacturonase (Pollard and Kieser, 1959; Malan, 1961). The remaining sulphite-binding compound, L-xylosone, is produced from the dehydroascorbic acid of the juice in the presence of sulphur dioxide (Whiting and Coggins, 1960a), a yeast being unnecessary for the reaction.

The characteristics of the chosen culture should remain unchanged over long periods of cultivation, i.e. it should form mutant or respiration-deficient cells only rarely. Stevens (1966) tested the flocculation of a large number of isolates from an ale yeast culture, using a method developed by Chester (1963): yeasts with stable Chester values rarely mutated. Petite-cells can be detected by plating the culture on the media of Czarnecki and Engel (1959), Nagai (1963, 1965) or Kleyn and Vacano (1963). The giant-colony characteristics of a large number of isolates from a culture will reveal the presence of morphological mutants (by the formation of sectors or pronounced papillae) and the basic colony type (Richards, 1967). New isolates from the stock culture, made using single-cell techniques, should have their industrial characteristics rigorously tested before being brought into use. It should be axiomatic, of course, that the culture is free of other strains and species (see also Beech *et al.*, 1968; Beech and Davenport, 1969b). Even trace contamination of a non-flocculent strain in a flocculent culture can be detected by the forcing test of Ellison and Doran (1961). The presence of other species can be tested for by plating on malt agar and selective media such as lysine agar (Walters and Thiselton, 1953), or possibly with serological techniques (Campbell and Brudzynski, 1966; Richards and Cowland, 1967). Freedom from bacteria can be ascertained by plating heavy suspensions on malt agar containing actidione and 8-OH-quinoline (Beech and Carr, 1960).

It should be possible to induce the culture to grow rapidly prior to use. Ideally it should not be exacting in its vitamin requirements (Goto, 1961; Taketa and Tsukahara, 1962), and should be capable of growing well between 10° and 15°, since cider vats are not attemperated, and also resistant to autolysis. The composition of the growth medium itself is important; our own tests have shown that liquid apple juice-yeast extract (Beech, 1957; Beech and Davenport, 1969a) is better than malt extract:yeast extract:peptone:glucose (MYPG; Wickerham, 1951) or competely synthetic media such as Wickerham's yeast–nitrogen base plus glucose. Most yeasts grow well at 25° but many, including Port 350R, show a pronounced lag phase at 15° before attaining their maximum growth rate.

Of perhaps greater importance is the fermentation efficiency of the culture. Thorne (1958, 1961) considered that this factor, which is the quotient of the fermentation velocity and the nitrogen content of the yeast, is characteristic of the strain. Juice treatment, as well as the chosen culture, can affect the rate of fermentation. Thus, sulphiting the juice gives the longest lag phase and the fastest rate of fermentation. Heating the juice, whether by flash pasteurization or for aroma stripping, reduces the lag phase and the rate of fermentation; enzymic depectinization has little effect on either parameter. Measurements of fermentation rates can be supplemented by pilot-scale fermentation trials using jacketed tubes of standard dimensions and inoculating with a known amount of yeast (Cook, 1963). The method also gives supplementary information on the size of the crop, degree of attenuation, etc. It might be thought that the larger the cell the more efficiently it would ferment. However, breeding experiments have shown (Emeis, 1964) that the maximum rate of fermentation in heterozygous yeasts was obtained with triploids and tetraploids; the rate being slower both for greater or lesser degrees of ploidy. The ethanol tolerance

(Ranganathan and Bhat, 1958) of the culture is also important; It should be sufficient to allow the yeast to grow and ferment the cider to dryness; but it should not be so great as to cause infection problems in the final product, should the bottling technique prove inadequate.

The ideal flocculation pattern of a cider yeast would be one in which it was largely in suspension for most of the fermentation but settled out almost completely in the dry cider; Port 350R is a good example. However, some yeasts, such as Bordeaux 88 XIII (*Sacch. oviformis*) and NCYC 1119 (*Sacch. cerevisiae*), ferment almost as rapidly, but remain intensely flocculent throughout the fermentation. Very flocculent slow-fermenting yeasts are unsuitable for fermenting fresh juice, but are invaluable for naturally-conditioned draught or champagne ciders. On the other hand, a cloudy naturally-conditioned draught cider, such as is favoured in certain areas, is best produced by a non-flocculent yeast, e.g. *Sacch. cerevisiae* var. *turbidans* (Wiles, 1950) or NCYC 202 (*Sacch. cerevisiae*). The methods used for testing the flocculence of brewing yeasts are not entirely suitable for cider yeasts. We have found Hough's (1957) aggregation test at pH 3·5 and 5·0, and Gilliland's (1951) test to be most useful. Recently, in a search for cultures suitable for continuous tower fermentations, we discovered that bubbling nitrogen gently through yeast suspended in a column of sterile dry cider disclosed immediately whether a yeast was flocculent or otherwise. A comparison of the results obtained with these flocculation tests is given in Table XXIII.

However, no matter how perfect a yeast is in these respects, it is useless unless it produces an acceptable flavour in the cider. Normally

TABLE XXIII. *Flocculation of Yeasts as Measured with Different Tests (Reproduce with permission from Beech, 1967a)*

[*Degree of response indicated by the system appropriate to each test*]

Yeast culture	Chain formation	Gilliland's test	Hough's test pH 3·5	pH 5·0	Bubble test
Saccharomyces cerevisiae var. *ellipsoideus*, Burgundy	—	I	—	—	
Saccharomyces oviformis, 88 XIII	—	I	+	+	
Johannisberg	+	I	∓	—	Non-flocc
Saccharomyces oviformis, AWY 723	+	I	+	+	
Saccharomyces cerevisiae, AWY 350R	—	II	+ +	+ +	Mod-flocc
Saccharomyces cerevisiae var. *ellipsoideus*, UCD 505	+	II	+ +	+ +	
Saccharomyces cerevisiae, NCYC 1119	+	III	+ + +	+ + +	Very flocc

this would be assessed by the quality control panel of the factory, but potential yeasts can be screened quickly beforehand by gas-liquid chromatography measurement of the higher alcohols or fusel oils they produce in standardized small-scale ciders. Ciders with suitable fusel oil patterns would still need to be tasted, since their flavour can be modified by other volatile compounds, such as esters, aldehydes and ketones also produced by the yeast (Lawrence, 1964; Gilliland and Harrison, 1966). The general range of fusel oils in ciders and perries, given by Kieser et al. (1964) and Pollard et al. (1965), is shown in Table XXIV. Only n-butanol (and possibly n-hexanol) is characteristic of the juice; all others are formed during the synthesis of amino acids by yeasts (SentheShanmuganathan and Elsden, 1958; Ingraham and Guymon, 1960; Crowell et al., 1961; Lewis, 1964). The normal range of amino acids in cider apple juices, equivalent to 10 mg N/100 ml or less (Burroughs, 1957), is insufficient to produce significant amounts of fusel oil by deamination, as occurs in beer (Hough and Stevens, 1961). The formation of fusel oils in beer has been reviewed by Stevens (1960) and Hudson and Stevens (1960), in spirits by Pfenniger (1963), and in wine fermentations by Webb (1967). The taste thresholds of fusel oils in beer and wine have been measured by Harrison (1963) and Rankine (1967b) respectively.

TABLE XXIV. *Fusel Oil Components in Apple Juices and Ciders* (*Reproduced with permission from Pollard, Kieser and Beech, 1966*)

Constituent	Ranges (p.p.m.) in	
	Apple juices	Ciders
n-Propanol	0·2–2	4–47
n-Butanol	3–24	4–32
Isobutanol		14–74
Isopentanol	0–1	42–410
2-Methylbutan-1-ol	0·1–2	16–39
n-Hexanol	1–2	2–17
2-Phenylethanol		7–258
Tyrosol		0–20

The amount of fusel oil components formed by a yeast will vary with different juices and fermentation conditions (see Sections IV.B.3b, p. 125), but several yeasts can be compared by allowing them to ferment batches of the same juice, as shown in Table XXV. The yeasts can be arranged in order according to the amounts they produce. *Saccharomyces oviformis*, Port 350R and Fendant form small quantities of all components, although under suitable circumstances Port 350R can form considerable quantities of *n*-propanol: *Sacch. uvarum* and Johannisberg consistently produce large quantities of fusel oils, the latter yeast producing more isopentanol than *Sacch. uvarum*.

TABLE XXV. *Fusel Oil Components of Ciders Prepared from Batches of the same Apple Juice*

Yeast culture	Fusel oil component (p.p.m.)			
	Isobutanol	Isopentanol	2-Phenylethanol	Total
Mixed Juice 1. CV's Sweet Coppin and Stoke Red				
Saccharomyces oviformis	14	81	18	140
Saccharomyces cerevisiae, Port 350R	14	120	20	185
Saccharomyces cerevisiae var. *ellipsoideus*, Burgundy	17	117	22	185
Saccharomyces cerevisiae, Montrachet	25	187	100	340
Saccharomyces uvarum, GEI	29	206	143	407
Johannisberg	55	270	145	501
Mixed Juice 2. CV's Sweet Coppin and Cox's Orange Pippin				
Fendant	38	131	36	205
Saccharomyces uvarum, GEI	97	342	258	697
Johannisberg	262	410	226	898

3. Biochemical Changes during Fermentation

a. Formation of organic acids. Graff and Bidan (1950) and Jacquin and Tavernier (1951) described the organic acid and higher alcohol production of yeasts isolated from French ciders, using classical methods of analysis. Carr and Whiting (1956) showed that yeasts formed succinic acid during the fermentation of both untreated and sulphited juices, the amount increasing with increasing pH value. Beech (1959b), surveying acid changes in a wide range of pilot-scale sulphited juice fermentations found that, over the pH range 3·4 to 4·0, yeasts formed malic and pyruvic acids in addition to succinic. The work was further extended by Whiting and Coggins (1960b), who stated that malic and quinic were the principal acids of cider apple juices, with traces of citramalic, chlorogenic, *p*-coumarylquinic, shikimic, citric, mucic, benzoic, gluconic and 2-methyl-2,3-dihydroxybutyric acids. Glycollic acid, identified in Swiss cider-apple juices (Tanner and Rentschler, 1954), was not detected in English juices. None of the acids decreased in amount during the fermentation of sulphited juices, but malic acid increased by yeast action, sometimes by more than 50% of the original value. New acids were also formed, including succinic, lactic, fumaric, gluconic, dihydroxybutyric, α-hydroxybutyric, and mono-, di- and tri-galacturonic acids were produced by the breakdown of pectin. In a further paper, Whiting and Coggins (1960c) showed that traces of three keto acids (pyruvic, α-ketoglutaric and oxalacetic) were also present in cider-apple juices. The first two increased considerably during the fermentation of sulphited juices. Some tentative work with pure yeast fermentations of apple juice has shown that *Sacch. oviformis* and Burgundy yeast can cause a slight diminution of malate, while Johannisberg causes an augmentation. The metabolism of organic acids by pure cultures of wine yeasts has been examined in great detail by Rankine (Section IV.B.2c, p. 118). Many cultures of *Saccharomyces* sp. were able to decompose between 3% and 45% of the malic acid, while *Schizosacch. malidevorans* decomposed it completely (Rankine, 1966). Species of this genus have not been found so far in English cider fermentations. The optimum conditions for the formation of malic acid by three wine yeasts were detailed by Drawert *et al.* (1965). Lactic acid can also be formed by yeasts; *Sacch. veronae* characteristically produces L(+)-lactic acid (Peynaud *et al.*, 1967) while other yeasts form only small amounts of D(−)-lactic with traces, if any at all, of the L(+)-isomer. In bacteria-free wine fermentations, Thoukis *et al.* (1965) found that 90% of the increase in non-volatile acidity could be ascribed to the formation of succinic acid, and 10% to lactic acid. Changes in organic acids in cider and wine fermentations would seem to have many features in common.

b. Formation of fusel oils. The amount of higher alcohols or fusel oils in ciders varies not only with the composition of the yeast flora, but also with the juice treatment and fermentation conditions.

Ciders made from sulphited juice have lower fusel oil contents than those made from the corresponding non-sulphited juice (Table XXVI).

TABLE XXVI. *Effects of pH Level and Sulphiting on the Fusel Oil Contents of Ciders*

Juice treatment pH	Fusel oil component (p.p.m.)			
	n-Propanol		Total fusels	
	Sulphited	Non-sulphited	Sulphited	Non-sulphited
3·0	25	83	102	174
3·5	26		105	
4·0	29		101	
4·5	51	88	135	150

TABLE XXVII. *Effect of Juice Nitrogen Content on the Fusel Oil Content of Ciders*

Cultivar	Juice nitrogen (mg/100 ml)	Fusel oil component (p.p.m.)					
		n-Propanol	Isobutanol	Isopentanol	2-Phenyl-ethanol	Total
(a) Ciders made from sulphited juices (100 p.p.m.) yeasted with *Saccharomyces uvarum*							
Sweet Coppin	4·9	6	47	228	147	428
Sweet Coppin/Stoke Red	5·2	6	29	206	143	384
Yarlington Mill/Cox's Orange Pippin	26·2	26	127	193	89	435
Sweet Coppin/Cox's Orange Pippin	30·8	18	85	160	44	307
(b) Ciders made from sulphited juices (150 p.p.m.) yeasted with *Saccharomyces cerevisiae,* Port 350R							
Tremlett's Bitter	4·8	10	30	82	49	171
Dabinett	5·4	11	26	112	16	165
Yarlington Mill	5·9	12	23	102	16	153
Sweet Coppin	6·4	11	26	117	11	165

However, sulphiting is unlikely to be abandoned for this reason since ciders from non-sulphited juices are liable to have oxidative or microbial spoilage taints. The pH value of the juice has an effect only above pH 4·0, whether the juice has been sulphited initially or not. Rankine (1967b) found increases in isobutanol and isopentanol with increased pH values in wines. The amounts of aliphatic fusel oils produced in naturally-fermenting sulphited apple juices over the temperature range 15° to 25° appeared to be reasonably uniform, decreasing sharply outside these temperatures (Pollard et al., 1966). Otsuka et al. (1963) found maximum fusel and minimum aldehyde and volatile acid in wines fermented at 25°. Similar results have been obtained with Russian wines by Valuǐko (1960), and for American wines by Ough et al. (1966) and Ough and Amerine (1967). There is considerable variation in the fusel oil content of ciders made by adding the same yeast to different juices, as shown in Tables XXV and XXVII. In Table XXV, the nitrogen contents of Mixed Juices 1 and 2 were 5·2 mg and 18·9 mg N/100 ml respectively. It would appear that, for either Sacch. uvarum or Johannisberg, the fusel oil content of the cider increases with increasing juice nitrogen. However, this is not borne out in Table XXVII, where isopentanol and 2-phenylethanol decrease with increasing juice nitrogen. We can only conclude that nutritional factors other than nitrogen control the levels of fusel oils in ciders. Äyrapää (1968) found that with ammonium sulphate as sole source of nitrogen, fusel oil formation, particularly of isobutanol, was at a maximum between 10 and 20 mg N/litre, decreasing at higher levels of nitrogen. When nitrogen was supplied as a complex mixture of amino acids, widely different fusel oil patterns were obtained according to the species and strain of brewing yeast.

Fermentation of apple juice after concentration, storage and subsequent dilution is becoming increasingly important in cider factories. It has been found that the cider from concentrated juice always has a lower fusel oil content than cider from fresh juice (Table XXVIII). The two higher alcohols derived from the juice disappear during the initial processing. Thus, n-hexanol is lost during aroma stripping, and n-butanol during the concentration of the stripped juice, as can be shown by fermenting the concentrates after dilution, not with water but with condensates normally discarded from these two stages. Even allowing for the loss of these two components, the total higher alcohol content of the cider from diluted concentrate is always less than that from juice. The lower level of fusel oils in ciders from concentrates is of concern to cidermakers, and the reasons for this difference are still being sought*. Not

---

* Our latest results indicate that clarification after enzymic depectinization causes the greatest loss in fusel oil content when the juice is fermented.

F. W. BEECH AND R. R. DAVENPORT

TABLE XXVIII. *Effect of Juice Concentration on Fusel Oil Content of Ciders*

Cultivar	Yeast	Medium	Fusel oil component (p.p.m.)					
			n-Propanol	Iso-butanol	n-Butanol	Iso-pentanol	n-Hexanol	2-Phenyl-ethanol
Sweet Coppin Cox's Orange Pippin	*Saccharomyces uvarum*	Juice	18	85	11	160	3	44
		concentrate	20	60		98		44
Sweet Coppin Stoke Red	Montrachet	Juice	3	23	16	185	6	80
		concentrate	4	22		140		52
	Johannisberg	Juice	6	42	15	270	6	144
		concentrate	4	44		162		86

only do the individual stages in processing have an effect, but the order in which they are performed is also important.

When carried out singly, prior to yeasting, any of the following operations gave ciders with similar higher alcohol contents: sulphiting, flash pasteurization, stripping, enzymic depectinization (enzyming). Of the combined treatments, heating then enzyming the juice or cider caused no change in fusels, whereas enzyming before heating reduced the higher alcohol content. Sulphiting before any heating and enzyming also caused a reduction (mainly in 2-phenylethanol). The values could be restored by enzyming the cider rather than the juice, but this is not practicable when enzymed juices are required for concentration. Concentrating the juice caused the greatest reduction of any treatment. Preliminary experiments indicate that freeze-concentration of stripped juice is no better in this respect than heat concentration.

TABLE XXIX. *Fusel Oil Contents of Dry Ciders Produced by Different Fermentation Systems at 25° with A W Y 350R (Reproduced with permission from Beech, 1967a)*

Fusel oil	Fusel oil content in cider (p.p.m.)			
	Anaerobic		Aerobic	
	Static batch	Continuous (heterogeneous)	Stirred batch	Continuous (homogeneous)
2-Phenylethanol	7	7	31	20
n-Propanol	200	250	62	25
Isobutanol	17	13	98	25
Isopentanol	67	98	197	53

Finally, the method of fermentation also has an effect on the fusel oil content of the ciders. Beech (1967a) compared anaerobic and aerobic systems for both batch and continous fermentations (Table XXIX). This illustrates the remarkable capacity of yeast Port 350R to synthesize n-propanol under anaerobic conditions, while the formation of the other higher alcohols is depressed. Aerobically, the amounts of fusel oil formed by this yeast vary with batch and continuous systems. Production of n-propanol is very much less in both; in the continuous system the amounts of all other components are small, whereas the levels of isobutanol and isopentanol are increased when the system is operated batchwise. Similarly, stirred wine fermentations have produced two to seven times more higher alcohols than anaerobic systems (Crowell and Guymon, 1963), the increase again being mainly with isobutanol and isopentanol. For yeast Port 350R, at least, the choice of fermentation system is critical, but this may not be so with yeasts able to form large amounts of all fusel oil components under normal fermentation conditions; cf. *Sacch. uvarum* or Johannisberg (Table XXV).

The heterogeneous or tower system of continuous fermentation (Klopper *et al.*, 1965; Royston, 1966) has been shown to give ciders similar in flavour and chemical composition to those produced by traditional methods (Beech, 1967a). This has been found on both the laboratory and pilot-plant scale. Residence times of four to six hours have been obtained, even though the nutrient status of apple juice is only approximately one tenth that of wort. A laboratory tower has been run satisfactorily for a month with freshly pressed juices used six hours after sulphiting. Normally, however, the greatest economic advantage would be gained from running the tower throughout the year on diluted apple juice concentrate. Also, the loss of fusel oils that seems inherent in the use of concentrate would tend to be counteracted by the higher fusel oil production with this heterogeneous, rather than the homogeneous system of continuous fermentation. Finally, it is essential to use a flocculating yeast in order to retain the optimum amount of yeast to the tower (Portno, 1967a). The bubble test, described in Section IV.B.2c (p. 118), carried out in dry cider at the operating temperature of the tower, appears to be very suitable for selecting a culture with the right characteristics.

C. FERMENTATION TREATMENTS

1. *Yeast Control*

With the decline of keeving for the production of naturally sweet ciders, a practice grew up in the 1930s of preparing such ciders by mechanical removal of the yeast crop during fermentation. The method was better suited for use in factories than the keeving process, but the problems were just the same and the process was finally abandoned in the U.K. for the same reasons (Beech, 1958). However, during the following research work on the process, insight was gained on the factors that limit the extent of normal fermentations.

The investigations of Challinor and Burroughs (1949) were based almost entirely on changes in the nitrogen content of the fermenting juice. Total nitrogen was determined on juice or cider direct, and the soluble nitrogen on a centrifuged sample. The differences between the soluble nitrogen of a fermenting cider and that of the juice from which it was made, was used as a measure of yeast growth at that stage of fermentation. The amount of yeast deposited in the lees was calculated from the decrease in total nitrogen during the fermentation, whilst yeast in suspension was given by the difference between total and soluble nitrogen of any particular sample. Checking the fermentation after different losses of specific gravity, enabled Burroughs and Challinor (1949) to find what loss in nitrogen (i.e. the amount of yeast growth) was necessary for the stabilization of the resulting sweet cider; the optimum time for centrifuging was after yeast growth had ceased (Burroughs and Challinor, 1951). The factors that prevented the stable ciders from re-fermenting were deficiencies of nitrogen and thiamine (Burroughs, 1952, 1953); phosphorus was not a limiting factor, as found also by Zubkova (1958), since this element was liberated rapidly back into the cider after the initial uptake. Elbert and Esselen (1959) found that the amount and rate of alcohol production in apple juice fermentations was related to the amino nitrogen content. Differences in thiamine were said to be unrelated to fermentation differences, a view that is not shared by other workers (Peynaud and Lafourcade, 1957; Okuda and Tanase, 1958; Hernandez, 1967). Ough and Amerine (1966) have produced equations for predicting the fermentation rates of grape juices. The equations, based on sugar content (°Brix), total nitrogen and pH value, gave standard errors of 0·07° Brix/h at 33°, but a similar degree of accuracy could be obtained from biotin and total nitrogen alone (Ough and Kunkee, 1968). They postulated the existence of another unknown factor affecting the fermentation rate that is fairly constant over the maturity range of the grapes tested.

2. *Treatment of "Stuck" Fermentations*

The converse of producing stable sweet ciders, is the treatment of normal fermentations that "stick"; that is, cease prematurely. There are numerous reasons for this phenomenon. Some such ciders may con-tain no fermentable sugars, but appear to have an appreciable residual specific gravity due to the presence of sorbitol, glycerol and traces of D-xylose, galactose, arabinose, ribose, rhamnose, sorbose, inositol and oligosaccharides. In certain years the sorbitol and organic acid content of some perries may be so high as to contribute over thirty degrees of specific gravity (Whiting, 1961). A simple and rapid test for reducing sugars, such as is used by diabetics, will show whether the residual gravity is due to these nonfermentables or not. Ciders made from non-sulphited juice may fail to ferment adequately because of a lack of efficient fermenting yeasts, or because the small amount of *Saccharo-myces* sp. present has been overwhelmed by growth of spoilage yeasts or bacteria. A heavy infection of *Acetobacter* in the wood of brewery vessels has caused the autolysis of yeast during fermentation (Comrie, 1951). One strain of *Acetobacter rancens* killed, not only a yeast used in naturally-conditioned bottled beer, but also yeasts in the genera *Saccharomyces, Hansenula, Pichia, Schizosaccharomyces, Torula, Candida* and *Brettanomyces*. The active principle was not identified, but was not acetic acid (Gilliland and Lacey, 1964, 1966). A similar effect has been reported from Japan (Kaneko and Yamamoto, 1966). Sulphited juices from factories with high standards of sanitation may fail to start fermenting unless yeasted deliberately. Normal fermentations of sulphited juices that cease prematurely do so because the yeasts die out. Usually it is because of the low nutrient status of the original juice, so that only a small working population develops. The number of new cells formed subsequently is inadequate to replace cells that die, until insufficient yeasts are left to complete the fermentation. In such cases addition of 50 mg ammonium sulphate/litre (250 g/1,000 gal) is required for every decrease of 0·01 in specific gravity; 0·2 mg thiamine/litre (1 g/1,000 gal) is sufficient, irrespective of the drop in specific gravity required (Burroughs and Pollard, 1954). The concentrations of other vitamins in cider is normally sufficient (Jacquet and Breton, 1966). Occasionally, the cider may also need aeration to encourage vigorous yeast growth (see also Rankine, 1955; Ricketts and Hough, 1961), and a new inoculum if the number of viable yeasts in the cider is less than 10^4/ml.

Very rarely fermentation may cease because of the presence of an inhibitory substance. Spraying fruit with captan just prior to harvesting usually prevents fermentation from starting; this is more likely to occur with grapes (Kasza, 1956) than with cider apples. Excessive tannin concentrations are said to be inhibitory to yeasts (Lüthi, 1957b), their inhibitory effect can be removed by fining the juice with gelatine prior to yeasting. Problems of inhibition usually arise during the fermentation of diluted concentrates that have been stored at elevated temperatures. Under such conditions the amino acids and vitamins of the concentrate are severely reduced (Lüthi, 1958; Petit and Godon, 1963; Príbela and Betušová, 1964) and inhibitors, such as hydroxymethylfurfural, are formed. Filtration after treatment with activated charcoal removes the inhibitors, and deficiencies in nutrients can be overcome by the addition of a yeast food containing ammonium sulphate and a vitamin mixture. Addition of a vigorously-fermenting yeast culture to the treated and aerated liquid then completes the process.

D. POST-FERMENTATION TREATMENTS

1. Clarification and Storage

Small casks of cider need processing fairly soon after fermentation ceases or infections of aerobic bacteria and yeasts begin to develop; usually these are Acetobacter xylinum and Pi. membranaefaciens (C. mycoderma) respectively. Ciders in large vats are less troubled by these disorders, particularly if they were made originally from properly sulphited juices. In fact, such bulks of cider benefit from a short holding period after fermentation, since it allows settlement of yeasts in suspension, particularly if the juice was inoculated with a flocculating yeast after sulphiting. The process is usually assisted by "racking" (i.e. decantation of the liquid from off the deposit or "lees"), centrifuging or coarse filtration into clean storage vats. Ciders made from non-sulphited juices, especially those with a high pH value, would certainly need careful filtration after fermentation because of their bacterial content (Tables XX and XXX). The importance of the correct juice treatment for reducing cider filtration costs is obvious. The microbiology of stored ciders has not been investigated systematically in recent years; results obtained before juice sulphiting was generally adopted have little bearing on present-day practices. The occurrence of the malo-lactic fermentation, and of such yeasts as *Hansenula* and *Brettanomyces* sp., has been mentioned in earlier sections.

TABLE XXX. *Effect of Clarification Treatments on the Microflora of Dry Ciders (Reproduced with permission from Beech, 1959b)*

Treatment	Viable organisms in ciders (cells/ml)			
	From untreated juice		From sulphited juice	
	Yeasts	Bacteria	Yeasts	Bacteria
Mixed contents of the vat	6×10^5	$1 \cdot 035 \times 10^7$	$2 \cdot 7 \times 10^7$	+
After centrifuging	40	$4 \cdot 3 \times 10^6$	$< 10^3$	
After centrifuging and pulp-filtering		$8 \cdot 65 \times 10^3$		

Normally the clarified ciders are blended to some extent before storage, but sometimes the general level of acid or tannin in the ciders is excessive. In such circumstances ciders can be diluted, sweetened, re-sulphited and yeasted. The re-fermented cider, with a better acid/tannin balance, is then used either in the blending programme or sold in the lower price range. Because of the increasing use of concentrated apple juice, it is more usual to make up a blend of cider, concentrate, sugar and water for sulphiting and yeasting. The final ciders then have more uniform flavours, with alcohol contents suitable for each type of product. Uniformity of flavour seems to go hand-in-hand with selling nationally. Perhaps as a reaction, small farm cider-makers are selling increasing quantities of ciders with distinctive flavours to passing motorists. Even these makers have been forced to modernize their packaging methods to ensure that the final products are stable micro-biologically.

2. *Packaging*

Just prior to being sold, ciders receive a final blending and any necessary additions of citric or DL-malic acid, sugar, colouring matter and sulphur dioxide. Their subsequent treatment is determined by the method of packaging.

a. Still ciders. A small amount of uncarbonated cider is sold non-sterile in wooden casks or plastic containers, but most of it receives some sterilizing treatment. This can be pasteurization of the filled and sealed containers or sterilization of the cider (by filtration or flash-pasteurization) and filling into sterile containers *via* a pre-sterilized bottling machine. The first treatment is most suitable for farm ciders since the product is always sterile, but its flavour may not be so fresh as that stabilized by the other two treatments. Not all manufacturers' EK-grade of filter pad retain coccoid bacteria of 0·8 μ diam, in which case it is necessary to use the EKS-grade, with consequent loss of throughput: membrane filters are an excellent alternative. Flash pasteurization (Brumstead and Glenister, 1963) at 85° for 30 sec, while destroying all vegetative forms of micro-organisms, may activate the growth of spores of any heat-resistant moulds present in the cider. Injection with about 0·5 volumes of carbon dioxide before sterilization has been found to prevent the growth of these and any other aerobic contaminants. Still ciders, especially in half or one gallon jars, must be free of yeasts since the contents are aerated each time some of the cider is poured. This process encourages the rapid growth even of minute traces of yeasts.

Non-microbiological problems can also be experienced with still ciders awaiting sale. Delicate flavours can be lost by oxidation in the bottle, while cider in jars will also darken for the same reason. Therefore, an anti-oxidant such as ascorbic acid or sulphur dioxide must be added; again a small amount of carbon dioxide is helpful. The process is aggravated if the cap lining is permeable to oxygen, or the cider contains more than 7 p.p.m. of iron. Both rigid and collapsible plastic containers are being used increasingly for still cider; not only must it be possible to sterilize the container, but care should be exercised in the choice of the plastic material. This should be impermeable to oxygen and not allow any of the carbon dioxide dissolved in the cider to diffuse away. Some plastics can impart an unpleasant flavour taint; they can be detected with a simple screening test, based on ultraviolet spectrophotometric examination of a synthetic solution left in contact with the sample (Beech, 1966).

b. Carbonated ciders. At one time it was quite common for draught or bottled cider to become "conditioned" due to the growth of yeasts. The source of this flora was contamination from the equipment or container, as shown below for cider processed in an old factory (Beech, 1959b; see also Lüthi, 1949).

Cider preparation stage	Viable organisms (cells/ml)	
	Yeasts	Bacteria
Before centrifuging	10,800	7,500
Leaving the centrifuge	25	5,000
Entering wooden vat	125	5,000
Leaving wooden vat	1,500	5,000
Sweetened in trade cask	12,500	not tested

Several types of draught cider are produced at present and most contain a small amount of carbon dioxide, either from controlled conditioning or by injection. The sugar content of cider for natural conditioning is calculated to give the required amount of gas when fully fermented. The composition of the sweetening syrup is adjusted to give slow fermentation while on "tap"; e.g. saccharin and simple dextrins are used for sweet ciders. Usually the nitrogen content of the cider is adjusted to a standard value by the addition of ammonium sulphate; thiamine is also added if a requirement is indicated by test fermentations. The choice of yeast culture is vital, slow-fermenting flocculent yeasts for a clear cider and non-flocculent for ciders to be sold with a pronounced haze (Section IV.B.2c, p. 118). Normally the filled and sealed containers are held in the factory for a few days to ensure that conditioning has begun. Successful operation of this process needs very careful control at all stages, and the latest trend is the change over to "keg" cider, i.e. sterile artificially-carbonated cider, dispensed by top pressure from metal containers. The process is identical with that used for beer, except that materials coming into contact with the cider must be more resistant to corrosion.

There are several ways of presenting bottled carbonated ciders. Conditioning in the bottle is virtually of historical interest only. Newly fermented ciders would be pulp-filtered in February, samples removed, sweetened and incubated in sealed half-filled champagne bottles. The first samples might ferment violently, but later samplings eventually showed when the bulk of cider could be safely sweetened and bottled. It would then ferment steadily in bottle, produce ample carbonation and, in the best examples, have only a small firm yeast deposit. If opened carefully, most of the cider could be poured out sparkling and free of haze. The process lasted while labour was cheap and the general public accepted yeast deposits in their cider. Cider champagnized in the bottle lasted until quite recently since it was more consistent in quality and free of sediment, having had the yeast deposit removed by disgorging before being finally sealed. Its place has now been taken by cider and perry carbonated by the Charmat (Tschenn, 1934; Tressler *et al.*, 1941; Cusick, 1950; Atkinson *et al.*, 1959) or pressure-tank process. The sterilized juice is yeasted with a culture suitable for fermenting under high pressure (Odintsova, 1956) and run into temperature-controlled pressure tanks. The first gases are vented to atmosphere, but eventually the tank is sealed and the pressure builds up sufficiently so that the final product will contain 3·5 to 5 volumes of carbon dioxide; any excess gas is discharged automatically. A continuous wine champagnizing process has been described by Russian workers (Merzhaman, 1961; see also Kunkee and Ough, 1966) but so far it has not been used for cider. In the batch process all subsequent operations are carried out under pressure; these include additions of sugar syrup, etc., clarification by centrifugation and filtration, and sterile filtration into a standard counter-pressure bottling machine. Full sterility precautions are required from the sterilizing filter onwards. These also apply to the greater bulk of ciders that are carbonated artificially before bottling. The technique for producing bottled ciders free from micro-organisms has been detailed by Beech (1955, 1967b) and Buckle (1967). It will not be described here since it is the same as that used for all sterile bottled beverages (Portno, 1967b, 1968). It differs only in that the bottled cider should contain a minimum of 30 p.p.m. of free sulphur dioxide, not only to prevent the growth of any microbial infection during bottling, but also as an anti-oxidant preventing flavour and colour changes. It will be impossible to achieve this level of free sulphur dioxide within the legal limit of 200 p.p.m. total SO_2 unless the cider is low in sulphite-binding compounds (Burroughs and Sparks, 1963). Thus, as shown in Sections III and IV, the stability of the final cider is determined by efficient control at every stage of cider-making, starting from the ripe fruit hanging on the tree.

Appendix

YEASTS ISOLATED FROM FRUIT BUDS, APPLES, APPLE JUICE AND CIDER, WITH BIBLIOGRAPHY

A = yeasts isolated at any stage of fruit development from the dormant flower bud to the mummified fruit; J = yeasts isolated from apple juice and concentrated apple juice; C = yeasts isolated from fermenting juice and stored cider.

Yeast	Source	
Apiculate yeasts—unidentified	A	
Aureobasidium pullulans	A	
Brettanomyces bruxellensis	J	C
Brettanomyces cidri		C

Yeasts	Source		
Brettanomyces claussenii			C
Brettanomyces claussenii var. A		J	C
Candida catenulata	A	J	
Candida guilliermondii/Pichia guilliermondii	A	J	
Candida humicola/Nadsonia slovaca	A		
Candida krusei	A	J	
Candida malicola (see *Aureobasidium pullulans*)			
Candida mycoderma/Pichia membranaefaciens	A	J	C
Candida mesenterica			C
Candida parapsilosis	A	J	
Candida parapsilosis var. *intermedia*	A		
Candida parapsilosis var. A	A		
Candida parapsilosis var. B	A		
Candida pulcherrima	A	J	C
Candida reukaufii	A		
Candida scottii	A		
Candida solani	A		
Candida tropicalis	A	J	
Candida utilis/Hansenula jadinii		J	
Candida sp. A	A		
Candida sp. B	A		
Candida sp. C		J	
Candida sp. D	A		
Candida sp. E	A	J	
Candida sp. F	A	J	
Candida sp. G		J	
Candida sp. X		J	
Candida/Hansenula sp. B	A		
Candida/Hansenula sp. C	A		
Citeromyces sp. A	A		
Cryptococcus albidus	A		
Cryptococcus diffluens	A		
Cryptococcus kutzingii	A		
Cryptococcus laurentii	A		
Cryptococcus neoformans	A	J	
Cryptococcus sp. A		J	
Debaryomyces castellii/Torulopsis dattila	A	J	
Debaryomyces hansenii/Torulopsis famata	A		
Debaryomyces kloeckeri (*Debaryomyces hansenii*)	A		C
D. nicotianae (*Debaryomyces hansenii*)	A	J	
Endomyces mali	A		
Hanseniaspora valbyensis/Kloeckera apiculata	A	J	
Hanseniaspora osmophila		J	
Hanseniaspora sp. A	A	J	
Hansenula anomala/Candida pelliculosa	A		
Hansenula beckii	A	J	
Hansenula beijerinckii/Torulopsis sp. A		J	
Hansenula californica	A		
Hansenula holstii/Candida silvicola		J	

Yeasts	Source		
Hansenula silvicola	A	J	
Hansenula silvicola var. B	A		
Hansenula suaveolens		J	
Hansenula subpelliculosa	A		
Hansenula sp. I	A		
Hansenula sp. II	A		
Hansenula sp. III	A		
Hansenula sp. X		J	
Kloeckera apiculata/Hansenula valbyensis	A	J	
Kloeckera apiculata var. A	A	J	
Kloeckera antillarum	A		
Kloeckera magna		J	
Kloeckera sp. A	A	J	C
Kloeckera sp. X	A		
Kloeckera sp. Y	A		
Pichia delftensis			C
Pichia farinosa	A		
Pichia fermentans	A	J	
Pichia guilliermondii/Candida guilliermondii	A	J	
Pichia membranaefaciens/Candida mycoderma	A	J	C
Pichia polymorpha	A		C
Pichia sylvestris		J	
Pichia sp. A	A	J	C
Pichia sp.		J	
Rhodotorula aurantiaca/Sporobolomyces salmonicolor	A		
Rhodotorula flava	A		
Rhodotorula glutinis/Sporobolomyces roseus	A		
Rhodotorula glutinis var. *rubescens*	A		
Rhodotorula glutinis var. *rufosa*	A		
Rhodotorula minuta		J	
Rhodotorula mucilaginosa/Sporobolomyces pararoseus	A	J	
Rhodotorula rubra/Sporobolomyces sp. D	A		
Rhodotorula sp. X	A		
Rhodotorula sp. Y	A		
Saccharomyces acidifaciens	A		
Saccharomyces bisporus	A		
Saccharomyces carlsbergensis	A		C
Saccharomyces cerevisiae/Candida robusta	A	J	C
Saccharomyces delbrueckii	A		
Saccharomyces elegans	A		
Saccharomyces florentinus	A	J	
Saccharomyces fructuum	A		
Saccharomyces fructuum var. A	A		
Saccharomyces heterogenicus		J	
Saccharomyces kluyveri		J	
Saccharomyces microellipsodes		J	
Saccharomyces oviformis			C
Saccharomyces rosei	A		
Saccharomyces rouxii		J	

Yeasts	Source		
Saccharomyces steineri			C
Saccharomyces uvarum	A	J	C
Saccharomyces willianus	A		
Saccharomycodes ludwigii	A	J	C
Saccharomycodes ludwigii var. *bisporus*			C
Sporobolomyces roseus/*Rhodotorula glutinis*	A		
Sporobolomyces salmonicolor/*Rhodotorula aurantiaca*	A		
Sporobolomyces sp. X/Fungus sp. I	A		
Sporobolomyces/*Rhodotorula* sp. B	A		
Torulopsis aeria	A		
Torulopsis anomala var. A	A		
Torulopsis bacillaris	A		
Torulopsis candida/*Debaryomyces subglobosus*	A	J	C
Torulopsis etchellsii	A		
Torulopsis famata/*Debaryomyces hansenii*	A	J	
Torulopsis famata/*Cryptococcus luteolus* var. A	A	J	
Torulopsis glabrata	A	J	
Torulopsis holmii/*Saccharomyces exiguus*	A	J	
Torulopsis ingeniosa		J	
Torulopsis inconspicua		J	
Torulopsis nitratophila	A		
Torulopsis nodaensis		J	
Torulopsis sp. A	A	J	
Torulopsis sp. B	A		
Torulopsis sp. C	A	J	
Torulopsis sp. R	A		
Torulopsis sp. W		J	
Torulopsis sp. X	A	J	
Torulopsis sp. Y		J	
Torulopsis sp. Z		J	
Trichosporon cutaneum	A		
Trichosporon pullulans	A		
Trichosporon sp.	A		

Bibliography: Beech (1953, 1957, 1959a, 1965); Beech *et al.* (1964); Bowen (1962); Capitain (1930); Clark *et al.* (1954); Davenport (1968); Fell and Phaff (1967); Legarkis (1961); Lewis (1910); Lodder and Kreger-van Rij (1952); Osterwalder (1924); Sasaki and Yoshida (1959).

References

Adams, A. M. (1961). *Rep. hort. Exp. Stn Prod. Lab. Vineland* for 1959–60, 79–82.

Adams, A. M. (1964). *Can. J. Microbiol.* **10**, 641–646.

Anh, T. H. (1966). *Revue Brass.* **77**, 365–370.

Anon. (1967). *Fd Inds S. Afr.* **19**, 55–57.

Asai, G. N. (1960). *Phytopathology* **50**, 535–541.

Atkinson, F. E., Bowen, J. F. and MacGregor, D. R. (1959). *Fd Technol.*, *Champaign* **13**, 673–678.

Äyrapää, T. (1968). *J. Inst. Brew.* **74**, 169–178.

Bab'eva, I. P. and Savel'eva, N. D. (1963). *Mikrobiologiya* **32**, 86–93.

Barker, B. T. P. (1950a). *Rep. agric. hort. Res. Stn Univ. Bristol* for 1949, 131–136.

Barker, B. T. P. (1950b). *Rep. agric. hort. Res. Stn Univ. Bristol* for 1949, 137–144.

Barker, B. T. P. (1951). *Rep. agric. hort. Res. Stn Univ. Bristol* for 1950, 178–187.

Beech, F. W. (1953). *Rep. agric. hort. Res. Stn Univ. Bristol* for 1952, 125–137.

Beech, F. W. (1955). *Rep. agric. hort. Res. Stn Univ. Bristol* for 1954, 179–186.

Beech, F. W. (1957). Ph.D. Thesis: University of Bristol.

Beech, F. W. (1958). *Soc. Chem. Ind. Monograph* No. 3, 37–51.

Beech, F. W. (1959a). *J. appl. Bact.* **21**, 257–266.

Beech, F. W. (1959b). *Rep. agric. hort. Res. Stn Univ. Bristol* for 1958, 154–160.

Beech, F. W. (1962). *Rep. agric. hort. Res. Stn Univ. Bristol* for 1961, 176–180.

Beech, F. W. (1963a). *Rep. agric. hort. Res. Stn Univ. Bristol* for 1962, 157–167.

Beech, F. W. (1963b). *Rep. agric. hort. Res. Stn Univ. Bristol* for 1962, 168–170.

Beech, F. W. (1965). *Antonie van Leeuwenhoek* **31**, 81–83.

Beech, F. W. (1966). *Rep. agric. hort. Res. Stn Univ. Bristol* for 1965, 60–61.

Beech, F. W. (1967a). *Rep. agric. hort. Res. Stn Univ. Bristol* for 1966, 227–238.

Beech, F. W. (1967b). *Rep. agric. hort. Res. Stn Univ. Bristol* for 1966, 239–245.

Beech, F. W. and Carr, J. G. (1953). *In* "Science and Fruit" (T. Wallace and R. W. Marsh, eds.), pp. 68–77. University of Bristol.

Beech, F. W. and Carr, J. G. (1960). *J. Sci. Fd Agric.* **11**, 35–40.

Beech, F. W. and Challinor, S. W. (1951). *Rep. agric. hort. Res. Stn Univ. Bristol* for 1950, 143–160.

Beech, F. W. and Davenport, R. R. (1969a). *In* "Methods in Microbiology" (J. R. Norris and D. W. Ribbons, eds.). Academic Press, London.

Beech, F. W. and Davenport, R. R. (1969b). *In* "Isolation Methods for Microbiologists, Part A" No. 3 Tech. Series Soc. appl. Bact. (D. A. Shapton and G. W. Gould, eds.). Academic Press, London.

Beech, F. W. and Pollard, A. (1966). *Rep. agric. hort. Res. Stn Univ. Bristol* for 1965, 259–264.

Beech, F. W., Burroughs, L. F. and Codner, R. C. (1952). *Rep. agric. hort. Res. Stn Univ. Bristol* for 1951, 149–159.

Beech, F. W., Kieser, M. E. and Pollard, A. (1964). *Rep. agric. hort. Res. Stn Univ. Bristol* for 1963, 147–149.

Beech, F. W., Davenport, R. R., Goswell, R. W. and Burnett, J. K. (1968). *In* "Identification Methods for Microbiologists Part B" No. 2 Tech. Series Soc. appl. Bact. (B. M. Gibbs and D. A. Shapton, eds.), pp. 151–175. Academic Press, London.

Boehi, Q. (1912). "Ein neues Verfahren zur Herstellung alkoholfrier Obst- und Traubenwein (Kohlensäurenverfahren)". Huber & Co., Frauenfeld.

Bowen, J. F. (1962). Ph.D. Thesis: University of Bristol.

Bowen, J. F. and Beech, F. W. (1964). *J. appl. Bact.* **27**, 333–341.

Bowen, J. F. and Beech, F. W. (1967). *J. appl. Bact.* **30**, 475–483.

Brumstead, D. D. and Glenister, P. R. (1963). *Brewers' Dig.* **38**, 49–52.

Brun, P. and Tarral, R. (1963). *Bull. Soc. scient. Hyg. aliment.* **51**, 238–255.

Buckle, F. J. (1967). *Process Biochem.* **2**, 37–40.

Burroughs, L. F. (1952). *Rep. agric. hort. Res. Stn Univ. Bristol* for 1951, 138–148.

Burroughs, L. F. (1953). *Rep. agric. hort. Res. Stn Univ. Bristol* for 1952, 110–125.
Burroughs, L. F. (1957). *J. Sci. Fd Agric.* **8**, 122–131.
Burroughs, L. F. (1959). *Rep. agric. hort. Res. Stn Univ. Bristol* for 1958, 164–168.
Burroughs, L. F. (1962). *Rep. agric. hort. Res. Stn Univ. Bristol* for 1961, 173–175.
Burroughs, L. F. (1964). *In* "4th Int. Symp. Fd Microbiol. Sweden" (N. Molin, ed.), pp. 133–137. Almqvist and Wiksell, Stockholm.
Burrough, L. F. and Challinor, S. W. (1949). *Rep. agric. hort. Res. Stn Univ. Bristol* for 1948, 207–215.
Burroughs, L. F. and Challinor, S. W. (1951). *Rep. agric. hort. Res. Stn Univ. Bristol* for 1950, 161–167.
Burroughs, L. F. and Pollard, A. (1954). *Rep. agric. hort. Res. Stn Univ. Bristol* for 1953, 184–188.
Burroughs, L. F. and Sparks, A. H. (1963). *Rep. agric. hort. Res. Stn Univ. Bristol* for 1962, 151–156.
Burroughs, L. F. and Sparks, A. H. (1964). *J. Sci. Fd Agric.* **16**, 176–185.
Burroughs, L. F. and Whiting, G. C. (1961). *Rep. agric. hort. Res. Stn Univ. Bristol* for 1960, 144–147.
Campbell, I. and Brudzynski, A. (1966). *J. Inst. Brew.* **72**, 556–560.
Capitain, R. (1930). Ph.D. Thesis: University of Lyons.
Carr, J. G. (1953). *Rep. agric. hort. Res. Stn Univ. Bristol* for 1952, 144–150.
Carr, J. G. (1957). *J. Inst. Brew.* **63**, 436–440.
Carr, J. G. (1958). *In* "Proc. Symp. Fruit Juice Concentrates Bristol", pp. 383–390. Juris-Verlag, Zürich.
Carr, J. G. (1959a). *Rep. agric. hort. Res. Stn Univ. Bristol* for 1958, 160–163.
Carr, J. G. (1959b). *J. appl. Bact.* **21**, 267–271.
Carr, J. G. (1960). *J. appl. Bact.* **22**, 377–383.
Carr, J. G. (1964a). *Rep. agric. hort. Res. Stn Univ. Bristol* for 1963, 167–172.
Carr, J. G. (1964b). *Brewers' J. Hop Malt Trades Rev.* **100**, 244–247, 324–327, 390–393.
Carr, J. G. and Whiting, G. C. (1956). *Rep. agric. hort. Res. Stn Univ. Bristol* for 1955, 163–168.
Carter, M. V. (1961). *Rep. Rothamsted exp. Stn* for 1960, 125.
Castelli, T. (1957). *Am. J. Enol. Vitic.* **8**, 149–156.
Chalenko, D. K. and Korsakova, T. F. (1958). *Trudy tsent. nauchno-issled. Lab. vinodel. Prom.* **1**, 5–11.
Challinor, S. W. (1955). *J. appl. Bact.* **18**, 212–223.
Challinor, S. W. and Burroughs, L. F. (1949). *Rep. agric. hort. Res. Stn Univ. Bristol* for 1948, 182–206.
Charley, V. L. S., Hopkins, D. A. and Pollard, A. (1941). *Rep. agric. hort. Res. Stn Univ. Bristol* for 1940, 97–104.
Chester, V. E. (1963). *Proc. R. Soc. Ser. B*, **157**, 223–233.
Clark, D. S. and Wallace, R. H. (1954). *Can. J. Microbiol.* **1**, 275–276.
Clark, D. S., Wallace, R. H. and David, J. J. (1954). *Can. J. Microbiol.* **1**, 145–149.
Comrie, A. A. D. (1951). *Proc. Eur. Brew. Conv. Brighton*, 168–177.
Cook, A. H. (1963). *Proc. Eur. Brew. Conv. Brussels*, 477.
Cooke, W. B. (1959). *Mycopath. Mycol. appl.* **12**, 1–41.
Cooke, W. B. (1962). *Mycopath. Mycol. appl.* **17**, 1–43.
Crowell, E. A. and Guymon, J. F. (1963). *Am. J. Enol. Vitic.* **24**, 214–222.
Crowell, E. A., Guymon, J. F. and Ingraham, J. L. (1961). *Am. J. Enol. Vitic.* **12/13**, 111–116.

Cusick, J. I. (1950). *Fd Technol., Champaign* **4**, 329–331.
Czarnecki, H. T. and Engel, E. L. van (1959). *Brewers' Dig.* **34**, 52–56.
Daepp, H. V. and Lüthi, H. R. (1966). *Fruchtsaft-Ind.* **11**, 5–10.
Davenport, R. R. (1967). *Rep. agric. hort. Res. Stn Univ. Bristol* for 1966, 246–248.
Davenport, R. R. (1968). Thesis: Institute of Biology.
Do, J. Y. and Salunkhe, D. K. (1964). *Fd Technol., Champaign* **18**, 182–184.
Domercq, S. (1956). Ph.D. Thesis: University of Bordeaux.
Drawert, F., Rapp, A. and Ullrich, W. (1965). *Vitis* **5**, 20–23.
Elbert, E. M. and Esselen, W. B. (1959). *Fd Res.* **24**, 352–361.
Ellison, J. and Doran, A. H. (1961). *Proc. Eur. Brew. Conv. Vienna*, 224–234.
Emeis, C. C. (1964). *Proc. Eur. Brew. Conv. Brussels*, 362–369.
Evelyn, J. (1664). *In* "Pomona, or an Appendix concerning Fruit-Trees, In relation to Cider, the Making, and several ways of Ordering it", p. 32. J. Martyn and J. Allestry, London.
Fabian, F. W. (1933). *Fruit Prod. J. Am. Vineg. Ind.* **12**, 141–142.
Falk, M. and Giguère, P. A. (1958). *Can. J. Chem.* **36**, 1121–1125.
Fell, J. W. and Phaff, H. J. (1967). *Antonie van Leeuwenhoek* **33**, 464–472.
Fornachon, J. C. M. and Lloyd, B. (1965). *J. Sci. Fd Agric.* **16**, 710–716.
Gilliland, R. B. (1951). *Proc. Eur. Brew. Conv. Brighton*, 35–58.
Gilliland, R. B. and Harrison, G. A. F. (1966). *J. appl. Bact.* **29**, 244–252.
Gilliland, R. B. and Lacey, J. P. (1964). *Nature, Lond.* **202**, 727.
Gilliland, R. B. and Lacey, J. P. (1966). *J. Inst. Brew.* **72**, 291–303.
Goto, S. (1961). *Hakko Kogaku Zasshi* **39**, 705–709.
Graff, Y. (1959). *Inds aliment. agric.* **76**, 183–191.
Graff, Y. and Bidan, P. (1950). *Inds agric. aliment.* **66**, 247–253.
Guilliermond, A. (1920). "The Yeasts", translated and revised by F. W. Tanner. Chapman and Hall, London.
Harrison, G. A. F. (1963). *Proc. Eur. Brew. Conv. Brussels*, 247–256.
Hernandez, M. R. (1967). *Sem. Vitivinic.* **22**, 423–425.
Hough, J. S. (1957). *J. Inst. Brew.* **63**, 483–487.
Hough, J. S. and Stevens, R. (1961). *J. Inst. Brew.* **67**, 488–494.
Hudson, J. R. (1967). *Internat. Brewer Distiller*, 27–30.
Hudson, J. R. and Stevens, R. (1960). *J. Inst. Brew.* **66**, 471–474.
Ingraham, J. L. and Guymon, J. F. (1960). *Archs Biochem. Biophys.* **88**, 157–166.
Ingram, M. (1958). *In* "The Chemistry and Biology of Yeasts" (A. H. Cook, ed.), p. 623. Academic Press, New York.
Jacquet, J. and Breton, J. Le (1966). *C. r. hebd. Séanc. Acad. Agric. Fr.* **52**, 1054–1058.
Jacquin, P. and Tavernier, J. (1951). *Inds agric. aliment.* **68**, 599–607; (1952). **69**, 115–127.
Jakubowska, J. (1963). *Zesz. nauk. Politech. lódz.* No. 8, 5–25; from *Chemy Abst.* 1965 **63**, 4909b.
Jenny, J. (1940). *Landw. Jb. Schweiz.* **54**, 739–774.
Kaneko, T. and Yamamoto, Y. (1966). *Rep. Res. Labs Kirin Brew. Co.* Dec. 37–50.
Kasza, D. S. (1956). *N.Z. Jl Agric.* **93**, 561.
Kawakita, S. and Uden, N. van (1965). *J. gen. Microbiol.* **39**, 125–129.
Kayser, E. (1890). *Annls Inst. Pasteur* **4**, 321.
Keener, P. D. (1950). *Am. J. Bot.* **37**, 520–527.
Keener, P. D. (1951). *Am. J. Bot.* **38**, 105–110.
Kieser, M. E., Pollard, A. and Stone, A. M. (1949). *Rep. agric. hort. Res. Stn Univ. Bristol* for 1948, 228–234.

Kieser, M. E., Pollard, A., Stevens, P. M. and Tucknott, O. G. (1964). *Nature, Lond.* **204**, 887.

Kir'yalova, E. N. (1960). *Trudȳ veses. nauchno-issled. Inst. sel'-khoz. Mikrobiol.* **16**, 190–201.

Kleyn, J. G. and Vacano, L. N. (1963). *Am. Brew.* **96**, 26–34.

Klöcker, A. (1912). *Zentbl. Bakt. ParasitKde (Abt II)* **35**, 369.

Klöcker, A. (1913). *C. r. Trav. Lab. Carlsberg.* **10**, 207.

Klopper, W. J., Roberts, R. H., Royston, M. G. and Ault, R. G. (1965). *Proc. Eur. Brew. Conv. Stockholm,* 238–259.

Kodaira, R. and Uemura, T. (1961). *Kôso Kagaku Shinpojiumu* **14**, 130–134; from *Chemy Abst.* 1961 **55**, 13543b.

Kodaira, R., Ito, Y. and Uemura, T. (1958). *J. Agric. Chem. Soc., Japan* **32**, 49–54; from *Chemy Abst.* 1959 **53**, 644h.

Kramer, C. L., Pady, S. M., Rogerson, C. T. and Oiye, L. G. (1961). *Trans. Kans. Acad. Sci.* **62**, 184–199.

Kreger-van Rij, N. J. W. (1964). Ph.D. Thesis: University of Leiden.

Kringstad, H. and Rasch, S. (1966). *J. Inst. Brew.* **72**, 56–61.

Kunkee, R. E. and Ough, C. S. (1966). *Appl. Microbiol.* **14**, 643–648.

Last, F. T. and Deighton, F. C. (1965). *Trans. Br. mycol. Soc.* **48**, 83–99.

Lawrence, W. C. (1964). *Wallerstein Labs Commun.* **27**, 123–152A.

Leal, B. I. (1958). *Revta Ciencia apl.* **12**, 318–324; from *Chemy Abst.* 1959 **53**, 2532e.

Leal, B. I., Varela, V. A., Bravo Abad, F. and Llanguno, C. (1961). *Revta. Agroquim. Technol. alimentos* **1**, 11–17; from *Chemy Abst.* 1962 **56**, 14740a.

Legarkis, A. (1961). Ph.D. Thesis: University of Athens.

Lewin, L. M. and Smith, E. J. (1964). *Nature, Lond.* **203**, 867–868.

Lewis, C. E. (1910). *Maine agric. Stn Bull.* No. 178, 45.

Lewis, M. J. (1964). *Wallerstein Labs Commun.* **37**, 29 39.

Lewis, M. J. and Rainbow, C. (1963). *J. Inst. Brew.* **69**, 39–45.

Lodder, J. and Kreger-van Rij, N. J. W. (1952). "The Yeasts: a Taxonomic Study". North-Holland Publishing Co., Amsterdam.

Lund, A. (1954). Ph.D. Thesis: University of Copenhagen.

Lüthi, H. (1949). *Schweiz. Z. Obst- u. Weinb.* **58**, 379–384, 401–409.

Lüthi, H. (1950). *Schweiz Z. Obst- u. Weinb.* **59**, 321–325.

Lüthi, H. (1957a). *Rev. Ferment. Ind. aliment.* **12**, 15–21.

Lüthi, H. (1957b). *Mitt. Geb. Lebensmittelunters. u. Hyg.* **48**, 201–217.

Lüthi, H. (1958). *In* "Proc. Symp. Fruit Juice Concentrates, Bristol", pp. 391–401. Juris-Verlag, Zürich.

Lüthi, H. (1959a). *Landw. Jbr Schweiz.* **73**, 108–111.

Lüthi, H. (1959b). *Adv. Fd Res.* **9**, 221–284, Academic Press, New York.

Lüthi, H. and Hochstrasser, R. (1952). *Schweiz. Z. Obst- u. Weinb.* **61**, 301–307, 359–361.

Lüthi, H. and Vetsch, U. (1955). *Schweiz. Z. Obst- u. Weinb.* **64**, 404–409.

Malan, C. E. (1961). *Accad. It. Vite Vino, Siena Atti* **12**, 201–216; from *Chemy Abst.* 1962 **57**, 3873h.

Marcus, O. (1942). *Arch. Mikrobiol.* **13**, 1–44.

Marshall, C. R. and Walkley, V. T. (1951a). *Fd Res.* **16**, 448–456.

Marshall, C. R. and Walkley, V. T. (1951b). *Fd Res.* **16**, 515–521.

Marshall, C. R. and Walkley, V. T. (1952). *Fd Res.* **17**, 197–203.

Masuda, H., Shijo, N. and Muraki, H. (1964a). *Hakko Kogaku Zasshi* **42**, 379–382; from *Chemy Abst.* 1966 **64**, 20593b.

Masuda, H., Shijo, N. and Muraki, H. (1964b). *Hakko Kogaku Zasshi* **42**, 383–387; from *Chemy Abst.* 1966 **64**, 20593e.
Melas-Joannidis, Z., Carni-Catsadimas, I. and Verona, O. (1958). *Ann. Microbiol.* **8**, 118–137.
di Menna, M. E. (1955). *Trans. Br. mycol. Soc.* **38**, 119–129.
di Menna, M. E. (1957). *J. gen. Microbiol.* **17**, 678–688.
di Menna, M. E. (1959). *N.Z. Jl agric. Res.* **2**, 394–405.
di Menna, M. E. (1962). *J. gen. Microbiol.* **27**, 249–257.
Ménoret, Y. and Gautheret, R. J. (1962). *Inds aliment. agric.* **79**, 419–425.
Merzhaman, A. A. (1961). *Trudy krasnodar. Inst. pishch. Prom.* **22**, 95–104; from *Chemy Abst.* 1962 **57**, 17207e.
Miller, J. J. and Webb, N. S. (1954). *Soil Sci.* **77**, 197–204.
Milleville, H. P. and Eskew, R. K. (1944). "Recovery and utilization of natural apple flavour". AIC–63, U.S. Dept Agric., Washington.
Millis, N. F. (1951). Ph.D. Thesis: University of Bristol.
Millis, N. F. (1956). *J. gen. Microbiol.* **15**, 521–528.
Monties, B. and Barret, A. (1965). *Annls Technol. agric.* **14**, 167–172.
Mosiashvili, G. I. (1956). *Mikrobiologiya* **25**, 484–488.
Mosiashvili, G. I., Osipova, S. A. and Gigineishvili, L. A. (1961). *Vinod. Vinogr. Mold.* **21**, 10–14; from *Chemy Abst.* 1962 **57**, 1382a.
Mrak, E. M. and Phaff, H. J. (1948). *A. Rev. Microbiol.* **2**, 1–46. Ann. Revs Inc., California.
Müller-Thurgau, H. (1899). *Zentbl. Bakt. ParasitKde (Abt II)* **5**, 684–685.
Nagai, S. (1963). *J. Bact.* **86**, 299–302.
Nagai, S. (1965). *J. Bact.* **90**, 220–222.
Niethammer, A. (1942). *Arch. Mikrobiol.* **13**, 45–49.
Neubeck, C. E. (1959). *J. Ass. off. agric. Chem.* **42**, 374–382.
Odintsova, E. N. (1956). *Soviet Abstr. Chem.* No. 56579; from *Chemy Abst.* 1959 **53**, 3591f.
Ohara, Y., Kushida, T., Nonomura, H. and Maruyama, C. (1957). *Yamanashi Daigaku Hakko Kenkyusho Kenkyu Hokoko* No. 4, 29–33; from *Chemy Abst.* 1959 **53**, 9565c.
Okuda, K. and Tanase, O. (1958). *Rep. Res. Labs Kirin Brew. Co.* **1**, 7–11; from *Chemy Abst.* 1959 **53**, 19293c.
Osterwalder, A. (1912). *Zentbl. Bakt. ParasitKde (Abt II)* **33**, 257–272.
Osterwalder, A. (1924). *Zentbl. Bakt. ParasitKde (Abt II)* **60**, 481–528.
Otsuka, K., Hara, S. and Imai, S. (1963). *Nippon Jozo Kyokai Zasshi* **58**, 631–635; from *Chemy Abst.* 1965 **63**, 9023e.
Ough, C. S. and Amerine, M. A. (1966). *Am. J. Enol. Vitic.* **17**, 163–173.
Ough, C. S. and Amerine, M. A. (1967). *Am. J. Enol. Vitic.* **18**, 157–164.
Ough, C. S., Guymon, J. F. and Crowell, E. A. (1966). *Fd Res.* **31**, 620–625.
Ough, C. S. and Kunkee, R. E. (1968). *Appl. Microbiol.* **16**, 572–576.
Parle, J. N. (1963a). *J. gen. Microbiol.* **31**, 1–11.
Parle, J. N. (1963b). *J. gen. Microbiol.* **31**, 13–22.
Pearce, E. B. and Barker, B. T. P. (1908). *J. agric. Sci., Camb.* **3**, 55–79.
Peter, A. (1964). *Fruchtsaft-Ind.* **9**, 5–14.
Petit, L. and Godon, B. (1963). *C. r. hebd. Séanc. Acad. Agric. Fr.* **257**, 1993–1995.
Peynaud, E. (1956). *Inds aliment. agric.* **73**, 253–257.
Peynaud, E. and Domercq, S. (1954). *Xe Congrès int. Inds aliment.* Madrid.
Peynaud, E. and Domercq, S. (1955). *C. r. hebd. Séanc. Acad. Agric. Fr.* **41**, 103.
Peynaud, E. and Lafourcade, S. (1957). *Inds aliment. agric.* **12**, 897–904.

Peynaud, E., Lafon-Lafourcade, S. and Guimberteau, G. (1967). *Antonie van Leeuwenhoek* **33**, 49–55.

Pfenniger, H. (1963). *Z. Lebensmittelunters-u.-Forsch.* **119**, 401–415; **120**, 100–116, 117–126.

Phaff, H. J., Miller, M. W. and Mrak, E. M. (1966). *In* "The Life of Yeasts", p. 108. University Press, Harvard.

Pollard, A. and Beech, F. W. (1957). "Cider-making". Rupert Hart-Davis, London.

Pollard, A. and Beech, F. W. (1963). *In* "Perry Pears" (L. C. Luckwill and A. Pollard, eds.), pp. 195–203. University of Bristol.

Pollard, A. and Beech, F. W. (1966a). *Process Biochem.* **1**, 229–233, 238.

Pollard, A. and Beech, F. W. (1966b). *Rep. agric. hort. Res. Stn Univ. Bristol* for 1965, 259–264.

Pollard, A. and Kieser, M. E. (1959). *J. Sci. Fd Agric.* **10**, 253–260.

Pollard, A., Beech, F. W. and Burroughs, L. F. (*in press*). *In* "Encyclopedia on Food and Food Science" (G. Borgström, ed.) Vol. 2. Interscience Publishers (John Wiley and Sons Inc.), New York.

Pollard, A., Kieser, M. E. and Beech, F. W. (1966). *J. appl. Bact.* **29**, 253–259.

Pollard, A., Kieser, M. E., Stevens, P. M. and Tucknott, O. G. (1965). *J. Sci. Fd Agric.* **16**, 384–389.

Portno, A. D. (1966). *J. Inst. Brew.* **72**, 193–196.

Portno, A. D. (1967a). *J. Inst. Brew.* **73**, 43–50.

Portno, A. D. (1967b). *J. Inst. Brew.* **73**, 512–514.

Portno, A. D. (1968). *J. Inst. Brew.* **74**, 291–300.

Príbella, A. and Betušová, M. (1964). *Fruchtsaft-Ind.* **9**, 15–25.

Ranganathan, B. and Bhat, J. V. (1958). *J. Indian Inst. Sci.* **40**, 105–110.

Rankine, B. C. (1955). *Aust. J. appl. Sci.* **6**, 408 413, 414 420, 421–425.

Rankine, B. C. (1963). *J. Sci. Fd Agric.* **14**, 79–91.

Rankine, B. C. (1964). *J. Sci. Fd Agric.* **15**, 872–877.

Rankine, B. C. (1966). *J. Sci. Fd Agric.* **17**, 312–316.

Rankine, B. C. (1967a). *J. Sci. Fd Agric.* **18**, 41–44.

Rankine, B. C. (1967b). *J. Sci. Fd Agric.* **18**, 583–589.

Rankine, B. C. (1968). *J. Sci. Fd. Agric.* **19**, 624–627.

Rankine, B. C. and Lloyd, B. (1963). *J. Sci. Fd Agric.* **14**, 793–798.

Rehm, H. J. and Wittmann, H. (1963). *Z. Lebensmittelunters-u.-Forsch.* **120**, 465–478.

Richards, M. (1967). *J. Inst. Brew.* **73**, 162–166.

Richards, M. and Cowland, T. W. (1967). *J. Inst. Brew.* **73**, 552–558.

Ricketts, R. W. and Hough, J. S. (1961). *J. Inst. Brew.* **67**, 29–32.

Rogerson, C. T. (1958). *Trans. Kans. Acad. Sci.* **61**, 155–162.

Romwalter, A. and Király, A. von (1939). *Arch. Mikrobiol.* **10**, 87–91.

Royston, M. G. (1966). *Process Biochem.* **1**, 215–221.

Ruyle, E. H., Pearce, W. E. and Hays, G. L. (1946). *Fd Res.* **11**, 274–279.

Sainclivier, M. (1951). *Inds. agric. aliment.* **68**, 277–278.

Samish, Z., Etinger-Tulczynska, R. and Blick, M. (1963). *Fd Res.* **28**, 259–266.

Sasaki, Y. and Yoshida, T. (1959). *J. Fac. Agric. Hokkaido* (*imp.*) *Univ.* **51**, 194–220.

Schanderl, H. (1957). *In* "Yeasts" (W. Roman, ed.), pp. 127–139. Dr. Junk, The Hague.

Schanderl, H. and Staudenmayer, T. (1964). *Mitt. höh. Bundeslehr-u. Vers Anst. Wein- Obst- u. Gartenb. Ser. A*, **14**, 267–281.

SentheShanmuganathan, S. and Elsden, S. R. (1958). *Biochem. J.* **69**, 210–218.
Smith, E. S., Bowen, J. F. and MacGregor, D. R. (1962). *Fd Technol., Champaign* **16**, 93–95.
Spencer, J. F. T. and Phaff, H. J. (1963). *Yeast Newsl.* **12**, 8.
Stevens, R. (1960). *J. Inst. Brew.* **66**, 453–471.
Stevens, T. J. (1966). *J. Inst. Brew.* **72**, 369–373.
Stewart, E. D., Hinz, C. and Brenner, M. W. (1962). *Proc. Am. Soc. Brew. Chem.* 40–46.
Swindells, R. and Robbins, R. H. (1966). *Process Biochem.* **1**, 457–460, 469.
Taketa, M. and Tsukahara, T. (1962). *Nippon Jozo Kyokai Zasshi*, **57**, 1109–1111; from *Chemy Abst.* 1964 **61**, 6081g.
Tanner, F. W. (1944). "Microbiology of Food" 2nd Ed. Garrard Press, Champaign, Illinois.
Tanner, H. and Rentschler, H. (1954). *Mitt. Geb. Lebensmittelunters-u.-Hyg.* **45**, 305–311.
Tavernier, J. and Jacquin, P. (1946). *Chim. Ind.* **56**, 104–113.
Thorne, R. S. W. (1958). *J. Inst. Brew.* **64**, 411–421.
Thorne, R. S. W. (1961). *Brewers' Dig.* **36**, 38–40, 43.
Thoukis, G., Ueda, M. and Wright, D. (1965). *Am. J. Enol. Vitic.* **16**, 1–8.
Tressler, D. K., Celmer, R. F. and Beavens, E. A. (1941). *J. ind. Engng Chem.* **33**, 1027–1031.
Tschenn, C. (1934). *Fruit Prod. J. Am. Vineg. Ind.* 111.
Valuĭko, G. G. (1960). *Trudy vses. nauchno-issled. Inst. Vinod. Vinogr.* **9**, 113–144; from *Chemy Abst.* 1961 **55**, 13759i.
Vas, G. and Ingram, M. (1949). *Fd Mf.* **24**, 414–416.
Vasatko, J. and Príbella, A. (1965). *Izv. vyssh. ucheb. Zaved.pishch. Tekhnol.* No. 6, 17–20; from *Chemy. Abst.* 1966, **64**, 9930h.
Veselov, I. Y., Kann, A. G. and Gracheva, I. M. (1963). *Mikrobiologiya* **32**, 610–615.
Wainwright, T. (1962). *Biochem. J.* **83**, 39P.
Walters, L. S. and Thiselton, M. R. (1953). *J. Inst. Brew.* **59**, 401–404.
Warcollier, G. and Moal, A. Le (1935). *Annls Falsif. Fraudes* **28**, 517–534.
Webb, A. D. (1967). *Biotechnol. Bioengng* **9**, 305–319.
Whiting, G. C. (1961). *Rep. agric. hort. Res. Stn Univ. Bristol* for 1960, 135–139.
Whiting, G. C. and Coggins, R. A. (1960a). *Nature, Lond.* **185**, 843–844.
Whiting, G. C. and Coggins, R. A. (1960b). *J. Sci. Fd Agric.* **11**, 337–344.
Whiting, G. C. and Coggins, R. A. (1960c). *J. Sci. Fd Agric.* **11**, 705–709.
Wickerham, L. J. (1951). "Taxonomy of Yeasts", Tech. Bull. No. 1029. U.S. Dept. Agric., Washington.
Wiles, A. E. (1950). *J. Inst. Brew.* **56**, 183–193.
Williams, R. R. and Child, R. D. (1966). *Rep. agric. hort. Res. Stn Univ. Bristol* for 1965, 71–89.
Williams, A. J., Wallace, R. H. and Clark, D. S. (1956). *Can. J. Microbiol.* **2** 645–648.
Yates, A. R., Seaman, A. and Woodbine, M. (1967). *Can. J. Microbiol.* **13**, 1120–1123.
Zambonelli, C. (1964). *Annali Microbiol.* **14**, 129–141.
Zubkova, R. D. (1958). *Trudy Inst. Mikrobiol. Virus., Alma-Ata* **2**, 143–154; from *Chemy Abst.* 1959 **53**, 8530c.
Zyl, J. A. van (1962). *Tech. Serv. Sci. Bull.* No. 381, Vitic Ser. No. 1, pp. 44. Dept. Agric., South Africa.
Zyl, J. A. van, Vries, M. J. de and Zeeman, A. S. (1963). *S. Afr. J. agric. Sci.* **6**, 165–180.

Chapter 4

Brewer's Yeasts

C. RAINBOW

Bass Charrington Limited, Burton-on-Trent, England

I. Introduction

Beer is essentially the product resulting from the alcoholic fermentation of brewer's hopped malt extract (brewer's wort) and the primary products of that fermentation, ethanol and carbon dioxide, are essential constituents of all beer. Nevertheless, not all yeasts (even those belonging to the species *Saccharomyces cerevisiae* and *Sacch. carlsbergensis*, to which all brewing strains belong) produce palatable acceptable beers by their fermentative action on malt wort.

The ability to ferment sugars in malt wort to ethanol and carbon dioxide as its major end-products of metabolic activity is therefore not the only factor determining the suitability of a yeast strain for brewing. Indeed, the critical factor may be ability to form, in subtly balanced proportions, quantitatively minor metabolic products such as acids, esters, higher alcohols and ketones. Such products arise, in some cases, from carbohydrate and, in others, from nitrogen metabolism. On the other hand, a yeast may be unsuitable for brewing because one or more of these minor metabolic products is formed in excessive amounts, either in the absolute sense, or relative to one another. Again, other yeasts may be unsuitable by virtue of excessive production of such undesirable trace products as mercaptans.

Still another factor important in determining suitability of a yeast strain for brewing is its technological behaviour; in particular, its flocculence, or tendency to separate from suspension in fermenting wort in a longer or shorter time. On this property, which is discussed in detail below (Section II.B, p. 149), depend important beer characteristics.

In what follows, therefore, the biochemistry of the formation of ethanol and carbon dioxide from fermentable sugars by the classical fermentation cycle, which is a pathway common to all brewer's yeasts and has been treated adequately many times elsewhere (see Volume 2) will be assumed, and this chapter will be concerned chiefly with those other aspects of yeasts which distinguish them as industrially suitable for beer production. For this purpose, brewer's yeasts must be discussed largely in terms of their interaction with their normal environment, brewer's hopped wort. It may be interjected here that the hop constituents of brewer's wort, which are responsible for the bitter flavour of beer, do not seem to influence the biochemical properties of yeast, although they tend to become concentrated in foam and therefore may exert surface effects on top yeasts accumulating in heads.

II. Technological Behaviour

A. TOP AND BOTTOM YEASTS

The many kinds of beer available may be classified into two main groups according to the type of yeast used and its behaviour at the end of fermentation. Some brewer's yeasts typically settle out to the bottom of the fermenting vessel: these are bottom yeasts, strains of *Saccharo-*

myces carlsbergensis, used for the production of lager beers. By contrast, the yeasts used to produce ale are strains of *Sacch. cerevisiae*, which accumulate as a yeasty head at the surface of the fermenting wort. Thus, the term *ale* now usually signifies a top fermentation beer.

Top fermentation beers are typical of Great Britain, whereas lager beers are characteristic of most of the rest of the world. Lager beers include pale types (e.g. Pilsen) and dark types (e.g. Munich). Ales also include pale and dark types and, additionally, the very dark stouts (porters). In either case, dark beers are brewed from malt grists containing malt which has been coloured by kilning at relatively high temperatures to induce melanoidin formation.

Apart from differences ascribable to the yeast strains used in their preparation, lager beers and ales acquire distinctions resulting from differences in the malts, hops and brewing waters used, and in the brewing and fermentation procedures. These factors lead to differences in the brewer's wort, which provides the substrates presented to the yeast for fermentation, although it is likely that such differences are less important as determinants of lager beer or ale character, flavour and aroma than are the respective yeast types applied to ferment the wort.

B. FLOCCULATION

1. *Technological Importance*

As lager fermentations proceed, the lager yeast, at first largely dispersed in suspension, and maintained thus by the evolution of the carbon dioxide, begins to form small aggregates of cells which fall out of suspension and collect at the bottom of the fermenting vessel. Flocculation phenomena also occur with top fermentation ale yeasts: in this case, cell aggregates, formed from a relatively large number of cells clumped together, appear to entrain bubbles of carbon dioxide, so that buoyant complexes of cells and gas are formed and carried to the surface of the fermenting liquid, where they collect to form a stable triphasic heterogeneous structure composed of gas, liquid and cells.

This aggregation of cells and their separation from suspension towards the end of fermentation is known as flocculation. Reviewing this subject, Comrie (1952) defined flocculation as follows: "A yeast is said to flocculate when the cells adhere in clumps and it tends to separate rapidly from the liquid in which it is suspended."

The property of flocculation forms the basis on which yeast for future inoculation ("pitching") is selected and collected. This is especially true in British ale breweries, where the yeast, having risen to the surface, is usually removed by a process of skimming. Since the most flocculent cells form the first, and the least flocculent cells the last heads, ale brewers are able to select yeast fractions with appropriate flocculating properties for future pitching. Frequently, middle skimmings, that is cells with powers of flocculation intermediate between those of the early (most flocculent) and the last (least flocculent) cells, are selected for reasons which will now be discussed.

Individual strains of brewer's top and bottom yeasts differ considerably in flocculating power from extremes of poorly flocculative ("powdery" yeast; *Staubhefe*) to highly flocculative ("clumping" yeast; *Bruchhefe*). Because highly flocculative yeasts separate early from suspension in fermenting wort, they tend to yield beers which are less fully fermented (less well "attenuated", in brewing parlance), sweeter, and having more "palate-fullness". They carry the advantage that prompt removal of yeast tends to avoid certain harsh flavours ("yeast bite") contributed by yeast if it is allowed to remain in mass contact with beer. Such flavours are probably caused by yeast excretion products, or even by autolytic changes occurring in the yeast mass, particularly when its vital activities slow down at the end of fermentation. However, the use of highly flocculent yeasts carries disadvantages. Thus, on transferring the beer from the fermenting vessel ("racking") for maturing and conditioning, the beer may have become so "cleansed" of yeast that there may be too few cells in suspension to maintain an adequate rate of secondary fermentation to accomplish these changes without having recourse to the addition of active yeast, perhaps in the form of fermenting wort. A second disadvantage of the use of highly flocculent yeasts for primary fermentation is that their early separation leaves a beer which is relatively rich in nutrient substances (carbohydrates, amino acids, salts, growth factors) on which beer spoilage micro-organisms can thrive. Such beers are thus prone to biological instability.

From the foregoing, the brewing behaviour of powdery yeasts will be apparent by contrast. In summary, they produce a more attenuated beer than do flocculating yeasts, tending to yield drier beers with better biological stability. Sufficient yeast usually remains in suspension at the end of primary fermentation to ensure adequate secondary fermentation for conditioning and maturing but, with very powdery yeasts, beer clarification is slow and there is danger of the beer acquiring undesirable yeasty flavours.

The importance of flocculation to the brewer is reflected in the number of attempts to measure flocculation, and to classify yeasts according to their flocculence. Thus, Gilliland (1951) classifies them:

Class I: completely dispersed at all stages of fermentation.

Class II: initially completely dispersed, forming small loose clumps towards the end of fermentation.

Class III: initially completely dispersed, flocculating towards the end of fermentation into "caseous" masses.

Class IV: begins to flocculate soon after the start of fermentation because of the failure of newly formed cells to separate.

Hough (1957) applied a more complex scheme of differentiating categories of yeast flocculence, in which isolates from pitching yeasts were tested for ability to form:

(a) A film of cells at the air-liquid surface of an aqueous suspension of yeast (head formation).

(b) Aggregates in calcium chloride solution buffered at pH 3·5.
 (i) Some strains aggregating in this test disperse on adding maltose.
 (ii) Other strains not aggregating in this test precipitate when ethanol is added.

(c) Aggregates in calcium chloride solution buffered at pH 5·0.

(d) Aggregates in calcium chloride solution buffered at pH 5·0 when an appropriate second strain is added (mutual flocculation).

(e) Chains of cells in malt extract liquid medium (chain formation).

Yeasts with the properties of Class II in Gilliland's scheme appear to possess the best compromise in flocculation properties to determine their frequent selection as brewing yeasts in ale breweries where a simple skimming system is adopted for collecting yeast, although there are, in Britain, a few cases of special systems of fermentation, characteristic of certain localities, for which more powdery or more flocculent yeasts are well suited. Finally, for certain systems of continuous fermentation in towers which are being developed (Klopper *et al.*, 1965), it seems that operational success may be entirely dependent on the selection of a suitable, essentially highly flocculent, yeast.

2. *Mechanism of Flocculation*

The importance of flocculation in brewing technology is quite apparent, but the mechanism of the phenomenon is still unexplained. The subject has been reviewed by Comrie (1952), Jansen (1958) and by Rainbow (1966a), on whose review much of the following discussion is based.

a. Physicochemical aspects. In aqueous suspensions at the pH values of worts and beers (3·8–5·6), brewer's yeasts migrate to the anode in electrophoresis experiments, thus behaving as negatively charged colloids. At more acid pH values (2·3), reversal of the charge may take place. Both flocculent and non-flocculent yeasts behave in this way, but the former generally carry the lower charge (Jansen and Mendlik, 1951). Flocculated yeasts are deflocculated by washing in distilled water and reflocculated by addition of bi- or poly-valent cations, especially Ca^{2+}.

These observations have been explained by suggesting that, when the charge on the cells reaches a certain minimum, their mutual repulsion becomes so insignificant as to enable them to clump together under the influence of other forces.

However, Eddy and Rudin (1958b) have found strains of top and bottom yeasts with electrophoretic mobilities independent of their flocculating characteristics, so that flocculence and surface charge are not necessarily interrelated, and a simple concept of flocculation related to surface charge cannot be upheld.

b. *Effect of environment*. Apart from the effects of pH level and cations already mentioned, there is a formidable list of substances variously reported as influencing flocculation (Gilliland, 1955; Rohrer, 1950; Kudo, 1952; Kijima, 1954). Perhaps the most important of these substances are carbohydrates and proteins. Fermentable sugars are known to prevent flocculation, or to cause yeast to redisperse (Lindquist, 1953). Proteins have been reported to cause flocculation (St. Johnston, 1953) in some cases, or to prevent it in others (Lindquist, 1953; Kijima, 1954); but, since flocculation can occur in defined glucose-ammonium salts medium, complex wort constituents do not necessarily play a part in flocculation (Eddy, 1955a).

Recently, Mill (1964b) has reported that Ca^{2+} is essential to flocculation, its effect being antagonized by Na^+. He also found that treating flocculated yeast with 1,2-epoxypropane (an agent for esterifying carboxyl groups) or with formamide (an agent which increases the dielectric constant) caused its dispersal. On the other hand, treatments which reduced hydrogen bonding (urea, or heating to 50–60°), or which reduced the dielectric constant of the medium, aggregated non-flocculent yeast. Mill considers that flocculated cells are linked by salt bridges, Ca^{2+} ions joining two carboxyl groups at the surfaces of two cells, the structure thus formed being stabilized by hydrogen-bonding between complementary carbohydrate hydrogen and hydroxyl groups in the cell surfaces.

c. *Flocculation and genetics*. There is good evidence that flocculence is genetically controlled in *Sacch. cerevisiae*. Thorne (1951) showed that flocculence was dominant (F) to non-flocculence (f), the mutation rate $F \rightarrow f$ being high. Flocculence may thus be a racial character of the yeast cell, but Thorne (1952) did not exclude environmental factors from playing a part. Gilliland (1951, 1955) also considered flocculence to be genetically controlled, but there was no clear evidence whether it, or non-flocculence, was dominant; nor whether flocculation was induced by such environmental factors as fall in pH value, precipitation of protein on cells, attainment of a specific Ca^{2+}:phosphate ratio, or accumulation of yeast metabolic products. Further, deficiencies of fermentable sugars, utilizable nitrogen or growth factors were not primary causes of flocculation.

d. Flocculation and metabolism. Rainbow (1966) considers that the different aspects of flocculation already discussed may not be mutually inconsistent and that, if genetic control of flocculence is the key, the flocculence thus determined will be affected in subsidiary ways by many environmental factors. Since enzymes are genetically controlled, it is reasonable to consider flocculence in relation to yeast metabolism.

Certain observations implicate metabolism in the mechanism of flocculation. Thus, ability to flocculate develops in yeast cells towards the end of fermentation, i.e. at a certain stage of metabolism and physiological age when the composition of the medium falls within certain limits. Eddy (1955a, b) reports that a strain, which was non-flocculent during the first 26 h of fermentation, gradually acquired the power to flocculate. Similarly, Mill (1964a) finds that potential flocculence increases rapidly in the last half of the log phase of yeast growth. The accumulation of some substance in the medium was not responsible for potential flocculence, but exhaustion of a nitrogenous constituent might be, since the onset of flocculation was delayed by ammonium salts, basic amino acids, glutamine, asparagine, γ-aminobutyrate and urea. No flocculence developed in the absence of glucose. Mill considers that the yeast cell walls contain a nitrogen compound which determines non-flocculence. When this substance is synthesized insufficiently because of lack of nitrogen and/or a source of energy, potential flocculence develops.

Another observation relating flocculation with metabolism is that fermentable sugars, especially maltose, protect the yeast cell against flocculation, whereas non-fermentable sugars do not. This suggests that flocculation is contingent upon the retardation of energy-yielding metabolism, a conclusion which is difficult to reconcile with that of Mill (1964a).

Harris (1959) has advanced a metabolic theory of flocculation, as yet untested experimentally, involving the NAD/NADP system, which takes into account: (i) the stereo-specificity of flocculation in relation to yeast strain; (ii) mutual flocculation of yeast strains; (iii) the requirement for polyvalent ions; and (iv) the association of flocculation with the end of fermentation. Kijima (1963) also relates flocculation to metabolism in that he finds that the fermentation system of yeast consists of parts sensitive (α) and insensitive (β) to uranyl ions, and that there is a close inverse relationship between flocculence and the contribution of the α system. The nature of the α and β systems awaits discovery.

e. Flocculation and cell structure. Since the outer layers of the yeast cell must be concerned in flocculation phenomena, studies of cell wall structure are particularly pertinent. Lindquist (1953) studied the influence on flocculation of environmental conditions and treatments directly affecting the cell surface. He considers that flocculation phenomena are determined by the composition and physicochemical properties of wort and the yeast cell wall. He visualizes the cell wall as containing positive and negative ionogenic groups. The latter, which predominate, are contributed by mannan and nucleic acids, while a glucosamine polymer provides the cationic groups. Under favourable conditions, flocculation would result by these polar groups attracting oppositely charged groups on other cell surfaces. Lindquist feels that a low level of net charge is not essential to flocculation, but accessibility of the positive ionogenic groups, which may be masked, perhaps by protein, is essential. This view is not inconsistent with that of Mill (1964a) quoted above.

That the property of flocculence resides in the cell wall was clearly demonstrated by Eddy and Rudin (1958a), who showed that the flocculation characteristics of isolated yeast cell walls were related to those of the corresponding whole cells. Treatment of the walls with papain destroyed their flocculence, with simultaneous solubilization of a protein–mannan complex, which may thus have a direct role in flocculation. That some yeasts flocculate in pairs (mutual flocculation) indicates that at least two types of chemical groups are involved (cf. Lindquist, 1953), one or both of which might be carried by the protein–mannan complex. Eddy (1958) sees the yeast cell as a solid glucan matrix to which glucan and mannan are bound through protein-containing carboxyl groups, which are responsible for the change in cell charge from positive to negative over the range pH 3–7. The mannan seems to carry phosphate groups. Eddy considers flocculating power not as being a function of cell charge, although, since papain solubilizes the protein–mannan complex with loss of flocculence, the cell wall constituent carrying the charge is involved in flocculation.

Masschelein and Devreux (1957) found no qualitative difference between the glucan and mannan of the powdery and flocculative yeasts used by Thorne and by Gilliland. In both types the glucan fraction remained constant throughout the growth period, but the mannan fraction in flocculative yeast first decreased, increased to a maximum and then decreased to a lower level than that attained in the powdery yeast, in which the content slowly increased during 120 h. They also associated flocculence with the cell walls, since the flocculence of isolated walls resembled that of whole cells. Flocculence appeared to be minimal when mannan and charge density were maximum, so that a relationship to physiological age and cell wall composition was apparent.

Later, Masschelein *et al.* (1963) found that the period of deflocculation during growth coincided with mannan synthesis, and that flocculent yeast possessed an intracellular mechanism by which mannan was rapidly utilized. They regard powdery yeast as a flocculating yeast in a permanent state of deflocculation, and consider that the protein : mannan ratio determines the physiological state of the cell, and that mannan may regulate the intensity of flocculation according to the extent to which it masks the active groups of the fraction specifically carrying the flocculence character. In favour of this hypothesis are the following points : (i) minimum mannan content corresponded to maximum flocculation; (ii) the period of deflocculation was eliminated by growth in the presence of sodium fluoride, an inhibitor of mannan synthesis *via* phosphoglucomutase; (iii) at the end of fermentation, cell wall material from deflocculated yeast contained less mannan than that from powdery yeast; (iv) rapid flocculation resulted when mannan was dissolved by treating powdery yeast with alkali; (v) endo-fermentation in flocculent yeasts in the presence of 2,4-dinitrophenol exceeded that in powdery yeasts, which do not possess the intracellular biochemical mechanism allowing rapid utilization of reserve polysaccharide. Masschelein and coworkers thus see flocculence as a cyclic phenomenon, with phases of deflocculation and flocculation alternating in successive fermentations, the relative durations of the phases depending on the synthesis and degradation of mannan and, in consequence, on the protein : mannan ratio in the cell wall.

A picture is therefore arising relating flocculence to yeast cell wall structure, but while current theories of the mechanism have common features, there are considerable differences to be resolved. Possibly, electron microscope studies of the yeast cell wall, such as those reported by Northcote (1963), will help materially in this resolution. At present, it seems that flocculence is a property of the cell wall, and that it is associated with the relative proportions of certain protein and polysaccharide constituents. Some workers consider that the mannan constituent is of primary importance, but that its role may be masked by the protein constituent (Lindquist, 1953; Mill, 1964a, b). In another view (Masschelein *et al.*, 1963), the mannan is considered to play the masking role in a system in which the mannan : protein ratio is critical. There is growing evidence that, whether flocculence occurs or not is contingent on the balance between biosynthesis and biodegradation of certain cell wall protein and carbohydrate components : these processes, being mediated by enzymes, are genetically controlled.

In all this, the role of the environment is probably exerted through its effect on cell metabolism and in supplying ions for floc formation. The effects of other environmental constituents, for example colloidal wort components, on flocculation are probably of a subsidiary nature, exerted either by local action at the cell surface, or indirectly through effects on cell metabolism.

III. Nutrition and Excretion

A. GENERAL

Brewer's yeasts, and indeed baker's yeasts (see Chapter 7, p. 349) and other species of *Saccharomyces*, grow well on simple media based on fermentable sugar as source of energy and carbon fragments for biosynthesis; ammonium salts as source of nitrogen; mineral salts; and one or more growth factors. One such medium applied in nutritional studies (Cutts and Rainbow, 1950) is given in Table I. This medium contains lactate as buffer to maintain the pH value around 4·0 during yeast growth and fermentation, although the lactate may not be a metabolically inert buffer in that it may contribute carbon fragments to the yeast cell. In the absence of buffer, the pH can fall to values of 2·8 or lower as a result of yeast growth.

TABLE I. *Defined Medium for Studying Yeast Growth and Nutrition (Cutts and Rainbow, 1950)*

Component	Concentration (g/litre)	Component	Concentration (mg/litre)
Glucose	40	D-Biotin	2
$(NH_4)_2HPO_4$	4·0	Ca D-pantothenate	1,000
$MgSO_4.7H_2O$	2·0	Thiamine HCl	1,000
KH_2PO_4	1·0	Pyridoxine HCl	1,000
$CaCl_2.6H_2O$	0·25	*myo*-Inositol	10,000
Lactic acid (syrupy)	10	*p*-Aminobenzoic acid	100
		KI	100
Trace elements solution, 1·0 ml/litre			
KOH solution to pH 5·0–5·2			

Composition of trace elements solution (g/litre): H_3BO_4, 1·0; $MnSO_4.4H_2O$, 0·4; $ZnSO_4.7H_2O$, 0·4; $CuSO_4.5H_2O$, 0·45; $FeSO_4.7H_2O$, 2·5; ammonium molybdate, 0·2

B. CARBOHYDRATE REQUIREMENTS

Brewer's yeasts grow on and ferment the hexose sugars D-glucose, D-mannose and D-fructose: D-galactose is not fermented until the yeast is specially adapted. Pentose sugars are not fermented.

Of the disaccharide sugars, sucrose and D-maltose are fermented by all strains, presumably following hydrolysis by yeast invertase (β-D-fructofuranoside fructohydrolase; see Chapter 10, p. 529) and maltase (α-D-glucoside glucohydrolase) respectively. In brewer's yeasts, invertase is a constitutive enzyme located in the outer layers of the cell, and washed suspensions of brewer's yeast readily convert solutions of sucrose to glucose and fructose. Maltose, however, is normally only fermented after a period of induction in its presence. Harris and Millin (1963) showed that a yeast grown in the absence of maltose or maltotriose lacked both the permeases required for the transport of these substrates into the cell, and also lacked maltase, although it could ferment glucose. Fermentation of maltose only took place after the yeast had been transferred to a maltose medium and, even then, after the elapse of an appreciable period. During this induction period, maltose permease and maltase were synthesized at the expense of the free amino acid pool. By contrast, the enzymes responsible for the entry and fermentation of hexose sugars are constitutive in brewer's yeast. At the end of primary fermentation, brewer's yeasts are de-adapted to maltose, and on being reseeded into a medium such as brewer's wort, containing maltose as its chief fermentable sugar, a period of re-adaptation to maltose is required before maltose is once more utilized for growth and fermentation. The sequence of carbohydrate utilization in brewer's wort is discussed below.

Lactose is not fermented by brewer's yeasts. On the other hand, melibiose fermentation may be applied as a diagnostic test to distinguish *Sacch. carlsbergensis* which ferments melibiose, from *Sacch. cerevisiae* which does not, although the test is not invariably successful.

Of the trisaccharides, brewer's yeast ferments maltotriose. As with maltose, maltotriose is only fermented after an induction period required for the synthesis of its appropriate permease (Harris and Millin, 1963). Having entered the cell, maltotriose is hydrolysed by maltase to glucose, which can then be fermented. The fructose and glucose moieties of raffinose are completely fermented by those yeasts which contain melibiase (*Sacch. carlsbergensis*): in the case of *Sacch. cerevisiae*, fructose is split off by the action of invertase and fermented, but the melibiose residue cannot be fermented. Here again, this fermentation reaction is applied for diagnostic purposes to distinguish top from bottom yeasts.

Brewer's yeasts are unable to utilize isomaltose or maltosaccharides more complex than maltotriose. These sugars are therefore left unchanged during the fermentation of brewer's wort.

C. NITROGEN REQUIREMENTS

In an otherwise suitable medium, most yeasts, including brewer's, grow well on ammonium salts as the sole source of nitrogen. Yeasts therefore possess great biosynthetic abilities. In the manufacture of baker's yeast this fact is applied (see Chapter 7), but the ability to utilize ammonia is of more academic interest in considering brewer's yeasts, the normal habitat of which is brewer's malt wort, the predominant yeast nitrogen nutrients in which are amino acids and low molecular weight peptides. MacWilliam (1968) and Jones and Rainbow (1966) give values for the nitrogen distribution of brewer's worts (Table II).

TABLE II. *Nitrogen Distribution of Brewery Worts and Beers (From Jones and Rainbow, 1966)*

Wort or beer	Gravity of wort	Total N	α-Amino N		Proline N		HMW–N	
			(μg/ ml)	(% total N)	(μg/ ml)	(% total N)	(μg/ ml)	(% total N)
Wort A	1042·0	698	201	28·8	37	5·3	123	17·6
Wort B	1049·0	898	236	26·3	—	—	147	16·4
Beer B	—	608	60	9·9	—	—	138	22·7
Wort C	1041·5	586	195	33·3	42	7·2	—	—
Beer C	—	359	31	8·6	37	10·3	—	—

Beers B and C were brewed respectively from Worts B and C. HMW–N = nitrogen in material of molecular weight greater than about 4,000

Brewer's yeasts do not secrete proteolytic enzymes, so that the polypeptide constituents of wort survive fermentation substantially and pass into the finished beer. However, Damlé and Thorne (1949) showed that brewer's yeast could grow on low molecular weight peptides, and there is evidence from Table II that these are taken up from wort by yeast, since the loss of α-amino nitrogen on fermentation accounts only partially for the loss of total nitrogen. The limits imposed by molecular size on availability of peptides to yeast is not yet known.

D. GROWTH FACTOR REQUIREMENTS

Observations showing that yeasts fail to grow when a small inoculum is seeded into defined glucose–ammonia–salts medium date from Pasteur's time, but the first clear demonstration that certain yeasts would not grow in such a medium unless small amounts of organic extracts were added was made by Wildiers (1901). He proposed the term "bios" for the factor or factors responsible. Later it was shown that bios was composite, and Eastcott (1928) identified the so-called

Bios I as *myo*-inositol. Miller *et al.* (1933) then fractionated Bios II into Bios IIA and IIB, the former being identified by Williams (1939) as pantothenic acid and synthesized by Williams and Major (1940). Bios IIB was first isolated by Kögl and Tönnis (1936), who named it biotin. The active isomer (D-biotin) is one of the most physiologically active substances known.

Most strains of brewer's yeast (and in the author's experience, possibly all strains of ale yeasts) have an absolute requirement for exogenous biotin. Many strains also require D-pantothenate (for which β-alanine will often substitute): rather less frequently, *myo*-inositol is essential or beneficial, while there are a few examples of brewer's yeasts which have an absolute requirement for *p*-aminobenzoate (Rainbow, 1948; Cutts and Rainbow, 1950). Exogenous supplies of pyridoxine and thiamine are rarely, if ever, essential requirements for brewer's yeasts, although their rates of biosynthesis are often insufficient to maintain optimum growth rates in simple defined media. The presence of these growth factors therefore usually stimulates growth in the early stages (Lewis and Rainbow, 1963). Indeed, this observation was applied in the microbiological assay of pyridoxine with *Sacch. carlsbergensis* 4228 by Atkin *et al.* (1943). The author knows of no reported cases of brewer's yeasts in commercial use being exacting to nicotinic acid, riboflavin, individual amino acids, purines or pyrimidines.

TABLE III. *Growth Factor Contents of Infusion Worts and Top Fermentation Beers*

Growth factor	Gravity	Wort (μg/ml)	Beer (μg/ml)	Reference
D-Biotin	1043	0·0050	0·0029	Lynes and Norris (1948)
Thiamine	1036	0·32	0·03	Hopkins and Wiener
	1031	0·40	0·34	(1944)
Riboflavin	1036	0·36	0·37	Hopkins and Wiener
	1033·5	0·58	0·25	(1944)
Pyridoxine	1031–1054	—	0·25–0·87	Hopkins and Pennington
	1035·5	0·45	0·64	(1947)
D-Pantothenic acid	1031–1054	—	0·42–0·87	Hopkins *et al.* (1948)
Nicotinic acid	1036	8·6	7·1	Norris (1945)

Values for biotin, pyridoxine and nicotinic acid are total (free + combined) contents: for the others, values represent the content of free growth factors

Brewer's wort is a rich source of growth factors and, in most cases, yeast depletes wort of these substances, so that the resultant beer contains substantially less of them (Table III). An exception to this is pantothenate, which may be present in some beers in greater concentrations than it is in the worts from which they were brewed (Hopkins *et al.*, 1948). Hopkins and his coworkers therefore consider that the pantothenate content of a beer may depend on the need shown by the particular yeast for an exogenous source. The riboflavin content of beers also seems to be as great as that of the parent worts, and some yeasts may enrich wort with riboflavin during fermentation (C. Rainbow, unpublished work). There appear to be no published values for the inositol content of ale worts, but MacWilliam (1968) quotes values of 18 to 60 μg/ml for lager worts.

The biochemical role of these growth factors, exerted by virtue of their roles as moieties of coenzymes, is dealt with in most textbooks of biochemistry and need not be reiterated here. However, perhaps the role of inositol needs comment. The relatively great concentrations of inositol required for good yeast growth indicate that it may have a structural rather than a coenzymic function. Possibly it is required by the yeast cell in the form of phosphoinositides in the lipid membranes of the cell. It is interesting that instances of bacteria showing requirements for exogenous inositol have not, so far as this author is aware, been reported. Possibly, all bacteria can synthesize their own inositol, but it is also possible that inositol is a cell constituent characteristic of cells other than those of bacteria.

E. MINERAL REQUIREMENTS

The mineral requirements of brewer's yeast resemble those of most other living cells. Thus, sources of iron, potassium, magnesium, zinc, manganese, copper and possibly other metal ions are required for the functioning of vital metabolic enzymes. In addition, yeast requires rather larger amounts of $(SO_4)^{2-}$ and $(PO_4)^{3-}$ ions. From inorganic sources of the former, yeast synthesizes all its organic sulphur compounds: this subject is discussed in Section IV.E (p. 204).

The utilization, uptake and storage of phosphate by brewer's yeast have been studied by Markham *et al.* (1966) and Markham and Byrne (1967). In experiments using a defined medium, these workers showed that yeast can continue to grow at the expense of stored phosphate even after the medium had become entirely depleted in phosphate. With 5% glucose as the carbon source, yeast yield increased with inorganic phosphate in the medium up to 60 μg P/ml. At higher concentrations of phosphate (up to 125 μg P/ml), the yeast removed the phosphate from the medium and presumably stored it, so that when *Sacch. cerevisiae* was cultured in a medium containing 65 μg P/ml and transferred to a defined medium containing 5% glucose and 65 μg P/ml, growth was limited only by exhaustion of the sugar and not by the phosphate.

F. NUTRITION OF YEAST IN BREWER'S WORT

1. *Carbohydrates*

The utilization of wort carbohydrates by brewer's yeast was studied by Phillips (1955). Although he did not measure their disappearance from wort quantitatively, he showed by paper chromatography that, with brewery strains of *Sacch. cerevisiae* and *Sacch. carlsbergensis*, sugars were removed from wort in the order sucrose (first), monosaccharides, maltose and maltotriose (last), while maltotetraose and more complex dextrins were not attacked. Unpublished observations by the author on commercial fermentations with top yeast confirm these conclusions (Table IV).

TABLE IV. *Utilization of Carbohydrates during Fermentation of Wort by Top Fermentation Brewer's Yeast (C. Rainbow, unpublished results)*

Time of fermentation (h)	Gravity of wort	Amount of carbohydrate in wort (mg/ml)					
		D	MT	M	G	S	F
0	1042·3	19·3	11·9	43·1	11·4	3·2	2·2
22	1038·8	19·0	13·2	43·4	8·7	1·1	2·8
28	1036·2	18·6	11·6	40·7	2·8	0	2·2
46	1024·1	18·7	10·6	25·5	0	0	0·2
52	1019·6	17·4	10·7	17·8	0·4	0	0

Determinations by the anthrone reaction after separation of individual sugars by quantitative paper chromatography. D = dextrins (maltotetraose and higher saccharides); MT = maltotriose; M = maltose; G = glucose; S = sucrose; F = fructose

Harris and Thompson (1960a) found that yeast cells readily take up monosaccharides, even those which are not utilized for growth, but they largely fail to absorb di- and trisaccharides. Their experiments showed that cells of brewer's yeast which had been de-adapted to maltose and maltotriose fermentation initially took up these (^{14}C-labelled) sugars only to a limited extent, whereas adapted cells took up maltose, but not the related disaccharide lactose. As re-adaptation of the cells proceeded, uptake of ^{14}C-maltose increased. They suggest that maltose enters the yeast cell, not by simple diffusion, but by an adaptive process linked to maltose utilization. Subsequently (Harris and Thompson, 1960b), they demonstrated the presence of a maltotriose-specific permease system in brewer's yeast. Later work in this field by Harris and Millin (1963) has already been mentioned (see Section III.B, p. 156).

These observations concerning the sequential uptake of carbohydrate by brewer's yeast, particularly as concerns the uptake of maltose, are obviously important in brewing, in which the substrate for fermentation (brewer's wort) contains maltose as its most abundant carbohydrate. The relatively small amounts of sucrose, glucose and fructose in malt wort acquire considerable importance in providing readily-utilizable substrates for the early stages of yeast growth, during which adaptation to maltose and then to maltotriose may proceed.

Maltotriose is the last major fermentable sugar of malt wort to be attacked by yeast. Substantial amounts may persist in beers at the end of primary fermentation, after which it may play an important part in the slow conditioning and maturing process, whether carried out in bulk, in cask, or (in a few cases in which beers are still conditioned by live yeast in bottle) in bottle.

2. *Nitrogen*

Absorption of amino acids from wort. Detailed analyses of wort and beer amino acids were first attempted by Barton-Wright (1951), who applied the technique of microbiological assay. More recently, Pierce and his coworkers (Jones and Pierce, 1963, 1964; Jones *et al.*, 1965) have exploited the technique of amino acid analysis by column chromatography, using an automatic analyser, to determine the content of individual amino acids in worts and beers (Table V). This work led to the discovery that, under brewery conditions, brewer's yeasts removed amino acids from wort in an orderly manner, different amino acids being removed at different speeds. Consequently, amino acids could be classified according to the order of their removal from wort (Table VI).

The sequence of absorption was almost independent of the conditions of fermentation and of the strain of yeast, but under aerobic conditions the process of removal was carried further, until proline was also largely absorbed. Jones and Pierce (1964) describe these classes of amino acids as follows:

Group A. Immediately absorbed and almost removed from wort after 20 h of fermentation.

Group B. Not removed rapidly, but absorbed gradually during fermentation.

Group C. Absorbed after an appreciable lag.

Group D. Represented only by proline, which was only slightly absorbed from wort after 60 h of fermentation.

In a further study with laboratory fermentations with a top yeast, Jones *et al.* (1965) measured the amino acid composition of yeast sampled at frequent intervals during fermentation, and related it to the amino acid composition of the fermenting wort at the same stages, so that uptake from wort of each amino acid could be correlated with the

TABLE V. *Uptake of Amino Acids from Wort by* Saccharomyces cerevisiae *under Brewery Conditions. From Jones and Pierce, 1964*

Original wort gravity approximately 1045

Amino acid	Time of fermentation (h)				
	0	19	28	40	63
			µg α-amino N/ml		
Alanine	12·5	12·6	9·4	1·4	0·2
Ammonia	22·0	22·0	11·6	1·6	1·7
Arginine	8·7	0	0	0	0
Aspartic acid	6·0	2·4	0·2	0·2	0·2
Glutamic acid	7·4	0·9	0	0	0
Glycine	5·1	4·8	4·3	2·3	0·2
Histidine	3·6	2·7	1·8	0·5	0
Isoleucine	5·2	4·2	2·0	0·1	0·4
Leucine	12·9	9·3	3·7	0·2	0·4
Lysine	8·7	2·4	0·6	0·5	0·2
Methionine	2·4	1·0	0·2	0	0
Phenylalanine	7·6	7·0	4·7	0·8	0
Proline	44·1	43·7	43·2	40·9	38·0
Serine + amides[a]	16·3	7·0	0·4	0·3	0·2
Threonine	5·2	1·4	0·1	0·1	0·2
Tryptophan	3·0	2·7	2·3	1·9	0·5
Tyrosine	5·5	5·2	4·5	2·6	0
Valine	12·1	11·2	8·4	2·4	0·4
Total	188·3	130·5	97·4	55·8	42·6

[a] Serine + glutamine + asparagine expressed as serine amino nitrogen.

TABLE VI. *Order of Absorption of Amino Acids from Wort by Brewer's Yeast. After Jones and Pierce (1964)*

Group A Immediately absorbed	Group B Absorbed gradually during fermentation	Group C Absorbed after a lag	Group D Only slowly absorbed after 60 h
Arginine	Histidine	α-Alanine	Proline
Asparagine	Isoleucine	Ammonia	
Aspartate	Leucine	Glycine	
Glutamate	Methionine	Phenylalanine	
Glutamine	Valine	Tryptophan	
Lysine		Tyrosine	
Serine			
Threonine			

corresponding amounts incorporated into the intracellular yeast pool and into yeast protein. In the early stages of fermentation more nitrogen was incorporated as amino acids into yeast than was absorbed from wort. Indeed, in this period there was evidence that glycine, α-alanine, proline and amides were excreted by the yeast. Peptides absorbed from wort may be the source of this nitrogen absorbed by yeast which cannot be accounted for by loss of wort amino acids. At this stage the yeast pool contains considerable concentrations of the readily-assimilable amino acids, glutamic and aspartic acids and their amides, lysine and arginine. However, it is more surprising that α-alanine was a major constituent of the yeast amino acid pool, even at a stage when it was not being absorbed from the fermenting wort. Over the ensuing major period of yeast growth, there was a good balance between total amino acid content of yeast and total amino acid uptake from wort. Nevertheless, values for individual amino acids in the yeast (protein + pool) did not balance the loss of these amino acids from wort at all stages and, towards the end of fermentation, the yeast amino acid content increased; presumably at the expense of wort peptide nitrogen, even after free wort amino acids had been exhausted.

Jones and coworkers conclude that, when growing in wort, not only does yeast synthesize readily-assimilable amino acids after their elimination from wort early in fermentation, as well as those amino acids which are poorly assimilated or only so after a considerable lag period, but that it synthesizes many amino acids before the medium becomes deficient in them.

G. SHOCK EXCRETION

The nitrogen nutrition of brewer's yeast is complicated by the fact, recognized since the time of Pasteur, that yeast may release nitrogen compounds into the medium in which growth or fermentation has taken place. This phenomenon has recently been studied by Phaff, Lewis and their co-workers. Delisle and Phaff (1961) showed that such "excretion" of nitrogenous substances by yeast was not the result of autolysis, but was a physiological phenomenon normal to healthy yeast cells. Lewis and Phaff (1963) studied the phenomenon in young but mature and fully-grown cells of a non-flocculent and a flocculent strain of *Sacch. carlsbergensis*, in an English top brewing strain of *Sacch. cerevisiae* and in a baking strain of *Sacch. cerevisiae*. All the strains gave qualitatively similar results. In suspension in water, amino acids and material absorbing at 260 mμ (presumed to be nucleotides) were slowly lost into the medium in the same ratio as they occurred within the cells, so that the process appeared to be non-selective. Aeration and, to a smaller extent, the presence of the sulphates of magnesium, calcium, potassium and sodium increased the rate of release. The released amino acids were re-absorbed on adding glucose, whereas the release of nucleotide material was accelerated. Neither maltose nor galactose would replace glucose in inducing these effects unless the cells were adapted to these sugars. In the presence of both glucose and assimilable nitrogen, the excretion of nucleotide material was much enhanced.

Pursuant of these observations, Lewis and Phaff (1964) found that, when washed cells of their strains of *Sacch. carlsbergensis* and baker's yeast were suspended in a glucose solution, amino acids were suddenly released and then rapidly re-absorbed within two to three hours. They termed this phenomenon "shock excretion". The amount of amino acids released into the suspension medium depended on the yeast strain, but was independent of osmotic pressure: it increased with size of the intracellular amino acid pool and with temperature, but shock excretion did not occur unless sufficient metabolizable carbohydrate was present.

The inference from the above that shock excretion is a dynamic process, perhaps dependent on energy-generating metabolic processes, was studied by Lewis and Phaff (1965) in experiments in which rates of fermentation and shock excretion of amino acids were compared in a strain of flocculent brewer's bottom yeast. The effects of metabolic inhibitors were also studied. Amino acid excretion increased more rapidly with rise in temperature than did rate of fermentation, and it was not inhibited by recognized uncouplers of ATP synthesis applied at concentrations sufficient to inhibit the energy-requiring process of re-absorption of these amino acids. Shock excretion was therefore apparently not an energy-requiring process. However, since inhibitors of glucose dissimilation (iodoacetate, fluoride) and of entry of glucose into the cell (uranyl acetate) prevented amino acid excretion, the authors concluded that shock excretion "may be the result of changes in the state of the cell contingent upon the continuous flow of a fermentable sugar across the cell membrane".

The most recent paper in this series (Lewis and Stephanopoulos, 1967) adduces evidence that changes in membrane permeability, implicit in the results described in the paper mentioned previously, are concerned in the shock excretion phenomenon. In this work the same flocculent brewer's bottom yeast was used. The α-amino nitrogen released was entirely material of low molecular weight, the composition of which was almost identical with that of the amino acids of the intracellular pool (Table VII). Leakage was again non-specific, since leakage of ultraviolet-absorbing compounds, inorganic phosphate and metal ions accompanied that of amino acids during shock excretion. Experiments with ^{14}C-labelled α-amino acids showed that, while amino acids were being rapidly excreted, the yeast failed to absorb amino acids present in the medium, although glucose was rapidly taken up and metabolized.

TABLE VII. *Relative Composition of the Internal Free Amino Acid Pool of Saccharomyces carlsbergensis and the Amino Acid Fraction of the Medium during Shock Excretion. From Lewis and Stephanopoulos (1967)*

Amino acid	Relative amino acid composition		
	Internal, before adding glucose	External, 10 min after adding 10% glucose	External, 60 min after adding 10% glucose and 10^{-3} M-arsenite
Alanine	100	100	100
Glutamate	70	76	68
Glycine	50	48	51
Serine	42	46	40
Lysine	40	42	40
Arginine	20	21	22
Aspartate	17	17	18
Threonine	15	12	14
Histidine	12	12	12

In each case, the concentrations of amino acids are referred to alanine = 100. Amino acids not listed were present in quantities smaller than that of histidine

Shock excretion was inhibited by Ca^{2+}, but increased by *n*-butanol, which is known to damage the cytoplasmic membrane of yeast (Rose, 1963; Dixon and Rose, 1964). Further evidence implicating changes in membrane permeability with shock excretion included the observation that the adenosine triphosphatase activity of intact cells exposed to glucose was greater than that of cells exposed to water: this enzyme appears to be located in the cell membrane (Bolton and Eddy, 1962). Again, cells of the yeast treated with glucose were less able to retain sorbose than water-treated cells. Lewis and Stephanopoulos suggest a mechanism of shock excretion in which mature yeast cells in glucose solution utilize glucose, the penetration of which into the cell causes leakiness of the cytoplasmic membrane. Simultaneously, there is a temporary shortage of the energy necessary for amino acid uptake and retention (Taylor, 1947), so that amino acid leakage takes place.

The significance of shock excretion in technical brewing has not yet been assessed, but in comparing young cells and mature cells harvested from the same medium, Lewis and Stephanopoulos (1967) reported that young cells (18 h) released considerably less α-amino nitrogen, and absorbed ^{14}C-amino acid (α-alanine) more rapidly and with less lag than did mature cells (30 and 72 h). Since the yeast cultures used for pitching commercial brewery batch fermentations are usually mature cells taken from a previous brew, or from a pure culture plant, it seems likely that shock excretion may occur on pitching brewer's wort, which contains glucose derived enzymically by amylolysis of malt starch, and may also contain glucose added as brewing sugar. The extent to which shock excretion occurs at pitching may thus be a factor in determining how rapidly and intensely fermentation sets in, and in detecting defective fermentations.

Again, at the end of primary fermentation, when the raw beers pass on for maturing, it is common practice to "prime" the beer with added fermentable sugar. In Britain in particular, priming may be carried out with starch-based sugars or invert sugar, both of which contain glucose as a major component. If the concentrations of glucose thus added to the beer are sufficient to induce, even to a minor extent, shock excretion, small amounts of microbial nutrients will be released from the yeast left in suspension in the beer to effect the maturing process, so that the beer will become a little more prone to microbial spoilage and may even acquire undesirable "yeasty" flavours.

Whether shock excretion is a factor affecting continuous fermentations is a matter of even less well-founded conjecture. However, in the system of continuous fermentation in towers (see Section VI.A.2, p. 213), in which wort is caused to flow upwards through high concentrations of yeast which is proliferating only minimally, the cells may well be in a physiologically mature state and thus likely to be susceptible to shock excretion. By a similar token, shock excretion may be less important in accelerated batch fermentations, in which fresh wort is added to cells in the logarithmic phase of growth.

IV. Metabolism

In conventional batch brewing, yeast is pitched into aerated hopped malt wort. Initially, therefore, the yeast has access to dissolved free oxygen. Oxygen availability in the initial stages of fermentation is sought by the brewer because it stimulates yeast growth, cell replication is encouraged, and sufficient cells are produced to carry out a vigorous fermentation. This initial phase of oxygen-stimulated growth is also important in encouraging enzyme protein synthesis, which is essential before maltose, the predominant fermentable sugar of brewer's wort, can be absorbed and metabolized.

The growth of yeast during this early relatively aerobic phase, coupled with the formation of carbon dioxide by respiration, soon results in the depletion of free oxygen in the wort and the accumulation of metabolic carbon dioxide, both in the fermenting liquid and at the surface, where it displaces air and accumulates by virtue of its greater density. Thus, conditions of virtual anaerobiosis are set up and the major energy-yielding metabolic pathway of the yeast, the Embden-Meyerhof-Parnas (EMP) pathway, is soon established.

This pathway is the frequent subject of detailed treatment and further discussion is not merited here. Suffice it to say that, while the formation of pyruvate from fermentable sugar by the EMP pathway, followed by its decarboxylation and reduction to ethanol and carbon dioxide, is the general feature of yeast metabolism as it concerns beer manufacture, the special flavour characteristics of beers are endowed by small amounts of other substances. In some cases these arise as a result of quantitatively minor biochemical transformations stemming from pyruvate; in other cases it is by-products of nitrogen metabolism which contribute important flavour and aroma characteristics to beer. In the following treatment of the metabolism of brewer's yeasts, emphasis will therefore be placed on these quantitatively minor metabolic pathways and the products to which they give rise.

A. ESTER FORMATION

Early work on ester formation by yeasts is briefly reviewed by Prescott and Dunn (1959), but recent knowledge of ester formation by brewer's yeasts, acting under anaerobic conditions comparable to those encountered in commercial brewing, arises in a large measure from the work of Nordström (for review, see Nordström, 1964f).

1. General Factors Affecting Ester Formation

Nordström set out to decide between three possibilities of ester formation by brewer's yeast: (i) chemical esterification of acetic acid and ethanol after their formation during fermentation; (ii) enzyme-catalysed esterification of fermentation acetic acid and ethanol in the medium; or (iii) intracellular ester formation and subsequent diffusion of the ester into the medium, the ester being either a primary metabolite formed by energy-requiring processes, or through endo-esterase activity.

Most of Nordström's experiments were performed with a top fermentation strain of Sacch. cerevisiae fermenting a defined glucose medium under anaerobic conditions. Nordström (1961) came to the conclusion that ethyl acetate was formed intracellularly (Alternative iii above) on the following evidence.

(a) During the period of most intense ethyl acetate formation, the velocity constant of formation was about 10^3 times greater than that determined by kinetic measurements of the chemical esterification of ethanol and acetic acid in admixture in citrate buffer, or in filtered fermented medium.

(b) At pH 4·5 (the pH optimum for ethyl acetate formation during fermentation) the estimated equilibrium concentrations of ethyl acetate were considerably higher than those determined in the kinetic study of chemical esterification.

(c) The velocity constant of ethyl acetate formation during fermentation increased greatly as the pH value decreased: if the ester were formed chemically, production would increase at a lower pH value until equilibrium was reached. However, the pH relation found differed entirely from this.

(d) Ethyl acetate formation during fermentation was independent of acetic acid added to the medium. This finding, resembling that of Peel (1951) and of Tabachnick and Joslyn (1953a, b) for *Hansenula* sp., indicated that the acetate moiety of the ester was not derived from exogenous acetate. Furthermore, ester formation was constant as well as independent of the concentration of exogenous acetic acid, so that it did not occur through endo-esterase action.

A kinetic study of ethyl acetate formation during fermentation (Nordström, 1962a) showed that, while added acetaldehyde did not affect ethyl acetate or acetic acid production (although some acetaldehyde was metabolized to glycerol), added ethanol stimulated ester formation, but depressed that of acetic acid. The relationship between velocity of ethyl acetate formation and ethanol concentration was satisfied by the Briggs and Haldane (1925) equation for a monomolecular enzyme reaction with ethanol as substrate. At pH 4·6 the respective values of ethyl acetate formation and Haldane constant were 0·71 m-mole ethyl acetate/mole ethanol and 1·23 M, the corresponding values at pH 5·0 being 0·55 m-mole/mole and 1·10 M respectively. The technically important point is made by Nordström that, since the ethanol concentration remains high during the whole fermentation period in continuous fermentation, more ethyl acetate should be formed in this type of fermentation than in batch fermentation. Some experimental evidence has been obtained which confirms this theory.

Evidence of coenzyme A (CoA) participation in the anaerobic formation of ethyl acetate, such as is known to occur under aerobic conditions with *Hansenula* species (Peel, 1951; Tabachnick and Joslyn, 1953a, b), was next sought by Nordström (1962b). Nitrogen was supplied in various forms, because the source of nitrogen influences the quantity of the fermentation by-products (Ribéreau-Gayon *et al.*, 1959; Stevens, 1960). Acetic acid formation decreased with added pantothenate (or β-alanine), presumably because more acetate was transformed *via* acetyl-CoA. At the same time there was a close relationship between added pantothenate and ethyl acetate formed, indicative of a direct function of CoA in ester synthesis. Conversely, when the supply of pantothenate was restricted, or when reactions involving CoA and acetyl-CoA formation by oxidative decarboxylation of α-oxo acids (Alvarez *et al.*, 1958) were inhibited by arsenite or diazine green, acetic acid formation increased, and ethyl acetate yields decreased considerably.

With low concentrations of pantothenate, there was less ethyl acetate formation in media containing DL-α-alanine or L-asparagine than in medium containing ammonia, although yeast growth was scarcely affected. This effect of amino acids was counteracted by further addition of pantothenate, but the effect was not observed in wort, which contains sufficient pantothenate to ensure constant yields of the ester. Nordström suggested that the effect of the added amino acids was due to their acid-generating potential, rather than as an effect on cell synthesis (and therefore energy requirement), since growth was unaffected by the additions. To the present writer, it seems arguable, however, that extra CoA might be required to metabolize the amino acids, thus creating an added demand for CoA. Point is added to Nordström's hypothesis by his demonstration that addition of butyric acid caused diminished ethyl acetate formation: simultaneously, ethyl butyrate formation increased, the acids appearing to compete mutually in ester formation. He concludes that ester formation requires energy supplied through the participation of CoA.

Nordström (1963a) used the scheme for acetyl-CoA metabolism in yeast, shown in Fig. 1, to predict the effects on ester formation of influencing the availability of CoA, and then tested his predictions experimentally. Thus, restriction of biotin should limit consumption of CoA for lipid synthesis *via* malonyl-CoA and the supply of TCA cycle intermediates, which are important as bio-precursors in nucleic acid and protein synthesis under anaerobic conditions. On any of these scores, biotin restriction should render more acetyl CoA available for ester formation. Again, if the consumption of acetyl-CoA in growth (Fig. 1, Reactions IVc, Va and Vb) were restricted by inhibition with 2,4-dinitrophenol (which restricts growth by creating ATP deficiency), more acetyl-CoA (formed *via* Reaction II) should be available for ester synthesis (Reaction III).

In Nordström's experiments biotin limitation did not give results capable of unequivocal interpretation: at such concentrations of biotin as permitted maximum growth (the test strain of *Sacch. cerevisiae* was biotin-exacting) ester production was diminished, as was predicted.

Inhibition of the TCA cycle by added malonate also produced the anticipated stimulation of ester formation; and lastly, the addition of low concentrations of 2,4-dinitrophenol to the medium did stimulate ethyl acetate formation, an effect which was enhanced by the addition also of lipoate and counteracted by the sulphydryl poison, arsenite. Nordström considered that these results indicated that oxidative decarboxylation of pyruvate was important in the anaerobic formation of acetyl-CoA by *Sacch. cerevisiae*, ester formation being stimulated when other reactions requiring acetyl-CoA (e.g. the TCA cycle and the biotin-dependent carboxylation of acetyl-CoA to malonyl-CoA) were inhibited.

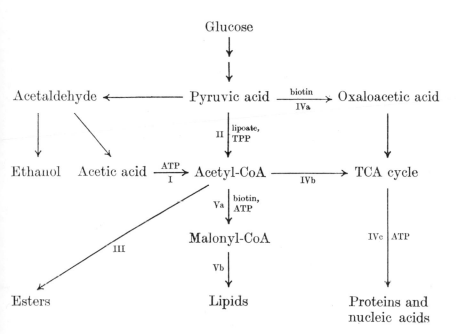

FIG. 1. Simplified scheme for metabolism of acetyl-CoA.
From Nordström (1963a).

There appears to be a correlation between yeast growth and ester formation: in general, ester formation decreases when growth is inhibited (Nordström, 1964b). Whether growth was limited by restricting supplies of mineral salts (Mg^{2+}, KH_2PO_4) or assimilable nitrogen, or by adding relatively high concentrations of Cu^{2+}, ester formation was depressed more rapidly than growth. The effects of added biotin or thiamine in inhibiting ethyl acetate formation, or of added pantothenate,

2,4-dinitrophenol or iodoacetate in stimulating it, are regarded by Nordström as being caused by the actions of these substances on the formation or consumption of acetyl-CoA.

2. *Formation of Esters from Acids*

Acids seem to compete in ester synthesis by yeast: ethyl acetate formation was diminished by adding butyric acid to the growth medium and ethyl butyrate appeared. Using the same top-brewing strain of yeast and the same conditions of anaerobiosis in a defined medium as had characterized his other experiments, Nordström (1963b) studied the relationship between rate of formation of ethyl butyrate (or valerate) and added butyric (or valeric) acid concentration. The relationship proved to be expressible by a modified Briggs-Haldane equation for monomolecular enzymic reactions. Nordström presented a theory of the kinetics of substrates in competition (presumably for CoA), in which the formation of ethyl butyrate and valerate from the respective acids by yeast satisfies the theory that the acids behave as substrates competing in a monomolecular enzyme reaction. The presence in beer and other fermented beverages of small amounts of these esters and the corresponding fatty acids is consistent with these views.

Nordström (1964a) next studied the formation of the ethyl esters of lower fatty acids added to his defined medium. The ethyl esters of butyric (C_4) up to pelargonic (C_9) and of isocaproic (C_6) acids were formed; the correlation between ester formation and acid concentration conformed with the kinetics of a monomolecular reaction, the Michaelis constants decreasing with increasing molecular weight of the fatty acid. All these acids behaved as competitive inhibitors of ester formation from the other acids. The ethyl esters of propionic, isobutyric and isovaleric acids were not formed in this experimental system. All the ester-forming fatty acids were toxic to the yeast, possibly because the fatty-acyl-CoA compounds formed inhibited essential metabolic reactions, such as lipid synthesis, which required acyl-CoA. Nordström offers as a scheme for ester formation in these experiments that given in Fig. 2.

Nordström (1964c) showed that the conclusions he had reached from his earlier experiments were equally valid for 20 other yeast strains, including haploid and diploid strains of both top and bottom yeasts. In fermentations in the medium supplemented with a number of lower fatty acids, the relative amounts of ethyl esters formed were essentially the same for all strains, but no strain formed ethyl esters of propionic, isobutyric or isovaleric acids. While the total formation of volatile esters varied relatively largely according to strain, there was no marked difference in amount of ester produced which was ascribable to top

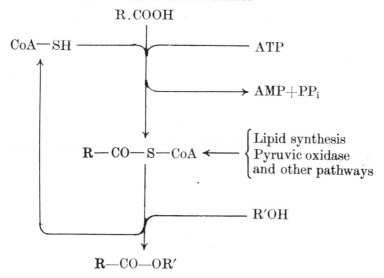

FIG. 2. Scheme for the formation of esters. From Nordström (1964a).

yeast (mean value 127 μM) on the one hand, or to bottom yeast (mean value 110 μM) on the other.

Later, Nordström (1966) analysed the medium after the complete exhaustion of glucose by fermentation with yeasts representing a number of genera and species, and confirmed that there was a quantitative difference between strains in ability to form esters. The distribution between the ethyl esters of butyric, valeric and isocaproic acids was, however, similar for all species, and it agreed well with that calculated from kinetic data. The total formation of volatile esters was not related to esters formed from acids added to the medium. Some values for total ester formation by various yeasts grown anaerobically in a defined

TABLE VIII. *Total Formation of Volatile Esters and of Ethyl Valerate by Various Yeasts Growing in a Medium Containing 2 mM Added Valeric Acid. From Nordström (1966)*

Yeast	Final concentration of esters (μM)	
	Ethyl valerate	Total volatile esters
Saccharomyces spp.	6–56	98–172
Saccharomyces cerevisiae (top yeast)	32	160
Saccharomycodes ludwigii	22	261
Kloeckera apiculata	4	414

glucose medium supplemented with valeric acid are given in Table VIII.

Although addition of lower fatty acids (up to hendecanoic, C_{11}) to the medium inhibited ethyl acetate formation by *Sacch. cerevisiae*, perhaps by diverting CoA from ethyl acetate synthesis (Nordström, 1964a), higher fatty acids were stimulatory (Nordström, 1964d), possibly because they are able to enter the cells, participate in lipid synthesis and thus spare CoA, which would otherwise be required for lengthening the carbon chain of small carbon fragments in lipid synthesis. The diversion of CoA from essential metabolic activities requiring acyl-CoA by fatty acids, especially those of intermediate chain length (C_8–C_{14}), may also explain the toxicity of such acids to yeast, since there is close parallelism between inhibition of ester synthesis and yeast growth (Fig. 3). The growth inhibition caused by the lower (toxic) acids was partially relieved by addition of higher fatty acids, amino acids or, still more effectively, by both.

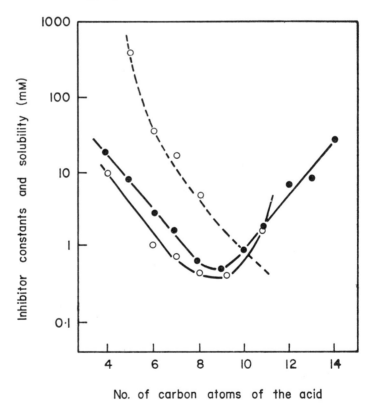

FIG. 3. Solubility (– – –○– – –), inhibitor constant of the formation of ethyl acetate (—○—) and inhibitor constant of yeast growth (—●—) for saturated, straight-chain fatty acids. From Nordström (1964d).

3. *Formation of Esters from Alcohols*

In a study of ester formation by his top strain of *Sacch. cerevisiae*, undertaken from the angle of the alcohol component, Nordström (1964e) showed that addition of alcohols to the medium prompted the formation of the appropriate esters in all cases studied, but isobutanol, isopentanol and tertiary alcohols were poorer substrates than were the straight-chain alcohols from methanol to heptanol. As with the acids, the relationship between ester formation and alcohol concentration could be expressed by a modified Briggs-Haldane equation for mono-molecular enzymic reactions. The maximum rate and Michaelis constant of ester formation decreased with increasing molecular weight of the alcohol.

As with the acids, the various alcohols competed in ester synthesis but, since ethanol dominated among them, and acetic acid similarly among the volatile acids, most of the volatile esters were ethyl esters or acetates, with ethyl acetate predominating. The distribution of the various esters depended on the availability of alcohols and acids in the fermentation medium. Next to ethyl acetate, isopentyl (isoamyl) acetate is the most abundant ester in fermented beverages, occurring to the extent of a few moles per cent of the alcohol. From his kinetic measurements, Nordström calculated values of about 200 for the ratio between ethyl acetate and isopentyl acetate; these are higher than other values (10–30) quoted in the literature. However, the kinetic studies showed that ester formation was an energy-requiring process, since the concentrations of the acetates of the higher alcohols greatly exceeded the equilibrium concentrations of the hydrolysis reactions. Since only minor concentrations of alcohols other than ethanol are present in beers and wines, esters other than ethyl esters can only occur in low concentrations.

Fermenting beer is thus a system for ester formation containing many competing substrates. However, the concentration of most alcohols is too low to exert an inhibitory effect on ester formation from other alcohols, so that a simplified kinetic equation of ester formation was derived:

$$v = \frac{dx}{dS_1} = \frac{VS}{K(B + S_1/K_1)}$$

where v = velocity; x = concentration of the ester under study; V = maximum velocity; S and S_1 = concentrations of alcohol under study; K and K_1 = Michaelis constants of the equations involving S and S_1; and B = system constant. This equation formulates the dependence of ester formation on that of ethanol; i.e. ester formation is linked to fermentation. Whether glucose, sucrose or maltose was used as the substrate for fermentation made little difference to ester formation.

4. Control of Ester Formation in Brewing

Nordström (1965) considered the possibilities of control of ester formation in brewing. In general, total volatile esters increased with temperature over the range used in brewery fermentations. The application of lower fermentation temperatures thus tends to yield beers with lower ester contents. Volatile esters were little affected by pH value, which therefore offers little scope for control.

Since the distribution of a number of added alcohols between acetates was the same for a number of yeast strains tested by Nordström, there seems to be no possibility of affecting the relative proportions of acetate esters by genetic means, other than by affecting the formation of the corresponding alcohols. This does depend on strain, over which selection could therefore be exerted. There is a connection between ability to form fusel alcohols and ability to break down amino acids. Thus, an isoleucine- and valine-deficient strain of Sacch. cerevisiae was unable to form isobutanol and 2-methylbutanol (Ingraham et al., 1961); this mutant had diminished ability to form volatile fusel alcohols even in wort (Drews et al., 1964).

Perhaps the most likely control methods centre on choice of yeast strain as discussed above, or on methods of control of yeast growth, following Nordström's (1964b) observation that ester formation was reduced when growth was inhibited. Nordström considers that control of nitrogen supply seems the most likely means of ester control. Added assimilable nitrogen stimulates ester production, so that dilution of wort nitrogen with brewing sugars, rice, maize or other starchy brewing adjuncts tends to give beers with low ester contents and with blander less characteristic flavours. However, this picture is complicated by the tendency of fusel alcohol formation to diminish with increasing nitrogen concentration above a threshold value (Äyräpää, 1963; Nordström, 1965; see also Section IV.B, p. 179), so that this effect would act contrarily to that already described. Thus, Nordström (1965) found that isopentanol decreased as the concentration of amino acids increased, while the formation of isopentyl acetate was less affected by changes of concentration of assimilable nitrogen than was that of ethyl acetate or ethyl caprylate (Fig. 4).

As mentioned earlier, any factor affecting biosynthesis and consumption of acyl-CoA also affects ester formation. Such factors include exogenous supply of D-biotin and pantothenate, but Nordström considers that control of esters by these means is unlikely, because the content of these growth factors in brewer's wort is likely amply to exceed the requirements of the yeast used for fermentation.

The special importance that control of ester formation is likely to have in continuous fermentation systems, because of the presence in the environment of relatively high alcohol concentrations at all times, has already been discussed.

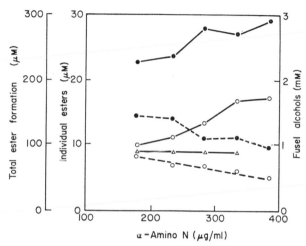

FIG. 4. Effect of reducing the α-amino nitrogen content of wort on the formation of volatile esters (—●—), ethyl caprylate (—○—), isopentyl acetate (—△—), isobutanol (---○---) and isopentanol + 2-methylbutanol (---●---) from Nordström (1965).

5. Mechanism of Ester Formation

a. Origin of the acid moiety. As a result of the work described above, Nordström (1965) rejects direct esterification:

$$R.COOH + R'OH \rightleftharpoons RCO.OR' + H_2O \qquad (1)$$

as a major reaction for ester formation in anaerobic fermentations by *Sacch. cerevisiae*, during which the equilibrium concentration expressed by Reaction 1 can be greatly exceeded. Indeed, this reaction proceeds very slowly at the pH values and temperatures under which beer is prepared and stored. Rather, he considers that esters are formed by alcoholysis of acyl-CoA compounds:

$$R.CO \sim S.CoA + R'.OH \rightleftharpoons R.COOR' + Co \sim ASH \qquad (2)$$

Ester formation thus depends on the supply of alcohols and acyl-CoA compounds. Since many brewing yeast strains are dependent on exogenous supplies of pantothenate, the formation of esters during fermentation also depends on wort pantothenate. However, beer wort contains larger amounts of this growth factor than are likely to limit yeast growth (Hopkins *et al.*, 1948), so that ester formation will not be conditioned by, or controllable by, wort pantothenate content.

The acid moiety of volatile esters in beer may arise in three ways (Nordström, 1965):

(i) by activation of fatty acids:

$$R.COOH + ATP + CoA \sim SH \longrightarrow$$
$$R.CO \sim SCoA + AMP + PP_i + H_2O \quad (3)$$

(ii) from α-oxo acids by oxidative decarboxylation:

$$R.CO.COOH + NAD + CoA \sim SH \longrightarrow$$
$$R.CO \sim SCoA + NADH_2 + CO_2 \quad (4)$$

(iii) from intermediates of higher fatty acid synthesis:

$$R.CO \sim SCoA + HOOC.CH_2.CO \sim SCoA + 2NADH_2 \longrightarrow$$
$$R.CH_2.CH_2.CO \sim SCoA +$$
$$CoA \sim SH + 2NAD + CO_2 + H_2O \quad (5)$$

In wort, the concentration of the acids required for Reaction 3 is so low, and so little of these acids, other than acetic, is formed by yeast (Äyräpää, *et al.*, 1961; Clarke *et al.*, 1962; Nordström, 1963c) that the reaction seems not to be an important factor in volatile ester formation in brewing. Probably Reaction 4 is responsible for much of the acetyl-CoA formed (Nordström, 1963a). However, since 2-oxobutyric acid and higher homologues are not apparently transformed by Reaction 4, the ethyl esters of propionic, isobutyric and isovaleric acids may not be formed as by-products of valine, leucine and isoleucine metabolism, unless indeed Reaction 4 occurs so slowly as to generate concentrations of these esters which are lower than the level of detection in beer (about $0 \cdot 01$ part/10^6). Minor quantities of ethyl esters were detected by Nordström (1964a) in his experimental fermentations in unsupplemented medium (Table IX), and are present in beer and wine.

TABLE IX. *Formation of Acids and their Ethyl Esters during Fermentation of Defined Medium (Nordström, 1964a)*

Acid radical	Final concentration of product (μM)	
	Free acid	Ethyl ester
Acetate (C_2)	3700	210
Butyrate (C_4)	20	1
Caproate (C_6)	20	2
Caprylate (C_8)	30	9
Caprate (C_{10})	0	8

These esters and the higher fatty acids of yeast lipids mostly contain an even number of carbon atoms, so that it seems feasible, as Nordström suggests, that these small quantities of volatile esters and the corresponding acids arise from small losses of lower acyl-CoA compounds during the synthesis of higher fatty acids by yeast, i.e. as leakage products from Reaction 5. Thus, when valeric acid was added to the medium, ethyl valerate increased with increasing temperature, whereas ethyl caprylate formation reached a maximum at 20°. If activation of the corresponding acids was responsible for the synthesis of these esters (Reaction 3), Nordström argues that ethyl caprylate and ethyl valerate formation would have followed the same pattern with respect to change of temperature. Since they did not, he suggests that Reaction 5 was responsible for caprylate formation.

b. Origin of the alcohol moiety. Fusel alcohols provide the alcohol moieties of the esters of fermentation. They arise mainly as by-products of yeast nitrogen metabolism; their origin from this source is described in the following Section. Fusel alcohols may arise also *via* carbohydrate metabolism (see Section IV.C.1, p. 190).

B. NITROGEN METABOLISM

1. *Early Work*

Ehrlich (1906, 1907a, b, 1911, 1912) took the first important steps to elucidate the problem of the nitrogen metabolism of yeast. He showed that the growth of yeast on certain amino acids led to the formation of the fusel fraction characteristic of yeast fermentations. He represented the sequence of changes as follows:

$$R.CHNH_2.COOH + H_2O \longrightarrow R.CHOH.COOH + NH_3 \qquad (1)$$
$$\underset{\alpha\text{-amino acid}}{} \qquad \underset{\alpha\text{-hydroxy acid}}{}$$

$$R.CHOH.COOH \longrightarrow R.CH_2OH + CO_2 \qquad (2)$$
$$\underset{\text{fusel alcohol}}{}$$

The ammonia formed in Reaction 1 was utilized for growth, while the α-hydroxy acid was decarboxylated to yield a primary alcohol (a fusel oil constituent) containing one atom of carbon fewer than the parent amino acid. Taking L-leucine as an example, the Ehrlich sequence becomes:

$$(CH_3)_2CH.CH_2.CHNH_2.COOH + H_2O \longrightarrow$$
$$\underset{\text{L-leucine}}{}$$

$$(CH_3)_2CH.CH_2.CHOH.COOH + NH_3$$
$$\underset{\alpha\text{-hydroxyisocaproic acid}}{}$$

$$(CH_3)_2CH.CH_2.CHOH.COOH \longrightarrow (CH_3)_2CH.CH_2.CH_2OH + CO_2$$
$$\underset{\text{isoamyl alcohol}}{}$$

In the conversion of phenylglycine to benzyl alcohol by yeast, Neubauer and Fromherz (1911) found evidence of the intermediate formation of phenylglyoxylic acid:

phenylglycine	phenylglyoxylic acid	benzyl alcohol

They therefore proposed the following sequence of reactions:

$$R.CHNH_2.COOH + O \longrightarrow R.CO.COOH + NH_3$$
$$R.CO.COOH \longrightarrow R.CHO + CO_2$$
$$R.CHO + 2H \longrightarrow R.CH_2OH$$

The corresponding α-hydroxy acid ($R.CHOH.COOH$) did not appear to be an intermediate, indicating that deamination of the amino acid was not effected by hydrolysis in the manner suggested by Ehrlich.

The decarboxylation step in the sequence received confirmation by the discovery of Neuberg and Hildesheimer (1911) of a carboxylase which they believed to be the catalyst responsible. This has been amply confirmed, and it is now known that the enzyme decarboxylates a number of 2-oxo acids.

The proposed reduction step received confirmation from a number of reports (Harden, 1932) that many aldehydes are reduced by fermenting yeast to the corresponding alcohols.

2. *Thorne's Work on the Ehrlich Mechanism*

The next main work on yeast nitrogen metabolism was that of Thorne. When he entered the field it was considered that, during growth on amino acids, yeast deaminated them, utilizing the liberated ammonia for growth and discarding the nitrogen-free carbon residue, after decarboxylating and reducing it, as fusel alcohol.

Thorne (1937) first satisfied himself that the Ehrlich formation of fusel alcohols took place with a top strain of *Sacch. cerevisiae* fermenting in a synthetic medium under conditions more closely resembling those of the brewery than had Ehrlich's. This proved to be the case: from fermentations in which tyrosine, tryptophan, phenylalanine or valine was supplied as the source of nitrogen for growth, the corresponding fusel alcohols (tyrosol, tryptophol, β-phenylethanol, isobutanol) were isolated in yields respectively of 80%, 78%, 67% and 60% of theoretical.

Two other amino acids, glutamic and arginine, tested by Thorne, yielded the acid and not the corresponding Ehrlich alcohol. Glutamic acid gave nearly theoretical yields of succinic acid. In this case, Thorne considered that the first two stages of metabolism took place as already described, with the formation of the appropriate α-oxo acid (2-oxoglutaric acid) and its decarboxylation to the aldehyde, which was then oxidized to succinic acid, instead of being reduced to γ-hydroxybutyric acid, as would be anticipated from the normal operation of the Ehrlich fusel alcohol mechanism:

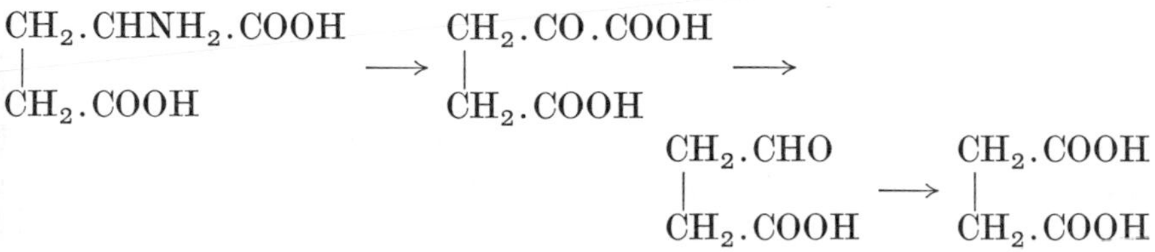

$$
\begin{array}{ll}
CH_2.CHNH_2.COOH & CH_2.CO.COOH \\
| & \longrightarrow \quad | \qquad\qquad \longrightarrow \\
CH_2.COOH & CH_2.COOH
\end{array}
$$

$$
\begin{array}{ll}
CH_2.CHO & CH_2.COOH \\
| & \longrightarrow \quad | \\
CH_2.COOH & CH_2.COOH
\end{array}
$$

Arginine also behaved anomalously in Thorne's experiments, succinic acid and γ-butylene glycol being obtained in yields of 60% and 19% respectively. Thorne proposed the following pathway to explain his findings:

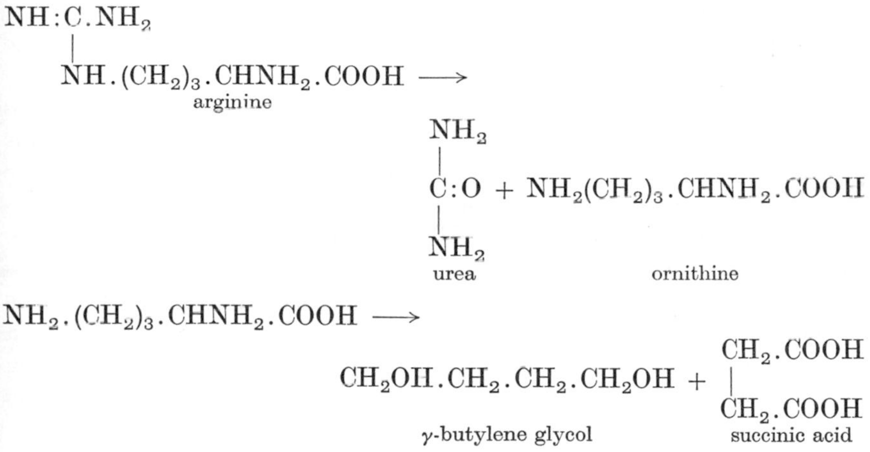

$$
\begin{array}{l}
NH:C.NH_2 \\
| \\
NH.(CH_2)_3.CHNH_2.COOH \longrightarrow \\
\qquad\qquad \text{arginine}
\end{array}
$$

$$
\begin{array}{l}
\qquad\qquad NH_2 \\
\qquad\qquad | \\
\qquad\qquad C:O + NH_2(CH_2)_3.CHNH_2.COOH \\
\qquad\qquad | \\
\qquad\qquad NH_2 \\
\qquad\qquad \text{urea} \qquad\qquad\qquad \text{ornithine}
\end{array}
$$

$$
NH_2.(CH_2)_3.CHNH_2.COOH \longrightarrow
$$

$$
\begin{array}{ll}
& CH_2.COOH \\
CH_2OH.CH_2.CH_2.CH_2OH + & | \\
& CH_2.COOH \\
\quad \text{γ-butylene glycol} & \text{succinic acid}
\end{array}
$$

Arginine metabolism by yeast has been little studied since Thorne's work, but in the light of some recent work it seems likely that the succinic acid may have arisen through glutamic acid, which is known to be metabolically derivable from arginine in many biological systems and through the TCA cycle. In fact, the pathway of arginine breakdown in yeast probably does not involve arginase as Thorne's scheme indicates; but rather arginine may be transformed to ornithine by arginine deiminase, an enzyme which has been identified in yeasts (Roche and Lacombe, 1952; Korzenovsky, 1955):

$$NH:C.NH_2$$
$$|$$
$$NH.(CH_2)_3.CHNH_2.COOH + 2H_2O \longrightarrow$$
$$NH_2.(CH_2)_3.CHNH_2.COOH + CO_2 + 2NH_3$$

3. The Ehrlich Mechanism and Transamination

Although Ehrlich had commented that he could not find free ammonia during amino acid deamination by yeast, some 50 years passed before the mechanism by which α-amino nitrogen was assimilated anaerobically by yeast in the Ehrlich pathway was established and the essential role played by transamination in that assimilation was elucidated. The experiments which led to this realization were performed by SentheShanmuganathan and Elsden (1958) and Senthe-Shanmuganathan (1960a, b) with a baker's strain of *Sacch. cerevisiae* operating under laboratory conditions unlike those involved in brewery fermentations. Despite this, there seems little reason to believe that the picture revealed would not be also true for brewer's strains growing under brewery conditions.

The findings of SentheShanmuganathan and Elsden (1958) may be summarized as follows:

(i) Washed cell suspensions of the yeast formed tyrosol from L-tyrosine under anaerobic conditions. Glucose, which was essential for this transformation, probably supplied energy for the transport of tyrosine into the cell, since azide and 2,4-dinitrophenol inhibited tyrosol formation by whole cells, but not by cell-free extracts, in which there is no question of an energy-requiring transport process.

(ii) Cell extracts formed glutamate, *p*-hydroxyphenylacetaldehyde (PHPA) and carbon dioxide from 2-oxoglutarate and L-tyrosine in a reaction which was stimulated by pyridoxal phosphate. This is strong presumptive evidence that enzymic transamination was involved in the conversion of tyrosine to tyrosol.

(iii) Cell extracts decarboxylated *p*-hydroxyphenylpyruvate to products mainly accounted for as PHPA.

(iv) Cell extracts oxidized reduced NAD on adding PHPA.

These results are entirely explicable by the following overall reaction catalysed by transaminase, carboxylase and alcohol dehydrogenase:

L-tyrosine + 2-oxoglutaric acid + $NADH_2 \longrightarrow$
tyrosol + L-glutamic acid + CO_2 + NAD

The individual reactions making up the overall reaction may be represented:

$$
\begin{array}{c}
\underset{\substack{| \\ \text{CH}_2.\text{CHNH}_2.\text{COOH} \\ \text{tyrosine}}}{\text{C}_6\text{H}_4.\text{OH}}
\; + \;
\underset{\substack{| \\ \text{CO}.\text{COOH} \\ \text{2-oxoglutaric acid}}}{\text{CH}_2.\text{CH}_2.\text{COOH}}
\; \longrightarrow
\end{array}
$$

$$
\underset{\substack{| \\ \text{CH}_2.\text{CO}.\text{COOH} \\ p\text{-hydroxy-} \\ \text{phenylpyruvic acid}}}{\text{C}_6\text{H}_4.\text{OH}}
\; + \;
\underset{\substack{| \\ \text{CHNH}_2.\text{COOH} \\ \text{glutamic acid}}}{\text{CH}_2.\text{CH}_2.\text{COOH}}
\tag{1}
$$

$$
\underset{\substack{| \\ \text{CH}_2.\text{CO}.\text{COOH}}}{\text{C}_6\text{H}_4.\text{OH}}
\; \longrightarrow \;
\underset{\substack{| \\ \text{CH}_2.\text{CHO} \\ p\text{-hydroxyphenyl-} \\ \text{acetaldehyde}}}{\text{C}_6\text{H}_4.\text{OH}}
\; + \; \text{CO}_2
\tag{2}
$$

$$
\underset{\substack{| \\ \text{CH}_2.\text{CHO}}}{\text{C}_6\text{H}_4.\text{OH}}
\; + \text{NADH}_2 \; \longrightarrow \;
\underset{\substack{| \\ \text{CH}_2.\text{CH}_2\text{OH} \\ p\text{-hydroxyphenyl} \\ \text{ethanol (tyrosol)}}}{\text{C}_6\text{H}_4.\text{OH}}
\; + \text{NAD}
\tag{3}
$$

Subsequently, SentheShanmuganathan (1960b) purified the tyrosine —2-oxoglutarate transaminase from the yeast, showing that it required pyridoxal phosphate as coenzyme, and that it was highly specific. SentheShanmuganathan (1960a) found transaminases for aspartate, norleucine, leucine, isoleucine, valine, norvaline, methionine, phenylalanine and tyrosine in the same yeast, cell extracts of which catalysed the transfer of the amino group from these amino acids to 2-oxoglutaric acid. The reactions involving aspartic acid, leucine, phenylalanine, tyrosine and valine were stimulated by pyridoxal phosphate. In addition, the cell extracts catalysed the release of carbon dioxide when incubated with 2-oxoglutarate and either leucine, norleucine, valine or norvaline. This indicated that the corresponding oxo acid formed from the amino acid was decarboxylated by the extracts. Confirmation was obtained by showing that 2-oxoisovalerate, 2-oxovalerate, 2-oxoisocaproate and 2-oxocaproate were rapidly decarboxylated by the cell extracts, and by purified yeast carboxylase, to the corresponding aldehydes. By comparison, the decarboxylation of p-hydroxyphenylpyruvate (derived from tyrosine by transamination) by purified yeast carboxylase occurred only slowly.

Finally SentheShanmuganathan (1960b) showed that cell extracts oxidized NADH_2 on addition of p-hydroxyphenylacetaldehyde, butyraldehyde, isobutyraldehyde, valeraldehyde or isovaleraldehyde, and that crystalline yeast alcohol dehydrogenase behaved similarly. The reverse reaction of reduction of NAD in the presence of the corresponding alcohols was also demonstrated.

These studies show that the mechanism of fusel alcohol formation in yeast involves successive stages of enzymic transformation by transaminase, carboxylase and alcohol dehydrogenase. Taking L-leucine as the example, the modern view of the Ehrlich mechanism is as follows:

$$(CH_3)_2CH.CH_2.CHNH_2.COOH + \begin{array}{c} CO.COOH \\ | \\ CH_2.CH_2.COOH \end{array} \xrightleftharpoons{\text{transaminase}}$$

L-leucine   2-oxoglutaric acid

$$(CH_3)_2CH.CH_2.CO.COOH + \begin{array}{c} CHNH_2.COOH \\ | \\ CH_2.CH_2.COOH \end{array}$$

2-oxoisocaproic acid   L-glutamic acid

$$(CH_3)_2CH.CH_2.CO.COOH \xrightarrow{\text{carboxylase}} (CH_3)_2CH.CH_2.CHO + CO_2$$

isovaleraldehyde

$$(CH_3)_2CH.CH_2.CHO + NADH_2 \xrightleftharpoons{\text{alcohol dehydrogenase}}$$

$$(CH_3)_2CH.CH_2.CH_2OH + NAD$$

isopentyl (isoamyl) alcohol

This scheme explains, not only the origin of the fusel fractions of liquids fermented by yeast, but also why no free ammonia is liberated during growth on amino acids. Furthermore, it brings out the vital role of transamination in yeast nitrogen metabolism. Indeed, from the key amino acid L-glutamic, the yeast cell can synthesize many other amino acids, given the appropriate 2-oxo acids, the carbon chains of which are generated from carbohydrate metabolism. This ability must be most important to *Sacch. cerevisiae*, which can synthesize all its cell nitrogen compounds (except for one or two growth factors) from ammonia as the sole source of nitrogen.

Lewis and Rainbow (1963) have also drawn attention to the importance of transamination in yeast nitrogen metabolism. They showed that brewer's yeast growing on L-glutamate as sole source of nitrogen liberated much 2-oxoglutarate into the medium, the amount being roughly proportional to that of the glutamate taken up. Much less 2-oxoglutarate was released during growth on other amino acids; the amount, in general, being least with those amino acids for which transaminases were demonstrated in cell extracts of the yeast. No 2-oxoglutarate was detected in the medium during growth on ammonia, urea or methionine, and production from L-glutamate was largely suppressed when the medium was supplemented with L-aspartate or ammonia. These experiments indicate that the glutamate–2-oxoglutarate transamination system is important in yeast metabolism, and that 2-oxoglutarate generated from glutamate is readily consumed in transamination reactions with other amino acids or by amination with ammonia when the latter is the source of nitrogen.

4. *Assimilation of Intact Amino Acids*

The belief that yeast assimilated nitrogen from wort amino acids by Ehrlich deamination focused attention on the nutrient value of individual amino acids and led to Thorne's important studies carried out between 1937 and 1949. He studied the growth and fermentation of brewer's yeasts, first in media containing ammonium salts or single amino acids as sole source of nitrogen (Thorne, 1937, 1941), and then in a medium containing binary (Thorne, 1944, 1945) and ternary (Thorne, 1946) mixtures of these nitrogen nutrients. He found that, in binary mixtures, yeast growth and fermentation were enhanced by about 20%, while addition of a third nitrogen source gave a further mean enhancement of 8%.

Taking the view widely held at that time that Ehrlich deamination was the normal mechanism by which amino acids were assimilated by yeast, ammonia should be superior to any amino acid as a yeast nitrogen nutrient. Although this was generally so when single amino acids were supplied, aspartate and asparagine were exceptions, and it was demonstrably false when mixtures of many amino acids, such as occur in brewer's wort, were supplied in Thorne's experiments. To explain this, Thorne (1949a, b) advanced the hypothesis that amino acids are assimilated intact by yeast and incorporated into its cell proteins, deamination only occurring when the cell is obliged to synthesize particular amino acids as they become deficient in the medium, and also to synthesize other nitrogen compounds required by the cell. Thorne (1949a) stated his hypothesis as follows: "When yeast is supplied with a diverse mixture of amino acids they will tend to be assimilated intact and at a higher rate than would be possible if they had to undergo deamination."

The product of such deamination was considered to be ammonia, which was utilized by the yeast as fast as it was formed, for biosynthetic purposes. These views must now be modified in the light of: (i) the new knowledge of the role of transamination as the concept of deamination leading to ammonia formation; and (ii) evidence (e.g. Jones *et al.*, 1965) that yeast synthesizes many amino acids before the medium becomes deficient in them (see Section III.F.2, p. 162).

In support of his hypothesis, Thorne (1949b) adduces the following evidence:

(i) The superiority of amino acid mixtures over ammonia as yeast nitrogen nutrients increases with increasing complexity of the mixture. Compared with growth on the corresponding single amino acids, the average enhancements on mixtures of 2, 3 and 8 amino acids were 20%, 28% and 54% respectively.

(ii) Certain amino acids on which yeast did not grow (e.g. lysine) were rapidly assimilated from complex mixtures.

(iii) More carbohydrate proved to be available for fermentation in medium containing mixed amino acids (casein hydrolysate) than in medium containing single amino acids or ammonia as the nitrogen source.

(iv) Growth of a pyridoxine-requiring strain of *Sacch. carlsbergensis* was less sensitive to pyridoxine deficiency in medium containing a complex mixture of amino acids than it was in medium containing single amino acids, in which the demands for pyridoxine for transaminations would be greater.

Thorne's final views on yeast nitrogen assimilation from wort are summarized in Fig. 5.

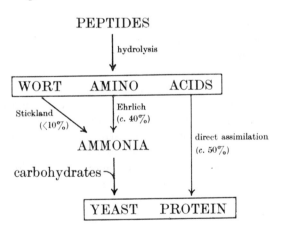

FIG. 5. Scheme of nitrogen assimilation from wort by yeast. After Thorne (1949a).

5. *Does Brewer's Yeast Operate the Stickland Reaction?*

The enhancement of growth and fermentation which occurred when yeast was grown on certain binary mixtures of nitrogen sources (Thorne, 1944; 1946a, b) led Thorne to the belief that the Stickland mechanism operated in yeasts. This mechanism, demonstrated by Stickland (1934, 1935) to operate in the anaerobe *Clostridium sporogenes*, involves the mutual oxidation and reduction of appropriate pairs of amino acids, one donating hydrogen to the other (the acceptor):

$$R'.CHNH_2.COOH + 2H_2O \longrightarrow R'.COOH + NH_3 + CO_2 + 4H$$

$$R''.CHNH_2.COOH + 2H \longrightarrow R''.CH_2.COOH + NH_3$$

The net reaction is thus:

$$\begin{matrix} R'.CHNH_2.COOH \\ + \\ 2R''.CHNH_2.COOH \end{matrix} + 2H_2O \longrightarrow \begin{matrix} R'.COOH \\ + \\ 2R''.CH_2.COOH \end{matrix} + CO_2 + 3NH_3$$

Given appropriate Stickland pairs of amino acids (e.g. alanine + glycine; phenylalanine + glycine; leucine + glycine, the H-donor being named first), Thorne obtained considerable enhancement of growth and fermentation and, with alanine + glycine mixtures, he also demonstrated the formation of substantial amounts of acetic acid, which is the only end-product of the Stickland reaction involving these amino acids:

$$\begin{matrix} CH_3.CHNH_2.COOH \\ + \\ 2CH_2.NH_2.COOH \end{matrix} + 2H_2O \longrightarrow \begin{matrix} CH_3.COOH \\ + \\ 2CH_3.COOH \end{matrix} + CO_2 + 3NH_3$$

So great has been the influence of Thorne's work that it is commonly accepted that yeast possesses this means of assimilating nitrogen. In fact, direct evidence that the Stickland reaction operates in yeast is lacking, and recent work by Lewis and Rainbow (1965) indicates that the evidence of enhanced yeast growth when pairs of amino acids are supplied as nitrogen source should not be interpreted as evidence of its operation. Working with Thorne's strains of brewer's *Sacch. cerevisiae*, Lewis and Rainbow showed that:

(i) Growth enhancements were recorded in only four out of nine Stickland amino acid pairs tested.

(ii) When an enhancement was observed, glycine was always one of the amino acid constituents.

(iii) Although they are both H-acceptors in the Stickland reaction, proline + glycine gave enhanced growth. Conversely, certain Stickland pairs (alanine + proline, leucine + proline, valine + glycine, aspartic acid + glycine, glutamic acid + glycine) did not (Table X).

(iv) From the above, growth enhancements appeared to depend on glycine as one constituent of the amino acid pair, but not all amino acids recognized as H-donors in the Stickland reaction promoted growth enhancements with glycine as H-acceptor.

TABLE X. *Growth of* Saccharomyces cerevisiae *on Mixed Nitrogen Nutrients. From Lewis and Rainbow (1965)*

Growth tests at 28° in defined medium containing 300 μg of nitrogen as ammonia, proline or α-amino acid per ml. S = Stickland pair of amino acids. Growth enhancement calculated from growth values for equimolar mixtures of the sources

Nitrogen nutrients	Yeast growth (mg dry matter/ml)					Growth enhancement
	Ratio of nitrogen supplies (a/b)					
	0/100	25/75	50/50	75/25	100/0	(%)
(a) L-α-Alanine ⎱ S (b) Glycine ⎰	0·13	1·44	2·07	1·63	1·03	78
(a) L-Aspartic acid ⎱ S (b) Glycine ⎰	0·13	2·44	3·00	3·00	2·90	0
(a) L-Glutamic acid ⎱ S (b) Glycine ⎰	0·13	2·52	3·10	3·00	3·00	0
(a) L-Histidine ⎱ S (b) Glycine ⎰	0·13	0·18	0·29	0·14	0·07	45
(a) L-Leucine ⎱ S (b) Glycine ⎰	0·13	1·60	2·45	1·90	1·56	45
(a) L-Phenylalanine ⎱ S (b) Glycine ⎰	0·13	0·58	1·42	1·05	0·83	48
(a) L-Valine ⎱ S (b) Glycine ⎰	0·13	1·77	2·52	2·90	3·10	0
(a) L-α-Alanine ⎱ S (b) L-Proline ⎰	0·68	0·98	0·99	0·98	1·03	0
(a) L-Leucine ⎱ S (b) L-Proline ⎰	0·68	1·42	1·75	1·75	1·58	0
(a) L-α-Alanine ⎱ (b) L-Leucine ⎰	1·58	1·75	1·53	1·36	1·03	0
(a) L-Proline ⎱ (b) Glycine ⎰	0·13	0·58	0·98	0·80	0·68	21
(a) Ammonia ⎱ (b) Glycine ⎰	0·05	1·15	1·32	1·44	1·06	20
(a) Ammonia ⎱ (b) L-α-Alanine ⎰	0·63	1·30	1·42	1·42	1·06	0

(v) Growth enhancements did not occur, under otherwise appropriate conditions, when exogenous pyridoxine and thiamine were supplied (Table XI): these growth factors were not included in Thorne's experimental medium.

(vi) Glycine was taken up by yeast during growth on amino acid pairs, whether they were Stickland pairs promoting enhanced growth,

TABLE XI. *Effect of Pyridoxine and Thiamine on Enhancement of Yeast Growth. From Lewis and Rainbow (1965)*

Growth tests at 28° for 72 h in basal defined medium to which was added a total of 300 μg of nitrogen as glycine and/or L-α-alanine. Growth enhancement calculated from growth values for equimolar mixtures of glycine + alanine. Pyridoxine HCl and thiamine HCl added as 1 μg of each per ml

Additions to medium	Yeast growth (mg dry matter/ml)			Growth enhancement
	Ratio of nitrogen supplies (glycine/alanine)			
	100/0	50/50	0/100	(%)
None	0·05	2·00	1·03	85
Thiamine	0·07	1·84	1·50	17
Pyridoxine	0·07	2·06	1·75	13
Pyridoxine + thiamine	0·10	2·33	2·30	0

Stickland pairs which did not do so, or non-Stickland pairs which did so. Glycine uptake was therefore not dependent on the presence of an H-donor amino acid.

(vii) The test yeast slowly adapted itself to grow on a medium containing glycine as sole source of nitrogen, but growth on glycine was stimulated by "sparker" concentrations, e.g. 10 μg N (as NH_3)/ml, of readily assimilable nitrogen in the form of ammonia, glutamate or aspartate (Fig. 6). Calculations of growth enhancements based on the insignificant growth levels obtained in 72 h on glycine alone are therefore misleading, since they fail to account for the contribution made to growth by glycine in the presence of readily assimilable amino acids.

(viii) The argument expounded in the previous paragraph was confirmed by comparing the growth behaviour of cells adapted to grow on glycine with that of unadapted cells in a medium (containing glycine + L-α-alanine) suited for the observation of growth enhancement. In this experiment, the unadapted cells gave a growth enhancement of 74%, while the adapted cells gave none. Thus, yeast adapts itself to growth on glycine, the process being accelerated, as might be expected, by the simultaneous presence of readily assimilable nitrogen. Consequently, growth enhancement effects when yeast is grown on mixtures of glycine and an amino acid on which it will grow readily as sole source of nitrogen, are attributable to utilization of glycine consequent upon adaptation to that amino acid.

(ix) Under conditions in which enhancement could occur (e.g. in an α-alanine + glycine medium), volatile acid formation (measured in terms of mg volatile acid per mg yeast formed) was not greater than that formed in the medium containing alanine as sole source of nitrogen.

Therefore, enhanced growth did not occur at the expense of a volatile acid-forming mechanism.

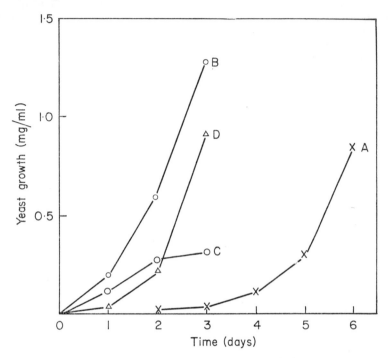

FIG. 6. Effect of "sparking" with ammonia on the lag phase which precedes growth of *Saccharomyces cerevisiae* in a glycine medium. The basal medium was supplemented as follows: Curve A, glycine; Curve B, glycine + ammonia; Curve C, ammonia. Curve D represents growth ascribable to glycine in the glycine–ammonia medium and is the difference between Curves B and C. 300 µg of nitrogen per ml was added as the glycine supplement and 10 µg of nitrogen per ml as the ammonia supplement. From Lewis and Rainbow (1965).

Lewis and Rainbow conclude from their experiments that "growth enhancements do not necessarily indicate the operation of the Stickland reaction in Yeast 37 (Thorne's strain of *Sacch. cerevisiae*) and, until more direct evidence for its operation in this and in other species of *Saccharomyces* is available, its contribution to nitrogen assimilation by yeast must not be accepted".

C. FORMATION OF FUSEL ALCOHOLS

1. *Fusel Alcohols, Fermentation and Nitrogen Metabolism*

Fusel alcohols certainly arise from the operation of the Ehrlich deamination of amino acids by yeast. However, it is now certain that this is not the only source of fusel alcohols of fermentation.

Äyräpää has made important contributions to the study of this subject. Initially he selected β-phenylethanol for study. He prepared this substance from beer by extraction with methanol-chloroform, purified it by column chromatography and identified it by gas chromatography, elementary analysis, ultraviolet spectrum and melting point of its 3,5-dinitrobenzoate (Äyräpää, 1961). It occurred in Äyräpää's beers in concentrations ranging from 10 to 40 μg/ml and, having an "intense flowery aromatic aroma" (sometimes likened to that of roses), it evidently contributes to beer aroma and flavour. Äyräpää (1962) has described a relatively simple spectrophotometric method for determining β-phenylethanol after it has been extracted from beer, and separated by chromatography on alumina-silicic acid using diethyl ether as the eluant.

In experiments with *Sacch. carlsbergensis*, grown anaerobically in a synthetic medium containing mixed amino acids in proportions resembling those of brewer's wort, β-phenylethanol, like the aliphatic fusel alcohols, was formed not only in the presence of the corresponding amino acids, but also when those amino acids were omitted, or even when ammonium sulphate provided the sole source of nitrogen (Äyräpää, 1963). The amount formed depended mainly on the ratio of assimilable nitrogen to sugar in the medium. High yields were obtained when the medium was poor in nitrogen (124 μg N/ml), so that fermentation was slower than in wort and the yeast became poorer in nitrogen. As

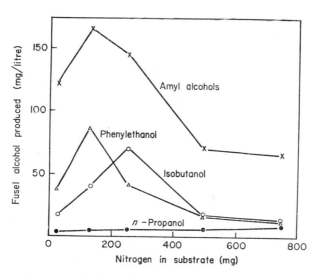

Fig. 7. Formation of fusel alcohols at various concentrations of an amino acid mixture simulating wort, but without phenylalanine. From Äyräpää (1963).

the concentration of assimilable nitrogen in the medium increased, so the yield of β-phenylethanol decreased (Fig. 7).

A similar relationship applied to the aliphatic fusel alcohols, and it was also valid when ammonium sulphate, mixed amino acids as in wort, or mixed amino acids lacking valine, leucine, isoleucine, phenylalanine and tyrosine, provided the source of nitrogen. Surprisingly, with ammonium sulphate as sole source of nitrogen, the yields of phenylethanol and amyl alcohols were only 20–40% less than in the mixed amino acid medium. Additions of phenylalanine to the mixed amino acid medium caused increased production of phenylethanol, but not proportionately and, at the most, only about 30% of the amino acid appeared as the alcohol. At the same time, the phenylalanine additions partially inhibited aliphatic fusel alcohol formation.

Äyräpää's results are consistent with those of Hough and Stevens (1961) and Enebo (1957) that production of fusel alcohols during fermentation followed the uptake of sugar, but was not correlated to uptake or excretion of nitrogen (Fig. 8).

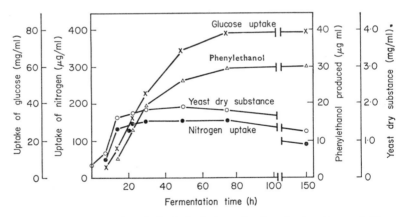

Fig. 8. Course of phenylethyl alcohol production during fermentation. From Äyräpää (1963).

Other results indicating that fusel alcohol formation did not relate to Ehrlich amino acid deamination were those of Jones et al. (1965), who showed that higher alcohols (and also ethyl acetate) were produced throughout fermentation and even early in fermentation in wort, before rapidly assimilable amino acids had been eliminated. Because of this, and the fact that fusel alcohols are formed in defined media containing ammonia as sole source of nitrogen, it seems that fusel constituents must be derived by deamination of amino acids synthesized intracellularly from the oxo acid pool, or directly by decarboxylation and reduction of these oxo acids.

Äyräpää considers that the formation of phenylethanol is analogous to that of aliphatic fusel alcohols, especially in being related to the level of nitrogenous nutrients. He explains the main features of the dependence of fusel alcohol formation on the level of assimilable nitrogen against a background illustrated in Fig. 9. From this scheme, it may be inferred that factors influencing the supply to, or withdrawal from, the oxo acid pool also influence the formation of fusel alcohols.

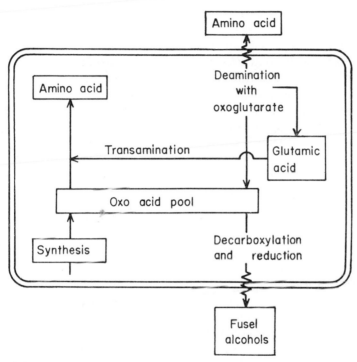

FIG. 9. Schematic presentation of the hypothesis of fusel alcohol formation. After Äyräpää (1963).

The observed relationship between assimilable nitrogen in the medium and fusel formation depends on two factors. Firstly, on the amount of nitrogen available for transamination and amination. The oxo acids resulting from nitrogen insufficiency are converted to the corresponding alcohols. Such oxo acids must arise from carbohydrate metabolism, which must also be the source of oxo acids and fusel alcohols when yeast is grown on ammonium salts as sole source of nitrogen. Secondly, the relationship depends on feed-back inhibition of amino acids on those reactions which synthesize their individual carbon chains. Äyräpää comments that the general form of the curves for amyl alcohols and isobutanol in Fig. 7 could be explained by feed-back inhibition, but not the curve for phenylethanol, since phenylalanine was absent from the medium.

Äyräpää's work neither established the influence of feed-back inhibition on, nor the quantitative contribution of the Ehrlich mechanism to, fusel alcohol formation. However, he points out that the relationship between fusel formation and level of nitrogenous nutrients is similar irrespective of whether amino acids or ammonium salts are the nitrogen source, so that feed-back inhibition by exogenous amino acids may play a subordinate role to that of the quantity of nitrogen available for amination and transamination. At nitrogen levels sufficient to allow development of biosynthetic pathways, but insufficient effectively to engage the oxo acids in transamination reactions, increased fusel alcohol formation from the unengaged oxo acids may be expected. By contrast, with increased amounts of exogenous nitrogen nutrients, transamination becomes increasingly possible; the oxo acids are used up and less fusel alcohol is formed. Only when an amino acid is present in excess of the needs of the yeast cell may it be anticipated that it will yield the corresponding fusel alcohol by deamination in the Ehrlich pathway.

Äyräpää (1965) attempted to assess the relative contributions to fusel alcohol formation of: (i) the Ehrlich mechanism by transamination or deamination of exogenous amino acids; and (ii) the mechanism *via* oxo acids synthesized from carbohydrate. As before, he used the same strain of *Sacch. carlsbergensis* grown anaerobically at 25° in a medium containing an amino acid mixture simulating brewery wort and supplemented with ^{14}C-phenylalanine. Even at low amino acid levels, insufficient to support much growth, most of the phenylalanine was converted to phenylethanol. Nevertheless, some phenylethanol was produced by the synthetic pathway. Even in media containing high amino acid concentrations (where assimilable nitrogen was in excess) some formation of the alcohol by synthesis was observed, although it was diminished in these circumstances and, in one experiment, synthesis was completely blocked by high concentrations of phenylalanine.

Äyräpää (1963) had shown that when amino acids capable of yielding higher alcohols by the Ehrlich mechanism were included in moderate amounts in the mixture presented to yeast, the amounts of the corresponding fusel alcohols in the fermented mixture were little greater than when those amino acids were absent. He therefore considers that the synthetic formation of higher alcohols may always be of major importance. However, as his later work with phenylalanine showed, suppression (perhaps by end-product inhibition) of the synthetic mechanism by high concentrations of the corresponding amino acids in the medium is possible. He therefore proposes the alternative, more likely hypothesis that the suppression of synthetic mechanisms may be compensated by the formation of the same alcohols by the Ehrlich pathway. Assuming that the difference between total phenylethanol formed and that formed by the synthetic pathway (as established in his experiments) represents the alcohol arising by the Ehrlich mechanism, Äyräpää's (1965) experiments indicate that phenylethanol formed by the latter mechanism was least at the highest amino acid concentrations (800 μg N/ml), maximum formation occurring at intermediate concentrations within the range studied (100–800 μg N/ml). Thus, transamination appeared to be substantially reduced when mixed amino acids were supplied in large excess: indeed, the fraction of phenylalanine transformed to phenylethanol bore an inverse relationship to the total nitrogen content of the medium, but depended little on the contribution made by phenylalanine to the total nitrogen content.

Since nutritional factors influence the formation of phenylethanol and isopentanol similarly (Äyräpää, 1963) and the general features of regulation of aromatic amino acid pathways resemble those of leucine, valine and isoleucine (Stadtman, 1963), Äyräpää inclines to the view that conclusions derived from his studies on phenylethanol will apply to fusel alcohols in general, and that both the Ehrlich and synthetic pathways operate in their formation during the fermentation of brewery worts.

2. Mechanism of Fusel Alcohol Formation from Pyruvate

The foregoing describes evidence that the Ehrlich mechanism is not the only pathway of fusel alcohol formation by yeasts. The evidence may be summarized as follows (Ingraham, 1965):

(i) In fermentations in complex media, the composition of the fusel alcohol formed does not correlate well with that of the amino acids of the unfermented medium.

(ii) Although amino acids are quickly taken up by yeast, the rate of fusel oil formation is no more rapid during the period of rapid absorption than it is later, when the medium is essentially free from amino acids.

(iii) Certain fusel alcohols (e.g. n-propanol, a major component) and n-butanol correspond to α-aminobutyrate and norvaline respectively, neither of which occur naturally.

(iv) Fusel alcohols are products of fermentations by resting cells in media containing no amino acids, and they are formed during fermentations in defined media in which ammonia is the sole source of nitrogen.

There is, therefore, strong evidence that fusel alcohols can be formed by yeast from carbon fragments generated by carbohydrate catabolism, i.e. by the so-called synthetic pathway of fusel alcohol formation. These fragments are likely to be 2-oxo acids, which are intermediates in amino acid synthesis by virtue of being able to engage in transamination with glutamate.

Stevens (1960) and Webb and Ingraham (1963) have reviewed the subject of the formation of fusel alcohols. They give the following pathway leading to isobutanol:

$$
\begin{array}{ccccc}
\mathrm{CH_3} & & \mathrm{CH_3} & & \mathrm{CH_3} \\
| & \xrightarrow{\text{Acetyl-CoA}} & | & \xrightarrow[\text{rearrangement}]{\text{reductive}} & | \\
\mathrm{CO} & & \mathrm{C(OH)COOH} & & \mathrm{CH_3C.OH} \\
| & & | & & | \\
\mathrm{COOH} & & \mathrm{COCH_3} & & \mathrm{HC(OH)COOH} \\
\text{pyruvic acid} & & \text{α-acetolactic acid} & & \text{2,3-dihydroxy-} \\
& & & & \text{isovaleric acid}
\end{array}
$$

$$\downarrow -\mathrm{H_2O}$$

$$
\begin{array}{ccccc}
\mathrm{CH_3\ CH_3} & & \mathrm{CH_3\ CH_3} & & \mathrm{CH_3\ CH_3} \\
\diagdown\diagup & \xleftarrow[\text{dehydrogenase}]{\text{alcohol}} & \diagdown\diagup & \xleftarrow{-\mathrm{CO_2}} & \diagdown\diagup \\
\mathrm{CH} & & \mathrm{CH} & & \mathrm{CH} \\
| & & | & & | \\
\mathrm{CH_2OH} & & \mathrm{CHO} & & \mathrm{CO.COOH} \\
\text{isobutanol} & & \text{isobutyraldehyde} & & \text{2-oxoisovaleric acid}
\end{array}
$$

$$\downarrow$$

$$
\begin{array}{c}
\mathrm{CH_3\ CH_3} \\
\diagdown\diagup \\
\mathrm{CH} \\
| \\
\mathrm{CHNH_2.COOH} \\
\text{valine}
\end{array}
$$

The steps in this scheme from 2-oxoisovaleric acid are readily recognizable as being common with those in the Ehrlich pathway from valine to isobutanol. With 2-oxobutyrate instead of pyruvate, the corresponding sequence leads to isoleucine and thence, through 2-oxo-3-methylvalerate to active amyl alcohol (Webb and Ingraham, 1963):

$$\text{2-oxobutyrate} \longrightarrow \text{2-acetohydroxybutyrate}$$

$$\downarrow$$

$$\text{2,3-dihydroxy-3-methylvalerate}$$

$$\downarrow$$

$$\text{2-oxo-3-methylvalerate} \rightleftharpoons \text{L-isoleucine}$$

$$\downarrow -\mathrm{CO_2}$$

$$\text{active pentanol} \rightleftharpoons \text{methylbutyraldehyde}$$

2-Oxobutyrate is not only a precursor of active pentanol and isoleucine, but on decarboxylation and reduction in the Ehrlich pathway it yields *n*-propanol, for which there is no amino acid precursor. Similarly, 2-oxoisovalerate is common in the sequences leading to L-valine, as described above, and also to L-leucine (Webb and Ingraham, 1963):

$$\text{2-oxoisovalerate} \xrightarrow{\text{CoA}} \text{3-carboxy-3-hydroxyisocaproate}$$

$$\downarrow$$

$$\text{2-hydroxy-3-carboxyisocaproate}$$

$$\downarrow {\scriptstyle -CO_2}$$

$$\text{isovaleraldehyde} \xleftarrow{-CO_2} \text{2-oxoisocaproate} \rightleftharpoons \text{L-leucine}$$

$$\Updownarrow$$

$$\text{isopentanol}$$

The origin of fusel alcohols by the synthetic pathway has been illuminated by studies with mutant yeasts. Ingraham and Guymon (1960) found that mutant yeasts, which could not synthesize certain amino acids, produced fusel fractions lacking in the corresponding alcohols. Thus, isoleucine- and leucine-requiring yeasts produced little or no active pentanol and isopentanol respectively. It therefore seemed that fusel alcohols in the non-mutant yeasts arose from intermediates in amino acid biosynthesis which were absent from the mutant yeasts.

3. *Fusel Alcohols in Beer*

The fusel alcohol content of beer is undoubtedly an important, but by no means the only, factor contributing to beer flavour. According to Hudson and Stevens (1960), top fermentation beers usually contain more fusel alcohols than do bottom fermentation beers, the chief fusel components consisting primarily of isopentanol, active pentanol and isobutanol. Of 56 ales examined, 35 which had been fermented by a skimming system had a mean fusel alcohol content of 101 p.p.m., as against 62 p.p.m., for 12 beers fermented by the Yorkshire Stone Square system, and 98 p.p.m. for 5 beers fermented by the Burton Union system. Four experimental continuously-fermented beers had contents ranging from 52 to 104 p.p.m.

Although there was no general correlation between original gravity of the beers and their fusel content, high gravity beers did contain high levels of fusel alcohols: thus, a beer containing 232 p.p.m. was a very high gravity stout. Values for the seven lager beers examined by Hudson and Stevens lay between 45 and 83 p.p.m., with a mean of 54 p.p.m.

There is therefore presumptive evidence that the fusel content of beers will depend on yeast strain, conditions of fermentation and wort composition. All these factors differ according as to whether a lager or ale is brewed and, even in ale brewing, yeast strain and fermentation conditions differ according to the particular system in use. Hough and Stevens (1961) studied the effects of some of these factors on fusel alcohol formation. They found that, in the earlier stages of fermentation, fusel formation increased linearly with attenuation; but in the final stages, it increased more rapidly than did attenuation. The amount of fusel oil produced by each of nine yeast strains fermenting the same batch of wort at 18° varied from 51 to 88 p.p.m. In each case, fusel alcohol continued to be formed during secondary fermentation, but it did not arise from yeast autolysing in the beer. In fermentations at 25°, they confirmed the observations of Hudson and Stevens (1960) that top yeasts generally form more fusel than do bottom yeasts, but there was no difference in this respect when top and bottom yeasts fermented the same wort at 10°, at which temperature fusel formation by top yeasts was depressed and that by bottom yeasts slightly enhanced.

Hough and Stevens (1961) studied the effect of wort composition by making additions of L-leucine or sugar to the wort and by carrying out fermentations in defined medium. Fusel alcohol formation increased in relation to the quantity of leucine added up to 223 μg/ml, although only 50 moles per cent of the added leucine was converted to isopentanol. When sugar was added to wort, fusel formation was depressed in proportion to the concentration of sugar added. This, and the observation that the fusel contents of beers produced from all-malt wort were proportional to the concentration of wort solids, led Hough and Stevens to conclude that the amount of fusel alcohols formed depended on malt constituents (presumably amino acids) other than sugar, and that fusel alcohols in beer arose chiefly from the Ehrlich pathway. Nevertheless, n-propanol, isobutanol and 2- and 3-methylbutanols were identified in fermented defined medium lacking amino acids, so that a pathway or pathways of fusel alcohol formation, other than the Ehrlich mechanism, must exist.

Concerning conditions of fermentation, Hough and Stevens found that: (i) more fusel alcohols were produced under anaerobic than under aerobic conditions; (ii) rather less fusel alcohol was produced by fermenting the same wort in conventional batch conditions compared with stirred batch or accumulated batch fermentations (Pollock, 1961); and (iii) beers produced by a system of continuous fermentation (Hough and Ricketts, 1960; Hough and Rudin, 1961) contained only slightly more fusel alcohols than did conventional beers.

In summary, these experiments serve to show that yeast strain, conditions of fermentation, and wort composition are all factors affecting fusel alcohol formation during beer fermentations, thereby modifying beer flavour, aroma and character. Not only will the fusel alcohol content *per se* influence flavour and aroma, but its constituents will participate in ester formation. As already described, the concentration of such esters in beers is very small: nevertheless, their contribution to flavour and aroma can be disproportionate to their concentration, and cannot therefore be disregarded. There can be no doubt that the brewer's constant striving for consistency with respect to yeast strain, wort composition and conditions of fermentation as essential to flavour consistency rests on a firm theoretical basis.

D. DIACETYL AND ACETOIN FORMATION

1. *Vicinyl Diketones in Beer*

Acetoin, diacetyl and 2,3-pentanedione are normal beer constituents which probably make an appreciable contribution to beer flavour and aroma, especially to that of ales. However, in concentrations exceeding threshold values, they are spoilage constituents. The threshold of tolerance varies according to the beer: the fuller flavoured beers are able to "carry" greater concentrations than are lager beers. Thus, 28 American beers contained 0·20–0·46 µg of vicinyl diketones/ml, 0·53–2·23 µg/ml being found in off-flavour beers (West *et al.*, 1952). Similarly, Drews *et al.* (1962) maintained that commercial American beers were unacceptable when the diacetyl content exceeded 0·45 µg/ml, while the upper threshold of acceptability for Pilsner or pale lagers was only 0·20–0·25 µg/ml.

It is now known that both diacetyl and 2,3-pentanedione (and possibly other carbonyl compounds) contribute to the sweetish, buttery or honey-like off-flavour which is commonly referred to as diacetyl flavour. Excessive acetoin is said to endow beer with a musty flavour.

The usual colorimetric method of determining diacetyl, involving the reaction in alkaline solution with creatine and α-naphthol (Brenner *et al.*, 1963), is not specific, but determines vicinyl diketones; that is, both diacetyl and 2,3-pentanedione, which are the chief representatives of this class of substance in beer. Since both substances contribute to diacetyl off-flavour in beer, this lack of specificity is not entirely disadvantageous. Harrison *et al.* (1965) were able to distinguish between diacetyl and 2,3-pentanedione by a gas chromatographic method, using electron capture for detection. The values they quote (Table XII) show that ales and stouts generally contain more vicinyl diketones than do lagers, and that there is a close correspondence between the total diacetyl + 2,3-pentanedione concentrations and those determined as vicinyl diketones colorimetrically.

TABLE XII. *Vicinyl Diketone Contents of Commercial Beers. After Harrison* et al. (*1965*)

Values for diacetyl and 2,3-pentanedione determined by gas-liquid chromatography; for total vicinyl diketones (VDK) by an absorptiometric method

Beer	Concentration of product (p.p.m.)		
	Diacetyl	2,3-Pentanedione	Total VDK
Lager A	0·02	0·01	0·04
Lager B	0·06	0·05	0·07
Lager C	0·03	0·02	0·12
Lager D	0·08	0·02	0·10
Ale E	0·06	0·02	0·09
Ale F	0·15	0·04	0·19
Ale G	0·30	0·20	0·49
Ale H	0·28	0·20	0·57
Stout I	0·04	0·01	0·09
Stout J	0·07	0·02	0·11
Stout K	0·58	0·26	0·85

2. *Formation of Diacetyl and Related Substances during Fermentation*

Several workers have observed that an increase in vicinyl diketones takes place during the initial, most vigorous stage of fermentation. Thereafter, under normal circumstances, the content decreases as fermentation is prolonged, until low levels of diacetyl and pentanedione are reached (Katô and Nishikawa, 1960–1; Harrison *et al.*, 1965). Vicinyl diketones are therefore normal minor products of yeast fermentation: they arise during the early stages and appear subsequently to be metabolized by the yeast as fermentation is prolonged. The ability of yeast to metabolize diacetyl is evident from the fact that its removal is most rapid during the period of active yeast propagation, and is directly related to the amount of yeast in suspension and hence to its flocculence (Burger *et al.*, 1957).

A more refined, but as yet comparatively untested, possibility of removing excessive vicinyl diketones from beer consists in treatment with an $NADH_2$-requiring diacetyl reductase prepared from *Klebsiella* (*Aerobacter*) *aerogenes* (Bavisotto *et al.*, 1964):

$$
\begin{array}{ccc}
CH_3.CO & \xrightarrow[\text{diacetyl reductase}]{NADH_2} & CH_3.CHOH \\
| & \xleftarrow{\quad NAD \quad} & | \\
CH_3.CO & & CH_3.CO \\
\text{diacetyl} & & \text{acetoin}
\end{array}
$$

The enzyme also catalysed the reduction of 2,3-pentanedione, albeit less effectively.

3. *Effect of Yeast Strain on Vicinyl Diketone Formation*

The properties of the yeast strain, especially its fermentative power and flocculence (Anh, 1965), is the most important factor affecting vicinyl diketone formation during fermentation. The respiratory deficient yeasts known as "verdants" and "petites" were found by Czarnecki and van Engel (1959) to produce diacetyl. In particular, the "petites", which lack succinic dehydrogenase and cytochrome oxidase, may be responsible for excessive concentrations of diacetyl in beer, and they suggest that pitching yeasts should be examined regularly for the presence of these mutants.

The ability of a yeast strain to take up valine from wort may also be important in determining its power of forming diacetyl. This question is discussed in Section IV.D.6 (p. 202). Kringstad and Rasch (1966) consider that the method of propagating the yeast strain also influences the formation of diacetyl and acetoin during fermentation. They found that yeast (presumably *Sacch. carlsbergensis*) propagated in static culture in brewer's hopped wort at 15° or 17°, yielded more diacetyl than it did in stirred culture. However, at 10° more diacetyl was formed in the stirred cultures, while yields in aerated cultures were similar to those in static cultures. In acetoin formation, the stirred yeast produced most and the static cultures the least acetoin, intermediate amounts being formed in the aerated cultures.

4. *Effect of Aeration on Diacetyl Formation*

In apparent disagreement with Kringstad and Rasch, other workers find that access of oxygen stimulates vicinyl diketone formation by brewer's yeast. Burger *et al.* (1957) consider that beer unduly exposed to air during processing (e.g. by prolonged holding in fermenters), so that it has an Indicator Time Test value (Gray and Stone, 1939) of 500–1,200, tends to have a high diacetyl content, while beers with values greater than 2,500 contain little or no diacetyl. Similarly, Anh (1965) reports that wort aeration and access of air to beer, arising from stirring or transference, predispose yeast to form diacetyl, the amount of which may be related to the redox condition of beer as indicated by its Indicator Time Test value.

On this subject, Portno's (1966) work indicates that diacetyl production is not affected by the concentration of oxygen in wort at pitching, whereas beer diacetyl concentrations, when air was passed through the wort during fermentations, were increased five- or sixfold in comparison with a control in which carbon dioxide was substituted for oxygen. The increase in diacetyl was proportional to the increased growth of yeast. When air was passed through the fermentation head-space, instead of through the fermenting liquid, beer diacetyl and yeast growth were again increased, but only twofold compared with the appropriate carbon dioxide control. These experiments indicate a relationship between oxygen, yeast growth and diacetyl formation, and show that lower diacetyl contents may be expected in beers brewed in closed, as compared with open, fermenters. Access of oxygen during fermentation appeared to enhance the beer diacetyl content by stimulating biosynthesis, so that the amino acid-carbohydrate balance was altered. Diacetyl did not arise from the direct oxidation of acetoin. The diacetyl content of beer during storage and conditioning was not affected by access of oxygen.

5. *Other Factors Affecting Vicinyl Diketone and Acetoin Formation*

Factors other than yeast strain and access to oxygen have been implicated in diacetyl formation in beer fermentations. Voerkelius (1961) reports observations suggesting that the formation of diacetyl and acetoin is influenced by incomplete conversion in the mash, high fermentation temperatures, the premature checking of fermentation and especially by weak fermentations. In his opinion excessive acetoin and diacetyl are avoided by eliminating all factors tending to cause sluggish fermentations, whether they be errors in mashing, temperature shock during fermentation, high nitrate and nitrite contents of the brewing water, or any other factor.

6. *Mechanism of Diacetyl Formation*

The mechanism of diacetyl formation by brewer's yeasts remains to be completely elucidated, but sufficient is known from studies involving brewer's and baker's yeasts and other micro-organisms to be able to offer a working hypothesis for future testing. Present hypotheses of the biosynthesis of diacetyl by yeast largely devolve on observations relating its formation to α-acetolactate or to L-valine. The chief relevant observations are as follows:

(i) In *Sacch. cerevisiae* and other organisms, acetolactate is a precursor of acetoin (itself a possible precursor of diacetyl) and valine (Dirscherl and Höfermann, 1954).

(ii) Cell-free preparations from *Sacch. cerevisiae* (baker's yeast) convert α-acetolactate to 2-oxoisovalerate, and thence to valine, in a system requiring NADP (Strassman *et al.*, 1958).

(iii) Cell extracts of baker's yeast convert pyruvate to α-acetolactate (Lewis and Weinhouse, 1958).

Taking up at this point, Owades *et al.* (1959) found that diacetyl formation during fermentation was strongly depressed by addition of L-valine, but not by any other amino acid or other nitrogen compound, and also by α-acetolactate. They considered that the evidence favoured the view that diacetyl was formed as a by-product of valine synthesis from pyruvate *via* acetolactate, and suggested that any treatment tending to supplant anaerobic by aerobic pathways (e.g. by pitching into aerated wort or by increasing contact with air by transference during fermentation) may predispose yeast to form diacetyl. Furthermore, wort low in valine may, by adaptive change, enhance the ability of the yeast to synthesize valine and, therefore, diacetyl. Since yeasts vary in ability to absorb valine from wort, they suggest that brewery yeasts should be screened for this ability, and those with poor powers of valine absorption rejected.

Owades *et al.* propose the following pathway of diacetyl formation in yeast:

$$CH_3.CO.COOH + CH_3.CHO$$

$$\downarrow$$

$$\underset{\text{acetolactic acid}}{\overset{\displaystyle CH_3.C(OH).COOH}{\underset{\displaystyle CO.CH_3}{|}}} \xrightarrow{-CO_2} \underset{\text{acetylmethylcarbinol (acetoin)}}{CH_3.CO.CHOH.CH_3} \xrightarrow{-2H} \underset{\text{diacetyl}}{\overset{\displaystyle CO.CH_3}{\underset{\displaystyle CO.CH_3}{|}}}$$

$$\downarrow$$

$$(CH_3)_2C(OH).CHOH.COOH \longrightarrow \underset{\text{2-oxoisovaleric acid}}{(CH_3)_2CH.CO.COOH} \longrightarrow$$

$$\underset{\text{valine}}{(CH_3)_2CH.CHNH_2.COOH}$$

The formation of diacetyl by micro-organisms other than yeast may take place from pyruvic acid by a similar pathway. Thus, *Streptococcus diacetilactis* possesses the enzymes to effect the sequence: citrate → oxaloacetate → pyruvate → α-acetolactate → 2,3-butanediol → acetylmethyl carbinol → diacetyl (Seitz *et al.*, 1963).

Reviewing the subject of diacetyl in brewing, Anh (1965) lists the evidence for the route of diacetyl formation postulated by Owades *et al.* as follows: (i) α-acetolactate is a precursor of valine and acetoin in *Sacch. cerevisiae*; (ii) pyruvate is a precursor of α-acetolactate in baker's yeast; (iii) valine, but not other amino acids, added to yeast fermentations depresses diacetyl formation, possibly by depressing synthesis of acetolactate; (iv) diacetyl formation is decreased by addition of α-acetolactate, acting perhaps by end-product inhibition of a step between pyruvate and acetolactate; and (v) diacetyl formation in primary fermentation reaches a maximum during the phase of yeast growth, i.e. when valine is being synthesized for cell material; thereafter, the diacetyl concentration decreases as a result of its metabolism by the yeast.

Of this evidence, the least satisfactory point is that of the action of acetolactate in depressing diacetyl levels. Looking at the scheme of Owades *et al.* as it stands, there is no obvious reason for the observed effect of acetolactate and Anh's suggestion (iv above) is not convincing.

E. SULPHUR COMPOUNDS IN BEER

1. *Hydrogen Sulphide*

The sulphur compounds required by brewer's yeasts for its life processes may be supplied entirely as sulphate ion, although, in wort, sulphur-containing amino acids are also utilized. The only exception to this is the requirement of brewer's yeast for quantitatively minor amounts of preformed D-biotin. An indication of the concentration of sulphate ion required by yeast is provided by Ekström and Sandegren (1951), who found that the sulphur content of the medium could be decreased to 5 μg/ml, supplied as Na_2SO_4, without affecting the yield of *Sacch. carlsbergensis*.

Presumably, sulphate undergoes a reductive sequence through sulphite and sulphide, before the sulphur enters organic combination into cysteine, cystine, methionine and other essential sulphur-containing cell constituents. It is therefore to be expected that small quantities of sulphite and sulphide, derived by reduction of sulphate, should appear in fermenting liquids.

In addition, traces of mercaptans are found in fermenting worts and beers. Together with sulphite, sulphide and mercaptans, mercaptals (Knorr, 1965), thioaldehydes and thioketones (Hashimoto and Kuroiwa, 1966) all contribute to the aroma and flavour of beer; and, in conditions in which excessive quantities of one or more of these constituents are produced by yeast, unpalatable beer can result. Brenner *et al.* (1953)

TABLE XIII. *Hydrogen Sulphide Contents of Commercial American Beers and Brewing Materials*

Material	Hydrogen sulphide content	Reference
Malt	(μg/kg) 190	Brenner *et al.* (1953)
Brewery worts	(μg/ml) 13–37	
Fermenting wort	up to 250	
Beers in bottle	0–64	
Wort at pitching	approx. 69	Kleber and Lampl (1956)
Fermentation gases	51–900	
Finished beer	approx. 48	
Ale	30	Maw (1965)
Lager	13	

consider that, while 5 µg H_2S/litre modifies beer aroma but not flavour, more than 100 µg/litre is required before an odour of hydrogen sulphide is detectable in beer. Table XIII shows some values for the hydrogen sulphide content of (American) beers.

The concentrations of hydrogen sulphide in beers are, then, very small. Much of that which is generated during fermentation is purged from the liquid by the action of the carbon dioxide evolved, as is evident from the high hydrogen sulphide content reported for fermentation gases. There are reports that top yeasts produce more than bottom yeasts (Macher, 1953; Maw, 1965). During fermentation production seems to reach a maximum while fermentation is vigorous (Ricketts and Coutts, 1951; Brenner *et al.*, 1953; Kleber and Lampl, 1956) and relatively intense evolution of hydrogen sulphide is considered to be a sign of strongly fermenting yeasts by Macher (1953), who found that the sulphate ion was reduced only with difficulty, requiring a low redox potential and low pH value, while thiosulphate was readily reduced to hydrogen sulphide and mercaptans.

The reduction of sulphite to hydrogen sulphide during fermentation was observed by Ricketts and Coutts (1951), who showed the enzymic nature of the process by demonstrating its inhibition with iodoacetate. In fact, hydrogen sulphide formation during normal fermentations was inhibited by a number of enzyme poisons, including sulphydryl inhibitors, sodium fluoride, azide and 2,4-dinitrophenol, but aldehyde and ketone reagents were not inhibitors.

Masschelein *et al.* (1961) commented on the contribution that hydrogen sulphide in wort might make to the content in beer. Cold aeration of the wort, followed by filtration at 0°, reduced the concentration of hydrogen sulphide in wort, and thereby had an important beneficial effect on the rate of maturation of the beer. They also noted that: (i) flocculent yeasts leave less hydrogen sulphide in beer than do powdery yeasts; (ii) prompt removal of yeast at the end of fermentation leads to lower hydrogen sulphide levels in the beer; and (iii) the sulphur dioxide content of beer increases during fermentation.

According to Stewart *et al.* (1962), many yeasts form hydrogen sulphide in artificial media deficient in pantothenate: 50–100 µg of pantothenic acid/litre, or 500 µg of β-alanine/litre, was sufficient effectively to depress the rate and amount of hydrogen sulphide production. However, these studies do not reflect hydrogen sulphide formation in fermenting worts, in which pantothenate concentrations are much higher (Hopkins *et al.*, 1948). A possible explanation of the observations of Stewart *et al.* is that, under conditions of pantothenate deficiency, hydrogen sulphide represents a metabolic end-product derived from a sulphur-containing moiety (e.g. mercaptoethanolamine) of CoA synthesis. A yeast able to synthesize such an intermediate, but unable to use it to complete the CoA molecule because of lack of the pantothenate moiety, might catabolize it with release of hydrogen sulphide.

2. Sulphydryl and Mercaptans

In 43 American beers, Brenner et al. (1954) found up to 70 μg or more of mercaptan/litre, but 80% of the beers contained 30 μg/litre or less. On exposing the beer to light in white or green bottles, this content increased by 30 μg/litre and also acquired the distinctive unpleasant aroma and flavour of "sun-struck" beers. Table XIV gives values for the total sulphydryl content of American worts and beer.

TABLE XIV. *Total Sulphydryl Contents of Worts and Beers. From Brenner et al. (1957, 1958)*

Material	Total -SH (μg/ml)		Remarks
	Range	Mean	
Laboratory malt worts	0·2–1·1	0·56	Represents about 1·5% of total soluble organic S of all-malt worts
Copper worts	0·26–1·66	0·77 ⎫	Pilot brews
Cold worts	0·16–1·60	0·75 ⎭	
74 Commercial beers and 6 ales	0·41–2·99	1·21	No marked differences between beers and ales
Beer	⎧ 0·87	—	Not exposed to light
	⎨ 1·06	—	Exposed to light for 1 day
	⎩ 1·28	—	Exposed to light for 13 days

The work of Brenner et al. (1958) also showed that the deterioration of the non-biological (protein-haze) stability of beers stored at room temperature over a period of 30 days was associated with a decrease in the total sulphydryl content of the beer, the magnitude of the decrease being the important factor, and not the initial sulphydryl content.

The volatile mercaptans of beer may comprise only about one-hundredth of the total beer sulphydryl, which, in turn, is only about one-fiftieth of the total organic sulphur of beer (Brenner et al., 1958). The volatile mercaptan content of lagers is stated by Maw (1965) to average 25 μg/litre. In commercial brewing, the mercaptan level fell during fermentation and increased during the post-fermentation storage period (Brenner et al., 1955).

Apart from their direct contribution to the aroma and flavour of beer, mercaptans produced photo-chemically from sulphur compounds normally present in beer have been implicated in the development of the "sun-struck" flavour of beer exposed to light in white or green bottles.

A critical note on the subject of mercaptans in beer is struck by Hashimoto and Kuroiwa (1966), who found no mercaptans in beer and considered that they were artefacts formed from S-amino acids or proteins by heating before or during distillation in the methods used for mercaptan determination. They produced gas chromatographic evidence of the presence in fermenting wort of three new volatiles, which they tentatively identified as thioformaldehyde, dithioformaldehyde and thioacetone.

The presence in beer of sugar mercaptals has been reported by Knorr (1965), but their origin and effects on aroma and flavour are not yet known.

3. *Sulphur Dioxide*

While the sulphur dioxide content of some beers may consist mainly of sulphur dioxide added as preservative, especially to unpasteurized draught beers in Great Britain, some sulphur dioxide is formed biochemically, presumably by reduction of sulphate ion, during fermentation. The levels thus attained are usually about 10–30 μg/ml, i.e. concentrations which are lower than can be detected by flavour tests (Brenner et al., 1955).

V. Biological Composition of Brewer's Yeasts

Many brewers pitch their fermentations with cultures of yeast which have passed from brew to brew for years, or even decades, part of the yeast grown in one brew being used to inoculate a subsequent brew, and so on. Under these circumstances, the pitching yeast becomes contaminated with micro-organisms derived from the air, from brewery vessels, plant, equipment or mains, from drops of condensate or from other sources, so that, especially in British brewing, the pitching yeast is not a pure culture. Even in those breweries where pure culture yeasts are grown up for pitching, a degree of contamination occurs as the yeast is transferred from brew to brew. Consequently, it becomes advisable to discard the yeast after a few generations (frequently fewer than ten) and replace it with fresh yeast grown up from a selected single-cell strain.

It may be interjected here that microbial contamination may not be the only reason for periodic replacement of yeast in a pure culture brewery: it sometimes happens that yeast properties change adversely with respect to fermentation behaviour or flocculence. Brewers often refer to such a state of affairs as "yeast weakness", a term which is usually applied to almost any adverse change in yeast properties. In breweries, pure cultures therefore do not remain uncontaminated for long, and the cultures in those breweries which have maintained a yeast for a prolonged period without changing it tend to carry a fairly constant proportion of microbial contaminants, the balance of which may become disturbed from time to time and lead to difficulties of microbiological stability of the beer.

The nature of these contaminants is circumscribed by their environment; namely that of fermenting brewer's wort and beer. Microorganisms which fail to grow in these media are unlikely to appear in the yeast crops formed during fermentation, which provide the inocula for further fermentations. Because of the chemical and physico-chemical properties of beer, the contaminant flora of brewer's yeasts and beers is thus restricted to a comparatively few species. Beer in particular is a medium ill suited to bacterial growth because: (i) it is impoverished with respect to bacterial nutrients of all classes, substantial quantities of which are removed by yeast during primary fermentation; (ii) as a result of primary fermentation, beer contains bacterial inhibitors (ethanol, fusel alcohols, etc.) and acquires low values of pH and redox potential; and (iii) it contains hop antiseptics, especially the isohumulones, which are active against gram-positive bacteria.

Typically the contaminant microflora of brewer's yeasts consists of wild yeasts, acetic acid bacteria, lactic acid bacteria and *Flavobacterium proteus*. Less frequently, *Zymomonas anaerobia* may be found.

a. Acetic acid bacteria. Certain species from the genera *Acetobacter* and *Acetomonas* are beer spoilage organisms (Rainbow, 1966b): under aerobic conditions they oxidize dilute solutions of ethanol to acetic acid. During primary fermentation, while anaerobic conditions prevail, their activity is probably negligible; but, as fermentation subsides and there is opportunity for air to gain access to the beer, they may commence to acetify the alcohol of the beer. The danger of acetic spoilage of beer is greatest in traditional cask draught beer, after air has entered the cask to replace beer which has been drawn off. In addition to spoilage by acetification, some strains of acetic acid bacteria, by virtue of their ability to secrete capsular polysaccharide (dextran) material, spoil beer by turning it "ropy". Today, beer spoilage by acetic acid bacteria is less common than of old: not only are high standards of hygiene maintained in modern breweries, but the handling of beer at all stages is conducted under conditions in which access of air is prevented or minimized. In addition, new systems of processing (including pasteurization and sterile filtration) and dispensing enable a high degree of cleanliness to be maintained, and access to air is prevented.

b. Lactic acid bacteria. Brewer's yeast may harbour members of two genera of lactic acid bacteria (Shimwell, 1949): (i) strains of hetero-fermentative rods (e.g. *Lactobacillus pastorianus*, *L. brevis*); and (ii) strains of lactic cocci belonging to the genus *Pediococcus* (or, as some prefer to consider it, *Streptococcus*). These gram-positive strains may be passed into beer from the pitching yeast and cause spoilage by turbidity, off-flavours, acid formation and (for some capsule-forming pediococci) ropiness. The beer-spoilage lactic acid bacteria are serious contaminants, since they are anaerobic or micro-aerophilic, and they have acquired a degree of tolerance to hop antiseptics, which are usually effective against gram-positive bacteria.

c. *Flavobacterium proteus.* This is the "short fat rod" of brewer's yeast, first systematically described by Shimwell and Grimes (1936). Later, Shimwell (1964) proposed placing it in a new genus as *Obesumbacterium proteus.* It is commonly present in pitching yeast, in which its relative numbers appear to remain fairly constant with time. It is not a beer-spoilage organism, because it fails to grow as soon as the fermenting wort attains a pH value lower than about 4·3.

d. *Zymomonas anaerobia* (Shimwell, 1950). If this gram-negative anaerobic rod gains access to beer containing glucose or fructose, it can create serious spoilage characterized by a rapid development of cloudi-ness, and a stench reminiscent of hydrogen sulphide and rotten apples. Fortunately, the organism is of comparatively rare occurrence in beers.

e. *Wild yeasts.* Strains of yeast belonging to species other than *Sacch. cerevisiae* or *Sacch. carlsbergensis* are obviously contaminants. However, it is less apparent that strains of these species themselves may be "wild" in the brewing context. Thus, *Sacch. carlsbergensis* (bottom yeast) and *Sacch. cerevisiae* var. *ellipsoideus* (wine yeasts) are not infrequently beer spoilage contaminants harboured by top fermentation pitching yeasts. Conversely, *Sacch. cerevisiae* is equally undesirable in bottom fermenta-tion breweries.

Even less apparent is the fact that, in a brewery using a strain with particular fermentation and flocculation properties, a strain of the same species, but having properties differing appreciably from those of the culture yeast, will be a "wild" yeast and may well create difficulties of fermentation, yeast separation, clarification and fining of the beer, and even uncharacteristic flavours and aromas.

f. *Detection of wild yeasts.* From the foregoing it is apparent that the detection of wild yeasts in the body of the culture yeast is an important item of brewing laboratory control. Such detection is not an easy matter because the analyst is usually seeking a few contaminant cells in the presence of a large number of morphologically similar culture cells, so that the microscope is of little assistance.

Use has been made of the fact that some wild yeasts utilize L-lysine as

8—Y. 3

sole source of nitrogen (Walters and Thiselton, 1953), whereas *Sacch. cerevisiae* and *Sacch. carlsbergensis* do not. However, this test has limitations in that, as already indicated, strains of *Sacch. cerevisiae* and *Sacch. carlsbergensis* are commonly "wild" in the brewery context. Further, the test may well indicate the presence of species of *Kloeckera* and other yeasts, many of which fail to grow in beer and are therefore of no interest as beer spoilage organisms.

Until recently, the problem of identifying and (what is equally important) enumerating small numbers of wild species of *Saccharomyces* in brewing pitching yeasts was most difficult, often too time-consuming to be of value for control purposes, or even impossible. However, advances towards the solution of this important brewing control problem have now resulted from applications of an immuno-fluorescent technique of antigenic analysis.

Antigenic analysis of the cell wall structure of *Sacch. cerevisiae* by Campbell and Allan (1964) revealed two distinct antigenic types among 19 strains. These types correspond to recognized morphological types; namely, *Sacch. cerevisiae* (rounded or slightly oval cells) and *Sacch. cerevisiae* var. *ellipsoideus* (elongated cells). In these experiments, serum was prepared from two strains of *Sacch. cerevisiae* and tested in agglutination tests against the other strains. The tests were repeated after absorption of the sera: (i) by other individual strains of *Sacch. cerevisiae*; and (ii) individually by two strains of *Sacch. cerevisiae* var. *ellipsoideus*. Absorption of the serum by *Sacch. cerevisiae* removed all anti-yeast agglutinins, but absorption by either of the var. *ellipsoideus* strains, while removing agglutinins to itself, left agglutinins to all *Sacch. cerevisiae* strains. The tests revealed three distinct antigens on all 12 strains of *Sacch. cerevisiae*, but only one was carried by the seven var. *ellipsoideus* strains. *Saccharomyces cerevisiae* was also antigenically sufficiently different from ten other *Saccharomyces* species examined to believe that the agglutination tests might provide a method of yeast identification.

The agglutination test was next extended to include species of *Sacch. carlsbergensis* (Campbell and Brudzynski, 1966). Serum agglutinating three out of 12 strains of *Sacch. carlsbergensis*, but not organisms of ten other *Saccharomyces* species, was obtained. Thereby, two groups of *Sacch. carlsbergensis* were differentiated serologically, although the groups were not distinguished by morphological and physiological properties. The nine strains of *Sacch. carlsbergensis* not agglutinated by the serum were serologically identical with *Sacch. cerevisiae* var. *ellipsoideus* and *Sacch. diastaticus*, from which they were distinguishable by fermentation tests with melibiose and starch. The results suggested that three serologically distinct groups of brewing yeasts are recognizable: (a) *Sacch. cerevisiae*; (b) *the Sacch. ellipsoideus-diastaticus* group, including some *Sacch. carlsbergensis* strains; and (c) other *Sacch. carlsbergensis* strains. Of these, Groups a and c are serologically unique, while Group b shares the antigenic structure of various species of *Saccharomyces*.

Although Campbell and his co-workers have thus shown how to distinguish *Sacch. cerevisiae*, *Sacch. cerevisiae* var. *ellipsoideus* and certain strains of *Sacch. carlsbergensis*, the quantitative problem of detecting very small numbers of cells of strains in the presence of large numbers of cells of a culture strain remains considerable. What the ale brewer wishes to know is how many cells of *Sacch. carlsbergensis* or *Sacch. cerevisiae* var. *ellipsoideus* there are in every million cells of his culture *Sacch. cerevisiae*. Conversely, the lager brewer wants to know the extent of contamination of his *Sacch. carlsbergensis* with *Sacch. cerevisiae* and var. *ellipsoideus*. The problem is not only one of detection, but also one of enumeration of a few in the presence of the very many.

Steps towards the resolution of this problem have been taken by Richards and Cowland (1967). They prepared antisera against *Sacch. cerevisiae* var. *ellipsoideus* and other wild yeasts and removed antibodies reacting with ale pitching yeast by repeated absorption with washed cells of the pitching yeast to obtain a specific rabbit anti-yeast serum (treated). This was used to prepare a goat antiserum, from which γ-globulin (containing the active antibody) was isolated and conjugated to fluorescein isothiocyanate to yield a fluorescent antibody. Thus, on treating washed cells of pitching yeast under appropriate conditions, first with anti-yeast serum (treated) specific for *Sacch. cerevisiae* var. *ellipsoideus* and then with fluorescein-conjugated goat antiserum reactive towards the anti-yeast, cells of *Sacch. cerevisiae* var. *ellipsoideus* will bind the specific fluorescent antiserum. On microscopic examination under ultraviolet illumination, such cells are revealed by their fluorescence against a background of non-fluorescent culture cells. Using a combination of two sera, all brewery contaminants of the genus *Saccharomyces* could be detected, provided: (i) the brewery strain used for cross-absorption was carefully selected; (ii) careful control was exercised over the cross-absorption itself; and (iii) the staining procedure was performed according to precisely defined details.

VI. Continuous Brewery Fermentations

To the brewer, the most important advantages of continuous fermentation are that present production can be achieved with much smaller plant, uniform quality of product can be maintained with lower labour costs, and the early slow stages of batch fermentation are eliminated so that the most rapid rate of batch working can be continually maintained or even exceeded (Hough and Rudin, 1959b).

A. SYSTEMS OF CONTINUOUS FERMENTATION

There are two basic types of system of continuous culture of microorganisms (Emeis, 1965):

(i) Turbidostatic systems, in which the cell concentration is kept constant by varying the flow-rate, which is controlled by a photoelectric device. The cell population may be maintained in the exponential phase of growth. These systems are particularly suitable for metabolic studies, and operate under physiological conditions quite different from those in chemostatic systems.

(ii) Chemostatic systems, in which the growth rate is regulated by the dilution rate and is kept well below the maximum value by maintaining concentrations of growth-limiting factors. The medium is introduced at a constant rate into the culture vessel and the cell population is maintained in the retardation phase of growth which precedes the stationary phase.

Continuous fermentation systems of commercial interest to brewers are chemostatic systems, forms of which will be briefly discussed in what follows. The efficiency of these systems is usually measured as the flow-rate of wort in unit volumes per day per unit volume capacity of the apparatus (Hough and Rudin, 1958); or by the number of times per day the vessel contents are completely changed (Hough, 1959). Normally this is about 0·1 times/d for conventional batch fermentations.

1. *Overflow Systems*

These are "open" continuous fermentations in one stirred and temperature-controlled vessel, or in a series of such vessels, into which wort is pumped at one end while the beer overflows at the other. Using such a system on a laboratory scale, Hough and Rudin (1958) found that a two-vessel was more efficient than a single-vessel system, but that a three-vessel system was no more advantageous. Within limits, Hough and Rudin (1959a) observed a simple relation between the rate of flow, the amount of yeast in suspension and the extent to which the wort was fermented. The flow-rate ensuring delivery of a beer at a predetermined gravity depended on yeast strain, temperature, wort quality, the number of fermenting vessels and the total volume of fermenting wort; and it could be determined by a few empirical measurements. For beers with attenuations 65–85% of that of the wort gravity, under certain conditions, the rate of wort inflow was linearly related to: (i) gravity of the effluent beer; (ii) time taken to achieve that gravity in the corresponding batch fermentation; (iii) temperature, in the range 15–25°; and (iv) for many yeast strains, original gravity between 1·030 and 1·100.

In these systems, yeast was maintained dispersed in wort by agitation, while fresh wort was introduced at the same rate as beer and suspended yeast were discharged, the concentration of yeast in the effluent beer being the same as that within the stirred vessel. In some systems, this yeast is returned to the system, so as to increase the yeast concentration in the fermenting vessel and so increase the rate of fermentation. Taking these points into consideration, Hough and Ricketts (1960) devised a V-tube apparatus which exploits the sedimenting properties of flocculent yeast to retain the yeast within the vessel, so that problems of yeast separation and recycling were minimized or eliminated and yeast accumulated within the fermenting vessel, giving high fermentation rates and low rates of yeast reproduction. The apparatus consists of a stirred vertical tube connected at its base to an unstirred tube set to it at an angle of 25 degrees of arc. The vertical tube is charged with pure flocculent yeast, and wort is pumped into the top. Here, the yeast is kept in suspension by stirring, but it settles to the bottom of the unstirred inclined tube, while clear beer flows off from a point near the top.

2. *The Tower System*

This system was described by Klopper *et al.* (1965). The tower consists essentially of a vertical tube charged with a high concentration of yeast. Wort enters at the bottom, passes through the plug of yeast, and beer overflows at the top, after passing through a section designed to disengage gas and separate yeast, which latter falls back into the tower. Because of the high yeast concentration maintained in the tower, fermentation rates are very high and, with selected yeast (see below), ale and lager worts have been fermented satisfactorily in four and eight hours respectively.

Details of the tower and its operation are given by Royston (1966a, b), who describes it as a heterogeneous, virtually "closed" fermenter, in which little yeast growth occurs. With normal brewery wort, this leads to the production of beer with a higher nitrogen content than usual, but a normal content can be obtained by using worts containing a low concentration of nitrogen-containing compounds. However, more recent experience (R. G. Ault, private communication) suggests that it may be desirable to operate the tower with a degree of aeration which, in so far as it promotes more yeast reproduction, can yield beers with normal nitrogen contents.

3. *Other Continuous Systems*

Two new laboratory-scale systems have been described by Portno (1967).

a. "Gradient tube" fermenter. This is designed to operate as a nearly perfect heterogeneous "open" system, incorporating a true fermentation gradient, which reproduces, under continuous conditions, the sequential changes of a batch fermentation. It is essentially a long narrow (0·25-in. bore) temperature-controlled tube, into which controlled proportions of pasteurized wort and yeast are fed, so that the fermentation proceeds as the mixture passes along the tube. A mixture of beer, gas and yeast is ultimately discharged into a separating vessel, from which settled yeast is recirculated to enable efficiencies greater than those usual for open systems to be attained.

b. Centrifugal continuous fermentation system. In closed fermenters, such as the tower, control of yeast concentration depends on the opposing factors of rate of upwards movement of liquid and the sedimentation characteristics of the yeast. In addition, there is a hazard that high concentrations of flocculent yeast may form cohesive plugs, which ferment inefficiently because of the limited surface of yeast exposed to the wort, and which may be expelled from the fermenter on a cushion of fermentation gas.

Portno's centrifugal fermenter attempts to overcome these difficulties and enables greater control to be exerted over the escape of yeast. It consists of a single stirred vessel from which beer escapes through passages radiating from a central spinning rotor. In escaping, yeast suspended in the beer is partially or entirely removed according to the speed of rotation of the rotor and its length of radius. The apparatus enables less flocculent yeasts to be used than would be acceptable for tower operation. It operates as a homogeneous fermenter in which, under steady conditions, the yeast remains in a constant metabolic state in a medium almost devoid of usable nutrients. In operating the system, an excessive degree of yeast retention (i.e. highly closed conditions) was reported to cause a continuous fall in rate of yeast growth, with concomitant cessation of fermentation and increase in α-amino nitrogen of the resultant beer. It therefore seems advisable to permit a regulated rate of yeast growth with this system, just as it may be with the tower.

B. SELECTION, BEHAVIOUR AND EFFICIENCY OF YEASTS

For all continuous systems, the primary considerations of yeast selection must be that it produces a palatable beer and that it ferments vigorously in the conditions of the chosen system. For practical reasons, for their stirred vessels Hough (1959) and Hough and Rudin (1959b) advise selection of a yeast which is not strongly disposed to head formation nor unusually flocculent, otherwise there is a tendency for yeast to accumulate and the maintenance of equilibrium conditions becomes difficult, necessitating frequent readjustments to increase the rate of wort inflow to maintain a constant degree of attenuation.

Rudin and Hough (1959) successfully employed 12 yeasts, including top and bottom yeasts isolated from commercial sources, for continuous fermentations in the apparatus of Hough and Rudin (1958). There were no substantial differences in behaviour between the top and bottom yeasts, and the quantity of yeast harvested was approximately the same.

The efficiency of some yeasts was independent of the number of vessels, but others were less efficient in single-vessel operation. In these experiments (Rudin and Hough, 1959) conditions in the first vessel of multi-vessel systems favoured yeast activity more than did those in the final vessel, where concentrations of metabolites were low and those of end-products high. When mixtures of strains were employed, one strain became dominant in some cases: while in others, even though the original proportions were maintained, the efficiency of the mixture was usually less than the average value for its individual components.

The effect of contaminants on efficiency depends on the particular contaminant, but efficiency is usually impaired (Hough, 1959; Hough and Rudin, 1959b; Rudin and Hough, 1959).

In tower operation, selection of a flocculent yeast is essential, and the characteristics of the fermentation depend *inter alia* on the equilibrium reached between the sedimentation of the yeast, the upward flow of wort (Klopper *et al.*, 1965; Royston, 1966a), the tendency of all cells to deflocculate in contact with unfermented wort, and the effect of fermentation gas on floc size (Royston, 1966b).

C. EFFECT OF AERATION

Ricketts and Hough (1961) found that, in a single-vessel system of continuous fermentation (Hough and Rudin, 1958), the rate of beer production was low under virtually anaerobic conditions. Increasing the aeration rate to 30 ml air/h caused a marked rise in yeast concentration and rate of beer production. Thereafter, little change was apparent with air-flows up to 500 ml/h, above which the yeast concentration and rate of beer production again increased to constant values attained at upwards of 25 litres/h. At the low aeration rates, the ethanol content of the beer resembled that of conventional fermentations, but at high rates the beer was low in ethanol. At aeration rates exceeding 500 ml/h, a sequence of regions was observed in which: (i) yeast concentration rose stepwise and the rate of beer production remained unaltered; (ii) at yet higher rates, the rate of beer production rose stepwise, while yeast concentration remained constant. These observations indicate that aeration had independent effects on rate of fermentation and on cell growth. Ricketts and Hough conclude that it is desirable to permit low rates of aeration in the system in view of the faster rate of beer production, for which the air dissolved in wort is insufficient.

Operating the V-shaped fermenter like that described by Hough and Ricketts (1960), Enari and Makinen (1961) concluded that oxygen was the factor limiting growth, and wort carbohydrate that limiting fermentation. When the same type of apparatus was operated with especially high concentrations of *Sacch. carlsbergensis*, restriction of air lowered the rate of beer production: even so, the efficiency of the system at high yeast concentrations under completely anaerobic conditions was greater than that of a system in which yeast escaped freely with the beer (Harris and Merritt, 1962b).

Cowland and Maule (1966) studied the effect of aeration on the growth and metabolism of *Sacch. cerevisiae* fermenting a complex synthetic medium (glucose-amino acids-salts-vitamins) in a single-stage open stirred continuous fermenter. Growth took place in anaerobic conditions, a steady state being maintained for 60 generations. At a low rate of aeration (oxygen tension = 0·2 mmHg) growth, glucose uptake and ethanol production reached maxima, but anaerobic metabolism continued even at the highest level of aeration studied (oxygen tension = 126 mmHg). However, even at a very low oxygen tension (0·05 mmHg), the Pasteur effect, as revealed by expressing the glucose uptake and ethanol production per unit of dry yeast present (i.e. on a specific basis), was pronounced. At this low oxygen tension ester formation was much lower than under anaerobic conditions. At higher levels of aeration, growth and rate of fermentation were reduced, ester formation was further depressed and acetoin and acetaldehyde were produced, although diacetyl formation never became significant. Fusel alcohol yields were depressed (to a rate one-half of that obtained under anaerobic conditions) only at high oxygen tensions and its production was more closely related to glucose uptake than to that of the related amino acids, which were completely absorbed from the medium at all the experimental rates of aeration.

This work was followed by a similar study of the effects of aeration during fermentations of hopped worts in the same system (Cowland, 1967). Oxygen continuously supplied to the ingoing wort stimulated the growth rate, so that yeast concentration and consumption of sugar and production of ethanol per unit weight of yeast increased markedly, without decreasing the proportion of sugar fermented to ethanol. Improved brewing performance therefore resulted, because ethanol formation was accelerated by the increase in cell population and the higher rate of yeast metabolism. The stimulation of growth also resulted in the uptake of wort nitrogen compounds more complex than the readily-assimilable amino acids which supply the needs of the yeast in the absence of air. By contrast, the specific rate of production of esters and fusel alcohols was not greatly affected by continuously aerating the ingoing wort.

When Cowland further increased the level of aeration by injecting air into the fermenter itself, growth was stimulated still more, but the specific rate of sugar utilization fell (i.e. the Pasteur effect operated), as did ester production, while fusel production underwent no further significant pattern of change. Additionally, considerable quantities of acetaldehyde and acetoin, neither of which were detected during fermentation under anaerobic conditions or under low oxygen tensions (0·05 mmHg), were found.

There is little published information on the effects of aeration on continuous fermentations in tower fermenters, but it appears that some aeration stimulates the growth of yeast in the tower, markedly improving the viability of the yeast, while depressing the ester content of the beer to levels nearer to those of conventionally brewed beers (R. G. Ault, private communication).

D. OTHER BIOCHEMICAL ASPECTS OF CONTINUOUS FERMENTATION

Reviewing biochemical aspects of continuous fermentation, Sandegren and Enebo (1961) report that: (i) both continuously-fermented and batch beers contained similar amounts of volatile acids, but the non-volatile acid content of continuous beers and their pH values were lower; (ii) beer made by the continuous process contained more acetaldehyde than batch beer and so tended to have a "green" flavour, and additionally, there might have been a tendency for the aldehyde to be converted to acetoin, although the oxidation of the latter to diacetyl was not likely in the reducing environment of continuous fermentation; (iii) the amounts of fusel alcohols and β-phenylethanol were higher in continuous beers; (iv) the relatively high ethanol content maintained throughout continuous fermentation tended towards beers with higher ester contents; (v) there was a possibility that beer flavour might be impaired by yeast excretion products, especially when yeast was recycled in continuous processes.

However, not all these points are necessarily valid. Thus, it has already been pointed out that the ester, fusel alcohol, acetaldehyde and acetoin content of beers produced by continuous processes is influenced greatly by aeration (Cowland and Maule, 1966; Cowland, 1967), so that the degree of aeration offers considerable scope in controlling levels of these flavour substances in continous beers.

The carbohydrates, total nitrogen and α-amino nitrogen of batch beers and beers brewed in their single-vessel continuous system were found by Hough and Rudin (1958) to be similar. Using the same system, Harris and Merritt (1961) reported that the pattern of carbohydrate and amino acid uptake by *Sacch. carlsbergensis* was essentially similar to that of stirred comparison batch fermentations. In the same apparatus, additions to wort of single amino acids (of which asparagine was the most effective), mixtures of amino acids, yeast autolysate or ammonia caused considerable increases in yeast growth and beer attenuation, but supplementing the wort with vitamins of the B-group complex, or with fermentable sugars did not (Harris and Merritt, 1962a, c).

Watson and Hough (1966) compared the nitrogen metabolism of *Sacch. cerevisiae* fermenting in two continuous systems: (i) an open system, consisting of a stirred vessel with free overflow of yeast with beer, in which no effort was made to build up high yeast concentrations; and (ii) a closed system, in which yeast was recycled, or trapped in the vessel by weirs or baffles.

Fermentations were performed in wort, in buffered synthetic medium and in synthetic medium in which the ammonium sulphate was replaced by casein hydrolysate + proline ("semi-synthetic medium"). The level of nitrogen in the synthetic medium severely restricted growth in the open system, but much higher yeast concentrations could be maintained using similar flow-rates in the closed system. The levels of ethanol, fusel alcohols (chiefly *n*-propanol, isobutanol and isopentanol), esters and cell carbohydrate were all two to five times greater in the closed system. Neither glucose nor ammonia was limiting in the open system, but either substance might limit growth in the closed system. In wort or the semi-synthetic medium, the level of assimilable nitrogen in the beer was greater in the closed system, and some nitrogen, particularly proline, was not utilized. Proline uptake was increased either by increasing the carbohydrate content of the medium, or by aerating to divert more carbohydrate from fermentation to growth. This observation recalls that of Cowland (1967), already mentioned, on the effect of aeration in extending the types of nitrogen compounds taken up by yeast in batch fermentation.

VII. Concluding Remarks

Lest it should appear that too much emphasis has been placed in this article on minor aspects of the metabolism of brewer's yeasts, a few words of explanation may not be amiss.

Brewer's yeasts are the vital agents in the production of an internationally popular drink, and the acceptability of beers depends largely on the relative amounts of yeast metabolic products formed only in small concentrations. Hence, in a contribution devoted to brewer's yeasts, the author has treated these matters at considerable length, while other quantitatively major considerations, such as those of the formation of ethanol and carbon dioxide, have been taken for granted as having received excellent treatment in biochemical, microbiological and technical texts on many occasions already.

The importance of the quantitatively minor yeast metabolic products, which may make or mar a beer, continues to stimulate workers to study their formation ever more closely. Such studies will benefit science in general and the technical brewer in particular. It will enable the academic scientist to fill in additional detail towards building up an integrated picture of cellular activity, which is a primary aim of biochemists. It will enable the technical brewer to exert greater control over beer flavour—something which he can do only inadequately at the moment by trying to repeat his brewing process precisely in each successive brew and then testing his success by submitting his product to subjective organoleptic evaluation. New knowledge of the minor components of beer can only serve to render more objective the assessment of beer character and flavour, and to render their formation more capable of effective control.

In batch fermentations, each phase of growth and fermentation has a characteristic metabolism: further studies of these phases will continue to illuminate temporal aspects of metabolism. Equally interesting, both academically and industrially, is metabolism at the instant of time which can be regarded as "frozen" in the steady-state conditions of some continuous fermentations. With the advent of continuous beer fermentations, the pressure and opportunity to study yeast metabolism under these new conditions is increasing, so that the continuous brewery of the future should be a splendid field for co-operation between academic and industrial scientists.

Acknowledgements

The author sincerely thanks those authors and publishers who have kindly allowed him to use their material, and the Directors of Bass Charrington Ltd. for permission to present this article.

References

Alvarez, A., Vanderwinkel, E. and Wiame, J. M. (1958). *Biochim. biophys. Acta* **28**, 333–341.

Anh, T. N. (1965). *Brasserie* **20**, 338–352.

Atkin, L., Schultz, A. S. and Frey, C. N. (1943). *Industr. Engng Chem. (Anal. ed.)* **15**, 141–144.

Äyräpää, T. (1961). *J. Inst. Brew.* **67**, 262–266.

Äyräpää, T. (1962). *J. Inst. Brew.* **68**, 504–508.

Äyräpää, T. (1963). *Proc. Eur. Brew. Conv., Brussels* pp. 276–287.

Äyräpää, T. (1965). *J. Inst. Brew.* **71**, 341–347.

Äyräpää, T., Holmberg, J. and Selmann-Persson, G. (1961). *Proc. Eur. Brew. Conv., Vienna* pp. 286–297.

Barton-Wright, E. C. (1951). *J. Inst. Brew.* **57**, 415–426.

Bavisotto, V. S., Shovers, J., Sandine, W. E. and Elliker, P. R. (1964). *Proc. A.M. Amer. Soc. Brew. Chem.* pp. 211–216.

Bolton, A. A. and Eddy, A. A. (1962). *Biochem. J.* **82**, 169–179.

Brenner, M. W., Blick, S. R., Frenkel, G. and Siebenberg, J. (1963). *Proc. Eur. Brew. Conv., Brussels* pp. 233–246.

Brenner, M. W., Jakob, G. and Owades, J. L. (1957). *J. Inst. Brew.* **63**, 408–414.

Brenner, M. W., Owades, J. L. and Golyzniak, R. (1953). *Proc. A.M. Amer. Soc. Brew. Chem.* pp. 83–98.

Brenner, M. W., Owades, J. L., Gutcho, M. and Golyzniak, R. (1954). *Proc. A.M. Amer. Soc. Brew. Chem.* pp. 88–97.

Brenner, M. W., Owades, J. L. and Fagio, T. (1955). *Proc. A.M. Amer. Soc. Brew. Chem.* pp. 125–132, 133–144.

Brenner, M. W., Schapiro, G. J. and Owades, J. L. (1958). *Proc. A.M. Amer. Soc. Brew. Chem.* pp. 104–111.

Briggs, H. E. and Haldane, J. B. S. (1925). *Biochem. J.* **19**, 388–389.

Burger, M., Glenister, P. R. and Becker, K. (1957). *Proc. A.M. Amer. Soc. Brew. Chem.* pp. 110–115.

Campbell, I. and Allan, A. M. (1964). *J. Inst. Brew.* **70**, 316–320.

Campbell, I. and Brudzynski, A. (1966). *J. Inst. Brew.* **72**, 556–560.

Clarke, B. J., Harold, F. V., Hildebrand, R. P. and Morieson, A. S. (1962). *J. Inst. Brew.* **68**, 179–187.

Comrie, A. A. D. (1952). *Wallerstein Labs Commun.* **15**, 339–345.

Cowland, T. W. (1967). *J. Inst. Brew.* **73**, 542–551.

Cowland, T. W. and Maule, D. R. (1966). *J. Inst. Brew.* **72**, 480–488.

Cutts, N. S. and Rainbow, C. (1950). *J. gen. Microbiol.* **4**, 150–155.

Czarnecki, H. T. and van Engel, E. L. (1959). *Brewers' Digest* pp. 52–56.

Damlé, W. R. and Thorne, R. S. W. (1949). *J. Inst. Brew.* **55**, 13–18.

Delisle, A. L. and Phaff, H. J. (1961). *Proc. A.M. Amer. Soc. Brew. Chem.* pp. 103–118.

Dirscherl, W. and Höfermann, H. (1954). *Biochem. Z.* **322**, 237–244.

Dixon, B. and Rose, A. H. (1964). *J. gen. Microbiol.* **35**, 411–419.

Drews, B., Specht, H. and Bärwald, G. (1964). *Mschr. Brau.* **17**, 101–116.

Drews, B., Specht, H., Ölscher, H. J. and Thürauf, F.-M. (1962). *Mschr. Brau.* **15**, 109–113.

Eastcott, E. V. (1928). *J. phys. Chem.* **32**, 1094–1111.

Eddy, A. A. (1955a). *Proc. Eur. Brew. Conv., Baden-Baden* pp. 65–68.

Eddy, A. A. (1955b). *J. Inst. Brew.* **61**, 318–320.

Eddy, A. A. (1958). *J. Inst. Brew.* **64**, 368.

Eddy, A. A. and Rudin, A. D. (1958a). *J. Inst. Brew.* **64**, 19–21.

Eddy, A. A. and Rudin, A. D. (1958b). *J. Inst. Brew.* **64**, 139–142.

Ehrlich, F. (1906). *Ber. dt. chem. Ges.* **39**, 4072–4075.

Ehrlich, F. (1907a). *Ber. dt. chem. Ges.* **40**, 1027–1047.
Ehrlich, F. (1907b). *Ber. dt. chem. Ges.* **40**, 2538–2562.
Ehrlich, F. (1911). *Ber. dt. chem. Ges.* **44**, 139–146.
Ehrlich, F. (1912). *Ber. dt. chem. Ges.* **45**, 883–889.
Ekström, D. and Sandegren, E. (1951). *Proc. Eur. Brew. Conv.*, Brighton pp. 178–186.
Emeis, C. C. (1965). *Mschr. Brau.* **18**, 224–228.
Enari, T.-M. and Makinen, V. (1961). *Brauwissenschaft* **14**, 253–256.
Enebo, L. (1957). *Proc. Eur. Brew. Conv.*, Copenhagen pp. 370–376.
Gilliland, R. B. (1951). *Proc. Eur. Brew. Conv.*, Brighton pp. 35–37.
Gilliland, R. B. (1955). *Brewers' Guild J.* **41**, 246–270.
Gray, P. P. and Stone, I. M. (1939). *J. Inst. Brew.* **45**, 253–263.
Harden, A. (1932). "Alcoholic Fermentation", 4th Edn. Longmans Green, London.
Harris, G. and Merritt, N. R. (1961). *J. Inst. Brew.* **67**, 482–487.
Harris, G. and Merritt, N. R. (1962a). *J. Inst. Brew.* **68**, 33–39.
Harris, G. and Merritt, N. R. (1962b). *J. Inst. Brew.* **68**, 241–244.
Harris, G. and Merritt, N. R. (1962c). *J. Inst. Brew.* **68**, 244–246.
Harris, G. and Millin, D. J. (1963). *Proc. Eur. Brew. Conv.*, Brussels pp. 400–411.
Harris, G. and Thompson, C. C. (1960a). *J. Inst. Brew.* **66**, 213–217.
Harris, G. and Thompson, C. C. (1960b). *J. Inst. Brew.* **66**, 293–297.
Harris, J. O. (1959). *J. Inst. Brew.* **65**, 5–6.
Harrison, G. A. F., Byrne, W. J. and Collins, E. (1965). *J. Inst. Brew.* **71**, 336–341.
Hashimoto, N. and Kuroiwa, Y. (1966). *Proc. A.M. Amer. Soc. Brew. Chem.* pp. 121–130.
Hopkins, R. H. and Pennington, R. J. (1947). *J. Inst. Brew.* **53**, 251–258.
Hopkins, R. H. and Wiener, S. (1944). *J. Inst. Brew.* **50**, 124–138.
Hopkins, R. H., Wiener, S. and Rainbow, C. (1948). *J. Inst. Brew.* **54**, 264–269.
Hough, J. S. (1957). *J. Inst. Brew.* **63**, 483–487.
Hough, J. S. (1959). *Brewers' Digest* pp. 39–44.
Hough, J. S. and Ricketts, R. W. (1960). *J. Inst. Brew.* **66**, 301–304.
Hough, J. S. and Rudin, A. D. (1958). *J. Inst. Brew.* **64**, 404–410.
Hough, J. S. and Rudin, A. D. (1959a). *Proc. Eur. Brew. Conv.*, Rome pp. 208–214.
Hough, J. S. and Rudin, A. D. (1959b). *Brewers' Guild J.* **45**, 298–301.
Hough, J. S. and Rudin, A. D. (1961). *J. Inst. Brew.* **67**, 404.
Hough, J. S. and Stevens, R. (1961). *J. Inst. Brew.* **67**, 488–494.
Hudson, J. R. and Stevens, R. (1960). *J. Inst. Brew.* **66**, 471–474.
Ingraham, J. L. (1965). *Tech. Quart., Master Brewers Assoc. Amer.* **2**, 85–87.
Ingraham, J. L. and Guymon, J. F. (1960). *Archs Biochem. Biophys.* **88**, 157–166.
Ingraham, J. L., Guymon, J. F. and Crowell, E. A. (1961). *Archs Biochem. Biophys.* **95**, 169–175.
Jansen, H. E. (1958). *In* "The Chemistry and Biology of Yeasts" (A. H. Cook, ed.), pp. 635–667. Academic Press Inc., New York.
Jansen, H. E. and Mendlik, F. (1951). *Proc. Eur. Brew. Conv.*, Brighton pp. 59–81.
Jones, M. and Pierce, J. S. (1963). *Proc. Eur. Brew. Conv.*, Brussels pp. 101–134.
Jones, M. and Pierce, J. S. (1964). *J. Inst. Brew.* **70**, 307–315.
Jones, M., Power, D. M. and Pierce, J. S. (1965). *Proc. Eur. Brew. Conv.*, Stockholm pp. 182–194.
Jones, M. O. and Rainbow, C. (1966). *Proc. A.M. Amer. Soc. Brew. Chem.* pp. 66–70.
Katô, H. and Nishikawa, N. (1960–1). *Bull. brew. Sci., Tokyo* **6**, 12–21.

Kijima, M. (1954). *J. Inst. Brew.* **60**, 223–227.

Kijima, M. (1963). *Rep. Res. Lab. Kirin Brewery Co.*, *Yokohama* **6**, 35–43.

Kleber, W. and Lampl, P. (1956). *Brauwissenschaft* pp. 66–69.

Klopper, W. J., Roberts, R. H., Royston, M. G. and Ault, R. G. (1965). *Proc. Eur. Brew. Conv.*, *Stockholm* pp. 238–254.

Knorr, K. (1965). *Proc. Eur. Brew. Conv.*, *Stockholm* pp. 343–351.

Kögl, F. and Tönnis, B. (1936). *Z. physiol. Chem.* **242**, 43–73.

Korzenovsky, M. (1955). In "Amino Acid Metabolism" (W. D. McElroy and B. Glass, eds.), p. 309. John Hopkins Press, Baltimore.

Kringstad, H. and Rasch, S. (1966). *J. Inst. Brew.* **72**, 56–61.

Kudo, S. (1952). *Proc. Japan Acad.* **28**, 93–96.

Lewis, K. F. and Weinhouse, S. (1958). *J. Am. chem. Soc.* **80**, 4913–4915.

Lewis, M. J. and Phaff, H. J. (1963). *Proc. A.M. Amer. Soc. Brew. Chem.* pp. 114, 123.

Lewis, M. J. and Phaff, H. J. (1964). *J. Bact.* **87**, 1389–1396.

Lewis, M. J. and Phaff, H. J. (1965). *J. Bact.* **89**, 960–966.

Lewis, M. J. and Rainbow, C. (1963). *J. Inst. Brew.* **69**, 39–45.

Lewis, M. J. and Rainbow, C. (1965). *J. Inst. Brew.* **71**, 150–156.

Lewis, M. J. and Stephanopoulos, D. (1967). *J. Bact.* **93**, 976–984.

Lindquist, W. (1953). *J. Inst. Brew.* **59**, 59–61.

Lynes, K. J. and Norris, F. W. (1948). *J. Inst. Brew.* **54**, 150–157, 207.

Macher, L. (1953). *Brauwissenschaft* pp. 54–57.

MacWilliam, I. C. (1968). *J. Inst. Brew.* **74**, 38–54.

Markham, E. and Byrne, W. J. (1967). *J. Inst. Brew.* **73**, 271–273.

Markham, E., Mills, A. K. and Byrne, W. J. (1966). *Proc. A.M. Amer. Soc. Brew. Chem.* pp. 76–85.

Masschelein, C. A. and Devreux, A. (1957). *Proc. Eur. Brew. Conv.*, *Copenhagen* pp. 194–211.

Masschelein, C. A., Ramos-Jeunehomme, C., Castiau, C. and Devreux, A. (1963). *J. Inst. Brew.* **69**, 332–338.

Masschelein, C. A., Ramos-Jeunehomme, C. and Devreux, A. (1961). *Proc. Eur. Brew. Conv.*, *Vienna* pp. 148–159.

Maw, G. A. (1965). *Wallerstein Labs Commun.* **28**, 49–70.

Mill, P. J. (1964a). *J. gen. Microbiol.* **35**, 53–60.

Mill, P. J. (1964b). *J. gen. Microbiol.* **35**, 61–68.

Miller, W. L., Eastcott, E. V. and Maconachie, J. E. (1933). *J. Am. chem. Soc.* **55**, 1502–1517.

Neubauer, O. and Fromherz, K. (1911). *Z. physiol. Chem.* **70**, 326–350.

Neuberg, C. and Hildesheimer, A. (1911). *Biochem. Z.* **31**, 170.

Nordström, K. (1961). *J. Inst. Brew.* **67**, 173–181.

Nordström, K. (1962a). *J. Inst. Brew.* **68**, 188–196.

Nordström, K. (1962b). *J. Inst. Brew.* **68**, 398–407.

Nordström, K. (1963a). *J. Inst. Brew.* **69**, 142–153.

Nordström, K. (1963b). *J. Inst. Brew.* **69**, 310–322.

Nordström, K. (1963c). *J. Inst. Brew.* **69**, 483–495.

Nordström, K. (1964a). *J. Inst. Brew.* **70**, 42–55.

Nordström, K. (1964b). *J. Inst. Brew.* **70**, 209–221.

Nordström, K. (1964c). *J. Inst. Brew.* **70**, 226–233.

Nordström, K. (1964d). *J. Inst. Brew.* **70**, 233–242.

Nordström, K. (1964e). *J. Inst. Brew.* **70**, 328–336.

Nordström, K. (1964f). *Svensk Kemisk Tidskr.* pp. 510–543.

Nordström, K. (1965). *Proc. Eur. Brew. Conv.*, *Stockholm* pp. 195–208.

Nordström, K. (1966). *J. Inst. Brew.* **72**, 38–40.

Norris, F. W. (1945). *J. Inst. Brew.* **51**, 177–184.

Northcote, D. H. (1963). *In* "Proceedings of the Symposium on the Chemistry and Biochemistry of Fungi and Yeasts", pp. 669–675. Butterworths, Dublin.

Owades, J. L., Maresca, L. and Rubin, G. (1959). *Proc. A.M. Amer. Soc. Brew. Chem.* pp. 22–26.

Peel, J. L. (1951). *Biochem. J.* **49**, 62–67.

Phillips, A. W. (1955). *J. Inst. Brew.* **61**, 122–126.

Pollock, J. R. A. (1961). *J. Inst. Brew.* **67**, 5–6.

Portno, A. D. (1966). *J. Inst. Brew.* **72**, 458–461.

Portno, A. D. (1967). *J. Inst. Brew.* **73**, 43–50.

Prescott, S. C. and Dunn, C. G. (1959). "Industrial Microbiology", 3rd Edn. McGraw-Hill, New York.

Rainbow, C. (1948). *Nature, Lond.* **162**, 572.

Rainbow, C. (1966a). *Process Biochem.* **1**, 489–492.

Rainbow, C. (1966b). *Wallerstein Labs Commun.* **29**, 5–15.

Ribéraud-Gayon, J., Peynaud, E. and Guimberteau, G. (1959). *C. r. hebd. Séanc. Acad. Sci. Paris*, **248**, 749–751.

Richards, M. and Cowland, T. W. (1967). *J. Inst. Brew.* **73**, 552–558.

Ricketts, J. and Coutts, M. W. (1951). *Amer. Brewer*, Aug. pp. 27–30, 72; Sept. pp. 27–30, 74–75.

Ricketts, R. W. and Hough, J. S. (1961). *J. Inst. Brew.* **67**, 29–32.

Roche, J. and Lacombe, G. (1952). *Biochim. biophys. Acta* **9**, 687–692.

Rohrer, R. (1950). *Schweiz. Brau.-Runds.* **61**, 1–2.

Rose, A. H. (1963). *J. gen. Microbiol.* **31**, 151–160.

Royston, M. G. (1966a). *Brewers' Guardian* **95**, 33–41.

Royston, M. G. (1966b). *Process Biochem.* **1**, 215–221.

Rudin, A. D. and Hough, J. S. (1959). *J. Inst. Brew.* **65**, 410–414.

Sandegren, E. and Enebo, L. (1961). *Wallerstein Labs Commun.* **24**, 269–279.

Seitz, E. W., Sandine, W. E., Elliker, P. R. and Day, E. A. (1963). *Can. J. Microbiol.* **9**, 431–441.

SentheShanmuganathan, S. (1960a). *Biochem. J.* **74**, 568–576.

SentheShanmuganathan, S. (1960b). *Biochem. J.* **77**, 619–625.

SentheShanmuganathan, S. and Elsdon, S. R. (1958). *Biochem. J.* **69**, 210–218.

Shimwell, J. L. (1949). *Wallerstein Labs Commun.* **12**, 71–88.

Shimwell, J. L. (1950). *J. Inst. Brew.* **56**, 179–182.

Shimwell, J. L. (1964). *J. Inst. Brew.* **70**, 247–248.

Shimwell, J. L. and Grimes, M. (1936). *J. Inst. Brew.* **42**, 348–350.

St. Johnston, J. H. (1953). *Wallerstein Labs Commun.* **16**, 39–70.

Stadtman, E. R. (1963). *Bact. Rev.* **27**, 170–181.

Stevens, R. (1960). *J. Inst. Brew.* **66**, 453–471.

Stewart, E. D., Hinz, C. and Brenner, M. W. (1962). *Proc. A.M. Amer. Soc. Brew. Chem.* pp. 40–46.

Stickland, L. H. (1934). *Biochem. J.* **28**, 1746–1759.

Stickland, L. H. (1935). *Biochem. J.* **29**, 288–290, 889–898.

Strassman, M., Shotton, J., Corsey, M. and Weinhouse, S. (1958). *J. Am. chem. Soc.* **80**, 1771–1772.

Tabachnick, J. A. and Joslyn, M. A. (1953a). *Plant Physiol.* **28**, 681–692.

Tabachnick, J. A. and Joslyn, M. A. (1953b). *J. Bact.* **65**, 1–9.

Taylor, E. S. (1947). *J. gen. Microbiol.* **1**, 86–90.

Thorne, R. S. W. (1937). *J. Inst. Brew.* **43**, 288–293.

Thorne, R. S. W. (1941). *J. Inst. Brew.* **47**, 255–272.

Thorne, R. S. W. (1944). *J. Inst. Brew.* **50**, 186–198.

Thorne, R. S. W. (1945). *J. Inst. Brew.* **51**, 6–17.

Thorne, R. S. W. (1946). *J. Inst. Brew.* **52**, 5–14.

Thorne, R. S. W. (1949a). *J. Inst. Brew.* **55**, 201–222.

Thorne, R. S. W. (1949b). *Nature, Lond.* **164**, 369.

Thorne, R. S. W. (1951). *Proc. Eur. Brew. Conv., Brighton* p. 21–33.

Thorne, R. S. W. (1952). *Wallerstein Labs Commun.* **15**, 201–211.

Voerkelius, G. (1961). *Brauwissenschaft* **14**, 389–397.

Walters, L. S. and Thiselton, M. R. (1953). *J. Inst. Brew.* **59**, 401–404.

Watson, T. G. and Hough, J. S. (1966). *J. Inst. Brew.* **72**, 547–555.

Webb, A. D. and Ingraham, J. L. (1963). *Adv. appl. Microbiol.* **5**, 317–353.

West, D. B., Lautenbach, A. L. and Becker, K. (1952). *Proc. A.M. Amer. Soc. Brew. Chem.* pp. 81–88.

Wildiers, E. (1901). *La Cellule* **18**, 313.

Williams, R. J. (1939). *J. Am. chem. Soc.* **61**, 454–457.

Williams, R. J. and Major, R. T. (1940). *Science, N.Y.* **91**, 246.

Chapter 5

Saké Yeast

K. KODAMA

Kodama Brewing Co. Ltd., Japan

I. Introduction

Saké is the traditional alcoholic beverage of Japan; it has been highly esteemed by the Japanese people for thousands of years, and has been the subject of poetic praise from early times. It is generally believed that the technique for brewing *saké* originated in China, but comparison of the production processes for *saké* and Chinese alcoholic beverages shows marked differences, especially in respect of the micro-organisms concerned.

The most important organisms in the *saké* brewing process are *koji* mould, belonging to the species *Aspergillus oryzae*, and the so-called *saké* yeasts. *Aspergillus oryzae* is cultivated on steamed rice to produce rice-*koji*, which is used mainly as a saccharifying agent; whereas in the Chinese alcoholic beverages, various moulds such as *Rhizopus*, *Mucor*, *Aspergillus*, *Penicillium*, *Absidia* or *Monascus* spp., and yeasts which grow spontaneously on steamed cereals produce *kyokushi*, the mixture being used not only as the saccharifying agent but also as the yeast inoculum.

In *saké* brewing the rice-*koji* and a larger quantity of steamed rice and water are added to an intermediate called *moto*, which acts as a yeast starter for the *moromi* main fermentation mash. *Moto* is classified into two types according to the procedure by which it is prepared; *ki-moto* or *yamahai-moto*, which is acidified by naturally-occurring lactic acid bacteria, and *sokujo-moto* in which lactic acid is added at the beginning of the process. In the *ki-moto* or *yamahai-moto* mash various aerobic bacteria, wild yeasts, lactic acid bacteria and *saké* yeast grow successively in this order, while in the *sokujo-moto* mash, *saké* yeast is almost the only micro-organism present. Accumulation of a high sugar concentration (over 20%) at an early stage in both *moto* processes, together with acidification produced either microbiologically or by the addition of acid, facilitates predominant growth of *saké* yeast in the later stages. In the *moromi* process, however, the presence of a high concentration of yeast cells used as inoculum (10^7–10^8/g) and the acidity of the mash due to the addition of acidified *moto*, permit enhanced growth of the *saké* yeast. The high initial yeast cell count is characteristically obtained by the *sandan-shikomi* procedure, by which rice-*koji*, steamed rice and water are added to the *moto* in three consecutive steps at definite time intervals. As the volume is increased in this way, there is a gradual increase in the population of *saké* yeast. Although these procedures for brewing *saké* were probably established on the basis of numerous trials and failures over the past centuries, they are now accepted as ecologically sound, and are used to control microbial growth in the mashing process. In particular the high concentration (40–45% w/v) of steamed rice used in the preparation of *moto* and *moromi* mashes contributes to sound brewing practice.

Saké yeast (in this context an ecological group of yeasts responsible for *saké* brewing) has a high resistance to unfavourable and violently changing conditions. High viscosity, and high concentrations of sugar, acid and alcohol in the mash, are all undesirable for the growth of *saké* yeast itself, but these conditions prevent the mash from being invaded by contaminating micro-organisms. Furthermore, by overcoming many of the above-mentioned difficulties, *saké* yeast gives the mash a good flavour and taste as well as a high alcohol concentration. On the basis of these characteristics, authentic strains of yeast suitable for *saké* brewing have been selected.

The research laboratory of the Japan Society of Brewing, attached to the Government Research Institute of Brewing, has long been engaged in the selection of excellent yeast strains and the distribution of strains to *saké* breweries, designating them by *Kyokai* numbers. Of these strains, *Kyokai* Nos. 6 and 7 (Fig. 1) are the most popular; they are used not only for commercial production but also for studies on *saké* yeast.

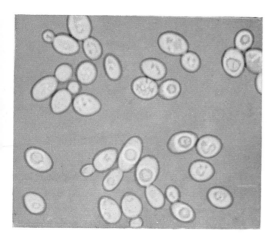

FIG. 1. Vegetative cells of *saké* yeast *Kyokai* No. 7 grown on *koji*-extract medium (x 1000).

II. Taxonomy and Ecology

In olden times, the *moto* mash was prepared by the addition of part of the *moto* saved from a previous batch, as yeast starter. It was considered at one time that the *saké* yeast was probably developed from the my-celia of *koji* mould, *Aspergillus oryzae*. Yabe (1895, 1897), however, isolated a yeast which was responsible for the *saké* fermentation and named it *Saccharomyces sake* Yabe. Based on his observations on its distribution in natural habitats, he first assumed that *saké* yeast occurred in rice straw, and might be carried into the *moto* mashes from straw which was used in those days in the construction of rice containers and mats for heat conservation in *koji* and *moto* preparation, and *moromi* fermentation processes. Following up these findings, the mor-phology, physiology and biochemistry of *saké* yeast have been exten-sively studied by many Japanese research workers (Kozai, 1900; Otani, 1903; Nakazawa, 1909; Takahashi *et al.*, 1913; Saito and Oda, 1932, 1934; Katsuya and Kitamura, 1949; Oda and Wakabayashi, 1954;

Tsukahara, 1961, 1962; Inoue *et al.*, 1962a; Kasahara, 1963; Ikemi, 1966). However, it is still difficult to distinguish taxonomically between *Saccharomyces sake* and related yeasts on the one hand, and *Sacch. cerevisae* Hansen on the other. Stelling-Dekker (1931) assumed that *saké* yeast was a subspecies of the species *Sacch. cerevisiae*, designating it *Sacch. cerevisiae* Hansen, Rasse *sake* Dekker. Later, Lodder and Kreger-van Rij (1952) brought *Sacch. sake* into the group *Sacch. cerevisiae*, after re-examining the properties of three strains of *saké* yeast.

The assimilation pattern of carbon compounds of two representative *saké* yeast strains, *Kyokai* Nos. 6 and 7, ascertained by using Wickerham's liquid medium test (1951), is presented in Table I (Kodama, 1966a). The results agree well with the pattern of the type culture of *Sacch. cerevisiae* (J. P. van der Walt, 1965, private communication). Therefore, according to the modern system of yeast taxonomy, which attaches considerable importance to assimilation of carbon compounds, it cannot be disputed that *saké* yeast is a strain of *Sacch. cerevisiae* Hansen.

TABLE I. *Assimilation of Carbon Compounds by* Saké *yeast*

Glucose	+	Melibiose	−	L-Rhamnose	−	Salicin	−
Galactose	+	Raffinose	+	Ethanol	±	Lactic acid	±
Sucrose	+	Melezitose	±	Glycerol	±	Succinic acid	±
Maltose	+	Inulin	±	Erythritol	−	Citric acid	−
Lactose	−	Soluble starch	−	Adonitol	−	Inositol	−
L-Sorbose	−	D-Xylose	−	D-Mannitol	−		
Cellobiose	−	L-Arabinose	−	D-Sorbitol	−		
Trehalose	±	D-Ribose	−	α-Methyl glucoside	−		

Further, recent numerical classification carried out by A. Kocková-Kratochvílová (1969, private communication), using 78 strains of *Sacch. cerevisiae* and other related species of *Saccharomyces*, also showed that two strains of *Kyokai*, Nos. 6 and 7, were taxometrically classified as members of the *Sacch. cerevisiae* group.

However, from a practical and ecological point of view, *saké* yeast can be differentiated from other types of *Sacch. cerevisiae* by additional properties: vitamin requirements (Section IV.A, p. 238), sugar and acid tolerance, osmophilic character, adaptability to anaerobic conditions and dominant growth on the culture fluids of *koji* mould (Takeda and Tsukahara, 1965c). These characteristics of *saké* yeast may be evaluated by consideration of its dominancy over other micro-organisms, including wild yeasts, in the *saké* process.

Since *saké* brewing is carried out under non-sterile conditions, the rice-*koji* and process water not being sterilized, and cleaned but unsterilized vessels being used for fermentation of the mash, there are many chances of contamination by wild micro-organisms during the brewing process. Nevertheless, as long as the process is controlled by techniques adequate to discourage invading micro-organisms, only a group of *saké* yeasts having the characteristics mentioned above dominates over the wild yeasts. Thus, as Saito (1950) reported, various types of wild *saké* yeast were found in the same brewery, or even in the same mash, where the classical type of *moto* (prepared by the action of spontaneously-occurring *saké* yeast) was used. Therefore, "*saké* yeast" is an ecological group of wild strains of *Sacch. cerevisiae* from which several favourable types for *saké* brewing have been selected and maintained as pure cultures.

In addition to taxonomic properties, ecological factors are also used in the grouping of *saké* yeasts. The following techniques for distinguishing the yeast used as inoculum from other wild strains may be employed for microbiological control in commercial brewing, and may also be applied in ecological studies.

a. TTC-agar-overlay technique. For the detection of various types of *saké* yeast belonging to the species *Sacch. cerevisiae*, including some yeasts of different genera which appear in the *saké* brewing process, the TTC (2,3,5–triphenyltetrazolium chloride)-agar overlay technique (Nagai, 1958, 1959), which was applied for this purpose by Akiyama and Furukawa (1963a), is useful. This technique was first devised for detecting respiratory-deficient mutants which could not reduce TTC, and thus were distinguished as white petite colonies. In *saké* mashes the respiratory-deficient mutant is usually found in the proportion of 1–2% (Akiyama and Furukawa, 1963b). *Kyokai* No. 6 and No. 7 can reduce TTC, and colonies of these strains turn red, whereas those of most wild *saké* yeast strains so far tested turn pink. The petite yeasts, and most film-forming yeasts, usually give pink or white colonies. When *Kyokai* No. 6 or No. 7 is used as inoculum, contaminating yeasts can easily be detected by a combination of this technique with the following method.

b. Selective medium technique. For detecting a very low population of *saké* yeast among an excess of dominant wild yeasts in rice-*koji* or the early stages of the *moto* preparation process, a special selective medium devised by Sugama *et al.* (1966) is convenient. This is Burkholder's medium modified by adding 2·5% ethyl acetate, which selectively suppresses most of the wild yeasts except for some strains of *Hansenula anomala*, but not *Kyokai* strains.

A modification of the Beerens and Castel (1958–9) method is also available, since in this partially anaerobic plate culture procedure using an alkaline pyrogallol solution, only *saké* yeast gives large colonies, while *H. anomala* and other aerobic yeasts hardly grow. On the other hand, for the differentiation of small numbers of *H. anomala* and other film-forming yeasts from a dense population of *saké* yeasts, selective media containing 1% KNO_3 as sole source of assimilable nitrogen, with 5% ethanol or 1% mannitol, sorbitol or glycerol as the only carbon source, are useful in the plate culture method (Kodama, 1960).

Furthermore, a natural mutant of *Kyokai* No. 7, which has been supplied to *saké* breweries as a starter yeast since 1962 shows, as compared with its parent strain or other normal *saké* yeasts, a much enhanced requirement for pantothenic acid in a medium containing ammonium sulphate as sole nitrogen source (see Section IV.A, p. 238). Its pantothenic acid requirement depends markedly on the culture temperature. When cultured at 35°, no growth occurs, while at a low temperature (15°), when there is a prolonged lag even in complete medium containing pantothenate, the temperature effect is nullified. Based on this characteristic of the mutant, the modified Burkholder's medium containing β-alanine (40 μg/l) instead of pantothenate, with incubation at 35°, is useful for its detection in the *moto* or *moromi* process. In this medium, small colonies of the mutant are found within one or two days further incubation at 20°, after the appearance of well developed large colonies of wild *saké* yeasts which appear in the first two days at 35° (Sugama *et al.*, 1965c).

Regarding the differences in morphological properties between strains of *saké* yeasts, most of the TTC-red yeasts are considered to be strains of *Kyokai* No. 6 or No. 7 because of their diverse distribution in *saké* breweries. Their ability to form ascospores is weak, but pellicle formation in liquid media, appearance of pseudomycelia in slide cultures (Fig. 2), and the rough surface of giant colonies on malt agar or gelatine media (Fig. 3) are prominent characteristics. In contrast, the TTC-pink yeasts in general form ascospores easily, but no pellicle or pseudomycelia, and the surface of giant colonies grown on each of these media is rather smooth (Takeda and Tsukahara, 1965a; Akiyama and Sugano, 1967).

Before the concept of pure cultures of micro-organisms had been developed, *saké* was brewed only by the wild yeasts which grew spontaneously during the preparation of *moto*. It is well known that the rice-*koji* is one of the habitats of wild yeasts, which include *Candida guilliermondii* and film-forming yeasts such as *Hansenula anomala* (in the majority), *Pichia membranaefaciens*, *Pi. farinosa*, *Debaryomyces kloeckeri*, *D. nicotianae*, *Torulopsis famata*, *C. fabianii* (to some extent) and *saké* yeasts (in the minority) (Kodama *et al.*, 1964; Takeda and Tsukahara, 1965a, 1966; Kodama *et al.*, 1966; Ouchi, 1966; Ouchi *et al.*, 1967). Of the total yeast cell count of 10^5–10^6/g in rice-*koji* (Murakami *et al.*, 1962), wild *saké* yeast comprises only about 10–10^2/g on average (Ouchi *et al.*, 1966b). In spite of the fact that *saké* yeast occupies only a small proportion of the yeast population in rice-*koji*, no competition is observed between this yeast and other wild-type yeasts because each yeast forms a separate colony which does not spread to cover the surface of the rice grains.

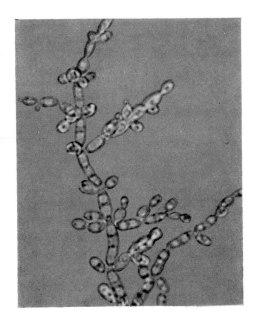

Fig. 3. Giant colony of *saké* yeast, *Kyokai* No. 7 strain.

FIG. 2. Pseudomycelia formed by *saké* yeast on potato agar (x 600).

However, the mycelia of *koji* mould develop on the grains until they touch the yeast colonies, and then the mould utilizes yeast nutrients, especially potassium, magnesium and phosphorus, suppressing further growth of the yeasts (Ouchi *et al.*, 1966a, b). In addition, in the later stage of *koji* preparation most of wild yeasts cannot grow further due to the rise of temperature in the *koji* material (38–40°) and the low humidity in the inoculation room (Sugama *et al.*, 1965b). Therefore, even when *saké* yeast (*Kyokai* No. 7) is inoculated at the initial stage of *koji* preparation, it increases only 10^2–10^3 times (Ouchi *et al.*, 1966a). This fact suggests that the process of *koji* preparation does not provide the *saké* yeast with suitable conditions for its growth. But, taking the vast surface area of the rice grains for the growth of *koji* mould (about 141 m²/100 kg of rice grains) into consideration, it is easily understandable that the rice-*koji* may act as a net in capturing wild *saké* yeast, which is rather sparsely distributed in nature (Sugama, 1967). Further, if careful attention is not paid to cleanliness and hygiene in the brewing process, especially during the preparation of *moto*, wooden equipment such as tubs, stirrers, hot-water casks, and worker's clothes and hands may act as carriers of both wild *saké* yeast and the yeast from previous batches of *moto* or *moromi*.

In the early stages of the *ki-moto* or *yamahai-moto* mash, wild *saké* yeast occurs in low concentration (10–10^2/g), but survives the unfavourable environment to grow gradually, mainly because of its tolerance to high viscosity, high concentrations of sugar, acid and alcohol, etc., and finally establishes its niche in the later stages. Such a dominant wild *saké* yeast often participates in the main fermentation of *moromi* mash, and is sometimes responsible for brewing *saké* with a unique character. In such a case, it may be called a "domesticated yeast".

At present, instead of domesticated *saké* yeast grown from chance sources, pre-cultures of authentic strains with known properties are used as seed for the preparation of *moto*. Both *Kyokai* No. 6 and No. 7 were isolated from mashes in breweries famous for producing excellent *saké*. These strains, having been successively cultured for many years by skilful craftsmanship in the brewery mashes containing rice-*koji* and steamed rice with a low (not more than 70%) polishing ratio (see Section V.C, p. 251), is presumed to have acquired certain superior properties, being especially suitable for better fermentation at a temperature sufficiently low for brewing *saké* with special bouquet and flavour. When these strains are used as seed in brewing, the aforementioned domesticated yeast is considered as a wild contaminant which should be suppressed as much as possible.

Akiyama *et al.* (1965) isolated a non-froth-forming yeast from *saké* mash which did not exhibit the frothy head formed by the normal *saké* yeast during main mash fermentation, since this yeast lacks a slime-like outer layer on the cell wall, which causes the formation of the froth, and can be observed by electron microscopy (Akiyama and Iwata, 1966). The visible difference between the non-froth forming yeast and *Kyokai* No. 7 consists mainly in the more pronounced smoothness of colonies formed by the former, which are coloured pink in the TTC-agar-overlay method; sugar fermentation and vitamin requirements are the same as those of *saké* yeast belonging to the species *Sacch. cerevisiae*. Yeast of this type had previously been isolated from the sediment in unpasteurized *saké* independently by Zenda (1916) and Takahashi (1916). The absence of a frothy head during the main mash fermentation by this yeast enables much more mash to be fermented than by the normal saké yeast in a vessel of the same capacity.

The effect of fermentation temperature on the appearance and chemical composition of mashes has been studied using various types of saké yeast (Hara et al., 1965, 1967). In these experiments, the maximum temperatures of mashes were controlled so as to reach 12°, 15° and 18° respectively 10 days after the commencement of mashing. Kyokai No. 7 was found to be suitable for fermentation at relatively low temperatures, while Kyokai No. 8 required higher temperatures. No scum was formed in mashes inoculated with Kyokai No. 7 or non-froth-forming yeasts, while thick scum was formed with Kyokai No. 8 regardless of the fermentation temperature. The acidity of the mash when the alcohol concentration had risen to 16% was influenced by the temperature and type of yeast. Acidity at a lower temperature (12°) was less than that produced at a higher temperature (18°), and the acidity of the mash with Kyokai No. 7 was less than with Kyokai No. 8. The formol nitrogen in the mash was not influenced by the kind of yeast, but higher fermentation temperatures tended to increase the nitrogen content in the last stage.

III. Mixed Populations

Since saké brewing is carried out without sterilization in an open fermentation system which allows the entry of various micro-organisms, the brewing technique is based primarily on the control of flora changes that occur during the fermentation of the mash, in such a way that the growth of the saké yeast used as inoculum is encouraged. Accordingly, mixed cultures of saké yeast with micro-organisms of various kinds, and particularly with other yeasts and lactic acid bacteria, require consideration from both ecological and practical aspects.

A. CULTURE OF SAKÉ YEAST STRAINS IN COMBINATION WITH OTHER SAKÉ
   OR WILD YEASTS

1. *Mixed Culture of Two Nutritional Variant Strains of* Saké *Yeast*

Two nutritionally different variants, Kyokai No. 6 and Strain 15-2-2, both of which had almost the same growth features when cultured singly in a defined medium, were grown in combination either in glucose-casein hydrolysate medium containing vitamins, or in koji-extract medium. The viable cell counts of the two strains were measured on media with and without pyridoxine. The results showed that Kyokai No. 6, which requires only pantothenic acid, prevailed markedly over Strain No. 15-2-2, regardless of the size and age of either inoculum. Strain No. 15-2-2 has substantial requirements for pantothenic acid and pyridoxine, and to a lesser extent for several other vitamins such as biotin and inositol. This strain, therefore, has more complex nutritional requirements than *Kyokai* No. 6. The ability of *Kyokai* No. 6 to predominate over the other yeasts was somewhat reduced when *koji* extract was used as the culture medium (Ito *et al.*, 1956). The ecological principle that micro-organisms which can grow in simplified media develop in preference to others which require more complicated media, even if the growth rate of the two organisms is similar, is thus applicable in this case.

2. *Mixed Culture of* Saccharomyces cerevisiae *Mutant Strain No. 328 with Wild* Saké *Yeast*

According to Akiyama and Umezu (1959), in an experimental brewing test carried out using a mutant of *Sacch. cerevisiae*, Strain No. 328 (R-type), which differed from other strains of this species in its reddish-purple colony colour when grown on *koji*-extract-agar, and in its requirement for adenine and arginine, the following phenomenon was observed in relation to the inoculated strain (R-type) and a wild *saké* yeast (W-type) which appeared as a contaminant during the course of brewing.

In the final stage of *moto* preparation, the ratio of inoculated R-type yeast to the contaminant was 1 : 1 at a total yeast cell count of $1·4 \times 10^8$/g, in spite of the fact that the original inoculum was 10^6/g. In *moromi* mash, thereafter, growth of the R-type yeast apparently occurred at an early stage, but it was overgrown by the contaminant W-type at the high-froth-forming stage, and later almost disappeared. Furthermore, in a mixed plate culture using *koji*-extract-agar medium containing adenine and arginine, the number of colonies of the R-type mutant and *Kyokai* No. 6 were in the same order, while that of the W-type was considerably greater than those of the other two yeasts after incubation for 48 h. Judging from this result, it may be concluded that, since various kinds of yeast can grow during the *saké* brewing process, deviation from the expected behaviour may be brought about by interaction between the inoculated yeast and the contaminants which grow spontaneously under various mash conditions.

3. *Mixed Culture of Non-froth-forming Yeast and* Kyokai No. 7

As explained above, the non-froth-forming yeast is one of the contaminants which occurs spontaneously in the *saké* brewing process. The two strains so far studied showed no appreciable difference in growth rate in glucose-yeast extract and *koji*-extract medium at various temperatures, or in their response to pantothenic acid, which supports yeast growth at a concentration of 25 ng/ml. However, when the content of pantothenate in the medium was increased to about 100 ng/ml, which is similar to that of pantothenate in saké mash but higher than required, incompatibility was observed in a mixed culture of the two yeasts by the TTC agar-overlay method. Several days later, the non-froth-forming strain dominated over Kyokai No. 7, even when the size of the inoculum of the latter was five times greater (5 × 10^3/ml) than that of the former (Akiyama et al., 1967). On the other hand, when the content of pantothenate was less than the amount required, incompatibility developed slowly.

When Kyokai No. 7 was grown in synthetic medium containing an excess of pantothenate (234 ng/ml), there was a lag in its uptake until the cell count reached about 10^7/ml. On the other hand, non-froth-forming yeast had no lag, and vigorous uptake commenced at the first stage of growth and lasted until the cell count reached the maximum (10^8/ml); pantothenate in the medium was then reduced to a very low concentration, as observed in the initial stage of the moromi fermentation (Fukui et al., 1955c; Nakamura et al., 1955). Further, when both yeasts were incubated singly in a pantothenate-deficient medium (10 ng/ml), the ratio of dead cells of Kyokai No. 7 tended to increase markedly, while that of non-froth-forming yeast was very low and almost constant. Therefore, it may be presumed that, in competition with non-froth-forming yeast in saké mash, the growth of Kyokai No. 7 may be restricted by a deficiency of pantothenate due to the difference of time lag in uptake of pantothenate between the two yeasts, resulting in more vigorous growth of the non-froth-forming strain than of Kyokai No. 7 (Akiyama and Shimazaki, 1967).

B. MIXED CULTURE OF SAKÉ YEAST AND SAKÉ LACTIC ACID BACTERIA

As is well known, most lactic acid bacteria have very complicated nutritional requirements, various B-group vitamins, amino acids and other substances being required for growth. In contrast, most saké yeasts require only one, or a few, B-group vitamins; accordingly saké yeast can synthesize vitamins other than those required for its own growth, as mentioned above. The fact that nutritive substances, including B-group vitamins, excreted by saké yeast cells serve to support the growth of saké lactic acid bacteria is clearly demonstrated in the following experiments.

Two strains of Lactobacillus, No. 333 and No. F7, and saké yeast Strain No. 15-2-2 were used. Strain No. 333, isolated from moto mash, has comparatively simple nutritional requirements, while Strain No. F7, isolated from acid-damaged moromi mash, requires very complex nutrients, including unknown growth factors. For example, only three

B-group vitamins (nicotinic acid, biotin and pantothenic acid) are required by Strain No. 333, while almost all are needed by Strain No. F7. For this experiment double culture tubes were used, the bottom of the inner tube being covered by a cellophane membrane, as devised by McVeigh and Brown (1954). Medium containing pantothenate and pyridoxine, but no other B-group vitamins, supported the growth of saké yeast Strain No. 15-2-2, but not that of two strains of *Lactobacillus*. This medium was placed in both the inner and outer tubes; saké yeast was inoculated into the inner tube and *Lactobacillus* into the outer. As the yeast grew in the inner tube, the growth of each *Lactobacillus* strain in the outer tube was increasingly enhanced. In view of the known components of the media used and other experimental results, it seems reasonable to conclude that several B-group vitamins such as nicotinic acid, biotin and folic acid, and nucleic acid bases are excreted by the saké yeast cells (Ito *et al.*, 1957). Most aerobic bacteria found in *moto* mash are also able to excrete these nutritive substances. Thus, commensalism may easily be established between various organisms in *ki-moto* or *yamahai-moto* mash.

Pronounced deterioration of the *moromi* mash occurs when both saké yeast and lactic acid bacteria are present under commensal conditions. This effect was demonstrated by mixing cultures of *Kyokai* No. 6 with each of the above-mentioned two lactic acid bacteria, using synthetic or *koji*-extract medium contained in a long glass tube. In this case the fermentative activity (Q_{CO2}) of the yeast cells harvested from the mixed broth by differential centrifugation was less than that of the single cultures, especially in the mixed culture with Strain No. 333, which not only dominated but killed most of the *Kyokai* No. 6 yeast. Recovery of the Q_{CO2} value of the yeast, which was observed when assimilable nitrogen (e.g. ammonium salts or casein hydrolysate) was supplied, indicated that viability was reduced by a deficiency of nutrients caused by utilization by the bacteria (Ito and Uemura, 1957).

This nutritive aberration of yeast in mixed culture with lactic acid bacteria was further studied by Momose *et al.* (1966) using *Kyokai* No. 7 and *Lactobacillus plantarum* Strain No. B-74, isolated from acid-damaged saké mash. When the mixed culture technique was applied either directly, or with a separating cellophane membrane, viable yeast cells, and the amounts of glutamic acid, threonine, serine and cystine in both the residual fluid and hot water extract of the harvested yeast cells were markedly reduced, especially in the case of the cultures separated by a cellophane membrane. Furthermore, the endogenous respiratory activity (Q_{O2}) of the yeast decreased markedly in the mixed cultures.

However, there were no effects of accumulated lactic acid or inositol deficiency on the decrease of Q_{O2} value. The decrease was partially recovered by incubating the yeast cells with amino acids, glutamic being particularly effective. From these results it is concluded that the reduction in viability of yeast in a mixed culture with lactic acid bacteria may be related to the uptake of certain amino acids, especially glutamic (Momose and Tonoike, 1968).

Recently, aggregation of yeast cells was observed in mixed cultures of Kyokai No. 7 with strains of saké lactic acid bacteria isolated from infected saké mash (Momose et al., 1968a). This phenomenon was not caused by accumulation of lactic acid, lowering of pH value, deficiency of inositol or any substance produced by lactic acid bacteria, but by contact of the yeast cells with cells of lactic acid bacteria. Further, this aggregation of yeast cells was observed when the cells obtained from single pure cultures of both yeast and lactic acid bacteria were mixed. It was observed that cells of saké yeast aggregated when mixed with Lactobacillus plantarum Strain No. B-74, but not with L. casei Strain No. B-83; those of wine, alcohol, baker's and brewer's yeasts aggregated when mixed with L. casei, but not with L. plantarum (Momose et al., 1968b).

Further study of this phenomenon by the moving boundary electrophoresis technique led to the conclusion that aggregation of yeast cells by lactic acid bacteria is due to electrostatic force exerted between the yeast and bacterial cell surfaces (Momose et al., 1969).

C. COMMENSALISM BETWEEN SAKÉ YEAST AND HANSENULA ANOMALA

Enhanced growth of Hansenula anomala is often observed in the early stages of ki-moto or yamahai-moto, preceding the development of saké yeast (Saito and Oda, 1932, 1934; Kodama et al., 1957). Hansenula anomala isolated from yamahai-moto has no vitamin requirement, but excretes pantothenic acid, pyridoxine and nicotinic acid during growth (Uemura, 1956; Kodama, 1965), while saké yeast requires pantothenic acid and several other vitamins. Thus, a culture filtrate of H. anomala, when added to medium deficient in pantothenate and pyridoxine, can support the growth of two saké yeast strains, Kyokai No. 6 and Strain No. 15-2-2, which require respectively pantothenic acid only, and pyridoxine plus pantothenic acid.

This fact, and other experimental results obtained by using McVeigh and Brown's double culture tube (see previous Section) showed that commensalism occurs between saké yeast and H. anomala, and that vitamins synthesized within the cells can be excreted into the culture fluid, while vitamins required by the yeast and supplied in the medium can only be excreted from the cells to a very limited extent (Uemura, 1956). The same commensalism was also observed between this strain of H. anomala and two saké lactic acid bacteria (Leuconostoc *mesenteroideus* and *Lactobacillus sake*), both of which grow spontaneously during the early stages of *ki-moto* or *yamahai-moto* (Kodama, 1965).

IV. Physiology

A. NUTRITIONAL REQUIREMENTS

Pantothenic acid is indispensable for the growth of all strains of *saké* yeast so far investigated. Takahashi (1954, 1956) first reported that *saké* yeast required this vitamin when the medium contained an amino acid mixture as the nitrogen source, but not when only inorganic ammonium salts were supplied. Fukui *et al.* (1955a) obtained the same results with *Kyokai* No. 7, but pointed out that even in the ammonium salt medium growth was promoted by addition of pantothenate together with several vitamins such as thiamine, biotin, inositol and pyridoxine. Ito *et al.* (1955) found a group of *saké* yeasts which required pyridoxine besides pantothenic acid among strains isolated from *moto* and *moromi* mashes from several breweries in Tōhoku district, but distribution of this group in the breweries was more restricted than that of the group which required pantothenic acid alone. Recently Sugama *et al.* (1965a) reported that pantothenic acid-requiring strains were in the majority (98% of cases), whereas strains requiring both pantothenate and pyridoxine comprised less than 2% of 415 *saké* yeast strains isolated from breweries in 1960 and 1961. According to Takeda and Tsukahara (1965b) the pantothenic acid and pyridoxine-requiring type was found mainly in the early stages of *moto* and *moromi* preparation, and seldom in the later stages.

In general, when an organic nitrogen source is provided, pantothenic acid and pyridoxine, alone or in combination, are essential while biotin, thiamine and inositol are supplementary for the growth of *saké* yeast. In respect of vitamin requirements, *saké* yeast is clearly distinguishable from baker's, distillery and wine yeasts, all of which require biotin besides pantothenic acid, and also from brewer's yeast, which requires biotin but not necessarily pantothenic acid (Takahashi, 1954; Nakanishi and Tsukahara, 1954; Takeda and Tsukahara, 1965b).

Furukawa and Akiyama (1962) obtained a mutant strain (*Kyokai* No. 7-1) by ultraviolet irradiation of *Kyokai* No. 7. This mutant requires pantothenic acid even in medium containing an inorganic nitrogen source. Since β-alanine cannot substitute for pantothenic acid in the case of this mutant, the enzyme system which catalyses the conversion of β-alanine to pantothenic acid appears to be lacking. Of the above-mentioned 415 strains, 13 were found to be of the same type as *Kyokai* No. 7-1. Most of these were isolated in 1961, and in that year 10 strains of this mutant type, *Kyokai* No. 7-1, were detected out of 14 strains recorded as *Kyokai* No. 7 in the Government Brewing Institute (Sugama *et al.*, 1965b). The coincidence of these findings might be accounted for by distribution to breweries of a naturally mutated strain of *Kyokai* No. 7 which had arisen in the stock culture. Since then, this mutant has been deliberately supplied to breweries.

As regards the interaction between amino acids and pantothenic acid for the growth of *saké* yeast, some of the individual amino acids so far tested, when used as sole nitrogen source, inhibited yeast growth in pantothenate-free medium. In particular, histidine inhibited growth completely: this inhibitory action was reversed by the addition of pantothenate (Takahashi, 1954). Using medium containing both pantothenate and alcohol, as is the case in *saké* mash, amino acids can be divided into two groups; (i) growth-promoting (glutamic acid, aspartic acid, arginine, proline, alanine); and (ii) growth-inhibiting (histidine, threonine, methionine, leucine, cystine). Mixed amino acids with opposing effects cancel each other to give normal yeast growth (Fukui *et al.*, 1956).

Pantothenic acid and thiamine differ in their effects on the growth and fermentation of *saké* yeast. Pantothenic acid promotes yeast growth but inhibits alcohol production, the depressant effect being restored by the addition of thiamine. As is well known, pantothenic acid and thiamine are precursors of coenzyme A (CoA) and cocarboxylase (TDP) respectively. Cells enriched with both CoA and TDP are obtained by the propagation of *saké* yeast in the presence of pantothenate and thiamine, which together stimulate both growth and fermentation. On the other hand, cells supplied with pantothenate alone give a higher value for the Meyerhof oxidation quotient than those with thiamine alone, and cells with both vitamins give an intermediate value. Accordingly, both pantothenic acid and thiamine are assumed to influence the Pasteur effect in yeast.

Based on the conception that a normal balance in the ratio of CoA to TDP in cells grown in the presence of both vitamins may be established, the inhibitory effect of pantothenic acid on alcohol production, which in turn is reversed by supplying thiamine, may be accounted for by an imbalance in the cells towards CoA. Adding thiamine in this case may cause the cell fermentability to rise to the level of normally balanced cells by the resultant enrichment in TDP (Fukui *et al.*, 1958a).

During *moromi* fermentation, free pantothenic acid, which is derived from rice-*koji* and steamed rice added to the mash, is in relatively high concentration (100–200 ng/ml) at an early stage, subsequently followed by a rapid decrease associated with stimulation of yeast growth (Fukui *et al.*, 1955c; Nakamura *et al.*, 1955). The concentration of amino acids in the early stage is low, because of active assimilation by the growing yeast, but the amount gradually increases later and the acids accumulate in the mash as the capacity of the yeast to take them up weakens because of the high alcohol concentration. Yeast in the early stage of the *moromi* fermentation is presumed to require thiamine for vigorous fermentation, but later pantothenic acid may reverse the inhibitory effect of accumulated amino acids on yeast growth, as already discussed. This can be corroborated by supplying thiamine to the *moromi* mash in the early stages, and pantothenate later (Fukui *et al.*, 1958b).

Evolution of hydrogen sulphide is sometimes observed during the *saké* brewing process. Kurono *et al.* (1935) pointed out that it is generated from cysteine by the desulphydrating action of yeast during the early stages of *moromi* fermentation. Particularly when the growth of yeast is markedly faster than the degradation of the rice components, this phenomenon is observed, probably because of the deficiency of pantothenic acid. In such a case the hydrogen sulphide once generated, usually disappears gradually, and hardly influences the quality of the resultant *saké*. Kodaira *et al.* (1958) and Kodaira and Uemura (1959) found that two strains of *saké* yeast, *Kyokai* No. 6 and Strain No. 15-2-2, when cultured in pantothenate-deficient ammonium sulphate medium produced hydrogen sulphide in the early logarithmic phase but less in the stationary phase, and that this effect was not brought about by a deficiency of other vitamins such as pyridoxine to which Strain No. 15-2-2 responds. The fact that the culture of *saké* yeast in pantothenate-deficient medium containing casein hydrolysate instead of ammonium sulphate abolished the evolution of hydrogen sulphide, led to research into the influence of amino acids. Methionine and leucine reduced the production of the gas, while alanine, glutamic and aspartic acids and cysteine caused an increase.

B. ACCUMULATION OF HIGH ALCOHOL CONCENTRATIONS IN *SAKÉ* MASH

It is a remarkable characteristic of *saké* brewing that high alcohol concentration, reaching about 20%, is produced from rice starch. This may be accounted for by the following factors: (i) successive mashing in three steps; (ii) saccharification of rice starch by the activity of *koji* enzyme and fermentation by *saké* yeast, which proceed simultaneously and so maintain a balance; (iii) the high concentration of the mash, which contains the solids of rice-*koji* and steamed rice (Yamasaki, 1958; Muto *et al.*, 1962); (iv) high yeast cell density (Nojiro, 1959; Muto, 1961); (v) factors contained in the mycelia of *koji* mould that favour alcohol tolerance of *saké* yeast (Inoue and Takaoka, 1953), comprising mainly calcium and magnesium pantothenates (Fukui *et al.*, 1955b).

Recently, special factors that influence the production of high concentrations of alcohol were investigated. Instead of enzymic saccharification of rice starch, a fermentation test with a yeast inoculum of 3×10^7/ml was carried out by stepwise addition of sucrose to the basal synthetic medium. Final concentrations of alcohol as high as 18–20% were reached when oryzenin, albumin or *koji* mould mycelia were added. Oryzenin, a major component of rice protein, serves as a detoxicant to adsorb trytophol, which is the most toxic to yeast of the higher alcohols derived from amino acids by yeast action. Furthermore, a high oxidation-reduction potential in the early stages and low later in the fermentation process, favoured the formation of a high concentration of alcohol. A fermentation test, first using cotton wool plugs and later a covering of paraffin, gave an alcohol concentration of 19·6% in synthetic medium.

Albumin and *koji* mould mycelia, in particular the latter, the water-insoluble mitochondrial fraction of *koji* were shown to play a part in the regulation of oxidation-reduction potential and stimulate yeast growth (Hongo *et al.*, 1967; Kawaharada *et al.*, 1967a, b). Yamashiro *et al.* (1966, 1967), pointed out the significance of the low temperature used in the fermentation process and of the rice-*koji* added to the *saké* mash, since low temperature helps to maintain the fermentative activity (Q_{CO2}) of the yeast, and the rice-*koji* was effective for increasing the yeast cell density. Further, comparison of the cells of *Kyokai* No. 7 isolated from fermented broth accumulating 10–12% alcohol ("fermenting cells"), with those from aerobic propagations ("aerobically-grown cells") revealed the following facts. The Q_{CO2} value and alcohol tolerance of "fermenting cells" were higher than those of "aerobically-grown cells", and the amounts of total nitrogen and vitamin B_6 were markedly higher in the former. The centrifuged supernatant of the fermented broth (with a concentration of about 10% alcohol) contained inhibitors of fermentation by "aerobically-grown cells" which "fermenting cells" tolerated.

The *moromi* yeast harvested from *saké* mash showed the same properties as the "fermenting cells". The inhibitory principles were adsorbed on charcoal and separated into volatile and non-volatile fractions. Gas chromatography of the volatile fraction before and after the active carbon treatment showed that the inhibitors included volatile fatty acids such as acetic, propionic and butyric. By addition of these acids to the medium, the fermentative activity of "aerobically-grown cells" was severely inhibited, especially in the presence of 10% alcohol, unlike that of "fermenting cells".

C. FORMATION OF ORGANIC ACIDS AND HIGHER ALCOHOLS

A characteristic of *saké* is the high content of succinic and lactic acids in relation to the total acids. Besides these, *saké* contains many other organic acids, as shown in Table X (see Section V.G, p. 266), which accumulate mainly during the *moromi* fermentation process, except for lactic acid which is derived chiefly from the *moto:* succinic acid is accumulated in the highest amount, followed by malic. Succinic and lactic acids are present in *saké* in almost the same concentrations (about 0·05%), while malic is less than half this; the total acids reach about 0·1% (Toyozawa *et al.*, 1960; Hayashida *et al.*, 1968).

Formation of succinic acid from glucose was observed by Kleinzeller (1941), using yeast cell suspensions under both aerobic and anaerobic conditions, only in bicarbonate solution, and not in phosphate buffer. However, according to Fujii and Uemura (1952), when *saké* yeast was used, succinic acid was formed aerobically from added glucose not only in bicarbonate solution but also in phosphate buffer. Furthermore, the ability to form succinic acid was enhanced when the yeast was subjected to anaerobic conditions in the deeper layers of the *moromi* mash. This was demonstrated by the anaerobic production of succinic acid from glucose in phosphate buffer with so-called *moromi* yeast, which was obtained by filtering *moromi* mash through gauze. The *moromi* yeast fraction, which was differentially sedimented by centrifuging at 1,000–1,500 r.p.m. for 5 min, was somewhat grey in colour and was covered with a layer of white starch granules. These were removed with a spatula, and the procedure was repeated several times until the iodine test was negative and the cells appeared homogeneous by microscopical examination. The yeast produced in this way can be treated as a pure culture, although it may be a mixture of several strains. This preparation was designated "*moromi* yeast" (Fujii and Uemura, 1953; Uemura and Fujii, 1954; Nakamura and Uemura, 1955).

The ability of *moromi* yeast to produce succinic acid from glucose under anaerobic conditions varied with the fermentation stage of the *moromi* mash, and the anaerobic activity seemed to be more vigorous than the aerobic throughout the *moromi* process. *Moromi* yeast was able to produce lactic acid anaerobically from added glucose, and also succinic acid from added glutamate when incubated in phosphate buffer. Furthermore, laboratory *saké* yeast strain *Kyokai* No. 6, cultured without shaking in deep *koji* extract or glucose-peptone-yeast extract medium, was able to produce succinic acid anaerobically from glucose when incubated in phosphate buffer. A comparative study of succinic acid formation by *saké* yeast, with and without pantothenate, demonstrated that pantothenate deficiency was responsible for the anaerobic production of the acid from glutamate but not from glucose. Judging from this and the other results, the anaerobic production of succinic acid from glucose may take place mainly through a pathway other than the TCA cycle in which CoA participates (Uemura and Fujii, 1954).

Saké yeast can metabolize amino acids because of its top fermenting characteristics, and formation of fusel oil from added amino acids is observed under anaerobic conditions. During anaerobic incubation of a cell suspension in tyrosine buffer solution containing a small amount of glucose, tyrosol (the deaminated alcohol) is almost stoichiometrically produced and excreted, whereas there is no excretion of ammonia. Ammonia produced in this way, part of which is utilized for the synthesis of nitrogenous cell constituents, remains mainly in the free state within the cell. In turn, "*moromi* yeast" has a weaker tyrosol-forming activity than pure *saké* yeast (Chihara and Uemura, 1957, 1958).

By the Ehrlich reaction amino acids are deaminated by yeast when they are supplied as sole nitrogen source (see Chapter 4, p. 147). Yamada (1932) has shown that higher alcohols produced by *saké* yeast have one more carbon atom than the aliphatic amino acids from which they are formed, and not one less as would be expected from the Ehrlich mechanism. Recently Yoshizawa (1963, 1964, 1965) has confirmed the participation of the acetoin condensation reaction in these processes. Following these reactions, the alcohols with one more carbon atom are formed from the corresponding amino acids.

D. INFLUENCE OF VARIOUS COMPOUNDS ON THE PHYSIOLOGY OF *SAKÉ* YEAST

1. *Lactic Acid*

The preventive action of lactic acid against contamination by harmful bacteria plays an important role in ensuring safe brewing of *saké*, even when using an open fermentation system; this procedure has been termed "protected fermentation". The influence of lactic acid on the physiology of *saké* yeast is consequently significant from the practical point of view. Generally, cultured and wild *saké* yeasts isolated from *moto* and *moromi* mashes have a strong tolerance to lactic acid. The maximum concentration of lactic acid in which *saké* yeast can grow was examined by using 41 yeast strains (Inoue *et al.*, 1962b). Before use, lactic acid was stored for a month as a 20% solution in order to obtain a stable concentration, in view of changes due to hydrolysis of lactide. After cultivation for 15 d at 27° the majority of the strains (about 70%) tolerated 2·0–2·5%, while several could tolerate 2·8%. It is noteworthy that *Kyokai* No. 6 had the strongest tolerance of all the strains tested. The culture temperature, the age of the pre-culture, and the composition of the medium affected the tolerance.

Tani *et al.* (1963a, b) reported that cell aggregation and an increase in the number of dead cells were observed in *saké* yeast grown in a medium containing 1·0–1·5% lactic acid, about the same concentration as that in the *moto* mash. The phenomenon caused by lactic acid appears to be analogous to that occurring in so-called "unbalanced growth" of inositol-requiring yeast when cultured in inositol-deficient medium. However, this action of lactic acid was observed even in a culture in synthetic medium containing sufficient inositol, and the degree of aggregation was less in organic nitrogen medium at a relatively low temperature (20°) than in inorganic nitrogen medium at a higher temperature (30°). Aggregation occurred in *koji* extract medium far later than in synthetic medium. In *moto* containing 0·7–1·0% lactic acid, however, no aggregation was observed, probably because of the high amino acid content and low temperature. The amount of inositol synthesized per unit weight of yeast cells was nearly the same regardless of the presence or absence of lactic acid, whereas the amount of inositol in the cells in the bound form was significantly lowered in the presence of this acid. Also, a marked reduction in content of NAD, CoA, cytochrome c, and several B-group vitamins such as thiamine, riboflavin and pantothenic acid was observed in the yeast cells cultured statically in the presence of lactic acid. Furthermore, lactic acid stimulated the disappearance of inositol supplied during the propagation of *saké* yeast.

From these results it is presumed that lactic acid blocks the transformation of inositol to its bound form (e.g. phospholipid inositol) and causes the unbalanced growth that is observed when the inositol-requiring yeast is cultured in inositol-deficient medium. In contrast, significant increases in yeast growth, as well as in the coenzyme and B-group vitamin content of the cells, were observed when the yeast was cultured aerobically in medium containing lactic acid. Under aerobic conditions, utilization of lactic acid occurred after most of the sugar had been metabolized. It can be inferred that the increase of coenzyme may enhance the respiratory system necessary for the oxidative utilization of lactic acid (Tani *et al.*, 1964, 1965).

2. *Alcohol*

In view of the fact that the alcohol concentration reaches a value as high as 19–20% in the mash after fermentation for 20–25 d, alcohol tolerance of *saké* yeast is an important property. There are several methods for measuring the alcohol tolerance of yeasts, and the definitions have been varied and rather confusing. Suto *et al.* (1951a) reported on the fermentative activity and alcohol tolerance of *Kyokai* No. 6, and expressed the tolerance in two ways: (i) as the ratio of the Q_{CO2} values in the presence and absence of 15% alcohol; (ii) as changes in the

Q_{CO_2} values when steeped in 15% alcohol at 30° for 14 h. In this case, the fermentative activity was measured by using a modified injection syringe with media containing an approximately constant amount of carbon source but varying amounts of organic nitrogen.

Too great an excess of assimilable nitrogen (0·2% nitrogen and 6·5% glucose) inhibited growth but slightly stimulated the fermentative activity. Regardless of the amounts of nitrogen supplied, each culture metabolized almost an equal amount of sugar after a long period of incubation. Fermentative activity in the early logarithmic phase (after incubation for 13 h) was high; alcohol tolerance was initially almost absent but increased considerably at a later stage (30 h). At lower nitrogen concentrations (0·01–0·1%), the higher the concentration of nitrogen the more resistant to 15% alcohol was the fermentative activity of the cells isolated after a long period of incubation (e.g. 140 h). In the early stages of the *moto* preparation process, the *moto* itself and the yeast obtained from it showed high fermentability, but the yeast lacked alcohol tolerance. In the last (resting) stage, changes in the components of *moto* and in the fermentative activity of the yeast decreased, until a stable equilibrium was apparently reached, although the nitrogen content of the cells continued to increase and they became resistant to 15% alcohol. It is interesting that the level of fermentability of resting *moto*, particularly that of *ki-moto* or *yamahai-moto*, which usually contains over 12% alcohol and 1·0% lactic acid, decreased only slightly in spite of incubation for another 20 d. This may have been due to the high nitrogen content of the yeast cells, which had grown under conditions of abundant assimilable nitrogen supply.

Alcohol tolerance was defined by Nojiro and Ouchi (1962) as the ratio of fermentative activities measured by the Warburg technique in the presence and absence of 18% alcohol. According to this definition, all yeast strains so far examined have an alcohol tolerance in the range 20–30%, and no differences have been observed between *saké* and other yeasts studied; including baker's, brewer's and wine yeasts. Furthermore, in all cases the fermentative activity is higher in young cells than in old, and the alcohol tolerance increases with the growth period. Changes in the tolerance of the yeast in fermenting *moromi* were investigated using the *moromi* yeast mentioned above. Fermentative activity both in the presence and absence of 18% alcohol was high during the vigorous fermenting stage. This was followed by a gradual decrease with time, during which the fermentability in 18% alcohol decreased more slowly than in its absence. Accordingly the alcohol tolerance ratio in both cases increased, reaching 16–32% during the main mash fermentation.

Inoue *et al.* (1962c) studied the conditions which affect the alcohol tolerance of *saké* yeast. In this case the critical concentration of alcohol which suppresses yeast growth was determined, and the following factors were found to be of importance: (i) pre-incubation period, 45 h being most suitable, shorter than 20 h or longer than 96 h reducing alcohol tolerance; (ii) temperature, a marked decrease in tolerance being observed over 27°; (iii) high initial glucose concentration, which has a similar effect, a suitable strength being 5%. Under these conditions the critical concentration of alcohol for the growth of *saké* yeast Strain SK-2-2 was shown to be 11·0–12·5% in medium containing 5% glucose at 27°.

The maximum (critical) concentrations of alcohol in combination with lactic acid which influence the growth of *saké* yeasts were further studied. Four strains were used which: (i) tolerated both alcohol and lactic acid well; (ii) tolerated both but less well; (iii) tolerated lactic acid well but alcohol less; (iv) tolerated alcohol well but lactic acid less. The yeasts were incubated in media containing alcohol (9–13%) and lactic acid (0·6–1·2%) at 23° and 27°, using various combinations of concentrations. It was shown that the maximum concentrations of alcohol and lactic acid which allowed yeast growth were greatly influenced by temperature. At 23° the yeasts were able to grow in 10–13% alcohol in the presence of 0·5% lactic acid, whereas at 27° they hardly grew in 9·5–11·5% alcohol. The strains with a high tolerance for both alcohol and lactic acid showed wide ranges of critical concentrations in the media containing both components, while those having poor tolerance had narrow ranges (Inoue *et al.*, 1962c).

3. *Sugar*

As already mentioned, a high concentration of sugar accumulates in the early stages of *moto* mash. This factor, together with the lactic acid effect, plays an important role in preventing growth of harmful bacteria. The *moto* process is clearly distinguishable from the *moromi* process in this respect. Soluble nitrogenous compounds, such as amino acids, also reach high levels early in the *moto* process, especially with the *ki-moto* system. In the *moto* process both growth and fermentation of *saké* yeast take place in strong medium (containing more than 25% of sugar), while in the *moromi* process the mash is rather weak (less than 10% of sugar) in respect of soluble substances.

In this connection, Suto *et al.* (1951b) reported on the physiological aspects of *saké* yeast grown on concentrated culture medium. *Koji* extract with a high sugar content of about 27%, and 1·5% soluble nitrogen, was filtered through a Berkefeld filter, and the sterile filtrate was used to prepare two series of concentrated and diluted media. Both basal media were adjusted to be rich or poor in nitrogen by adding either peptone or glucose. Each culture was incubated at 20° for 18 d. In the diluted medium (6–7% sugar) the yeast consumed nitrogen corresponding to the C:N ratio of the medium (g sugar/mg N), while in the concentrated medium (23–28% sugar) the yeast chiefly utilized the sugar, so that the C:N ratio of the materials metabolized did not correspond with the ratio in the medium. Further, in the former case the nitrogen content of the medium markedly affected the physiological properties of the yeast (the yield and the nitrogen content of the yeast, acid accumulation, fermentative activity, etc.), but in the latter this effect was slight.

These physiological properties of *saké* yeast which are observed in the concentrated medium are of significance in connection with the *saké* brewing process, in which alcohol fermentation is normally accompanied by a fairly large accumulation of acid, without being influenced by the amount of soluble nitrogenous material (0·10–0·15%) which accumulates during the course of brewing.

Kawabata and Kawano (1956) reported that when *Kyokai* No. 6 was cultured in medium containing 30% glucose (obtained by adding glucose to thin *koji* extract) large amounts of succinic, lactic, pyruvic and volatile acids accumulated, accompanied by a decrease in growth rate. The same phenomenon was also observed by Inoue *et al.* (1962d) when the growth of *Kyokai* No. 7 was restricted in a medium containing a high concentration of sugar (up to 40–45%). The acids formed in this case were shown to be mainly acetic, succinic and lactic, and the amount of acetic acid increased rapidly with higher sugar concentration. Therefore, slight "*kan-san-pai*" (sweet taste and acid damage), which occurs rarely during *moromi* fermentation, is probably due to acetic acid formation by *saké* yeast, caused by excessive accumulation of sugar, and may be followed by heavy "*kan-san-pai*" due to contamination by lactic acid bacteria (see Section V.F, p. 263).

V. Industrial Production of *Saké*

A. OUTLINE OF BREWING PROCESS

Saké is not only one of the most popular alcoholic drinks of the Japanese, but is also a very valuable factor in the Japanese national economy because of its importance as a source of indirect tax revenue. The annual consumption of various alcoholic beverages in Japan is given in Table II.

The raw materials used in brewing *saké* are rice and water. As the first step *koji*, a culture of *Aspergillus oryzae* on steamed rice, is prepared for the purpose of saccharification of the starch and decomposition of protein contained in the polished rice grains. Next, *moto*, a starter for *saké*

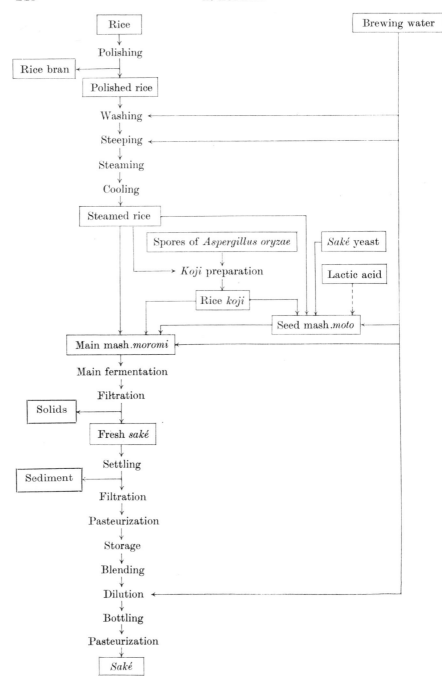

FIG. 4. Flow diagram of *Saké* brewing process.

TABLE II. *Annual Consumption of Alcoholic Beverages in Japan in 1967*

Beverage	Consumption (kl.)
Saké	1,293,421
Synthetic *saké*	53,331
Shochu	221,405
Beer	2,461,707
Whisky	108,742
Brandy	3,800
Fruit wine[a]	42,259

[a] Wine made from all varieties of fruit, including grapes.

yeast, is prepared by mashing rice, rice-*koji* and water in a small vessel. Since lactic acid added initially, or produced during the early stages of *moto* preparation, suppresses harmful bacterial contaminants, usually only the inoculated culture of *saké* yeast can grow in the *moto* mash. About 5–9% of the total amount of cleaned rice for mashing is used for the *moto* preparation. Finally, *moromi*, the main fermentation mash, is prepared by adding steamed rice, rice-*koji* and water to the *moto* in a large vessel, additions being made successively in three batches. Thereafter the main fermentation takes 20–25 d. When fermentation has ceased, the *moromi* mash is filtered to remove the solids, and the filtrate thus obtained is fresh *saké*. After about one month, the fresh *saké* is pasteurized and stored. All processes in *saké* brewing are carried out at room temperatures below 10°. Thus the brewing season is usually confined to winter, but breweries equipped with refrigeration and air-conditioning are increasing in numbers and are brewing *saké* throughout the year. A flow diagram of the *saké* brewing process is given in Fig. 4.

B. RAW MATERIALS

1. *Rice*

Ordinary rice grain is the principal raw material for brewing *saké*. As it is also the daily food of the Japanese, many varieties have been grown. Not all of these, however, are suitable for *saké* production, but only some of the indigenous types. The desirable properties of rice for brewing are: (i) the least possible value of the "unavailable polishing ratio" (see Section V.C, p. 251); (ii) soft kernel into which the mycelia of the *koji* mould can easily penetrate; (iii) a high rate and large percentage of water absorption during the water-steeping process, which are related to the elasticity of the steamed rice as felt by the touch of the hand, and usually suggest easy solubilization in the mash by the *koji* enzymes. Properties (ii) and (iii) can be explained in terms of the microscopic structure of rice grains as follows (Shibata, 1968). In cross section, the rice grain is seen to contain endosperm cells arranged in concentric circles, the density being more compact at the outer edge and less in the central region. This tendency is stronger in the large grain varieties, which are highly recommended as suitable rice for *saké* brewing. These properties are largely satisfied in cleaned rice with a low polishing ratio (70–75%), which indicates that a high proportion of the bran is polished away from the original rice grains (see Section V.C, p. 251)

The quality of rice is considerably influenced by the temperature during the ripening stage of its growth and the soil of the rice fields, as well as by the variety. Japanese rice ripened under average optimum temperature conditions (21·5°) provides the most suitable raw material for brewing rich bodied *saké*. Table III gives the general analysis of rice with various polishing ratios (Yamada *et al.* 1935).

TABLE III. *Analysis of Rice (percentages)*

Type of rice	Polishing ratio	Water	Protein	Fat	Ash	Starch
Whole	0	13·9	8·0	2·20	1·10	71·0
Cleaned {	90	13·6	7·4	0·67	0·40	75
	80	13·2	5·6	0·20	0·20	77
	50	10·1	5·5	0·08	0·17	80

2. *Water*

Process water is of fundamental importance, since *saké* contains more than 80% of water, and consequently the mineral composition exerts a strong influence on the quality, type and character of the *saké*. Some types of *saké* are associated with localities where the water supply has a special chemical composition. For instance, *"Miyamizu"* which is famous for brewing *"saké* of *Nada"* is characterized by its extremely high content of phosphate, which averages 2·275 mg/l. (Kanoh, 1953, 1961; Yamada *et al.*, 1955). Of the mineral components, potassium, phosphorus and magnesium are necessary for the growth of *koji* mould as well as *saké* yeast, and have a considerable effect on the intensity of fermentation in *moto* and *moromi* mashes, while calcium and chlorine accelerate the extraction of amylases from rice-*koji* (Tokuoka, 1941; Ito, 1952; Wakabayashi *et al.*, 1957; Shimizu, 1960; Ichikawa and Maeda, 1963). However, since the amounts of these minerals in water is usually much smaller than in rice, less attention need be paid to the minerals in water used during the growth of saké yeast, except in the case of the initial stage of moto and moromi mashes, when only a small quantity of mineral matter has been extracted from the rice. Water rich in potassium is desirable, because a fairly large proportion of the potassium component of rice is washed away during the washing and steeping processes (Nojiro and Aoki, 1957; Kanoh, 1962). On the other hand, the presence of iron is most undesirable in view of the fact that it causes browning of the colour and deterioration of taste (see Section V.G, p. 266).

The presence of ammonium and nitrate ions and organic substances in the water, though not particularly harmful in itself, suggests contamination by harmful micro-organisms. For industrial brewing, the water should be neutral or slightly alkaline, colourless, crystal clear, free from objectionable odour and taste, uncontaminated by harmful micro-organisms, and free from iron and excessive amounts of other minerals so that the residue on evaporation is white in colour. Usually, medium hard water (German hardness scale 3–7°) with a rather high chlorine content (30–70 mg/l.) is recommended. Soft water enriched with adequate amounts of calcium phosphate, potassium phosphate, magnesium sulphate and sodium chloride may be used, especially for brewing mild tasting saké.

C. PRETREATMENT PROCEDURES

1. Polishing (Milling) and Steeping

The first pre-treatment is polishing the rice grains. About 25–30% of the original weight of the rice grains is removed by means of a polishing machine. The ratio (percentage) by weight of cleaned rice after polishing to the original weight of grain used is known as the "polishing ratio". The lower the ratio, the more rice bran is removed, which indicates that the grain is more highly polished. For brewing refined saké of the best quality, the most suitable grain for the purpose is selected, and often polished to a ratio of 50%. The reason why, unlike rice for food, the low polishing ratio is required, is that the bran in the outer layers of the grain contains relatively larger amounts of protein, fat and ash, which not only affect the colour and flavour of the saké, but also may to some degree be converted into substances harmful to health; for example, fusel oil is produced from protein via amino acids (see Section IV.C, p. 241). In addition, excess of vitamins and nutrients in the outer part of the grain may promote the growth of harmful contaminants.

Toyozawa and Yonezaki (1959) pointed out that the loss by crushing or smashing of rice grains is far greater than by scattering of bran during the polishing process, and proposed the use of the parameter "unavailable polishing ratio", which indicates the difference between the calculated ratio (ratio by weight of 1,000 grains measured after and before polishing) and the practical ratio (ratio of total weights in one lot measured after and before polishing). This factor is of considerable value in controlling the polishing process. It varies with the quality of the rice, the polishing ratio, the technique of the polishing procedure, etc., ranging from 0·5% to 7·0%, and the smaller the ratio the better is the polishing efficiency. After polishing, the cleaned rice is washed with water to remove residual bran, and is then steeped in water for 1–20 h according to the rate of water absorption. The rice absorbs water up to 25–30% of its weight during the steeping process, and after steeping excess water is removed.

2. Steaming

The steeped rice is steamed to gelatinize the starch and denature the protein, so that the *koji* mould can grow easily with deep mycelial penetration and the rice can be thoroughly digested by the enzymic action during mashing. At the same time the rice is sterilized to a certain extent by the steaming process. The rice is heaped up in a *koshiki*, a large shallow cylindrical open vessel made of cedar wood, aluminium or stainless steel, with a special jet in the centre of the base through which steam is blown. This is placed on top of a large kettle filled with water, and steamed for about a half to one hour under atmospheric pressure. In contrast to the long duration of steaming in the classical procedure, it has been reported that steaming for 15–20 min is enough to gelatinize the starch and denature the protein (Oana, 1933; Yamada *et al.*, 1945; Akiyama and Takase, 1958; Akiyama *et al.*, 1963).

Instead of the older batchwise *koshiki*, new machinery for continuous rice steaming, using a metal (stainless-steel) conveyor, has been devised. The rice, which is placed on the conveyer in a relatively thin layer, is steamed for about 20–30 min while being transported. The following points were taken into consideration in its design (Imayasu *et al.*, 1963): (i) maximum resistance of the rice layer to steam heat occurs during the initial stage of steaming, and indicates that the latent heat of the steam is transferred to the rice, and the steam is converted to water on reaching the rice layer; (ii) resistance to the passage of steam is controlled by the thickness of the layer, the steam pressure and the rice polishing ratio; (iii) vertical leakage of steam from the conveyer is prevented by rollers and contact boards, and horizontally by guide boards and a flexible band placed between the conveyer and the walls.

D. PREPARATION OF KOJI

Koji is a culture of *koji* mould, *Aspergillus oryzae*, grown on and within the steamed rice grains. For the preparation of *koji*, *tane-koji* seed of *koji* mould has long been used in all Japanese breweries. The *tane-koji* is prepared by drying aged *koji*, the *Aspergillus* mould having been cultured on steamed rice at 34–36° for 5–6 d; this results in abundant spore formation. Several kinds of *tane-koji* have been developed from different strains of *Aspergillus* that have not only been handed down as heirlooms by every producer, but also obtained by natural or artificial mutation. Since olden times *tane-koji* has been distributed to all breweries by *tane-koji* producers, and the strains used for *tane-koji* have eventually formed a unique group of fungi, different from the wild types, as a result of the long history of development.

The scientific name for Japanese *koji* mould, *Aspergillus oryzae*, dates back to Ahlburg and Matsubara (1878) and Cohn (1883), but it was not until the work of Wehmer (1895) was published that *Asp. oryzae* was described in detail. Later, mycological studies on *koji* mould revealed that it is a large group of *Asp. flavus-oryzae*, including innumerable strains in which there are slight graded variations in morphological and physiological properties from strain to strain (Takahashi, 1912; Thom and Church, 1926; Sakaguchi, 1933; Saito, 1950; Nehira and Nomi, 1956a, b). However, Murakami *et al.* (1968) reported that most of the *koji* mould strains for *saké* brewing belong to *Asp. oryzae*, and not to *Asp. flavus*. The latter is clearly distinguishable from *Asp. oryzae* on the basis of the following mycological characteristics of the authentic type cultures of the two species and 314 Japanese industrial strains: (i) most of the *Asp.* oryzae strains do not produce kojic acid, in contrast to *Asp. flavus* strains, when cultured in *koji* extract; (ii) most *Asp. oryzae* strains cause the rice *koji* to turn brown, while no *Asp. flavus* do so, and those *Asp. oryzae* strains which do not cause the *koji* to turn brown are strictly confined to non-producers of kojic acid; (iii) morphological properties such as the colour of surface cultures, and the arrangement of sterigmata, etc., are of little value for distinguishing *Asp. oryzae* from *Asp. flavus*. It is worth noting that no aflatoxin-producing strains are found among the Japanese industrial strains of *koji* mould (Yokotsuka *et al.*, 1967; Murakami *et al.*, 1967; Kurata, 1968). Strains of *Asp. oryzae* can also be grouped by their physiological characteristics into three types: aerobic, anaerobic and intermediate. The aerobic strains most widely distributed in *tane-koji* for *saké* brewing have been classified mainly in a group of *Asp. oryzae* having large vesicles (Murakami, 1958; Murakami and Kwai, 1958; Murakami and Takagi, 1959).

Rice-*koji* contains various enzymes, mainly amylases and proteases. The former include α- and β-amylases, sometimes accompanied by maltase and isomaltase (Okazaki, 1950); the amylases are produced in greater amount when the *koji* mould is cultured at a relatively high temperature, e.g. 40° (Suzuki *et al.*, 1956). Rice-*koji* also contains two kinds of protease, acid and alkaline, the former being more abundant (Kageyama and Sugita, 1955), especially when the *koji* mould is grown at a rather lower temperature, e.g. 35° (Suzuki *et al.*, 1957). *Aspergillus oryzae* is characterized by its strong α-amylase activity, which is considered to be one of the reasons why the unique procedure of *saké* brewing can be carried out safely starting from mash containing as much as 35% (w/v) of rice starch (Kitahara and Murata, 1953).

During the long history of *saké* brewing, it has been found that the following mycological characteristics are of major importance in *koji* moulds: (i) rapid growth on and into the grains of steamed rice; (ii) production of abundant amylolytic and a little proteolytic activity; (iii) production of good fragrance, and accumulation of flavorous compounds; (iv) limited production of coloured substances. The abovementioned group of *Asp. oryzae* can be considered as representative of industrial strains possessing most of these characteristics. During *koji* preparation, various changes take place: hydrolytic enzymes, B-group vitamins, sugars, peptides and amino acids accumulate in the rice grains, and components or precursors of the special fragrance of *saké* are produced by fungal action. Also, the temperature of the material rises gradually due to the respiration of the *koji* mould; this should be controlled so as not to exceed a maximum of 42° by intermittent stirring which also supplies the material with air.

The ability of *koji* mould to grow under both aerobic and semianaerobic conditions, enables its mycelia not only to cover the surface of the rice grains but also partially to invade the inner tissue. The key point in *koji* preparation is the maintenance of a proper balance between the hardening of the steamed rice (as the moisture in the rice grains evaporates) and the growth rate of the mould.

The following is an example of conventional *koji* preparation. After the steamed rice has been cooled to about 35° by passing through cooling apparatus on a belt conveyer, it is transferred into the *koji-muro*, a large incubation room in which the temperature (26–28°) and humidity are controlled at suitable levels for the growth of the mould. Next, *tane-koji* is dispersed in the proportion of 60–100 g/1000 kg of the material and mixed with the steamed rice, and this mixture is left heaped in the centre of the floor. At this stage the temperature of the material is 31–32°. As the spores germinate and mycelia develop, the rice begins to smell mouldy. After incubation for about 10–12 h, during which time the temperature of the material rises by about 1°, the heap of rice grains is mixed, either by hand or by a simple machine, in order to maintain uniformity of growth, temperature and moisture content. After another 10–12 h, when the growth of mould mycelia on the grains can be distinctly seen with the naked eye as small white spots, and the temperature of the material has risen to 32–34°, it is dispensed into small flat wooden trays, each of which contains 1·5–2·0 kg of mouldy grains, and the trays are laid in piles on a shelf. Thereafter, at intervals of 6–8 h, the material in the trays is gently mixed by hand. Meanwhile, the trays are replaced one upon another to equilibrate the rising temperature of the material in the upper and lower trays. During the later stages of the *koji* preparation process, special caution must be taken to prevent over-heating of the material caused by respiration of the mould. After incubation for 40–48 h, the temperature of the material rises to 40–42°, white mycelia develop to cover and penetrate the grains, and a special odour, "*koji*-fragrance", is produced. At this stage the grains contain sufficient enzymes, vitamins and various nutritive components for mashing and growth of the *saké* yeast. Finally, the rice-*koji* is taken out of the room and spread on a clean cloth at a low temperature until it is required for the preparation of *moto* or *moromi*.

Koji can be classified into many types but, in general, the following two groups can be distinguished by particular features of the mycelial development on the rice grains.

i. *Sohaze-koji* has a white layer of mycelial growth covering the whole surface of the grains. It is usually employed for mashing at a comparatively high temperature (16–18°), and is suitable for brewing *saké* of common quality. Since this type is easily prepared without any special technique or precautions, it is now widely adopted, particularly for mass production.

ii. *Tsukihaze-koji* has several white spots of fungal colonies on the surface of the grains with abundant mycelia invading deeply into the interior of the grains. This type is prepared by using highly polished rice (low polishing ratio, 60–50%) and is used for slow but prolonged mashing at a low temperature (10–12°). It is suitable for brewing *saké* of superior quality with a fruit-like flavour. In contrast to *sohaze-koji*, the preparation of this type requires technical skill and various precautions.

Through long practical experience it has become apparent that the external appearance of *rice-koji* is an important criterion in *saké* brewing. As scientific studies on rice-*koji* advanced, it was found that the appearance usually corresponded to the enzyme activity. The quality of rice-*koji* can therefore be evaluated by the activities of enzymes such as α- and β-amylases, protease, and by the correct balance between the activities of these enzymes (Matsuyama, 1964; Nunokawa, 1964; Miyoshi, 1968).

In order to save labour, and to prepare *koji* of as uniform a quality as possible, the conventional method is continually being improved, and automatic methods are being devised for commercial application. Large trays with bottom inserts of wire mesh or wooden lattice which can hold 30–45 kg of material have come into use, instead of the numerous small trays previously used (Ogawa and Wakayama, 1961). Also a large wooden table with a rim round the edge, and fitted with coarse wire mesh or perforated metal plates capable of holding 300–500 kg of material has recently been used (Ito, 1959, 1962; Sakai *et al.*, 1960). In either method it is necessary to cover the surface of the material with a clean cloth to prevent the rice grains from drying too quickly, and to ensure normal growth of the *koji* mould during the first half of *koji* preparation.

Various types of *koji*-making machines have been devised, and are used for preparing large amounts in most modern breweries (Kawamura *et al.*, 1959a, b, 1960; Terui *et al.*, 1960; Terui and Shibasaki, 1960). In these machines, a large quantity of steamed rice is heaped thickly on a wire mesh or perforated metal plate fixed in a closed box, and clean air at the required humidity and temperature is passed through the layer of material from an air conditioner, automatic adjustment being applied according to the progress of the culture. By adequate control of both the temperature and moisture of the material by the conditioned air, *koji* with almost the same quality as is prepared by the conventional method is obtained, thus shortening the time needed for *koji*-preparation by about 6–8 h.

E. PREPARATION OF *MOTO*

In the brewing of *saké*, *moto* (which means "mother of *saké*" in Japanese) plays an important role as a starter for the yeast culture when carrying out fermentations in the *moromi* main mash. It therefore corresponds to the pitching yeast in beer brewing *Moto* is required to provide a pure and abundant yeast crop, and to supply sufficient lactic acid to prevent the multiplication of harmful wild micro-organisms during the preparation of *moto* and in the early stages of the main mash fermentation. The purity of the *moto*, measured as the ratio of the culture yeast cells to the total yeast cells at the end of the *moto* preparation process, is influenced by the following factors: (i) the ratio of inoculated culture yeast cells to wild yeast cells at the commencement of mashing; (ii) the time lag for the initiation of yeast growth at the given temperature and growth characteristics of the yeast; (iii) requirements (in terms of quality and quantity) and assimilability of nutrients; (iv) tolerance for the various components of the fermentation mash such as alcohol, lactic acid, nitrite and sugar, and for environmental conditions such as temperature, viscosity and osmotic pressure; (v) oxygen supply.

As already mentioned, *moto* can be divided into two groups according to the method of preparation. In the classical procedure lactic acid is produced in the mash by lactic acid bacteria derived from the rice-*koji*, whereas in the modern method lactic acid is added to the mash at the beginning of *moto* preparation, so that little growth of lactic acid bacteria and other harmful bacteria derived from the rice-*koji* and water supply takes place.

1. Yamahai-moto

Ki-moto is a representative example of classical *moto*, and though it has been handed down from generation to generation since olden times, its practice is still significant from the view point of modern microbiology. First, the raw materials (steamed rice, rice-*koji* and water) are mashed in several small shallow wooden vessels, and the mash is thoroughly kneaded by special wooden oars. This treatment, called "*yamaorosi*" in Japanese, promotes the degradation and solubilization of rice grains at a low temperature, prior to propagation of naturally-grown *saké* yeasts, and stimulates the growth of *saké* lactic acid bacteria. After this treatment the mashes from the individual vessels are mixed together in one container. This procedure is laborious and time-consuming, besides requiring some skill. Kagi *et al.* (1909) devised a modified form of *ki-moto* called *yamahai-moto*, which is based on the same microbiological principle as *ki-moto* and has practically replaced the *ki-moto*, because the procedure is simpler. In *yamahai-moto*, mashing is done in one vessel without the process of *yamaorosi*. Instead of *yamaorosi*, strong stirring and agitation are required, for which a small mechanical agitator is available (Kodama *et al.*, 1962; Kawakami, 1964). The proportions of raw materials in *yamahai-moto* vary somewhat in different breweries.

The following is an example of the conventional method for *moto* preparation (Kodama *et al.*, 1956; Kodama, 1963). Steamed rice (120 kg) is mixed with 60 kg of rice-*koji* and 200 l. of water in a vessel at an initial temperature of 13–14°, and kept for three to four days with intermittent stirring and agitation. During this period the rice grains are partially degraded and saccharified, and the temperature gradually falls to 7–8°. The mash is then warmed at a rate of 0·5–1·0°/d by an electric heater placed under the bottom of the vessel, or by placing a wooden or metal cask filled with hot water in the mash, and after warming for a further 10–15 d, the temperature reaches 14–15°; this completes the first half of the *moto* preparation. During this stage successive changes in the wild microflora occur because mashing starts at a pH value near neutral, and groups of organisms which require few nutrients are gradually replaced by others which have complicated requirements as compounds are dissolved from the rice-*koji* and steamed rice.

In the early stages, nitrate-reducing bacteria such as *Achromobacter*, *Flavobacterium*, *Pseudomonas* or *Micrococcus* spp. (derived from water or rice-*koji*) appear, followed by lactic acid bacteria including *Leuconostoc mesenteroides* var. *sake* and *Lactobacillus sake* (derived from rice-*koji*). These bacteria multiply to reach a maximum count of about 10^7-10^8/g, but successively disappear before fermentation by the *saké* yeast begins, due to the accumulation of a high concentration of sugar and acidification resulting from the growth of lactic acid bacteria (Katagiri and Kitahara, 1934; Saito and Oda, 1932; Oda, 1935; Saito, 1950; Kodama, 1959a, b; Kitahara, 1960; Kodama *et al.*, 1960; Ashizawa, 1961; Ashizawa and Saito, 1965).

During this stage changes in wild yeasts also occur. For the first few days TTC-pink and TTC-white yeasts appear (see Section II, p. 227). The TTC-pink colonies consist mainly of film-forming yeasts, especially *Hansenula anomala* and its variety *cifferii*, and of a minority of wild *saké* yeasts. The TTC-white colonies consist chiefly of *Candida guilliermondii*, and to a lesser extent *Pichia*, *Debaryomyces* and *Torulopsis* spp. Most of these yeasts are derived from the rice-*koji* (Saito and Oda, 1934; Kodama *et al.*, 1957; Kodama and Kyono, 1963; Kodama, 1966b; Akiyama and Sugano, 1967). The TTC-pink yeasts, which number about 10^3/g for about the first 10 d, rapidly decrease to almost zero within another two or three days due to the toxic effect of nitrite, produced in the concentration 5–7 mg/l. by nitrate-reducing bacteria from nitrate contained in or added to the water (Hanaoka, 1918; Zenda, 1920; Kanai and Iida, 1932; Ashizawa, 1961, 1962, 1963). Yeasts of the *H. anomala* group are easily detected by plating on a medium containing nitrate as sole nitrogen source. This aerobic yeast reaches a maximum count of 10^4-10^5/g on the surface layer of the mash, followed by a rapid decrease due to the anaerobic conditions in the viscous mash (Saito, 1950; Kodama, 1960). The TTC-white yeasts can still be detected after the TTC-pink yeasts have disappeared, but the count falls as the fermentation by *saké* yeast proceeds.

The TTC-red yeasts, which are considered to be the same as or very similar to *Kyokai* No. 7 can, in a few cases, be detected in the initial stages. When wild yeasts, especially wild *saké* yeasts of the TTC-pink type, which are frequent contaminants, have almost disappeared $(10-10^2$/g) because of the toxic effect of nitrite, and the temperature of the mash reaches about 15°, the pure culture of *saké* yeast (usually *Kyokai* No. 7) is added to give a count of 10^5-10^6/g. The mash filtrate at this stage has the following composition: density, 16·0–16·5 degrees

Baumé; reducing sugars, 26–28%; amino acids (as glycine) 0·5–0·8%; total acids (as lactic, 0·30–0·40%. After 2–3 d, when the temperature has risen to 17–19° and increased acidity (0·5–0·6%) has caused the removal of nitrite, the inoculated *saké* yeast grows fully and fermentation begins. When the yeast cell count has reached about 10^8/g, vigorous fermentation takes place, accompanied by a temperature rise to 20–23°. A few days later the mash is gradually cooled to prevent the yeast from dying or being weakened by the high concentrations of alcohol and acid. After resting for a further 5–7 d, the *moto* is used for the main fermentation of *moromi* mash. The filtrate of the final *moto* mash has the following composition: density, 4–6 degrees Baumé; alcohol, 12–15%; amino acids, 0·45–0·65%; total acids, 0·9 1·0%.

It is noteworthy that the amount of amino acids contained in *moto* of this classical type is two or three times higher than that of the modern variety, *sokujo moto* (Oana, 1931). Regarding the cause of the difference between the two types of *moto* in the amount of amino acids present, it has long been thought that in the former, sufficient breakdown of rice protein by *koji* proteases is attained under conditions in which the pH value changed slowly from almost neutral to slightly acid due to the growth of lactic acid bacteria; whereas in the latter, strongly acidic conditions, caused by the initial addition of lactic acid, inhibits the activity of *koji* proteases (Sugiyama and Nagahashi, 1932). However, it was found that *koji* proteases consist mainly of an acid protease system by the action of which rice protein in the *moto* mash is broken down to soluble compounds, but is not fully decomposed to amino acids under the strongly acidic conditions, as in the case of *sokujo-moto* (Kageyama, 1955). Therefore, the above concept appears still to be controversial. In this connection, Akiyama (1957, 1958a, b) assumed that in the early stages of *yamahai moto* preparation, when the pH value of the mash is almost neutral, the rice protein is degraded to an intermediate substance which is acid-soluble but water and tri-chloroacetic acid-insoluble, easily decomposed into amino acids under the acidic conditions, and therefore is presumably a precursor of amino acids. Further, by adding lactic acid to *moto* mash in small portions at intervals during the first stage of *moto* preparation, in order to lower the pH value gradually, the content of amino acids could be increased to almost the same as in *yamahai moto* (Akiyama, 1959).

On the other hand, Takeuchi and Shimada (1965a, b) and Takeuchi *et al.* (1967, 1969) pointed out that non-specific physico-chemical interactions between protein and starch gel might occur in *moto* mash; solubilization of some of the cereal protein under acidic aqueous conditions, or enzymatic hydrolysis of protein over a wide pH range is affected by the coexistence of protein and starch gel. However, no interference with rice protein solubilization by starch gel under acidic conditions occurs at the near-neutral pH values existing in the early stages of *moto* mash of the classical type. This phenomenon may be explained if protein and starch combine by electrostatic force. Further, enzymatic hydrolysis of protein adsorbed on starch gel is presumably inhibited by the operation of van der Waals forces.

To avoid interference with the hydrolysis of protein, it is necessary to rid the mash of gelatinized starch, which causes pastiness, as soon as possible by hydrolysing it to glucose. The procedure for hardening steamed rice (by cooling it to a low temperature before mashing) and kneading or braying the hardened rice (which have been practised since olden times) are effective, not only for saccharification of the degraded starch, but also for solubilization of the rice protein with resultant abundant formation of amino acids. Both the conceptions mentioned above provide strong clues towards the elucidation of the mechanism of the proteolysis that occurs during the preparation of *ki-moto* or *yamahai-moto*. The kneading procedure also helps to dissolve the nutrients contained in the rice-*koji* and steamed rice, and provides semi-anaerobic conditions in the mash which promote the growth of lactic acid bacteria in the early stages (Kodama, 1960).

The advantage of this classical method is that a healthy *moto*, containing a yeast inoculum of high purity, is obtained by the natural selection of wild yeasts. However, a drawback is that the procedure requires rather a long time (25–30 d) and hard physical labour, besides considerable technical skill.

2. Sokujo-moto (*Rapidly-processed* Moto)

Moto of this type, devised by Eda (1909), is most popular in modern *saké* brewing. It can be prepared in a relatively short time (12–15 d) by a fairly simple procedure, since the solubilization and saccharification of the mash proceed more quickly, due to the high initial mashing temperature (18–22°). In this method commercial lactic acid (75%, S.G. 1·21) is added to the mash (650–700 ml/100 l. water) to give a pH value of 3·6–3·8, in place of lactic acid produced by naturally-occurring bacteria. However, even though pure culture yeast is used as the inoculum, wild *saké* yeasts as well as the culture yeast can develop during the *moto* process, and often the wild types finally predominate (Akiyama and Sugano, 1967). This may be ascribed to the fact that the high mashing temperature and acidic conditions are close to the optimum for the multiplication of both culture and wild yeasts. In addition, as opposed to the behaviour in the classical process, no natural selection of wild yeasts by the toxic effect of nitrite occurs, because nitrate-reducing bacteria are inhibited by the presence of lactic acid. To prepare *moto* containing a high proportion of culture yeast, it is necessary to inoculate the yeast at a high level (10^5–10^6/g) at as early a stage as possible (Tsukahara, 1954). Also, in contrast to the classical method, gentle mixing rather than kneading should be used under the strongly acidic conditions during the early stage of the preparation, to produce a mash which is less pasty, and rich in sugar and other nutrients for the growth of *saké* yeast. Except for several points mentioned above, the procedure is almost the same as that of *yamahai-moto*.

3. Kōontoka-moto (*Hot-mashed* Moto)

This is a variety of *sokujo-moto*, but mashing is carried out at 56–58° for 5–7 h, which is not only the optimum condition for liquefaction and saccharification of rice starch by the *koji* enzymes, but also effectively sterilizes the vegetative cells of wild yeasts and bacteria. A high ratio of water to rice (150–160 l./100 kg) produces a thin mash and prevents excessive accumulation of sugar and high viscosity, both of which inhibit the growth of the *saké* yeast. A thin mash also makes it easier to raise the temperature to a suitable level for enzymic digestion, and to cool it for preparing *moto*.

The following example is typical: to 220 l. of water (out of a total of 270 l.) warmed to 60°, a mixture of 120 kg of steamed rice, cooled to about 55°, and 60 kg of rice-*koji* are added in a vertical mashing vessel made of stainless-steel and equipped with a water-jacket and agitators for heating and cooling. After incubation for about six hours, the mash is cooled by adding ice blocks equivalent to the remaining water (50 l.), and lactic acid is added at a temperature below 50°. The filtrate of the mash thus obtained has the following composition: density, 16·0–16·5 degrees Baumé; reducing sugars, 24·5–25·5%; amino acids, 0·16–0·20%; total acids, 0·30–0·35%. The mash is further cooled to about 20° and is transferred to another vessel for inoculation with pure culture yeast (10^5–10^6/g). From this stage on, the procedure is almost the same as that for *sokujo-moto*.

The advantage of this method is that excellent *moto*, containing inoculated yeast of high purity, is obtained more quickly (8–10 d) since most of the wild yeasts from the rice-*koji* are killed (Akiyama and Iwata, 1964) and the broth is rich in nutrients for the growth of *saké* yeast during the hot-mashing process.

4. Compressed Yeast Method

An attempt to use culture yeast in place of *moto* was made by Takahashi (1907). Later it was shown by Yamada *et al.* (1939) that an experimental brewing test could safely be carried out by using a *saké* yeast crop from a culture in Carlsberg vessels as a substitute for conventional *moto*. Recently, based on the same principle, a new procedure which may be called "the compressed yeast method" has been devised by Shimoide *et al.* (1964, 1965b). By this procedure compressed *saké* yeast harvested from an aerobic propagation is used.

The following is an example of this method. One hundred litres of medium, composed of molasses, urea, salts and 0·5% lactic acid, is placed in a closed stainless-steel vessel of 180 l. capacity equipped with coils for sterilization and cooling, an agitator, a compressor and an air filter. After the seed culture has been added, the propagation is continued for about 48 h at 28° with an aeration rate of 90–120 l./min and agitation at 250 r.p.m. The yeast crop is harvested by centrifugation and washing, and put through a filter press. In this way 3·5–4·0 kg of fresh yeast, with a water content of about 70%, is obtained from 100 l. of medium. Instead of the *moto*, about 300–500 g of compressed yeast is added to the first stage of the *moromi* mash per 10^3 kg of total polished rice, with lactic acid equivalent to that contained in the *moto*. According to Miyasaki *et al.* (1965), a medium containing malic acid in place of lactic acid can also be used for this purpose; this provides the advantages of satisfactory yeast yield, pH stability, and easy separation and washing of the yeast.

When this method is used for *saké* brewing, the nutrients for *saké* yeast, such as B-group vitamins and amino acids that are transferred from the *moto*, are initially deficient in the *moromi* fermentation process. However, the cells obtained by this method have high respiratory activity and grow sufficiently well for fermentation to proceed normally (Fukui *et al.*, 1966). Comparison of the *saké* brewed by using conventional *moto* with that produced with compressed yeast revealed no differences in the amino acids content or the quality by the sensory responses (Shimoide *et al.*, 1965a).

Regarding the role of *moto* in *saké* brewing, there appear to be two contradictory opinions. One view is that, because the role of *moto* is to provide sufficient pure active yeast for healthy and vigorous fermentation of the main mash under acidic conditions, the simplest method for preparation is to be preferred, provided that *saké* of good quality can be produced. In fact, statistical analysis of sensory tests on many brands of *saké* from different districts indicated that the method of preparing *moto* is not relevant to the character or quality of the *saké* produced therefrom (Kanto-Shinetsu Tax Administration Bureau, 1961, 1962, 1964). The alternative opinion is that, since the methods of *moto* preparation have a considerable influence on the delicacy of bouquet, flavour and taste of *saké*, individual brewers should choose a method by which *saké* with its own particular trait can be brewed.

F. *MOROMI* FERMENTATION

Moromi main mash is fermented in large open vessels of enamelled or glass-lined iron, or of stainless-steel, with capacities ranging from about 6 to 18 kl. The unit weight of polished rice used for mashing in one vessel has long been standardized at 1·5 tons, but recently in large breweries 3–5 tons has become accepted in correspondingly larger vessels. The main mash is composed of steamed rice, rice-*koji* and water. Table IV gives an example of the proportion of raw materials used for a typical *moromi* mash.

TABLE IV. *Proportion of Raw Materials in* Moromi *using Hot-mashed* Moto

	Total rice (kg)	Steamed rice (kg)	Rice-*koji* (kg)	Water (l.)
Moto mash	180	120	60	270
1st addition	510	360	150	500
2nd ,,	1020	810	210	1230
3rd ,,	1710	1350	360	2329
4th ,,	225	225	—	286
Total	3645	2865	780	4615

Steamed rice, rice-*koji* and water are added to the *moto* yeast starter (already prepared) in batches of increasing size which are mashed consecutively over a period of four days, as shown in Table IV. After adding the first small batch to the *moto*, an interval of one day is allowed for the propagation of the *saké* yeast, and before adding the second batch the yeast count has reached about 10^8/g, the same order as in the *moto*. The temperature of the first mash on the first day is 12–13°, and fermentation starts slowly on the second day. The second and third batches are added on the third and fourth days respectively. Meanwhile the mash temperature is lowered by adjusting the temperatures of the steamed rice and water, and is held at 7–8° on the fourth day. This successive mashing in three steps, accompanied by gradually lowered mash temperatures, plays an important role in restricting invasion by wild contaminants, since the high population of inoculated *saké* yeast and the concentration of lactic acid in the original *moto* is kept from being diluted in one step.

During the fermentation of *moromi*, various biochemical changes take place: (i) saccharification of rice starch to glucose by *koji*-amylase; (ii) fermentation, converting glucose to ethanol and carbon dioxide by the action of *saké* yeast; (iii) formation of organic acids (lactic, succinic, malic, etc.) from glucose by *saké* yeast and *koji* enzymes; (iv) decomposition of rice protein to peptides and amino acids by *koji* protease; (v) conversion of amino acids to higher alcohols by *saké* yeast; (vi) conversion of lipids to fatty acids and glycerol by *koji* lipase, and esterification of fatty acids with alcohols by *saké* yeast.

It is significant that both saccharification and alcohol fermentation occur simultaneously in *moromi* mash, whereas the processes are consecutive in *moto* mash. After adding the third batch, the mash is agitated, usually twice daily with a long wooden paddle, to assist the solubilization of the materials. Three to four days later the density of the mash reaches a maximum. The composition of a filtrate of the mash at this stage is: density, 7–8 degrees Baumé; alcohol concentration, 3–4%; total acids (as succinic acid) 0·06–0·07%. A froth resembling soap suds gradually spreads over the entire surface, and subsequently increases to a thick foam. A fresh fruit-like odour at this stage indicates a healthy fermentation.

The fermentation gradually becomes more vigorous with the rise in mash temperature, and a rather viscous froth rises to form *taka-awa* (deep layer of foam) which reaches to the brim of the vessel. For several days this has to be broken down with a small electric agitator. At this stage, the yeast cell count reaches its maximum, about $2·5 \times 10^8/\text{g}$ (Nojiro, 1959). As the alcohol and acid concentrations increase, the foam becomes less dense, and is easily dispersed. Later, when the density of the mash falls, the froth begins to recede, displaying numerous beautiful hemispheres, followed by various features including wrinkled scum, a smooth thick or thin covering, or no covering on the surface of the mash according to the type of *saké* yeast and the physical and nutritional conditions (Tsukahara *et al.*, 1964; Saito, 1965).

The temperature of the mash reaches a maximum of 13–18° by the sixth to the ninth day, and this is maintained for another 5–7 d, after which it decreases as the fermentation subsides. On the twentieth to twenty-fifth day, when the alcohol concentration in the mash has reached 17·5–19·5% and the fermentation has almost ceased, purified alcohol (30–40%) is usually added to the mash to adjust the final concentration to about 20–22%. If necessary, other additions such as glucose, lactic acid, succinic acid or sodium glutamate are made; the amounts, including alcohol, conforming to government standards. This addition of purified spirit and other substances to the mash was legally permitted in the 1940s to compensate for the fall in *saké* production due to shortage of rice, and even today this practice is continued, since correctly chosen amounts of these additives make the quality of *saké* more uniform and clean. The officially permitted amount of added alcohol is 220 l. (calculated as 100% ethanol) per 1,000 kg of unpolished rice, but addition of the maximum amount reduces the sweetness and body of the *saké*. However, at present sweet flavoured *saké* is generally preferred.

Accordingly, in order to sweeten the mash, 7–10% of the total amount of steamed rice is occasionally added during the final stage of the *moromi* process. By this means a certain amount of glucose produced from the starch by the saccharifying action of the rice-*koji* accumulates in the mash, because of the weakened fermentative activity of the yeast in the presence of over 15% alcohol. Further, for the purpose of obtaining *saké* with better body, some of the additional alcohol may be omitted or replaced by rice at a legally-defined exchange ratio. Usually no alcohol is added to the mash for exported *saké* or some of the so-called "de luxe" grades.

An example of the changes with time in various mash components during the *moromi* fermentation process is shown in Table V.

TABLE V. *Changes with Time in Various Components of Mash during* Moromi *Fermentation (without the fourth addition)*

Time from start (third addition) (d)	Gravity (degrees Baumé)	Reducing sugar (%)	Alcohol (% v/v)	Acid (as succinic) (%)
3–4	7–8	7–8	3–4	0·06–0·07
6	5–6	6–7	6–7	0·08–0·09
9	4–5	6–7	9–10	0·09–0·10
12	3–4	5–6	11–12	0·10–0·11
15	2–2·5	4–5	15–16	0·11–0·15
18	1–1·5	3–4	17·5–18·5	0·14–0·16
21	0–0·5	2–3	18·5–19·5	0·14–0·16

There would appear to be more contamination by wild *saké* yeast in the *moromi* than in the *moto* process. If wild *saké* yeast predominates over the *saké* culture yeast in the *moto* process, this effect is further accentuated in the *moromi* process where a greater bulk of material is mashed, and the culture yeast may be completely overgrown by the contaminant. It is important, therefore, to suppress wild *saké* yeast in the *moto* process as much as possible; the proportion of wild *saké* yeast should not exceed 10% of the total in the *moto* (Akiyama and Sugano, 1967). To maintain normal pure fermentation in the *moromi* process by the yeast used as inoculum, a sufficient amount of aerobically-cultivated yeast is valuable as a starter in place of the conventional *moto*, as explained above (Section V.E. 4, p. 261).

Deterioration due to contamination by *Lactobacillus* spp. rarely occurs during *moromi* fermentation. If, however, the growth of *saké* yeast is extremely retarded during the early stages of the *moromi* process (because of the use of over-aged *moto* in which the majority of the yeast cells are weakened or dead, or of an exceptionally low temperature in the *moromi* mash), lactic acid bacteria derived from rice-*koji* predominate over the yeast, and acid damage occurs when the viable cell count of the bacteria reaches more than $10^7/g$ (Momose *et al.*, 1965). In this case fermentation declines rapidly and may cease entirely, while accumulation of lactic acid and sugar continue progressively, resulting in an excessively sweet and acid mash, the so-called *kan-san-pai* (sweet and acid-damaged) *moromi*. Competition for the uptake of nutrients between *saké* yeast and lactic acid bacteria, as discussed above (Section III.B, p. 235), may be the basic cause. In order to prevent wild contaminants from obtaining a foothold through which they may invade the *saké* brewing process, and to ensure healthy fermentation throughout the brewing season, brewery hygiene must be of the highest standard. The importance of cleanliness and sanitation in all department of the brewery cannot be too strongly impressed upon the brewer, because only in a clean atmosphere and with clean equipment and utensils can sound and good *saké* be produced.

G. FILTRATION, PASTEURIZATION AND STORAGE

After standing for one or two days, the mash is poured into bags of about 5 l. capacity made of linen or synthetic fibre, which are placed in a large wooden or stainless-steel box and filtered under hydraulic pressure to remove the solids. An efficient automatic mash filter has recently been devised for commercial use (Yabuta, 1965; Imayasu, 1966). The slightly turbid *saké* filtrate obtained by using the conventional mash filter is clarified by settling, and five to ten days after filtration, by which time the fresh *saké* has become almost completely clear, the supernatant and sediment are separated. The solids contain starch, cellulose, protein, yeast cells and enzymes. During the precipitation process the fresh *saké* is gradually aged, probably because of the enzymatic decomposition of various components. If the sediment is kept for too long, especially at high room temperatures, without being separated from the supernatant, autolysis of the yeast cells, secondary fermentation or over-ripening of the *saké* may occur, resulting in a deterioration in quality.

After separation, the supernatant is pumped to another vessel through a filter of activated carbon, asbestos or cotton. This process is repeated once or twice, and the bottom layer of filtrate, which still contains a certain amount of sediment, is mixed with another mash. The clarified *saké* is then blended in order to ensure uniform quality. After standing for a further 30–40 d, pasteurization is carried out to kill or suppress yeasts and harmful micro-organisms, destroy enzymes, and adjust the maturity of the *saké*. The fresh *saké* is heated to 55–65° by passing through a coil of tinned copper, aluminium or stainless-steel immersed in a tank or tube containing hot water. Recently multi-tube or plate-type heat exchangers with high heat transfer efficiency have come into use.

A memorandum on *saké* brewing in an old diary preserved in a Buddhist temple in *Nara*, records that the same "pasteurization" process (invented by Pasteur in 1865) had already been in use since the *Eiroku period* (1558–1570) to sterilize fresh *saké* brewed during the warm season, and many descriptions of similar processes are found in books of later years (Sakaguchi, 1964). Immediately after pasteurization the *saké* is transferred to sealed vessels for storage. Closed vertical or horizontal vessels, made of enamelled or glass-lined steel, or stainless-steel, are now used for storage instead of the traditional wooden tubs. The maturation of pasteurized *saké* during storage is probably due to oxidation reactions and physico-chemical changes. The storage building should preferably be kept at 13–18°, consideration being given to the rate of maturing and the time of bottling.

The composition of *saké* differs somewhat from brewery to brewery, and even in the same brewery may vary between batches according to the polishing ratio of the rice, the proportion of raw materials used for mashing, including the amounts of additional alcohol and glucose, as well as various factors in the brewing process. Table VI gives average analytical values for several components of 45 kinds of *saké* (Hayashida *et al.*, 1968). In this table the values are calculated on the basis of an alcohol content of 15%.

TABLE VI. *Analysis of* Saké *(Averages of 45 types)*

Analysis	Content
Total sugar, as glucose (%)	4·20
Direct sugar, as glucose (%)	3·46
Acidity (ml N/100 ml)	1·52
Total organic acids (mg/100 ml)	115·22
Glutamic acid (mg/100 ml)	20·23
Total nitrogen (%)	0·0726
Formol nitrogen (%)	0·0288

Typical levels of sugars, B-group vitamins, organic acids and amino acids in *saké* are shown in Tables VII, VIII, IX and X respectively.

Differences in the yield of *saké* from polished rice may occur; these are normally recorded as l./10³ kg. The average yield from breweries

TABLE VII. *Sugars in* Saké

Type of sugar	Content (%)		
	Aso *et al.*, 1954	Fukinbara and Muramatsu, 1952	Masuda *et al.*, 1965
Total sugar	4·53	—	3·76
Direct sugar	3·77	—	2·28
Glucose	3·69	2·31	2·38
Sakebiose	0·08	⎱0·18	—
Kojibiose	0·08	⎰	—
Isomaltose	0·33	0·54	0·48
Panose	0·08	0·21	⎱0·23
Isomaltotriose	0·08	⎱0·32	⎰
Other oligosaccharides	0·28	⎰	

TABLE VIII. *B-group Vitamins in* Saké (*Assay method given in brackets*)

Vitamin	Content (ng/ml)	
	Akiyama, 1963	Tani, 1956
Thiamine	5·3–12·0	0
	(*Lactobacillus fermenti* 36)	(Thiochrome)
Riboflavin	13–69	2–84
	(*Lactobacillus casei E*)	(Lumiflavin)
Pantothenic acid	80–260	110–410
	(*Lactobacillus arabinosus* 17–5)	(*Saccharomyces carlsbergensis*)
Nicotinic acid	123–171	750–860
	(*Lactobacillus arabinosus* 17–5)	(*Lactobacillus arabinosus*)
Folic acid	0·53–1·28	
	(*Streptococcus faecalis R*)	
Biotin	0·05–0·08	0·5–14
	(*Saccharomyces carlsbergensis*)	(*Saccharomyces carlsbergensis*)
Pyridoxin	650–930	690–2270
	(*Saccharomyces carlsbergensis*)	(*Saccharomyces carlsbergensis*)
Inositol	3,600–8,400	11,000–31,000
	(*Saccharomyces carlsbergensis*)	(*Saccharomyces carlsbergensis*)

in the *Tōhoku* district in 1966 was 3,145 l./10^3 kg. Since stored *saké* may vary in flavour and taste from batch to batch, blending is necessary to achieve uniform brand quality. The blended *saké*, containing 20–22% alcohol, is diluted by adding water to 15·0–16·5% according to the class required (special, first or second). Filtering through a small amount of activated carbon after dilution improves the flavour and taste. Finally

TABLE IX. *Organic Acids in* Saké

Acid	Content (mg/100 ml)				
	Otaka and Yamanouchi, 1955 (silica gel and paper chromatography)	Ueda *et al.*, 1960 (silica-gel chromatography)	Matsui and Sato, 1963 (gas chromatography)	Matsui and Sato, 1966 (silica-gel chromatography)	Muto *et al.*, 1964 (gas chromatography)
Acetic		3·5–11·9	24·9	7·8–12·0	7·2–23·4
Propionic		0·5–0·6	1·0		
n-Butyric			1·4		
i-Butyric			1·0		
Oxalic	trace			0–2·7	
Malonic				0–7·3	
Succinic	54·0	49·4–61·5		73·2	28·3–44.9
Fumaric	1·0	1·7–3·4		18·6–33·7	
Aconitic		2·0–7·8			
Glycolic		4·7–6·4		7·6–13·7	
Lactic	55·0	37·1–52·1		34·2	13·5–36·9
Malic	19·0	19·4–39·0		22·8–44·2	9·4–32·2
α-Hydroxyglutaric					trace–1·5
Citric	12·0	3·5–5·4		17·3–48·0	trace–7·1
Pyruvic	1·0	0·5		0–6·2	
α-Ketoglutaric	trace	0·3–1·2		0–17·5	
Pyroglutamic	18·0			12·9–23·2	

TABLE X. *Amino Acids in* Saké

Amino Acid	Content (mg/100 ml)		
	Tamura *et al.*, 1952 (bioassay)	Omachi and Kawano, 1957 (chromatography)	Takahashi, 1965 (paper chromatography)
Alanine		38	25
Arginine	48	46	19
Aspartic acid	45	26	21
Cystine	3	trace	4
Glutamic acid	75	47	29
Glycine	36	30	31
Histidine	17	21	12
Isoleucine	21		13
Leucine	53	}105	25
Valine	50		21
Lysine	19	18	19
Methionine	9	trace	3
Ornithine			4
Phenylalanine	37	29	8
Proline	40	27	11
Serine	51	22	24
Threonine	22	18	4
Tryptophan	3	trace	7
Tyrosine	39	12	6
Total	538	349	286

it is bottled and re-pasteurized. The quality of *saké* sold as special and first class requires authorization by official appraisers. Besides alcohol percentage, the specific gravity, which is correlated with sweetness or body, is measured by a special hydrometer (*saké* meter) in which one scale unit indicates one-tenth of a degree Baumé, minus (−) and plus (+) indicating heavier and lighter than water respectively. The *saké* meter value of *saké* on the market differs somewhat from brand to brand. The average value of 63 kinds of *saké* (first class) was reported as − 5·7 (Sendai Tax Administration Bureau, 1968). In contrast to wine, *saké* is usually consumed within about a year of being brewed.

In regard to the colourants in *saké*, flavin and melanoidin were long considered to be the chief components (Kurono and Katsume, 1930; Teramoto, 1936). Recently, however, Ueno *et al.* (1966) reported that the percentage of colour (optical density at 420 nm) resulting from flavin compounds was small (1·2–6·0%). The marked increase of fluorescence in old *saké* (stored after pasteurization) is considered to be due to melanoidin, which has an intense fluorescence. This is formed by the Maillard reaction (Kurono and Katsume, 1930) in a similar way to the browning reaction in other foods; and harmane (blue fluorescence) and other fluorescent compounds (yellow fluorescent YS compound) derived from tryptophan (Takase and Murakami, 1966, 1967; Takase *et al.*, 1968a, b). Many investigators have studied the *saké*-colurants, involving iron (Eda *et al.*, 1929; Kawasaki, 1955; Yoshizawa, 1958; Yoshizawa and Makino, 1960; Osaka Tax Administration Bureau, 1960; Kobayashi and Akiyama, 1961), since the presence of iron in *saké* causes heavy colorization and deterioration of flavour and taste. Recently, Sato *et al.* (1967) showed that the greater part of the iron-containing *saké* colourants is composed of ferrichrysin, which belongs to the ferrichrome series (Neilands, 1967). In *saké*, ferrichrysin is also present in the colourless iron-free form, deferri-ferrichrysin, which is produced by *Aspergillus oryzae* on the steamed rice during *koji* preparation. Most of the ferrichrysin in *saké* is thought to be produced during the course of brewing from deferri-ferrichrysin in rice-*koji* containing iron present as a contaminant. At present excessively coloured *saké* is not acceptable for sale, but it is difficult to remove ferrichrysin from *saké* unless high concentrations of activated carbon or aluminium trichloride are used. Therefore, precautions should be taken to prevent contamination by iron throughout the course of brewing, and to use iron-free water.

Activated carbon at a concentration of 200–300 g/kl. is normally used for decolorizing *saké*. According to Sato *et al.* (1967) the total adsorption ratio of colourants in *saké* depends mainly on that of the melanoidin, and only to a small extent on that of riboflavin and ferrichrysin. Besides the decolorization effect, the treatment with activated carbon plays an important role in adjusting the maturity of unpasteurized *saké* by adsorbing residual enzymes and various other components.

Reports on the flavour components of *saké* have been published by Yamada *et al.* (1928), Higashi (1928), Yamamoto *et al.* (1953), Yamamoto (1961a), Yoshikawa *et al.* (1963), Sasaki (1964) and Komoda *et al.* (1966). Among these, Yamamoto (1961b, c) and Yamamoto *et al.* (1961) reported that ethyl esters of α-keto carbonic acids (e.g. pyruvic, α-keto-*n*-butyric, α-keto-isovaleric, α-keto-isocaproic) and of α-hydroxy carbonic acids (e.g. lactic, α-oxy-butyric, α-oxy-isovaleric, leucic), phenethyl acetate, ethyl caprylate and some of aromatic acids (e.g. ferulic, vanillic) and non-volatile carbonyl compounds (e.g. benzaldehyde, phenyl acetaldehyde, cinnamic aldehyde) are components which contribute to *saké* flavour. These components were separated by adsorption, elution or fractional distillation, and were identified by paper or gas chromatography. According to Komoda *et al.* (1966), the compounds associated with *saké* flavour collected from fermenting mash by passing the evolved carbon dioxide through a solvent trap, consist mainly of ethyl and isoamyl acetates, isoamyl, *active* amyl and isobutyl alcohols, ethyl caproate, ethyl caprylate, isobutyl acetate and ethyl butyrate. In the particular case of "*ginjoshu*", *saké* of the best quality with a unique bouquet and flavour, which is brewed from highly polished rice (polishing ratio, 50–45%) by an elaborate technique, at an extremely low mashing temperature, larger amounts of the ethyl esters of acetic, butyric, caproic and lauric acids are present (Yoshizawa, 1966). The combined components, the proportions of which may vary somewhat in *saké* from different breweries, produce a delicate bouquet and aroma which is an intrinsic characteristic of *saké*.

With regard to the relationship between the constituents and the taste of *saké*, alcohol and other volatile substances, glucose and its complexes, organic acids such as lactic and succinic, amino acids, amines and decomposition products of nucleic acids are known to be essential components, although it also contains various other metabolic products of *Asp. oryzae* and *Sacch. cerevisiae* which are supposed to influence the delicacy of flavour and taste.

It seems that high molecular weight substances, including peptides and mucopolysaccharides, contribute with other components to the smoothness and balance of the taste. When several fractions of high molecular weight, particularly nitrogen-containing substances such as peptides, were removed from *saké* by means of dialysis and electrophoresis, the taste deteriorated but was to a large extent restored by adding these fractions back to the treated *saké* (Takahashi *et al.*, 1955; Ando, 1957; Iida, 1960; Tajima, 1965; Otaka, 1958). It has also been reported that the complexity and smoothness of the taste are related to the buffering action of *saké* components (Tanaka, 1952; Tajima, 1962; Iwata *et al.*, 1964) and to the well-balanced composition of organic acids and sugars (Sato, 1957). At present, it seems that the factors responsible for the special taste and flavour are highly complicated, and not confined to the chemical components stated above.

Spoilage of *saké* is sometimes encountered. The so-called *hiochi* is a characteristic damage after pasteurization which causes turbidity and a disagreeable off-flavour and taste. Atkinson (1881) pointed out that this was due to the presence of bacilli. More recent microbiological studies have revealed that the *hiochi* organisms are lactic acid bacteria, most of which require certain peptides, amino acids, vitamins and bases for growth, and have a tolerance to 12–16% alcohol. According to Kitahara *et al.* (1957) the *hiochi* bacteria are classified into the following two groups, based on the criteria proposed by Takahashi (1907b) and Yamasaki (1926).

a. True-*hiochi*-bacteria

Addition of *saké* or *koji* extract is essential for the growth of this group. The optimum pH for growth is 5·0, in contrast to normal lactic acid bacteria true-*hiochi*-bacteria grow slightly at pH 6·0 but not at pH 7·0.

> *a*–I. Homo type: *Lactobacillus homohiochii*
> *a*–II. Hetero type: *Lactobacillus heterohiochii*

b. *Saké*-saprogenic lactic acid bacteria

Addition of *saké* or *koji* extract is not essential for the growth of this group.

> *b*–I. Homo type: *Lactobacillus plantarum*, *Lactobacillus acidophilus*.
> *b*–II. Hetero type: *Lactobacillus fermentum*, *Leuconostoc mesenteroides*.

Tamura (1956, 1958a, b) isolated a growth factor for the first group of *hiochi* bacteria in crystalline form from *saké* and from the culture fluid of *Asp. oryzae* and other micro-organisms, and named it hiochic acid. This was identified as mevalonic acid, the growth-promoting factor for *Lactobacillus acidophilus* ATCC 4963, isolated from distiller's solubles by Folkers *et al.* (1956). Tomiyasu (1933) showed that *hiochi* off-flavour is caused by diacetyl, which is present in contaminated *saké* in concentrations as high as 1·57 mg/100 ml, whereas the content in normal *saké* is less than 2×10^{-5} mg/100 ml. It seems that the *hiochi* off-flavour is produced mainly by true-*hiochi* bacteria; while a strong acidic odour, but not *hiochi* off-flavour, is produced by the second group of *saké* lactic acid bacteria.

So-called "*saké*-protein turbidity" occurs independently of the putre-faction mentioned above when fresh and clear *saké* turns slightly turbid by pasteurization. This turbidity, if observed in bottled *saké*, would cause the product to lose its market value. Ichikawa *et al.* (1955) re-ported that the turbid material consists mainly of protein and poly-saccharide, and it is presumed to be closely associated with the auto-lysis of *saké* yeast and *koji* mould, since the carbohydrate components of the turbid material are identical with those of the hydrolysates of these micro-organisms. On the other hand, according to Sugita and Kageyama (1957, 1958a, b) the precursor of the turbid material is the β-amylase of *koji* mould. The β-amylase activity of *saké* has been shown to be proportional to the turbidity after pasteurization (Suzuki *et al.*, 1958). The content of turbid material was of the same order as that of β-amylase (20–70 mg/l.) and the protein content was about 60%. It is characteristic of its amino acid composition that the aspartic acid con-tent is higher than that of glutamic acid, and the basic amino acids are lower than in typical protein and α-amylase. This particular amino acid composition is analogous to that of enzyme proteins, especially enzyme preparations from rice-*koji*. The turbid material also contained about 30% of polysaccharide, composed of mannose and galactose (Akiyama, 1962a). These investigations show that the main compon-ent of the turbid material in *saké* is β-amylase produced by the rice-*koji*.

Either physical or enzymic clarification methods can be used to re-move *saké*-protein turbidity by precipitation. For the former approach, protein precipitants of various kinds (e.g. egg white, wheat flour and gelatine in combination with persimmon tannin), while in the latter the bacterial protease preparation "Spitase F" (Inoue and Kawasaki, 1953) and the fungal protease preparation Taka-diastase "Klarin-S" are em-ployed respectively (Yamada and Akiyama, 1957; Akiyama, 1962b).

VI. Conclusions

Saké brewing has long been carried out largely on the basis of tradi-tional techniques which have been handed down and improved from generation to generation. Recent knowledge of the physiology, bio-chemistry and ecology of the micro-organisms involved enables us to confirm the fundamental soundness of the conventional methods of brewing. This scientific approach may also make possible further im-provements in the technology of *saké* brewing by eliminating certain irrational procedures. In fact, there are many modifications in *saké*

10—Y. 3

brewing techniques which are based on the personal experiences of older workers. For example, new equipment for continuous steaming and filtration of mash, automatic *koji* preparation and cultivation of yeast have come into use, and pneumatic systems and belt conveyors for transporting rice and rice-*koji* have replaced manual labour. Moreover, up to date research on the application of commercial enzyme preparations (obtained from various moulds and bacteria), and of submerged culture of *koji* mould substituted for classical rice-*koji*, as well as research on continuous multiple-stage brewing techniques and so on will, in the near future, bring about revolutionary changes in the technology of *saké* brewing; changes that Japanese breweries have never previously experienced, and which will result in the domestic breweries being replaced by large-scale enterprises.

On the other hand, the brewing of *saké* by traditional methods will still remain, and refined *saké* with its individual characteristics will undoubtedly be brewed by elaborate manual techniques in a few breweries using the classical procedures. These unique types of *saké* will continue to be highly appreciated by connoisseurs, and always admired by expert patrons just as the *chateau* wines are in European countries.

References

Ahlburg, H. and Matsubara, S. (1878). *Tokyo Ijishinshi* **24**, 6–12.
Akiyama, H. (1957). *J. agric. Chem. Soc. Japan* **31**, 913–918.
Akiyama, H. (1958a). *J. agric. Chem. Soc. Japan* **32**, 355–359.
Akiyama, H. (1958b). *J. agric. Chem. Soc. Japan* **32**, 526–529.
Akiyama, H. (1959). *J. agric. Chem. Soc. Japan* **33**, 1–6.
Akiyama, H. (1962a). *J. agric. Chem. Soc. Japan* **36**, 825–834.
Akiyama, H. (1962b). *J. agric. Chem. Soc. Japan* **36**, 903–907.
Akiyama, H. (1963). *J. Soc. Brew. Japan* **58**, 638–640.
Akiyama, H. and Furukawa, T. (1963a). *J. agric. Chem. Soc. Japan* **37**, 398–402.
Akiyama, H. and Furukawa, T. (1963b). *J. agric. Chem. Soc. Japan* **37**, 529–538.
Akiyama, H. and Iwata, C. (1964). *J. Soc. Brew. Japan* **59**, 520–521.
Akiyama, H. and Iwata, C. (1966). *J. Ferment. Technol., Osaka* **44**, 1–7.
Akiyama, H. and Shimazaki, T. (1967). *J. Ferment. Technol., Osaka* **45**, 383–387.
Akiyama, H. and Sugano, N. (1967). *J. Ferment. Technol., Osaka* **45**, 1093–1100.
Akiyama, H. and Takase, S. (1958). *J. Soc. Brew. Japan* **53**, 434–439.
Akiyama, H. and Umezu, K. (1959). *J. Ferment. Technol., Osaka* **37**, 586–591.
Akiyama, H., Iwata, C. and Naganawa, M. (1965). *J. Ferment. Technol., Osaka* **43**, 629–634.
Akiyama, H., Saito, Y. and Saito, K. (1967). *J. Ferment. Technol., Osaka* **45**, 377–382.
Akiyama, H., Yamamoto, A. and Suzuki, M. (1963). *Rep. Jap. Govt Inst. Brew.* **135**, 1–11.
Ando, T. (1957). *J. agric. Chem. Soc. Japan* **31**, 638–645.

Ashizawa, H. (1961). *J. Soc. Brew. Japan* **56**, 1135–1139.
Ashizawa, H. (1962). *J. Soc. Brew. Japan* **57**, 422–426.
Ashizawa, H. (1963). *J. Soc. Brew. Japan* **58**, 543–548.
Ashizawa, H. and Saito, Y. (1965). *J. Soc. Brew. Japan* **60**, 803–807.
Aso, K., Shibasaki, K. and Yamauchi, F. (1954). *J. Ferment. Technol., Osaka* **32**, 47–51.
Atkinson, R. W. (1881). *Memo. Tokyo imp. Univ. Sci. Dep.* **6**, 1–73.
Bearens, H. and Castel, M. M. (1958–9). *Ann. Inst. Pasteur Lille* **10**, 183–192.
Chihara, S. and Uemura, T. (1957). *J. agric. Chem. Soc. Japan* **31**, 268–272.
Chihara, S. and Uemura, T. (1958). *J. agric. Chem. Soc. Japan* **32**, 73–78.
Cohn, F. (1883). *Jahreb Schles. Gesellsch. Cultur.* (1883). Breslau 226.
Eda, K. (1909). *Rep. Jap. Govt Inst. Brew.* **25**, 22–67.
Eda, K., Oana, F. and Arimatsu, K. (1929). *Rep. Jap. Govt Inst. Brew.* **103**, 1–25.
Folkers, K., Wright, L. D., Skeggs, H. R., Aldrich, P. E., Hoffman, C. H. and Wolf, D. E. (1956). *J. Am. chem. Soc.* **78**, 4499.
Fujii, Y. and Uemura, T. (1952). *J. Ferment. Technol., Osaka* **30**, 423–427.
Fujii, Y. and Uemura, T. (1953). *J. Ferment. Technol., Osaka* **31**, 474–475.
Fukinbara, T. and Muramatsu, K. (1952). *J. agric. Chem. Soc. Japan* **26**, 583–589.
Fukui, S., Tani, Y. and Kishibe, T. (1955a). *J. Ferment. Technol., Osaka* **33**, 1–5.
Fukui, S., Tani, Y. and Kishibe, T. (1955b). *J. Ferment. Technol., Osaka* **33**, 59–64.
Fukui, S., Tani, Y. and Kishibe, T. (1955c). *J. Ferment. Technol., Osaka* **33**, 302–307.
Fukui, S., Tani, Y. and Kishibe, T. (1956). *J. Ferment. Technol., Osaka* **34**, 369–377.
Fukui, S., Tani, Y. and Kishibe, T. (1958a). *J. Ferment. Technol., Osaka* **36**, 131–141.
Fukui, S., Tani, Y. and Kishibe, T. (1958b). *J. Ferment. Technol., Osaka* **36**, 296–306.
Fukui, S., Tani, Y. and Shimoide, M. (1966). *J. Ferment. Technol., Osaka* **44**, 610–622.
Furukawa, T. and Akiyama, H. (1962). *J. agric. Chem. Soc. Japan* **36**, 354–358.
Hanaoka, M. (1918). *J. Soc. Brew. Japan* **13**, 1–22.
Hara, S., Otsuka, K., Yoshizawa, K., Hosoya, K., Kita, M. and Morinaga, K. (1965). *J. Ferment. Technol., Osaka* **43**, 873–880.
Hara, S., Takagi, U. and Otsuka, K. (1967). *J. Ferment. Technol., Osaka* **45**, 282–288.
Hayashida, M., Ueda, R. and Teramoto, S. (1968). *J. Ferment. Technol., Osaka* **46**, 77–91.
Higashi, T. (1928). *Sci. Pap. Inst. Phys. Chem. Japan* **7**, 506–526.
Hongo, M., Hayashida, S., Inoue, S. and Koizumi, R. (1967). *J. agric. Chem. Soc. Japan* **41**, 629–634.
Ichikawa, K. and Maeda, Y. (1963). *J. Ferment. Technol., Osaka* **41**, 530–541.
Ichikawa, K., Inoue, Y. and Kawasaki, T. (1955). *J. Soc. Brew. Japan* **50**, 278–292.
Iida, S. (1960). *J. Soc. Brew. Japan* **55**, 296–299.
Ikemi, M. (1966). *J. Soc. Brew. Japan* **61**, 347–351, 1183–1185.
Imayasu, S. (1966). *J. Soc. Brew. Japan* **61**, 106–112.
Imayasu, S., Kuriyama, K. and Ando, N. (1963). *J. Ferment. Technol., Osaka* **41**, 254–261.
Inoue, T. and Takaoka, Y. (1953). *In* "Lecture Meeting of the Society of Fermentation Technology, Japan".

Inoue, T., Takaoka, Y. and Hata, S. (1962a). *J. Ferment. Technol., Osaka* **40**, 237–251.

Inoue, T., Takaoka, Y. and Hata, S. (1962b). *J. Ferment. Technol., Osaka* **40**, 505–511.

Inoue, T., Takaoka, Y. and Hata, S. (1962c). *J. Ferment. Technol., Osaka* **40**, 511–521.

Inoue, T., Takaoka, Y. and Hata, S. (1962d). *J. Ferment. Technol., Osaka* **40**, 544–548.

Inoue, T., Takaoka, Y. and Hata, S. (1962e). *J. Ferment. Technol., Osaka* **40**, 602–609.

Inoue, Y. and Kawasaki, T. (1953). *J. Soc. Brew. Japan.* **48**, 425–431.

Ito, K. (1952). *J. Soc. Brew. Japan* **47**, 137–141.

Ito, K. (1959). *In* "Various Ways of *Saké* Brewing". *Ann. Brew. Ass. Symp.* pp. 59–60.

Ito, K. (1962). *J. Soc. Brew. Japan* **57**, 1172–1177.

Ito, Y. and Uemura, T. (1957). *J. agric. Chem. Soc. Japan* **31**, 783–786.

Ito, Y., Endo, I. and Uemura, T. (1956). *J. Ferment. Technol., Osaka* **34**, 18–22.

Ito, Y., Matsuki, M. and Uemura, T. (1957). *J. agric. Chem. Soc. Japan* **31**, 779–782.

Ito, Y., Nakamura, Y. and Uemura, T. (1955). *J. Ferment. Technol., Osaka* **33**, 421–426.

Iwata, C., Akiyama, H. and Suzuki, M. (1964). *J. Soc. Brew. Japan* **59**, 800–805.

Kageyama, K. (1955). *J. Ferment. Technol., Osaka* **33**, 53–59.

Kageyama, K. and Sugita, O. (1955). *J. Ferment. Technol., Osaka* **33**, 109–113.

Kagi, K., Otake, I., Moriyama, Y., Ando, F., Eda, K. and Yamamoto, T. (1909). *Rep. Jap. Govt Inst. Brew.* **29**, 1–83.

Kanai, H. and Iida, S. (1932). *Rep. Jap. Govt Inst. Brew.* **114**, 184–219.

Kanoh, S. (1953). *J. agric. Chem. Soc. Japan* **27**, 881–887.

Kanoh, S. (1961). *J. agric. Chem. Soc. Japan* **35**, 1304–1308.

Kanoh, S. (1962). *J. agric. Chem. Soc. Japan* **36**, 379–380, 489–494.

Kanto-Shinetsu Tax Administration Bureau (1961). *J. Soc. Brew. Japan* **56**, 1168–1174.

Kanto Shinetsu Tax Administration Bureau (1962). *J. Soc. Brew. Japan* **57**, 1111–1117.

Kanto Shinetsu Tax Administration Bureau (1964). *J. Soc. Brew. Japan* **59**, 157–160.

Kasahara, H. (1963). *J. Soc. Brew. Japan* **58**, 583–586.

Katagiri, H. and Kitahara, K. (1934). *J. agric. Chem. Soc. Japan* **10**, 942–969.

Katsuya, N. and Kitamura, S. (1949). *J. Soc. Brew. Japan* **44**, 7–22.

Kawabata, S. and Kawano, Y. (1953). *J. Ferment. Technol., Osaka* **31**, 434–438.

Kawabata, S. and Kawano, Y. (1956). *Rep. Synthetic Saké Corporation* **11**, 315–324.

Kawaharada, H., Hayashida, S. and Hongo, M. (1967a). *J. agric. Chem. Soc. Japan* **41**, 635–639.

Kawaharada, H., Koga, H., Hayashida, S. and Hongo, M. (1967b). *J. agric. Chem. Soc. Japan* **41**, 640–645.

Kawakami, H. (1964). *Ann. Brew. Ass. Japan* **19**, 1–21.

Kawamura, M., Yamashita, M., Umemura, Y. and Shibata, M. (1959a). *J. Soc. Brew. Japan* **54**, 581–585.

Kawamura, M., Yamashita, M., Umemura, Y. and Shibata, M. (1959b). *J. Soc. Brew. Japan* **54**, 581–585.

Kawamura, M., Yamashita, M., Umemura, Y. and Shibata, M. (1960). *J. Soc. Brew. Japan* **55**, 780–785.

Kawasaki, W. (1955). *J. Soc. Brew. Japan* **50**, 107–109.

Kitahara, K. (1960). *In* "Ecology of Micro-organisms". *Inst. appl. Microbiol. Japan Symp.* **11**, 42–53.

Kitahara, K. and Murata, U. (1953). *J. Ferment. Technol.*, *Osaka* **31**, 90–98.

Kitahara, K., Kaneko, T. and Goto, O. (1957). *J. agric. Chem. Soc. Japan* **31**, 556–564.

Kleinzeller, A. (1941). *Biochem. J.* **35**, 495–501.

Kobayashi, I. and Akiyama, H. (1961). *J. Soc. Brew. Japan* **56**, 171–175.

Kodaira, R. and Uemura, T. (1959). *J. agric. Chem. Soc. Japan* **33**, 621–631.

Kodaira, R., Ito, Y. and Uemura, T. (1958). *J. agric. Chem. Soc. Japan* **32**, 49–53.

Kodama, K. (1959a). *J. Soc. Brew. Japan* **54**, 138–142.

Kodama, K. (1959b). *J. Soc. Brew. Japan* **54**, 770–773.

Kodama, K. (1960). *In* "Ecology of Micro-organisms" *Inst. appl. Microbiol.*, *Japan Symp.* **11**, 54–70.

Kodama, K. (1963). *Ann. Brew. Ass. Japan* **18**, 64–77.

Kodama, K. (1965). *J. Ferment. Ass. Japan* **23**, 253–257.

Kodama, K. (1966a). *J. Soc. Brew. Japan* **61**, 318–319.

Kodama, K. (1966b). *J. Soc. Brew. Japan* **61**, 677–681.

Kodama, K. and Kyono, T. (1963). *J. Ferment. Technol.*, *Osaka* **41**, 113–116.

Kodama, K., Kyono, T. and Kodama, S., (1956). *Rep. Synthetic Saké Corporation Japan* **11**, 329–336.

Kodama, K., Kyono, T. and Kodama, S. (1957). *Rep. Synthetic Saké Corporation*, **14**, 554–562.

Kodama, K., Kyono, T. and Kodama, S. (1960). *Rep. Synthetic Saké Corporation*, *Japan* **20**, 919–923.

Kodama, K., Kyono, T. and Kodama, S. (1962). *Rep. Synthetic Saké Corporation*, *Japan* **27**, 428–435.

Kodama, K., Kyono. T. and Matsuyama, S. (1966). *J. Ferment. Technol.*, *Osaka* **44**, 8–13.

Kodama, K., Kyono, T., Iida, K. and Onoyama, N. (1964). *J. Ferment. Technol.*, *Osaka* **42**, 739–745.

Komoda, H., Mano, F. and Yamada, M. (1966). *J. agric. Chem. Soc. Japan* **40**, 127–134.

Kozai, Y. (1900). *Zentbl. Bakt. ParasitKde Abt. II* **6**, 385–405.

Kurata, H. (1968). *J. Fd Hygien. Soc. Japan* **9**, 23–34.

Kurono, K. and Katsume, H. (1930). *Rep. Jap. Govt Inst. Brew.* **106**, 30–41.

Kurono, K., Tatsui, S. and Washita, K. (1935). *Rep. Jap. Govt Inst. Brew.* **119**, 1–35.

Lodder, J. and Kreger-van Rij, N. J. W. (1952). "The Yeasts, a Taxonomic Study". North-Holland Publishing Co., Amsterdam.

Masuda, Y., Yamashita, K., Sasaki, S., Hattori, K. and Ito, Y. (1965). *J. Soc. Brew. Japan* **60**, 1114–1119.

Matsui, H. and Sato, S. (1963). *J. Soc. Brew. Japan* **58**, 734–738.

Matsui, H. and Sato, S. (1966). *J. Ferment. Technol.*, *Osaka* **44**, 14–19.

Matsuyama, M. (1964). *J. Soc. Brew. Japan* **59**, 672–675.

McVeigh, I. and Brown, W. H. (1954). *Bull. Torrey bot. Club* **81**, 218–233.

Miyasaki, R., Nagano, O., Yoshida, H. and Akaki, M. (1965). *J. Soc. Brew. Japan* **60**, 982–992.

Miyoshi, T. (1968). *J. Soc. Brew. Japan* **63**, 1255–1258.

Momose, H. and Tonoike, R. (1968). *J. Ferment. Technol., Osaka* **46**, 765–771.

Momose, H., Doi, S. and Tonoike, R. (1968a). *J. Soc. Brew. Japan* **63**, 682–685.

Momose, H., Ishii, H. and Tonoike, R. (1966). *J. Soc. Brew. Japan* **61**, 1037–1040.

Momose, H., Iwano, K. and Tonoike, R. (1969). *J. gen. appl. Microbiol.* **15**, 19–26.

Momose, H., Kobayashi, S., Koizumi, R. and Tonoike, R. (1965). *J. Soc. Brew. Japan* **60**, 539–542.

Momose, H., Okazaki, N. and Tonoike, R. (1968b). *J. Soc. Brew. Japan* **63**, 686–688.

Murakami, H. (1958). *J. agric. Chem. Soc. Japan* **32**, 91–95.

Murakami, H. and Kwai, M. (1958). *J. agric. Chem. Soc. Japan* **32**, 96–100.

Murakami, H. and Takagi, K. (1959). *J. agric. Chem. Soc. Japan* **33**, 905–909.

Murakami, H., Owaki, K. and Hashimoto, Y. (1962). *J. Soc. Brew. Japan* **57**, 1046–1049.

Murakami, H., Sagawa, H. and Takase, S. (1968). *J. gen. appl. Microbiol.* **14**, 251–262.

Murakami, H., Takase, S. and Ishii, T. (1967). *J. gen. appl. Microbiol.* **13**, 323–334.

Muto, H. (1961). *J. Soc. Brew. Japan* **56**, 189–181.

Muto, H., Tadenuma, M. and Furuichi, A. (1964). *Rep. Jap. Govt Inst. Brew.* **136**, 1–9.

Muto, H., Tadenuma, M. and Narumi, K. (1962). *J. Soc. Brew. Japan* **57**, 819–824.

Nagai, S. (1958). *Náturwissenschaften* **45**, 441–442.

Nagai, S. (1959). *Science, N.Y.* **130**, 1188–1189.

Nakamura, Y. and Uemura, T. (1955). *J. Ferment. Technol., Osaka* **33**, 203–208.

Nakamura, Y., Ito, Y. and Uemura, T. (1955). *J. Ferment. Technol., Osaka* **33**, 481–485.

Nakanishi, S. and Tsukahara, T. (1954). *J. Soc. Brew. Japan* **49**, 471–475.

Nakazawa, R. (1909). *Zentbl. Bakt. ParasitKde Abt. II* **22**, 529–540.

Nehira, T. and Nomi, R. (1956a). *J. Ferment. Technol., Osaka* **34**, 391–399.

Nehira, T. and Nomi, R. (1956b). *J. Ferment. Technol., Osaka* **34**, 423–428.

Neilands, J. B. (1967). *Science, N.Y.* **156**, 1443–1447.

Nojiro, K. (1959). *J. Soc. Brew. Japan* **54**, 658–661.

Nojiro, K. and Aoki, M. (1957). *J. Soc. Brew. Japan* **52**, 1008–1014.

Nojiro, K. and Ouchi, K. (1962). *J. Soc. Brew. Japan* **57**, 824–833.

Nunokawa, Y. (1964). *J. Soc. Brew. Japan* **59**, 1035–1039.

Oana, F. (1931). *Rep. Govt. Inst. Brew.* **111**, 124–159.

Oana, F. (1933). *Rep. Govt. Inst. Brew.* **117**, 101–113.

Oda, M. (1935). *J. Ferment. Technol., Osaka* **13**, 629–656.

Oda, M. and Wakabayashi, K. (1954). *J. Ferment. Technol., Osaka* **32**, 289–294.

Ogawa, C. and Wakayama, K. (1961). *J. Soc. Brew. Japan* **56**, 161–163.

Okazaki, H. (1950). *J. agric. Chem. Soc. Japan* **24**, 201–215.

Omachi, H. and Kawano, Y. (1957). *Rep. Synthetic Saké Corporation* **14**, 528–536.

Osaka Tax Administration Bureau (1960). *J. Soc. Brew. Japan* **55**, 226–230.

Otaka, Y. (1958). *J. Sci. Res. Inst.* **52**, 171–197.

Otaka, Y. and Yamanouchi, A. (1955). *J. agric. Chem. Soc. Japan* **29**, 880–883.

Otani, K. (1903). *J. Chem. Soc. Japan* **24**, 1052–1159.

Ouchi, K. (1966). *J. Soc. Brew. Japan* **61**, 161–164.

Ouchi, K., Ishido, T., Sugama, S. and Nojiro, K. (1966a). *J. Soc. Brew. Japan* **61**, 633–637.

Ouchi, K., Ishido, T., Sugama, S. and Nojiro, K. (1968a). *J. Soc. Brew. Japan* **63**, 186–190.

Ouchi, K., Nagai, T., Sugama, S. and Nojiro, K. (1966b). *J. Soc. Brew. Japan* **61**, 646.

Ouchi, K., Sugama, S. and Nojiro, K. (1967). *J. Ferment. Technol.*, *Osaka* **45**, 889–897.

Ouchi, K., Sugama, S. and Nojiro, K. (1968b). *J. Soc. Brew. Japan* **63**, 1300–1303.

Saito, K. (1950). *In* "The Mycology of *Saké* Brewing". Osaka.

Saito, K. and Oda, M. (1932). *J. Ferment. Technol.*, *Osaka* **10**, 787–802.

Saito, K. and Oda, M. (1934). *J. Ferment. Technol.*, *Osaka* **12**, 159–174.

Saito, T. (1965). *J. Soc. Brew. Japan* **60**, 898–900.

Sakaguchi, K. (1933). *Ann. Brew. Ass. Japan* **3**, 247–262.

Sakaguchi, K. (1964). *In* "*Nihon no sake*", pp. 181–182.

Sakai, S., Tsuzumi, H. and Nagayama, F. (1960). *J. Soc. Brew. Japan* **55**, 525–529.

Sasaki, K. (1964). *J. agric. Chem. Soc. Japan* **38**, 309–313.

Sato, S. (1957). *J. Soc. Brew. Japan* **52**, 281–384.

Sato, S., Tadenuma, M., Ueno, K. and Okuda, T. (1967). *J. Soc. Brew. Japan* **62**, 1230–1234.

Sendai Tax Administration Bureau (1968). *In* "Annual Report on *Saké* Brewing".

Shibata, T. (1968). *J. Soc. Brew. Japan* **63**, 545–547.

Shimizu, T. (1960). *Ann. Brew. Ass. Japan* **15**, 20–30.

Shimoide, M., Tani, Y. and Fukui, S. (1964). *J. Soc. Brew. Japan* **59**, 996–1000.

Shimoide, M., Tani, Y. and Fukui, S. (1965a). *J. Soc. Brew. Japan* **60**, 801–812.

Shimoide, M., Tani, Y., Sumino, K., Yamashiro, K. and Fukui, S. (1965b). *J. Soc. Brew. Japan* **60**, 443–446.

Stelling-Dekker N. M. (1931). "Die sporogenen Hefen". *Verh. K. Acad. Wet. Sect. II*, **28**, 1–547.

Sugama, S. (1967). *J. Soc. Brew. Japan* **62**, 927–935.

Sugama, S., Ouchi, K., Ninchoji, A. and Nojiro, K. (1966). *J. Soc. Brew. Japan* **61**, 164–169.

Sugama, S., Saheki, H. and Nojiro, K. (1965a). *J. Soc. Brew. Japan* **60**, 362–366.

Sugama, S., Shingaki, E. and Nojiro, K. (1965b). *J. Soc. Brew. Japan* **60**, 543–545.

Sugama, S., Yamakawa, K., Kataoka, G., Yamamura, H. and Nojiro, K. (1965c). *J. Soc. Brew. Japan* **60**, 453–456.

Sugita, O. and Kageyama, K. (1957). *J. Ferment. Technol.*, *Osaka* **35**, 347–350.

Sugita, O. and Kageyama, K. (1958a). *J. Ferment. Technol.*, *Osaka* **36**, 63–64.

Sugita, O. and Kagoyama, K. (1958b). *J. Ferment. Technol.*, *Osaka* **36**, 157–162.

Sugiyama, S. and Nagahashi, K. (1932). *Rep. Jap. Govt Inst. Brew.* **115**, 99–119.

Suto, T., Furusaka, C. and Uemura, T. (1951a). *J. Ferment. Technol.*, *Osaka* **29**, 105–109.

Suto, T., Takagi, K. and Uemura, T. (1951b). *J. Ferment. Technol.*, *Osaka* **29**, 447–452.

Suzuki, M., Kobayashi, N., Nunokawa, Y., Ito, M., Tanaka, K. and Sekizawa, S. (1956). *J. Soc. Brew. Japan* **51**, 906–914.

Suzuki, M., Nunokawa, Y., Hara, S., Baba, K. and Ito, Y. (1958). *J. Soc. Brew. Japan* **53**, 603–612.

Suzuki, M., Nunokawa, Y. and Hiroshima, G. (1957). *J. Soc. Brew. Japan* **52**, 470–477.

Tadenuma, M. and Sato, S. (1967). *Agric. Biol. Chem.* **31**, 1482–1489.

Tajima, O. (1962). *Rep. Synthetic Saké Corporation Japan* **28**, 520–523.

Tajima, O. (1965). *J. Soc. Brew. Japan* **60**, 20–22.

Takahashi, A. (1965). Doctorate thesis.

Takahashi, A. and Nose, T. (1955). *J. Ferment Technol.*, *Osaka* **35**, 167–170.

Takahashi, A., Sekine, K. and Nose, T. (1955). *Rep. Synthetic Saké Corporation Japan* **8**, 61–70.

Takahashi, G. (1916). *J. Soc. Brew. Japan* **11**, 15–24.

Takahashi, M. (1954). *J. agric. Chem. Soc. Japan* **28**, 395–398.

Takahashi, M. (1956). *J. agric. Chem. Soc. Japan* **30**, 140–145.

Takahashi, T. (1907a). *J. Soc. Brew. Japan* **1**, 20–26.

Takahashi, T. (1907b). *Bull. Coll. Agric. Tokyo imp. Univ.* **6**, 531–561.

Takahashi, T. (1912). *Rep. Jap. Govt Inst. Brew.* **42**, 361–416.

Takahashi, T., Eda, K., Yukawa, M. and Yamamoto, T. (1913). *Rep. Jap. Govt Inst. Brew.* **54**, 1–66.

Takase, S. and Murakami, H. (1966). *Agric. Biol. Chem.* **30**, 869–876.

Takase, S. and Murakami, H. (1967). *Agric. Biol. Chem.* **31**, 142–149.

Takase, S., Murakami, H., Sakai, T., Egashira, Y. and Makino, M. (1968a). *J. Soc. Brew. Japan* **63**, 881–888.

Takase, S., Sakai, T., Egashira, Y. and Murakami, H. (1968b). *J. Soc. Brew.* **63**, 783–788.

Takeda, M. and Tsukahara, T. (1965a). *J. Ferment. Technol., Osaka* **43**, 447–456.

Takeda, M. and Tsukahara, T. (1965b). *J. Ferment. Ass. Japan* **23**. 352–360.

Takeda, M. and Tsukahara, T. (1965c). *J. Ferment. Ass. Japan,* **23**, 453–456.

Takeda, M. and Tsukahara, T. (1966). *J. Ferment. Technol., Osaka* **44**, 367–371.

Takeuchi, I. and Shimada, K. (1965a). *J. agric. Chem. Soc. Japan* **39**, 83–88.

Takeuchi, I. and Shimada, K. (1965b). *J. agric. Chem. Soc. Japan* **39**, 89–94.

Takeuchi, I., Shimada, K. and Nakamura, S. (1967). *J. agric. Chem. Soc. Japan* **41**, 260–270.

Takeuchi, I., Shimada, K. and Nakamura, S. (1969). *J. Ferment. Technol., Osaka* **47**, 102–108.

Tamura, G. (1956). *J. gen. appl. Microbiol.* **2**, 431–434.

Tamura, G. (1958a). *J. agric. Chem. Soc. Japan* **32**, 707–712.

Tamura, G. (1958b). *J. agric. Chem. Soc. Japan* **32**, 783–798.

Tamura, G., Tsunoda, T., Kirimura, J. and Miyasawa, S. (1952). *J. agric. Chem. Soc. Japan* **26**, 480–483.

Tanaka, S. (1952). *J. Soc. Brew. Japan* **47**, 264–266.

Tani, Y. (1956). *J. Ferment. Technol., Osaka* **34**, 428–431.

Tani, Y., Shimoide, M., Sumino, K. and Fukui, S. (1963a). *J. Ferment. Technol., Osaka* **41**, 445–450.

Tani, Y., Sumino, K., Shimoide, M. and Fukui, S. (1963b). *J. Ferment. Technol., Osaka* **41**, 523–530.

Tani, Y., Sumino, K., Shimoide, M. and Fukui, S. (1964). *J. Ferment. Technol., Osake* **42**, 599–606.

Tani, Y., Sumino, K., Shimoide, M., Yamashiro, K. and Fukui, S. (1965). *J. Ferment. Technol., Osaka* **43**, 642–647.

Teramoto, S. (1936). *J. Ferment. Technol., Osaka* **14**, 671–679.

Terui, G. and Shibasaki, I. (1960). *J. Ferment. Technol., Osaka* **38**, 40–48.

Terui, G., Shibasaki, I., Mochizuki, T. and Takano, M. (1960). *J. Ferment. Technol., Osaka* **38**, 29–39.

Thom, C. and Church, M. (1926). *In* "The Aspergilli", pp. 198–207.

Tomiyasu, Y. (1933). *J. Ferment. Technol., Osaka* **10**, 515–518.

Tokuoka, U. (1941). *J. Ferment. Technol., Osaka* **19**, 791–809.

Toyozawa, M. and Yonezaki, H. (1959). *J. Ferment. Technol., Osaka* **37**, 64–68.

Toyozawa, M., Yonezaki, H., Ueda, R. and Hayashida, M. (1960). *J. Ferment. Technol., Osaka* **38**, 342–350.

Tsukahara, T. (1954). *J. agric. Chem. Soc. Japan* **28**, 405–408.

Tsukahara, T. (1956). *J. Ferment. Technol., Osaka* **34**, 122–125.

Tsukahara, T. (1961). *J. Soc. Brew. Japan* **56**, 888–890.

Tsukahara, T. (1962). *J. Soc. Brew. Japan* **57**, 117–123.

Tsukahara, T., Sakai, T., Maeda, K. and Miyasaka, T. (1964). *J. Soc. Brew. Japan* **59**, 358–363.

Ueda, R., Hayashida, M. and Kitagawa, E. (1960). *J. Ferment. Technol., Osaka* **38**, 337–342.

Uemura, T. (1956). *J. Ferment. Technol., Osaka* **34**, 117–121.

Uemura, T. and Fujii, Y. (1954). *Tohoku J. agric. Res.* **5**, 197–226.

Ueno, K., Yao, T., Tadenuma, M. and Sato, S. (1966). *J. Sov. Brew. Japan* **61**, 1169–1173.

Wakabayashi, K., Uozumi, M. and Tukahara, T. (1957). *J. Soc. Brew. Japan* **52**, 721–726.

Wehmer, C. (1895). *Zentbl. Bakt. ParasitKde Abt. II* **1**, 150–160.

Wickerham, L. J. (1951). *In* "Taxonomy of Yeasts", *Tech. Bull. U.S. Dep. Agric.* No. 1029. Washington, D.C.

Yabe, K. (1895). *Bull. Coll. Agric. Tokyo imp. Univ.* **2**, 219–220.

Yabe, K. (1897). *Bull. Coll. Agric. Tokyo imp. Univ.* **3**, 221–224.

Yabuta, N. (1965). *J. Soc. Brew. Japan* **60**, 686–691.

Yamada, M (1932). *Bull. chem. Soc. Japan* **8**, 97–100.

Yamada, M. and Akiyama, H. (1957). *J. agric. Chem. Soc. Japan* **31**, 127–132.

Yamada, M. and Hisano, K. (1957). *J. Soc. Brew. Japan* **50**, 762–763.

Yamada, M., Ishida, I. and Kobayashi, C. (1928). *Rep. Jap. Govt Inst. Brew.* **99**, 188–211.

Yamada, M., Ishimaru, K. and Masui, S. (1939). *Rep. Jap. Govt Inst. Brew.* **128**, 265–268.

Yamada, M., Katsume, H., Urano, T., Iwashita, N. and Shirai, M. (1935). *Rep. Jap. Govt Inst. Brew.* **122**, 157–158.

Yamada, M., Kobuyama, N., Yamasuga, K., Takagi, S. and Suzuki, D. (1945). *J. Soc. Brew. Japan* **40**, 44–45.

Yamada, M., Yoshizawa, K., Inouo, H. and Okada, T. (1955). *J. Soc. Brew. Japan* **50**, 760–761.

Yamamoto, A. (1961a). *J. agric. Chem. Soc. Japan* **35**, 616–623.

Yamamoto, A. (1961b). *J. agric. Chem. Soc. Japan* **35**, 711–715.

Yamamoto, A. (1961c). *J. agric. Chem. Soc. Japan* **35**, 819–823.

Yamamoto, A., Saaki, K. and Saruno, R. (1961). *J. agric. Chem. Soc. Japan* **35**, 715–719.

Yamamoto, G., Kanoh, S. and Sugama, S. (1953). *J. agric. Chem. Soc. Japan* **27**, 114–117.

Yamasaki, I. (1926). *J. agric. Chem. Soc. Japan* **5**, 377–387.

Yamasaki, I. (1958). *J. Ferment. Ass. Japan* **16**, 335–341.

Yamashiro, K., Shimoide, M., Tani, Y. and Fukui, S. (1966). *J. Ferment. Technol. Osaka* **44**, 602–609.

Yamashiro, K., Shimoide, M., Tani, Y. and Fukui, S. (1967). *J. Ferment. Technol., Osaka* **45**, 942–953.

Yokotsuka, T., Sasaki, M., Kikuchi, T., Asao, Y. and Nobehara, A. (1967). *J. agric. Chem. Soc. Japan* **41**, 32–38.

Yoshikawa, K., Okumura, U. and Teramoto, S. (1963). *J. Ferment. Technol.,* *Osaka* **41**, 357–362.

Yoshizawa, K. (1958). *J. Soc. Brew. Japan* **53**, 771–781.

Yoshizawa, K. (1963). *Agric. biol. Chem., Tokyo* **27**, 162–164.

Yoshizawa, K. (1964). *Agric. biol. Chem., Tokyo* **28**, 279–285.

Yoshizawa, K. (1965). *Agric. biol. Chem. Tokyo* **29**, 672–677.

Yoshizawa, K. (1966). *J. Soc. Brew. Japan* **61**, 481–485.

Yoshizawa, K. and Makino, R. (1960). *J. Soc. Brew. Japan* **55**, 54–59, 734–738.

Zenda, N. (1916). *Rep. Jap. Govt Inst. Brew.* **65**, 1–3.

Zenda, N. (1920). *J. Soc. Brew. Japan* **15**, 1–22.

Chapter 6

Yeasts in Distillery Practice

J. S. Harrison and J. C. J. Graham

Research Department, Distillers Company (Yeast) Limited,
Epsom, England

I. Introduction

Although methods for the concentration of spirit from weaker brews have been known since the time of the ancient Egyptians (Simmonds, 1919), the history of the use of special yeasts for the preparation of spirit prior to distillation dates only from about the beginning of the present century (Kervégant, 1946). Previously brewer's and wine yeasts were used. Many different forms of potable distilled spirit became accepted; in the first place these were characteristic of the regions where particular suitable raw materials were available. Thus, in France and neighbouring countries, where grapes have been grown since early times, the distilled product was called *eau de vie* by the French; in England, it was known as brandy, a word used as early as 1657, and derived from the Dutch *brandewijn* (distilled wine). The Scottish and Irish equivalents were whisky and whiskey respectively, from the Gaelic *uisge beatha*, water of life, another reminder of the regard in which concentrated alcoholic beverages were held. The corresponding Latin term *aqua vitae* had been used from 1547 for any form of potable spirits, and previous to that (1471) it was applied in alchemy to unrectified ethanol. The word *vodka* is also derived from the same root as the Russian for water. In the West Indies, where sugar cane was plentiful, the product was rum, a term used since 1654. More recently (1714) gin, or hollands, was made from fermented malt and grain; in Germany the similar product was called schnapps (1818). For many years increasing amounts of distilled spirit were used for industrial purposes, although more recently ethanol synthesized from petroleum is replacing fermentation spirit for these applications.

When it became apparent that benefits would be gained by manufacturing industrial spirit in the greatest possible yield at high concentration in the shortest time, for economic reasons, attempts were made to select suitable yeast strains from those available. In the potable-spirit industry, as early as 1903, Pairault (Arroyo, 1945) recommended the use of selected pure strains, chosen on the basis of adaptability to rum fermentation. It is now the common practice to select, adapt or breed special strains, which are maintained as pure cultures, for most of the specialized processes by which distilled spirit is manufactured. The virtues of particular yeasts for the production of the different forms of potable spirit are difficult to assess because of the subjective nature of the required properties. For the commercial production of industrial spirit, however, a quantitative estimation can be made on theoretical grounds of the total available carbon-containing compounds in the fermentation medium which can be converted to ethanol. As the manufacture of spirit for industrial purposes became more competitive, due to the development of large-scale chemical synthesis, much attention was given to this subject, and yields of ethanol very close to the theoretical maximum can now be obtained.

The economic importance of ethanol is considerable, the annual world output of potable spirit being equivalent to about twenty million hectolitres (1·5 million tons) of pure ethanol, and of industrial alcohol several million tons, of which about 1·3 million tons were made by fermentation processes in 1963 (World-Wide Survey of Fermentation Industries, 1966). Yeast therefore still plays a very significant part in producing ethanol for large-scale manufacturing purposes in spite of synthetic competition, and contributes indirectly considerable sums of money, of the order of thousands of millions of pounds sterling, to national exchequers in the form of excise duty. In many countries, for instance France, Scotland and the West Indies, spirituous liquors represent a high proportion of the export business.

The spirit distillation industry is seen to comprise a somewhat heterogeneous assortment of manufacturing processes linked by yeast as a factor common. As there is a great amount of overlap in the individual applications, an attempt has been made in the following pages to generalize the discussion of the fermentation processes as much as possible, in order to emphasize the essentially biological function of the yeast. Some necessary details are given, however, of the principal differences between the processes for the manufacture of different forms of spirit.

II. Yeast in Relation to Distillery Practice

A. GENERAL CONSIDERATIONS

Distillery spirit is available in many forms, varying from pure ethanol to complex potable spirits. The processes employed for its production are, however, all based on the same biochemical and physical principles. The manufacturing stages can be defined as:

(i) Preparation of medium.
(ii) Propagation of yeast inoculum.
(iii) Fermentation.
(iv) Treatment of fermented liquor.
(v) Distillation.
(vi) Maturing (this stage is only applicable in the case of certain potable spirits).

1. *Preparation of Medium*

As all industrial processes are based on the conversion of carbohydrates to ethanol and carbon dioxide, the medium (wort, must, mash, according to the particular industry) consists of an aqueous solution of assimilable sugars, or complexes which can be broken down during the process to simple sugars as, for example, by the action of enzymes on starches. The range of these raw materials is wide, and includes all the sources used for the preparation of non-distilled alcoholic beverages and for the manufacture of baker's and food yeasts. The simplest are the expressed juices of fruit or sugar cane, which can be fermented directly with little or no pretreatment. Next there are various by-products of other industries in the form of sugar solutions or concentrates, such as molasses from the manufacture of beet and cane sugar, and sulphite liquor from the making of paper. Another form of carbohydrate substrate is obtained by the chemical hydrolysis of wood, and finally almost all commercially available sources of starch from grain, roots, fruits, and so on are used in some part of the world for the preparation of distilled spirit.

According to the particular process to be employed, the medium may be treated to clarify it, or to remove undesirable substances, but almost universally some form of partial sterilization is practised in order that the yeast may be able to act without interference from infecting micro-organisms. Complete sterilization is seldom resorted to, although the medium may be held at boiling point for some time. When the action of enzymes originally present in the solution is required during the course of fermentation, lower temperatures must be used in order to avoid denaturation of the proteins. Alternatively, for instance in the manufacture of wine from which brandy is made, additions such as sulphur dioxide may be used (see Chapter 2, p. 5).

The essential function of the medium is to supply carbohydrate which is assimilable by the chosen yeast, along with smaller amounts of other nutrients and growth factors. In certain cases, such as the production of industrial alcohol, materials that are deficient in the original preparation may be added; in others, for example the manufacture of Scotch whisky, additives are not used. Some sugars, for instance glucose and fructose, are directly fermentable by all distillery yeasts; others, such as raffinose, can only be fermented by certain special yeasts (Lodder and Kreger-van Rij, 1952).

The concentration of carbohydrate in the medium is of importance because of its direct effect on the fermentation reactions; for instance, high osmotic pressure is detrimental to yeast but, equally important, a high concentration of ethanol in the later stages, resulting from an initially high sugar concentration, is inhibitory. On the other hand, the efficiency of the overall process depends to a considerable extent on the production of the maximum amount of ethanol in each fermentation vessel; high spirit concentration also has the virtue of involving the removal of the least possible amount of water during the distillation stage. In this respect distillery practice is different from that in the production of most other alcoholic beverages, where the final product contains a predetermined concentration of ethanol, mostly in the range

1–5%, requiring 2–10% sugar in the medium, as opposed to 10–15% sugar present initially in the fermentation stage in the distillery.

The initial pH value and the buffering capacity of the solution should be such that the yeast will function satisfactorily throughout the fermentation, during which acidic compounds are excreted by the yeast and often also by infecting micro-organisms; typically the commencing pH value is in the range 4·5 to 5·5. If known inhibitory substances are likely to be present, pre-treatment is employed to lessen their effect. For instance, sulphur dioxide, which may have been used as a preservative in the raw material, can be removed by passing steam and air through the liquid. Similarly, it may be advantageous to remove solids by centrifugation, filtration or simple settling.

2. *Propagation of Yeast Inoculum*

Because high initial concentrations of yeast are required for the production of spirit under distillery conditions, special methods are used to propagate the inoculum before it is added to the main fermentation medium. The practice for potable spirits in Great Britain and some other countries is to use selected yeast strains grown by aerobic methods as for baker's yeast (see Chapter 7, p. 349) using molasses as substrate (Stark, 1954; Simpson, 1966). Methods for the propagation of pure culture distiller's yeast by a continuous aerobic process have been described (Unger *et al.*, 1942). By such methods the large amounts of yeast required are produced rapidly and economically without expending valuable fermentation medium. In some cases, aerobically-grown baker's yeast itself is used directly (Stark, 1954), or brewer's yeast may be employed (Simpson, 1966). The magnitude of the problem can be judged from the fact that the output of a single grain distillery can total hundreds of thousands of gallons of spirit each week, representing millions of gallons of wort; this requires a supply of many tons of yeast, the inoculation rate usually being in the range 0·1–1·0% moist yeast based on the final volume of the medium (Butschek and Kautzmann, 1962).

An alternative method of propagation, used particularly when the medium can be satisfactorily sterilized, is the use of semi-aerobic starter, or "bub", stages which are later mixed with the main fermentation medium (Prescott and Dunn, 1959). By this procedure portions of sterilized wort of progressively larger size, containing the required substrates for yeast growth, are prepared. The first stage is inoculated with a culture of the chosen yeast strain. This is allowed to grow anaerobically, or with a limited supply of sterile air to increase the yield and growth rate, for a suitable time (e.g. 24 h, during which period the yeast multiplies 10–20 times) after which the whole culture is transferred to the next larger aliquot. The procedure is repeated in increasing volumes until sufficient yeast has been produced for seeding the main fermentation. This method is especially suitable for the preparation of industrial spirit from molasses, as no sugar is lost, other than that used for yeast growth, which is not allowed to become excessive by limitation of some essential growth requirement such as assimilable nitrogen. The early growth stages are sometimes carried out in so-called yeast machines, of which the Magné (1917) automatic system is typical.

3. *Fermentation*

In respect of yeast, the fermentation step is obviously of major importance in distillery practice, because it is here that the yeast actually functions. The weaker alcoholic strength and the presence of considerable concentrations of non-alcoholic substances in beer, cider and such products, as compared with distilled spirits mean that quantitatively the fermentation stage is relatively more important in the latter case, especially for industrial spirit. In spite of this, comparatively little has been published in direct relation to distillery procedure, particularly for potable spirits, because the industry is almost entirely traditionally based, but the principles of biochemistry and biophysics can quite satisfactorily be applied to explain the sequence of events occurring during fermentation; these principles apply equally well to all the various commercial processes.

It is known that anaerobic fermentation by yeast largely involves the conversion of the sugars, or similar substrates, to ethanol and carbon dioxide through a series of well known enzyme reactions (see Vol. 2, Chapter 7; Baldwin, 1963; Wilkinson and Rose, 1963). However, other reactions take place. Almost a hundred years ago Pasteur (1872) observed that only about 95% of the glucose utilized was converted to ethanol (48·4%) and carbon dioxide (46·6%), with 3·3% forming glycerol and 0·6% succinic acid. Apart from the yeast, which multiplies if adequate assimilable nitrogen, phosphate and other necessary growth factors are present, the other main product is fusel oil, a mixture of higher alcohols.

The overall process is best understood by considering in detail the growth of yeast under anaerobic conditions. From this the consequent phenomena, including the yield of ethanol and carbon dioxide, the amount of glycerol formed, the significance of other excretion products, and the order of magnitude of heat production, can be deduced.

a. Yeast growth. When yeast is grown under highly aerobic conditions the main energy requirements are supplied by oxidative reactions in the course of which free oxygen, normally atmospheric, finally combines with hydrogen (Utter *et al.*, 1968). In this way intracellular molecular re-arrangements of carbohydrate substrates take place, and unwanted by-products are excreted. Yeast organic matter, in a typical yield of 50% of the sugar used and with an empirical formula near to $C_6H_{10}O_3N$, is formed by the utilization, to a close approximation, of 4·5 O_2, the incorporation of one atom of nitrogen supplied as ammonium salts or amino acids, and the production of 5·5 CO_2 and 6·5 H_2O (Harrison, 1967). It has also been established that the complete oxidation by *Saccharomyces cerevisiae* of one glucose molecule through the oxidative phosphorylation system provides 28 molecules of ATP which are available for synthetic purposes (Kormančíková *et al.*, 1969), and an accepted value for the amount of yeast dry matter formed by the expenditure of one mole of ATP is 10·5 g (Bauchop and Elsden, 1960).

Under the anaerobic conditions of distillery fermentations yeast must still perform its essential functions, which are to maintain life and to reproduce; but the metabolism has, of necessity, to take a different and less efficient course than when free oxygen is available (Nord and Weiss, 1958). The two main essentials are: (i) an alternative to oxygen as a hydrogen acceptor; and (ii) a means of transferring phosphate from ADP to ATP to provide the mechanism for synthesis, a reaction which requires a high level of energy, about 10 kcal/mole (Jencks, 1968). During the anaerobic conversion of one molecule of glucose to ethanol and carbon dioxide, two active phosphate groups (2 ATP) are produced (from 2 ADP + P_i) in excess of the fermentation requirements (Baldwin, 1963). For these to be available anaerobically for synthetic reactions, by analogy with the oxidative mechanism, about 14 times as much glucose requires to be metabolized to produce a given amount of yeast dry matter as when the process is fully aerobic. When two molecules of pyruvate (corresponding to one mole of glucose) are deflected from the fermentative system to form new growth material *via* the tricarboxylic acid cycle and similar pathways, the formation of two molecules each of carbon dioxide (2 C atoms) and acetaldehyde (4 C atoms) is prevented. Consequently the acetaldehyde cannot act as a hydrogen acceptor to allow conversion of triose phosphate to diphosphoglycerate to take place at an earlier stage in the sequence of intracellular reactions. The excess hydrogen atoms are incorporated by the only other readily available acceptor, dihydroxyacetone phosphate, which does not normally act significantly in this way as it has a much lower affinity than acetaldehyde for $NADH_2$, the active agent in the transfer. The immediate reaction product is glycerol phosphate, which is converted by the loss of phosphate to glycerol, which is excreted (see Chapter 10, p. 529).

Part of the energy required is for the synthesis, *via* pyruvic acid, of cell material and excretion products which result from the various reactions in the chain. Nitrogen, for the formation of proteins, nucleic acids and so on, is introduced at a stage beyond pyruvic acid through the assimilation of amino acids or ammonium salts present in the medium, followed by transaminations and similar reactions (Harris, 1958). Meanwhile the necessary amounts of intracellular carbohydrates are synthesized much more simply from glucose via glucose phosphate and the elimination of water, the reactions involving, per mole of glucose, one molecule of UTP which is interconvertible with ATP (Trevelyan, 1958). As the total carbohydrate content of yeast is about the same as the combined content of protein and other organic non-carbohydrate compounds, the amount of glucose converted into yeast matter can be calculated. The amounts of the excretion products which result from the operation of the tricarboxylic acid cycle are dependent on various conditions, such as the pH value and the source of available nitrogen, and cannot therefore in general be predicted with precision.

(i) *Amino acids as nitrogen source.* The balance of reactions under practical conditions can, however, be illustrated by considering a typical laboratory fermentation balance sheet. When an amount of assimilable carbohydrate estimated by analysis as 10·10 g (calculated as glucose, derived from maltose, by the difference between initial and final values) in 100 ml of wort prepared from a malt mash, containing an adequate supply of growth factors, and mixed amino acids estimated by analysis to contain initially 15 mg of assimilable nitrogen, was fermented using an inoculum of 50 mg of yeast (as dry matter) under simulated distillery conditions (initial pH, 4·7; initial temperature, 22° rising to a maximum of 34° at 48 h; programmed agitation), the only products found by analysis of the wash in significant quantities were present in the amounts shown in Table I.

Taking the empirical formula of the amino acids in the medium as

TABLE I. *Analysis of Wash after Fermentation with Amino Acids as Nitrogen Source*

Product	Amount (mg)	Equivalent (millimoles)
Ethanol	5,130	111·52 C_2H_6O
Glycerol	216	2·35 $C_3H_8O_3$
Organic acids (as succinic[a])	54	0·46 $C_4H_6O_4$
Fusel oils (as butanol[b])	20	0·27 $C_4H_{10}O$
Yeast dry matter (final - initial)	208	1·30 $C_6H_{10}O_3N$[c]

[a] Succinic is the main organic acid present.

[b] Butanol has approximately the same elementary composition as the bulk oils.

[c] See text.

$C_4H_8O_2N$, and the 187·2 mg of yeast dry matter (excluding the inorganic constituents) as 1·30 millimoles $C_6H_{10}O_3N$ (Harrison, 1967), and assuming that the only extra products formed were carbon dioxide, equivalent to the ethanol plus that evolved in various decarboxylation reactions, and water released in polymerization and other reactions, the complete equation (in millimoles) can be balanced as follows, the weights of the compounds involved being given (in milligrams) below the formulae:

$$58\cdot10 \; C_6H_{12}O_6 \; + \; 1\cdot30 \; C_4H_8O_2N \; \longrightarrow \; 1\cdot30 \; C_6H_{10}O_3N \; + \; 2\cdot35 \; C_3H_8O_3$$
$$10{,}457 \qquad\qquad 133 \qquad\qquad\qquad 187 \qquad\qquad 216$$

$$+ \; 0\cdot46 \; C_4H_6O_4 \; + \; 0\cdot27 \; C_4H_{10}O \; + \; 0\cdot625 \; H_2O \; + \; 1\cdot485 \; CO_2$$
$$54 \qquad\qquad 20 \qquad\qquad 11 \qquad\qquad 65$$

$$+ \; 111\cdot52 \; CO_2 \; + \; 111\cdot52 \; C_2H_6O$$
$$4{,}907 \qquad\qquad 5{,}130$$

The compounds concerned in the tricarboxylic acid cycle and related reactions are succinic acid, higher alcohols, carbon dioxide (1·485 millimoles), water (0·625 millimoles) and half the yeast organic compounds, including all the nitrogen (93·6 mg, $10^{-3} \; C_{4\cdot33}H_{7\cdot22}O_{1\cdot01}N_{1\cdot30}$). The remaining organic yeast material is polysaccharide (0·578 millimoles $C_6H_{10}O_5$) which is formed by the polymerization of glucose units with the elimination of water (0·578 millimoles H_2O by calculation from the sugar used, compared with 0·625 millimoles needed to establish the above balance). The tricarboxylic acid cycle compounds require the passage of 8·74 millimoles C if they were all synthesized *via* pyruvic acid; thus liberating $2 \times 8\cdot74$, or 17·48 millimoles H, which would produce 2·91 millimoles of glycerol. This calculated value is somewhat higher than the 2·35 millimoles observed, due to the fact that some amino acids required for protein synthesis are incorporated into the cell intact (Thorne, 1949, 1950). The total calculated glucose required is 10·457 g, compared with 10·100 g observed experimentally, and the nitrogen in the yeast is estimated to be 18·2 mg, 15·0 mg assimilated nitrogen having been found by analysis.

According to the findings of Bauchop and Elsden (1960), the formation of 208 mg of yeast dry matter requires at least $208/10\cdot5 = 19\cdot8$ millimoles of ATP. During the fermentation of 10,457 mg of glucose $2 \times 10{,}457/180 = 116$ millimoles of ATP become available, which is much in excess of the growth requirement. Some of this may be required to supply maintenance energy (see next section), but inhibitory substances severely reduce the amount of yeast formed from a given amount of substrate (Kormančíková *et al.*, 1969). The concentrations of sugar used by Bauchop and Elsden and by Kormančíková *et al.* were very much lower than are used in distillery practice, and consequently the concentrations of inhibitory excretion products, such as ethanol, are much greater under commercial conditions.

Considering the assumptions made in the analytical procedures for measuring total carbohydrates and nitrogenous compounds, in the estimations based on equivalent aerobic yeast yields and of the composition of the nitrogen-containing yeast constituents, variability of the order found can be expected between the observed values and those calculated from theory. From the general agreement, it can be concluded that the theoretical biochemical approach is sound.

(ii) *Ammonium salts as nitrogen source.* When assimilable carbohydrate and nitrogen from natural sources was replaced by glucose and an ammonium salt respectively, in a synthetic medium with adequate supplies of minerals and growth factors, the analytically-determined products of fermentation, based on the observed utilization of 10·0 g of glucose, were as shown in Table II. The calculated balanced equation

TABLE II. *Analysis of Wash after Fermentation with Ammonium Salts as Nitrogen Source*

Product	Amount (mg)	Equivalent (millimoles)
Ethanol	4,563	99·20 C_2H_6O
Glycerol	518	5·63 $C_3H_8O_3$
Organic acids (as succinic)	140	1·19 $C_4H_6O_4$
Fusel oils (as butanol)	5	0·07 $C_4H_{10}O$
Yeast dry matter (final - initial)	190	1·20 $C_6H_{10}O_3N$

(in millimolar terms; milligrams below formulae) obtained from these values is:

$$54·90 \; C_6H_{12}O_6 + 1·20 \; NH_3 \longrightarrow 1·20 \; C_6H_{10}O_3N + 5·63 \; C_3H_8O_3$$
$$\quad\quad 9,882 \quad\quad\quad\quad 20 \quad\quad\quad\quad\quad\quad 173 \quad\quad\quad\quad\quad 518$$

$$+ \; 1·19 \; C_4H_6O_4 + 0·07 \; C_4H_{10}O + 1·15 \; H_2O + 2·66 \; CO_2$$
$$\quad\quad 140 \quad\quad\quad\quad 5 \quad\quad\quad\quad\quad 21 \quad\quad\quad\quad 117$$

$$+ \; 99·20 \; CO_2 + 99·20 \; C_2H_6O$$
$$\quad\quad 4,365 \quad\quad\quad 4,563$$

It will be seen that more glycerol was formed with ammonium salts as the nitrogen source than in the previous example when mixed amino acids were used. This is to be expected according to the findings of Nordström (1966) and, of course, is logical, as in the absence of amino acids which can be assimilated in the intact form, the incorporation of the ammonium ion requires the expenditure of more energy; hence the transfer of a greater amount of hydrogen, with consequent greater excretion of glycerol. Also, more organic acid is formed, but less fusel oil. The last effect is due to the fact that the supply of amino acids is closely connected with the formation of the higher alcohols which constitute fusel oil (see Section III, p. 323; and Chapter 4, p. 147), and the larger amount of organic acids is the indirect result of more pyruvic acid being deflected into the tricarboxylic acid cycle, as an intermediate step in the synthesis of amino acids, and eventually proteins. The total amount of yeast dry matter formed from ammonium salts is almost exactly the same, in relation to the carbon compounds metabolized, as when amino acids are supplied. This supports the theory that the intracellular reactions are required to provide a defined minimal basic supply of energy for the survival of the organism.

b. *Fermentation in the absence of assimilable nitrogen.* When no assimilable nitrogen is available, true growth cannot take place because proteins, nucleic acids and similar essential cellular components cannot be synthesized, except to replace natural losses by re-assimilation of decomposition products, although the polysaccharide content of the cells may increase somewhat. However, ethanol and carbon dioxide are produced, and in synthetic medium that was complete except for the omission of assimilable nitrogen, the products of fermentation, based on the observed usage of 10·0 g of glucose by 0·13 g of yeast (dry weight; same strain as in previous two examples) in 100 ml, were as given in

TABLE III. *Analysis of Wash after Fermentation in the Absence of Assimilable Nitrogen*

Product	Amount (mg)	Equivalent (millimoles)
Ethanol	4,723	102·63 C_2H_6O
Glycerol	425	4·62 $C_3H_8O_3$
Organic acids (as succinic)	150	1·27 $C_4H_6O_4$
Fusel oils (as butanol)	9	0·12 $C_4H_{10}O$
Yeast dry matter (final - initial)	38	0·24 $C_6H_{10}O_5$[a]

[a] Yeast carbohydrate, see text.

Table III. The balanced equation calculated (as above) from these values is:

$$55·11\ C_6H_{12}O_6 + 1·31\ H_2O \longrightarrow 0·24\ C_6H_{10}O_5 + 4·62\ C_3H_8O_3$$
$$\quad\ 9,921 \qquad\qquad 24 \qquad\qquad\qquad 38 \qquad\qquad 425$$
$$+\ 1·27\ C_4H_6O_4 + 0·12\ C_4H_{10}O + 1·93\ CO_2 + 102·63\ CO_2$$
$$\quad\ 150 \qquad\qquad 9 \qquad\qquad 85 \qquad\qquad 4,515$$
$$+\ 102·63\ C_2H_6O$$
$$\quad\ 4,723$$

In the three examples cited, the yields of ethanol, as percentages of the glucose used with amino acids, ammonia and no nitrogen source, were 96·0%, 90·5% and 93·1% respectively. As the amino acids have a sugar-sparing effect by contributing to yeast growth, the relations of ethanol carbon to the total carbon metabolized are somewhat closer, being 63·2%, 60·3% and 62·1% respectively, the theoretical proportion for complete conversion of glucose to ethanol being 66·7%. Much more glycerol and organic acids are produced in the absence of amino acids as the nitrogen source, but fusel oil formation is less, as would be expected. The observed variations in spirit yield are the result of differences in a complex of variables, such as the energy requirements for protein synthesis according to the type of nitrogen supply. However, no clear advantage in ethanol yield is gained by attempting to avoid completely the growth of yeast. In fact, the absence of cell proliferation implies that the fermentation rate, which under any given conditions is proportional to the yeast present (but see Section II.A.3.*f*, p. 298), remains constant, whereas in the presence of complete medium (i.e. containing assimilable carbohydrates, nitrogen, growth factors and minerals) it increases exponentially in the early stages. Graphs of ethanol production plotted against time are given in Fig. 1. The yeast inoculum

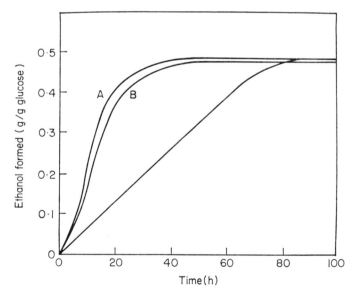

FIG. 1. Rate of fermentation of sugar by yeast under various nutrient conditions. Complete medium: A = with amino acids as assimilable nitrogen supply; B = with ammonium salts; C = with no supply of assimilable nitrogen. Sugar concentration, 10%; yeast inoculum, 0·016 g dry wt./g sugar; temperature, 30°.

in this experiment (Curve C) was pretreated by growing in nitrogen-free medium to decrease the nitrogen content to the minimum. The rate of fermentation at 30° was 1·0 g glucose metabolized (i.e. 0·47 g ethanol formed)/h.g yeast dry matter.

The reason that metabolism must proceed when yeast is in aqueous suspension if assimilable substrates are available, is that energy must be provided for essential functions, in particular the prevention of leakage of cell constituents into the surrounding liquid in the direction of the concentration gradient. Studies of maintenance energy have been made by Pirt (1965) and others (Hansford and Humphrey, 1966; Mor and Fiechter, 1968). Under aerobic conditions bacteria were found to require the utilization of about 0·1 g glucose/h.g dry matter for this purpose. In the example given in Fig. 1 (Curve C), when no nitrogen was present, the rate was about 0·24 g/h.g dry matter for yeast under anaerobic conditions. The amount of glucose required per gram hour might be expected to be greater than is needed aerobically because of the less efficient utilization of the substrate, although other factors are no doubt involved. When the sugar is exhausted, the yeast begins to die.

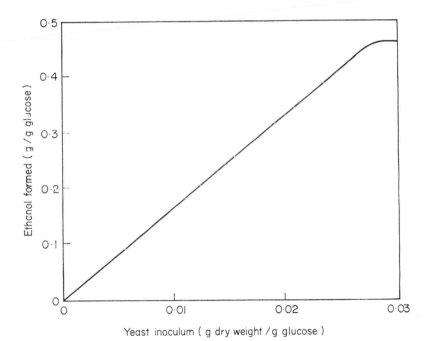

Yeast inoculum (g dry weight / g glucose)

FIG. 2. Relationship between total ethanol production and amount of yeast inoculum in fermentations of simple sugar solution. Yeast inoculum grown in nitrogen-deficient medium. Fermentation medium, 10% glucose in water; temperature, 30°; fermentation time, 200 h.

If a solution of glucose in water is used to replace a medium which is complete except for assimilable nitrogen, and the yeast inoculum is small, the metabolic reactions slow down and finally cease before all the sugar is used. The total sugar metabolized, and therefore the amount of ethanol formed, is proportional to the inoculum added initially, as shown in Fig. 2, where the amount of inoculum is plotted against the glucose consumed. This observation suggests that, in spite of the apparently excessive potential energy supply, some constituent or combination of constituents in the yeast is progressively destroyed, or diffuses from the cells and is lost to the intracellular reaction sites. The fermentation curve obtained under these conditions using an inoculum of the order employed commercially resembles Curve C of Fig. 1.

c. *Heat production.* In contrast to the situation with aerobic propagation of yeast (Harrison, 1967), heat produced by the anaerobic fermentation reactions is sufficiently small that cooling presents no problems when low concentrations of ethanol are formed at a relatively slow rate, as in the production of beer and wine. More attention to this question is necessary when high concentrations of sugars are fermented rapidly, as in distillery practice. Cooling systems are often not employed but, although this avoids the costs that would be entailed in fitting and operating special equipment, it has the disadvantage that the fermentation cannot be run at a predetermined optimum temperature. To overcome this difficulty, cooling coils are used in certain molasses fermentation procedures, such as the manufacture of industrial spirit (Hodge and Hildebrandt, 1954) and rum (Kervégant, 1946). Because spirit yield is related to temperature, other conditions being similar (White, 1954), a study of the economic advantages of using such a system more generally may be worthwhile in other large-scale processes. According to White (1954), more ethanol would be produced at a faster average rate if a controlled temperature around 40° was used with due precautions against evaporation.

From the equations derived in the previous section, the heat evolved under the given conditions can be calculated within reasonable limits of accuracy. Taking the simplest case, for ease of calculation, in which amino acids are not used as a source of nitrogen and part of the carbon requirement, about 9·7% of the glucose takes part in reactions other than the formation of ethanol and carbon dioxide (see Section II.A.3.a.ii, p. 292), and it can be shown that the transformation of sugar to yeast substance involves only a relatively small change in free energy (Harrison, 1967). A close approximation to the heat produced in the overall reaction can therefore be obtained by assuming that 90·3% of the glucose is fermented by the classical scheme. Thus, in 100 ml of 10% glucose medium, 9·03 g glucose are converted to ethanol and carbon dioxide with the formation of two high-energy phosphate groups per mole of glucose, each of which requires the expenditure of about 10 kcal (a quantity which varies somewhat with the pH value, temperature and other factors; Jencks, 1968). The free energy relations are approximately:

$$0.0502 \; C_6H_{12}O_6 + 0.1004 \; ADP + 0.1004 \; P_i \longrightarrow 0.1004 \; C_2H_6O$$
$$-10.89 \qquad\qquad\qquad\qquad\qquad\qquad\qquad\qquad -4.20$$

$$+ \; 0.1004 \; CO_2 + 0.1004 \; ATP$$
$$-9.48 \qquad\quad +1.00$$

Hence the free energy change is about -1.79 kcal, which provides sufficient heat to raise the temperature of 100 ml by $17.9°$.

In the absence of a special cooling system there are two main routes by which fermentation vessels lose heat: (i) through the walls by conduction and radiation; and (ii) by evaporation caused by the evolution of carbon dioxide gas saturated with water vapour. Obviously loss of heat from the walls of the vessel is conditioned by a number of factors (Webb, 1964). These include: (i) the thermal conductivity of the material of which the vessel is made, a parameter which is measured in kcal/m h deg; (ii) the surface area exposed; (iii) the temperature difference between the inside and outside; (iv) the thickness of the walls; (v) time. Because all these factors may have widely varying values, for instance according to whether the vessel is made of metal or wood, a practical solution under given conditions is to observe the rate of fall of temperature after the fermentation has ceased, and extrapolate back to zero time. A typical difference between the theoretical and actual maximum temperatures is $5°$, but it can vary considerably. Loss of heat by evaporation can be calculated for the example above as follows: the weight of carbon dioxide given off by 9.03 g of glucose is 4.42 g, giving a volume of 2.3 litres at $35°$. This removes about 0.1 g of water, which requires as energy for evaporation 0.06 kcal, representing a drop in the temperature of 100 ml of $0.6°$. Thus, instead of rising by $17.9°$, as would be the case if the whole of the 1.79 kcal were applied in a completely insulated calorimeter, the increase is reduced by about $5.0° + 0.6°$, to give an actual rise of the order of $12°$. From the initial temperature of $25°$, the fermentation wash therefore reaches about $37°$ when the initial glucose concentration is 10%. This figure falls within the range observed commercially.

As temperatures much higher than $37°$ are detrimental to the behaviour of the yeast (White, 1954), it is common practice, in the absence of a special cooling system, to avoid overheating either by starting the fermentation at an appropriately low temperature or, if this is impracticable, by using a weaker concentration of sugar so that less heat is produced. Adjustments on these lines are necessary in the case of large changes in ambient temperature because it is advantageous to employ as high a temperature as is feasible without losing spirit yield; otherwise the fermentation rate is slowed, with consequent inefficient use of the fermenter capacity.

d. Temperature effects. Like other biological processes, the rate of fermentation increases with rise in temperature to an optimum between 30° and 40°. The precise effects of temperature can, however, vary with the yeast strain. For example, in laboratory tests a particular yeast was shown to give a better spirit yield than a second strain in a molasses medium when the fermentation temperature was 33°, whereas at 38° the reverse was observed. This type of effect is probably related to different optimal temperatures for the growth and fermentative processes in the two yeasts (Farrell and Rose, 1967).

e. Sugar concentration. The concentration of assimilable sugar that can be fermented most efficiently depends to some extent on other components of the medium. For instance, blackstrap molasses is best fermented in a wort which contains 13–14% sugar, whereas high test molasses can be used satisfactorily with 16–17% sugar present initially. This difference is largely due to the higher concentration of mineral constituents and other non-assimilable compounds in the blackstrap molasses which have osmotic, and possibly other, effects. Similar behaviour may be expected in fermentation media other than molasses. Because different yeast strains are known to react differently under varying osmotic conditions (see Chapter 7, p. 349), it follows that the characteristics described may be affected by the particular type of yeast used. Furthermore, the maximum temperature reached during the fermentation period has an influence on the efficiency of fermentation in relation to sugar concentration, and must therefore be taken into consideration. One known effect of high mineral (i.e. ash) content in the medium is that more glycerol is formed owing to changes in the intracellular reaction equilibria (Nordström, 1966, 1968). The result of this is that some carbon is diverted from the ethanol production pathway.

f. Hydrogen ion concentration. When the buffering capacity of the medium is relatively high, the best spirit yields are generally obtained with an initial pH value in the range 4·5–4·7. If higher values are employed, more glycerol and organic acids are formed at the expense of ethanol. In more lightly buffered media, the optimum starting value is nearer pH 5·5; after eight to twelve hours this falls to about 3·5. If a more acid starting value is used, the final low level tends to slow the fermentation rate. This difficulty can be overcome to some extent by the use of larger yeast inocula.

g. Yeast inoculum. The amount of yeast added initially affects the fermentation in differing ways according to the particular strain, and to the nature of the medium. When the medium is relatively sterile a low inoculum can be used successfully, and is to be preferred providing the rate is adequate. With a very high inoculum, the large amount of nitrogenous material within the cells may allow considerable increase of yeast weight to take place by synthesis of polysaccharides such as glycogen, to produce yeast with a lower percentage cell-nitrogen content. In this way carbohydrates may be wasted in producing yeast rather than excreting ethanol. The reactions necessary for this behaviour also result in a somewhat increased formation of glycerol.

h. Addition of nutrients. In malt and grain worts there is sufficient assimilable nitrogen, phosphate and magnesium for maximal yeast growth to take place; additional amounts are therefore not necessary. When some recycling is practised, and spent wash is used as the make-up water for the preparation of new media, additional buffering capacity, nitrogen and phosphate are provided. In the case of molasses the situation is more variable; some samples contain sufficient of these nutrients to give adequate growth for the fermentation to be completed in an economical time, but the precise effect depends on the strain of yeast employed. With the majority of batches of blackstrap, high test and beet molasses, however, better results can be obtained by adding extra nitrogen in the form of ammonium salts or urea. It is rarely necessary to add phosphate or magnesium, except to the seed stages. Addition of too much nitrogen or phosphate may be detrimental. Some media may be lacking in growth factors. In certain cases this can be helpful, for instance in media which contain only just sufficient biotin to support yeast growth, the proliferation of lactic acid bacteria, which have a high requirement for biotin, is restricted, with beneficial effects on spirit production.

i. Effect of alcohols and aldehydes on yeast function. The maximum concentration of ethanol that can be obtained is controlled by such conditions as the sugar and mineral concentrations in the medium, and does not in practice usually reach a sufficient strength to influence the growth or fermentative activity of yeast to any marked degree. Ethanol concentrations up to 5% (v/v) have been shown to have little effect on most commonly used yeasts (see Chapter 2, p. 5). Higher alcohols, on the other hand, are much more toxic: Ingram (1955) states that 0·6% (v/v) of amyl alcohol completely inhibits yeast action, as does 1·5% isobutyl alcohol. Acetaldehyde has an inhibitory influence on anaerobic growth in concentrations as low as 0·1–0·2% (v/v). These effects are more pronounced at high fermentation temperatures.

j. Physical considerations. As a result of the evolution of carbon dioxide, the specific gravity of fermenting medium diminishes, an effect which is used in practice to follow the course of fermentation, to determine when the reactions have ceased, and to assess empirically the efficiency of the process. In fact, a number of other changes affect the specific gravity during distillery fermentations, apart from variations in temperature that are normally allowed for in the determination. For example, water and a small amount of ethanol are removed by evaporation in the carbon dioxide gas which leaves the vessel. When a 10% solution of carbohydrate is fermented, the volume of gas evolved is 2·7 litres/100 ml, assuming 4·7 g CO_2 at 35°. Saturated with water vapour at this temperature, 2·7 litres of gas will remove 0·11 ml of water; that is, decrease the volume by 0·11%. On the other hand, if the temperature of the liquid rises during the fermentation by 12°, there will be an expansion of 0·35 ml (i.e. 0·35%), which is partly compensated by the expansion of the material of which the vessel is made. Although these volume changes are relatively small they are significant on the large scale when efficiency is of primary importance.

Owing to the low partial pressure of ethanol in the solution, loss of spirit by evaporation in the evolved carbon dioxide is not excessive, but amounts to about 1% of the total spirit produced, a value which is highly dependent on the temperature and the concentration of ethanol.

If disaccharides, such as sucrose and maltose, or starches provide assimilable monosaccharides by hydrolysis, water is taken up, thus:

$$C_{12}H_{22}O_{11} + H_2O \rightarrow 2 C_6H_{12}O_6$$

As, in commercial practice, the content of sucrose in solution may be 20%, the water incorporated can be as much as 3·6 g/100 ml.

Another problem is supersaturation with carbon dioxide. This can easily occur, particularly in clear worts, and in any case, in deep commercial vessels the pressure at the bottom can approach twice atmospheric, with the resultant increase in gas solubility. These two effects can produce a temporary excess concentration of 0·1 g/100 ml, a specific gravity excess of 0·001, equivalent to an apparent 0·5–0·8% (w/w) ethanol concentration according to the total concentration. For this reason care must be taken when following the course of fermentation by specific gravity measurements, as a reduction in the final hours is often an indication of evolution of gas rather than of ethanol formation. Other physical problems are concerned with sampling techniques; these are dealt with in Section VI.A (p. 328).

k. *Infection.* Contaminating micro-organisms have effects on distillery fermentations which differ in some respects from those observed in other yeast processes. In the first place, no direct microbial danger to the consumer exists because the distillation process kills organisms in the fermented wash, leaving only cell residues in the still, and the concentration of ethanol in the distillate is usually sufficiently high to inhibit the development at a later stage of most micro-organisms. However, during fermentation foreign organisms can exert a number of influences; this they can do by: (i) using sugars which would otherwise be available to the yeast, and so lower the spirit yield; (ii) grow and excrete by-products, often late in the fermentation, on substrates which are not assimilable by the yeast; (iii) contribute substances which improve flavour; (iv) produce "off-flavours"; (v) excrete acids or other compounds which affect yeast performance and enzyme activity in the medium; (vi) provide enzymes which modify the fermentation conditions.

To utilize sufficient substrate to make a substantial difference to the spirit yield, the infection level would have to be numerically very high, 10^8 cells/ml or more; a state of affairs which is not likely to occur in modern distillery practice. Methods of sterilization of equipment and media are well understood (Sykes, 1965), and the design of plant is being continually improved from this point of view (Thorley, 1968). Much of the modern progress in this respect is directly due to the knowledge acquired in connection with the large scale manufacture of antibiotics, where a very high standard is required. Nevertheless, many bacteria multiply at a much faster rate than distillery yeasts, partly because of their inherent growth rate, but also because the conditions during certain stages of the fermentation process can be more favourable to their metabolic requirements than for the yeast. This is particularly so in the early stages, when the pH value is relatively high and ample sugar is present, and at the end, when the concentration of sugars which can be utilized by yeast is low, but substrates such as pentoses and glycerol, which can be metabolized by some micro-organisms, are available.

Probably the most universally damaging types from the distillers' point of view are the *Lactobacilli* and other lactic acid-forming bacteria such as *Streptococcus lactis*. These types tend to be present in many of the raw materials used in distillery practice, and may be difficult to eliminate completely. This is particularly true where enzymes in the original mash are required to modify starches during the fermentation period, thus rendering the application of high temperatures inadmissible. Fortunately many amylolytic enzymes will withstand a temperature of around 60°, which is sufficient to kill, or at least seriously damage, most of the troublesome bacteria. In general, in the case of malt mashes the effect of low pH values, caused by the presence of bacteria, on the amylolytic enzymes is more serious than the influence on the yeast, although continued exposure to highly acid media has a deleterious effect on yeast, particularly as the concentrations of assimilable substrates decline.

Because a large yeast inoculum is normally used, and the initial number of contaminating micro-organisms is kept as small as possible, many hours are required before the concentration of contaminants becomes serious, even if the growth rate of the contaminant is considerably greater than that of the yeast. Assuming reasonably clean working conditions, ethanol production is almost complete before the infection predominates, but as the bacteria can often metabolize substrates which are not assimilated by yeast, and their growth continues so long as substrate is available, very high levels can eventually be reached if the fermentation stage is allowed to continue beyond the time when ethanol production is complete. A number of possibilities arise under such conditions: for instance, ethanol may be lost through direct assimilation by certain types of organism such as *Acetobacter* spp. (Rose, 1968). Another consequence is that excessive amounts of organic acids or other compounds may be formed as by-products of bacterial metabolism: in certain cases these substances, either in their own right or as a consequence of later reactions with other compounds in the wort, may provide organoleptically attractive properties, for example esters (see Section III.B, p. 324). On the other hand, unpleasant effects may result from the action of bacteria on wash components. Examples are hydrogen sulphide produced by the reducing action of organisms on sulphur-containing compounds, and the metabolism by certain species of *Lactobacillus* of glycerol, formed during the fermentation by the yeast, to produce β-hydroxypropionaldehyde, which subsequently breaks down to acrolein when the wash is distilled.

4. *Treatment of Fermented Liquor*

In some distillery processes the fermented wash may be treated before being introduced into the still to remove solids which cause difficulties during distillation. Damage to the metal surfaces may result from burning of carbonaceous and mineral material with consequent oxidation, combined with deposition of an insulating coating which affects heat transfer. Also, the effect of over-heating of organic residues while volatiles are being distilled is to form undesirable breakdown products. These include furfural, formed by the effect of heat on pentoses.

The simplest treatment given to the wash is to allow it to stand for some time in a settling tank, or wash charger, and then to decant or pump off the clear liquid into the still. For the production of industrial spirit, various additives may be used which help to precipitate or coagulate unwanted substances. Where possible, however, the maximum possible amount of fermented material is passed to the still in order not to lose valuable spirit. If solids added later in the charging of the still produce unpleasant volatile contaminants, the corresponding distillate can be used as crude spirit or be rectified by further treatment.

Residues remaining attached to the still surfaces may be removed by hot caustic soda solution, or manually through openings in the sides of the distillation columns.

5. *Distillation*

The fermented wash, after suitable treatment, may be distilled in a number of different ways according to the requirements of the final product. These vary from crude separations, where the first and last runnings are collected independently but returned to later still charges with very little fractionation, to highly sophisticated and often multiple-stage distillation systems which produce virtually pure ethanol. The less efficient separations are used for the production of many types of potable spirit, where a complex mixture of congeners is required in order to obtain the characteristic flavour. A recent detailed illustrated description of the various types of still is given by Suomalainen *et al.* (1968).

a. Pot stills. Comparatively little fractionation of the components is obtained with pot stills of the normal type, which are generally of a characteristic shape based on old established tradition. A number of classical designs are used in different industries, the lower part of the still usually being bulbous, with a constriction above which there is a wider section before the bend to the condenser. The material is frequently copper. A certain amount of refluxing does, in fact, occur due to cooling of the vapours in the upper part of the still body, which may be fitted with a cooling system such as a jacket or internal coil for water circulation (Simpson, 1966, 1968). Heating may be by a number of alternative means, including the original form, a coal or coke fire, now sometimes fed by a mechanized stoking arrangement; otherwise oil or gas firing, steam coils or steam jackets may be used (Simpson, 1968). The condenser is often a copper coil immersed in a water bath, but other systems may be employed.

Such stills are normally operated batchwise. The still is charged, the contents are heated, and then some suitable means is used to determine when to make a "heads" or "first runnings" cut. This may be a turbidity test, or a specific gravity measurement by hydrometer. A middle fraction is then collected and, towards the end of the distillation, the distillate is again tested by hydrometer to determine the point at which to make a "tails" cut. The distillation is stopped when all the spirit has distilled. The spent wash is then run out of the still which, after washing down, is recharged. Malt whisky, rum and brandy are produced by this type of still, the particular significance being that a high proportion of the minor organoleptically active compounds formed by the yeast are retained in the distillate.

b. Coffey still. The Coffey (or "patent") still which was patented in 1830 (Manning, 1947), and remains in use in many British potable spirit distilleries, comprises first an "analysing" or "beer" column, and second a "rectifying" column. Its special feature is that it operates continuously, as opposed to the batch system used with the pot still, but this gives less complete separation of components than the rectifying still. The specific gravity of wash entering the stills is of the order of 0·998–1·040 (Suomalainen, 1968). Heating is by steam, introduced below a baffle near the bottom of the fractionating column, which contains a large number of perforated plates. Vapour condenses on the plates and an amount of condensate, depending on the operating conditions (e.g. steam pressure) returns to the lower levels. As the lower-boiling vapours collect at the top of the column, by tapping off liquid at the appropriate plates, fractions of selected composition can be achieved.

c. Rectifying stills. To obtain the purest ethanol, stills with highly efficient rectifying columns are used. Various designs of plates are fitted in order to effect the best rectification. The "bubble-cap" plate is probably the best known and the most widely used. The efficiency of these columns is such that the volatile impurities are concentrated in relatively narrow zones in the columns, and can be drawn off as required. Fusel oil, mainly consisting of a mixture of higher alcohols, is concentrated near the bottom of the rectifyng column, and can be removed in concentrations sufficiently high for sale without further treatment (Kreipe, 1962).

More sophisticated stills, working on the principle of extractive distillation, can produce ethanol of the highest purity. Azeotropic columns can also be included in multiple-column stills, so that absolute (99·7% v/v) ethanol can be made directly from fermented wash. (Suomalainen *et al.*, 1968).

B. INDUSTRIAL SPIRIT

Before the modern methods of synthesis of ethanol from petroleum oil fractions were developed, all spirit was made by fermentation. Strains of *Saccharomyces cerevisiae* have mainly been used as the active organism. In the technologically more advanced countries, ethanol for purposes such as power, fuel, chemical solvents and synthetic processes is now made by chemical means but, for perfumery and pharmacy, spirit of fermentation origin is often still preferred, and it is used, of course, for all potable applications. However, in regions where suitable cheap materials are obtainable, spirit for industrial use is still made in large quantities by fermentation. Many naturally occurring substances are employed, including cereals, fruits, roots, tubers and other types of materials like sugar cane, wood and the sap of trees. Where the carbohydrates are not present in an assimilable form, some pretreatment is required before the fermentation step.

The processes for the production of industrial spirit by fermentation include a high degree of rectification during distillation, so that the final products of the overall process are almost pure ethanol and carbon dioxide in yields which approach closely to the theoretical maximum for the sugar supplied (Macher, 1962).

1. *Raw Materials*

a. Molasses. One of the simplest processes uses various forms of molasses which, being of plant origin, contains most of the essential constituents for yeast growth in fairly concentrated form.

Two main sources of molasses are available, from sugar cane and sugar beet, which arise as residual products after sucrose has been separated by crystallization. These are sub-divided into blackstrap, refiner's and high test types, in order of increasing sugar purity, and correspondingly decreasing proportions of non-sugar constituents. The latter may be neutral as far as yeast fermentation is concerned, or either beneficial or detrimental. The beneficial constituents other than sugars, mainly sucrose, include minerals such as phosphates, magnesium, potassium and various trace elements which are necessary for growth, and growth factors, the more important of which are biotin, pantothenic acid, inositol, thiamine and pyridoxine (Becker and Weber, 1962; see also Section II.A.3.*h*, p. 299).

A number of constituents, such as betaine, take no part in the fermentation reactions, while others, for example glutamic acid, if present in high concentration, as is sometimes the case in beet molasses (White, 1954), may be considered to be either beneficial or detrimental, according to the purpose for which the product is used, on account of the influence of amino acids on the content of congeners in the distilled spirit (see Section III.A, p. 323). Harmful constituents include compounds produced by interaction between sugars and amino acids caused by heat during evaporation of the molasses (Ingram *et al.*, 1955). Hydroxymethylfurfural, for instance, is toxic to yeast under anaerobic conditions if present in high concentration (Stakhorskaya and Tokgarev, 1964). Compounds such as potassium imidodisulphonate are present on occasion as a consequence of the interaction between sulphur dioxide used during the sugar refining process and nitrites formed by bacterial action; these also can be inhibitory to yeast (see Chapter 7, p. 349).

(i) *Batch fermentation.* In the most common processes for fermenting molasses all the material is run into the vessel initially. To obtain the best results, extra assimilable nitrogen in the form of ammonia or urea is added, and when necessary extra phosphate, often as superphosphate. The pH value is normally adjusted to below 5·0 by addition of sulphuric acid, except when high-test molasses is used, in order to control the growth of infecting micro-organisms. Yeast is then added, and the fermentation commences as the yeast metabolizes monosaccharides that are present. It is important that the yeast should have adequate invertase activity (see Chapter 10, p. 529) to convert sucrose, the principal constituent of molasses, to glucose and fructose. Being associated with the cell wall, rather than the internal environment of the cell, invertase acts, in effect, almost as an extracellular enzyme and, if in sufficient concentration, rapidly converts all the sucrose in the fermentation medium to monosaccharides (Trevelyan, 1958). The fermentation then continues to completion and, after suitable treatment, if required, the wash is distilled. Typically the initial and final yeast counts in batch processes are about 10^7 and $1·5 \times 10^8$ cells/ml (equivalent approximately to 0·1 and 1·5% w/v compressed yeast) respectively.

(ii) *Semi-continuous processes.* Alternatively, semi-continuous processes can be employed. Various patents have been taken out (Boinot, 1936a, b) whereby the yeast is used repeatedly. By using a large yeast inoculum (1% or more of moist yeast) with the consequent rapid fermentation rate, and removing the yeast by centrifuging at intervals and re-using it, greater efficiency is claimed. The separation of the spent liquor removes many bacteria, and washing the yeast at an acid pH value and the presence of carbon dioxide before re-use is said further to decrease the level of infection (Boinot, 1941). In this way fermentation can, under suitable conditions, be continued semi-continuously for comparatively long periods. The duration depends, in practice, on the composition of the fermentation medium. By using beet molasses with a low biotin content, the growth of contaminating organisms is restricted, and the process can continue for as long as two months without trouble from infection. With blackstrap molasses, where a high spirit yield is the main consideration, sterile worts must be used, but in spite of this infection can build up in two to three weeks or even sooner unless extreme precautions are taken. Where high yield based on molasses is not especially required, but more importance is attached to a high rate of production, fermentation can be continued for several months by the use of high yeast concentrations; by this means infecting micro-organisms are kept in check. In this modification of the process, yeast growth is so great that periodically it is necessary to remove about half of the yeast present; normally this excess is sold as fodder yeast.

Acid washing of yeast does not effect a complete kill of infecting organisms. Too low a pH value causes changes in yeast function which become apparent during the early stages of the fermentation by the evolution of hydrogen sulphide; ammonium persulphate has been recommended as preferable to acid for the treatment (Bruch *et al.*, 1964). Penicillin and pentachlorophenol have been used in South America to repress infection, but with all these treatments resistant strains of micro-organisms can eventually develop.

The Usine de Melle-Boinot patents (Boinot, 1936a, b) describe methods of decreasing yeast growth, such as restricting the nitrogen supply, adding inhibitors and de-aerating with inert gases, which are said to result in an improvement in spirit yield. Adequate rate of spirit production is achieved by retaining a high concentration of yeast while restricting growth. Although semi-continuous methods are used extensively in some countries, notably France, the high capital and running costs have discouraged their use in the U.S.A. and in Great Britain (Hodge and Hildebrandt, 1954).

(iii) *Continuous processes*. Steady-state continuous processes may be used, under which conditions the gain in spirit yield can be 1–2% due to the low equilibrium concentration of sugars in the medium, as in the semi-continuous process. For this reason, in both types of process ethanol production ceases immediately the feed of substrate is stopped. Because the rate of addition of molasses affects the equilibrium concentration of the sugars, it is important in relation to the balance between ethanol yield and yeast growth. A similar effect has been shown in the case of continuous beer fermentation by Hough (1961) and Hough and Ricketts (1960). Slow feed rates result in low yeast growth.

According to Hospodka (1966) all ethanol produced from molasses in the U.S.S.R. is made by modifications of a continuous process (Malchenko and Krishtul, 1954). Infection troubles inevitably arise after some time. These are overcome to a large extent by care with sterilization of equipment and raw materials. In the Malchenko-Krishtul method antiseptics, including chlorinated lime, formaldehyde, pentachlorophenol and acids, are added to the molasses to suppress bacterial growth. By such means the steady-state runs can be continued for up to ten days. Yeast concentrations of at least 10% are used, and excess yeast is used for baking. This type of dual-purpose system, which can also be applied by batch techniques, is still in use on the continent of Europe, but in America and Great Britain it has generally been superseded by separate methods for the production of ethanol and carbon dioxide on the one hand, and baker's yeast (Chapter 7, p. 349) on the other, as both processes can be made more effective when operated with their special requirements in mind. Where the combined production schemes are still employed, the reason is often connected with local excise and legal arrangements.

The relative economics of the various continuous and other processes depends heavily on the exact balance between the extra cost involved in running such systems and the gain in efficiency obtained. In general, it might be observed that essentially simple production processes, such as are traditionally used in the manufacture of spirit by fermentation methods, and yield a product of low basic cost, are not suitable subjects for the application of highly sophisticated operating systems.

b. *Cane and beet juice.* The expressed juices obtained from sugar cane and sugar beet, and other sugar-containing by-products from sugar factories, are used directly in the countries of origin as substrates for the productions of ethanol. The problems involved are much the same as for molasses, except that storage of the raw material for any length of time is not possible on account of the greater infection risk with more dilute solutions of sugars than is the case with molasses, in which the concentration is sufficiently high to discourage the multiplication of micro-organisms.

c. *Tuberous roots.* A wide variety of starch-containing materials are used in distillery practice. These can be broadly divided into tuberous roots and cereals (Kreipe, 1962).

(i) *Potatoes.* These are the main source of carbohydrate for the manufacture of spirit in Germany. In a typical process, the potatoes are washed and placed in a *converter* where they are subjected to a steam pressure of two to three atmospheres for about 15 min. Fermentation is then carried out after the addition of 2–3% of malt, the temperature in the mash tun being 55° or less. The starch content of potatoes varies between 10% and 25% (Kerr, 1950).

(ii) *Cassava.* The tuberous roots of the cassava plant are used in the West Indies (Teixeira *et al.*, 1950). Mould bran (bran treated with *Aspergillus* spp.) gives a higher conversion of cassava starch than when malt is used. The difference is presumably connected with the molecular structure of starch from this source. Cassava root contains about 25% of starch and 5% of other sugars (Simmonds, 1919).

(iii) *Mangolds, sweet potatoes, yams, beetroot, artichokes.* These and other similar root vegetables and tubers are used in various countries for spirit production. The methods of treatment are similar to those for cassava. A particularly interesting case is that of artichoke starch, or inulin, which on hydrolysis produces an unusual breakdown pattern, fructose being formed from the inulin (Prescott and Dunn, 1959). Typical sugar content is 13% fructose and 16% total sugars. Specially selected yeasts are required in order to metabolize these carbohydrates, including strains of *Sacch. cerevisiae*, *Sacch. anamensis* and *Schizosacch. pombe* (Underkofler *et al.*, 1937).

d. Cereals. (i) *Maize.* One of the most commonly employed cereals is maize, which is grown in good yield in many parts of the world, and has a very high content of starch. It is particularly used in Eastern Europe and the U.S.A. The dry maize corns are cooked with water under steam pressure to gelatinize them. Too high a temperature tends to cause caramelization. If the starch paste is too viscous, it may be thinned by treatment with hydrochloric acid or by the amylolytic action of malt or fungal enzymes (Aschengreen, 1969; for further details see Section II.C.2.*a*, p. 313). When malt enzymes are employed, 10–15% of malt wort is added to the starch paste, which is left for some time to liquefy. The starch content of maize varies according to the variety and the climate of the region in which it is grown. The proportion of starch is of the order of 60% (de Clerck, 1957) to 68% (Kerr, 1950).

In the main, maize and other grains are treated batchwise. Although continuous methods of cooking and mashing have been developed (Stark, 1954), full-scale production of spirit by these techniques is not practised outside the U.S.S.R. according to Hospodka (1966). The reason for this is that complete sterilization of the raw materials by heat is contra-indicated because of the need to preserve the activity of the amylolytic enzymes in the medium.

(ii) *Barley, malt.* Barley is used as a raw material for preparing mashes for fermentation both in its natural form and also after germination, when the product is called malt (more stricly malted barley, as other cereals can be malted), and contains a strong amylolytic activity (see Section II.C.2.*a*, p. 313). The malt may be used in the moist state, as produced (green malt), or dried (kilned malt) when some 15% of the amylolytic activity may be lost (see Section II.E.1, p. 319). The starch content of barley is 55–65% dry weight (de Clerck, 1957).

(iii) *Wheat, oats, rye, sorghum, rice.* All these cereals are used in the fermentation industry. Wheat may be malted in a similar fashion to barley, with the production of amylolytic enzyme activity. The starch content is about 65% (de Clerck, 1957). Oats can be saccharified by the addition of 30% of barley malt: its starch content is about 53% (Kerr, 1950; de Clerck, 1957). Rye, like maize, is commonly used in Eastern Europe for the production of industrial and potable spirits. It contains a relatively large amount of starch, about 63% (Kerr, 1950; de Clerck, 1957). Sorghum, also known as millet or Guinea corn, is mainly grown in Africa; the starch content of some varieties is as high as 80% (Kerr, 1950). Rice is also used in large quantities for spirit production, more particularly in China and Japan; it contains about 70% starch (Kerr, 1950; de Clerck, 1957).

e. Sulphite liquor. Waste liquor from the sulphite process for wood pulp manufacture contains low concentrations of sugars which are extensively used for the production of ethanol (Butschek and Krause, 1962b). Patents were taken out for a fermentation process as early as 1878 (Mitscherlich, 1878). A variety of sugars are formed by hydrolysis of the lignins and hemicelluloses of wood; these include glucose, fructose, mannose, galactose, xylose and arabinose (McCarthy, 1954). Soft woods are preferred to hardwoods as they contain a higher proportion of complexed hexoses which, after hydrolysis, are capable of being fermented to ethanol by yeasts, although the pentoses can be utilized for the propagation of certain yeasts in the presence of oxygen. Up to forty gallons of ethanol can be made from 4,000 gal of waste liquor produced during the manufacture of one ton of dry sulphite pulp. Compounds inhibitory to yeast may be formed during the pulping process, particularly sugar-bisulphite addition compounds and furfural.

Bulked sulphite waste liquors are treated before fermentation by heat, acid and nutrient addition, and in other ways to make them sterile, to remove inhibitory substances, and to provide yeast nutrients such as assimilable nitrogen, minerals, growth factors, and so on, which are required for yeast metabolism. Sulphur dioxide is removed by passing air and steam through the solution, and acids are neutralized. Selected strains of *Sacch. cerevisiae* are used for this process (Johnson and Harris, 1948). In some cases the Usine de Melle-Boinot process (Boinot, 1936a, b) is used. Because of the low concentration of sugars in waste liquor, various means are employed to raise the substrate concentration, for instance, by addition of other sugar sources such as cellulose or starch hydrolysates (McCarthy, 1954), wood-waste hydrolysates and sulphite liquors (Hospodka, 1966). These materials are easily adaptable to such a process because the pretreatment produces a sterile solution, although various additives must be used to support yeast growth. By using very high yeast concentrations, with re-use of the yeast, short retention times can be attained. Andreyev (1958) and Kalyuzhnyi (1958, 1960) employed a specially selected quick-settling yeast which enabled them to increase the flow rate of the medium considerably. The stated amount of ethanol produced by 1 kg of yeast in 24 h is 7·5 kg, and the yield is 56·7 litres (45·0 kg) of ethanol per 100 kg sugar fermented.

f. Wood waste. In countries such as Canada, Sweden and Norway, where wood is plentiful, wood waste is hydrolysed by acid to give 65–88% of sugars fermentable by *Saccharomyces cerevisiae* for the manufacture of ethanol (Saeman and Andreasen, 1954; Conrad, 1962). Because the hydrolysate obtained in this way contains a relatively low concentration of sugars, a recycling system is used. Four potential sources of substances toxic to yeast occur in these solutions (Leonard and Hajny, 1945): (i) carbohydrate decomposition products, such as furfural formed from pentoses; (ii) decomposition products from lignin; (iii) extraneous substances in the wood; (iv) metal ions produced by corrosion of equipment by acids at high temperatures. In order to reduce toxic effects a number of treatments are available, including: (i) steam distillation, which also sterilizes the medium; (ii) addition of lime to precipitate certain compounds; (iii) protein precipitation; (iv) adjustment of pH value; (v) adsorption by surface-active materials; (vi) treatment with sulphur dioxide, which alters the oxidation-reduction potential.

The yeasts used for these processes require rather special properties as the fermentation conditions are such that some adaptation is required, and special strains of yeast which are able to metabolize the rather wide variety of sugars produced by the hydrolytic treatment. These include varying proportions of glucose, fructose, mannose, galactose, xylose and arabinose. Yeast growth is controlled during the fermentation by the regulated addition of urea as a source of assimilable nitrogen, and phosphate (e.g. monosodium phosphate), the yeast concentration being of the order of 1%. Continuous processes are in use for the fermentation of wood waste in some countries (McCarthy, 1954; Hospodka, 1966). Straw is also employed in certain parts of the world for spirit production. Hydrolytic treatments, similar to those applied to wood wastes are used, followed by fermentation by selected yeasts

g. *Fruit juices.* Juices expressed from many kinds of fruit are fermented when available in sufficient amounts (Malsch, 1962). Typical examples are apples, grapes, pineapples, plums, pears, dates, coconuts and bananas (Simmonds, 1919). The yeasts used for these processes may differ according to the particular sugars in the fruits. The best known example is probably the grape, the juice of which is fermented by the use of many different strains of yeast to produce the various types of wine (see Chapter 2, p. 5), from which brandy is distilled (Section II.F, p. 322). While special care is taken to select certain wines for the manufacture of potable products, unwanted low quality or spoiled wines may be used for the production of industrial spirit.

h. *Sap.* The sap of certain trees, notably the nipa and coco palms, is employed in some areas for the manufacture of spirit. Typically sap contains about 15% of available sugars (Simmonds, 1919). The stalks of the sorghum plant also contain a juice which has about the same content of sugars as sap.

i. *Whey.* Large quantities of whey, obtained during the processing of milk, are available in certain areas. Whey contains lactose, which can be fermented by selected yeasts to produce industrial spirit (Prescott and Dunn, 1959). Butschek and Neumann (1962) quote evidence that the lactose can be fermented to ethanol by *T. cremoris, T. sphaerica, T. lactosa, T. kefir, Sacch. fragilis, Sacch. lactis, Sacch. anamensis, C. pseudotropicalis, Monilia lactis, Zygosacch. lactis, Br. anomalis* and *Br. clausenii.*

2. *Uses of Industrial Spirit*

It is a universal practice to subject alcoholic products to taxation in various forms. Exceptions are made to varying degrees according to the use to which the spirit is put. High duty is payable on potable spirit in relation to the strength of the ethanol content, according to graded scales which differ in different countries. At the other extreme, spirit which has been denatured by the addition of various noxious substances in order to make it unfit for human consumption, may be used for commercial purposes either duty-free or at low rates (Prescott and Dunn, 1940; Stark, 1954). Laws to this effect were passed in Great Britain as early as 1855, but the use of denatured spirit was not legalized in the U.S.A. until 1906 (Stark, 1954). The additives employed for this purpose vary according to the use to which the spirit is to be used (Prescott and Dunn, 1959). When used for motor fuel, the ethanol must be carefully dehydrated (Pleeth, 1949; McCarthy, 1954). The principle uses of ethanol are for chemical synthesis and as a solvent, for which purposes fermentation spirit is in direct competition with the synthetic product; and in the perfumery industry and in pharmacy, where spirit made by fermentation may be preferred because of the different pattern of minor constituents (see Section III.B, p. 324).

C. VODKA AND GIN

Vodka and dry gin are both relatively pure colourless solutions of ethanol in water, although some other compounds are present in low concentrations. The methods of manufacture are broadly similar.

1. *Vodka*

Originally made in Russia, Poland and Finland, vodka is now also produced in considerable quantities in the U.S.A. and Great Britain. It is generally prepared from grain spirit (see Section II.C.2, p. 313), but molasses spirit is sometimes used (Warwicker, 1963). Russian vodka, which is largely made by fermenting rye mashes, is treated by passing the clear distilled spirit over a specially activated wood charcoal which, according to Maravin and Oshmyan (1960), catalyses the formation of ethyl esters of a number of fatty acids if the latter are present in the alcohol before treatment.

In the main, no additions are made to vodka, although in some Scandinavian countries a herbal extract is added. To comply with British law, however, additives must be used in order that the process of manufacture may be classified as compounding. The additives are usually odourless and tasteless, but can be detected chemically or physically. Much of the detail given in the following section in relation to the manufacture of gin applies also to the production of vodka which, similarly to London dry gin, is not matured before sale.

2. Gin

Gin, like vodka, is of comparatively simple composition, the spirit being, in fact, first made in rectified form. In the best practice materials of plant origin, known as botanicals and containing essential oils, particularly juniper berries, are added and the solution is redistilled to produce London dry gin, a term which relates to the manufacturing process rather than to the place of origin (Simpson, 1966). Other types of gin are made by direct addition of suitable essential oils to rectified spirit. Various sweetened gins are also manufactured.

a. *London dry gin.* For the production of typical London dry (or English) gin, cooked maize is frequently used as the main carbohydrate source, with malt as the saccharifying agent. Alternative sources, such as molasses, may be employed, as the official requirement is for spirit of high purity. In fact, the spirit is often bought from outside suppliers and put through rectifying stills by the gin manufacturers to remove higher alcohols and other contaminants. In a typical process (Simpson, 1968) maize provides about 80% of the carbohydrate, and malt (see Section II.E.1, p. 319) the remainder. The maize, usually first crushed to form a "grist", is heated by steam in pressure cookers, with stirring. The temperature is allowed to rise, over a period of 2·0–2·5 h, to 138° or thereabouts at which it is held for 0·5–1·0 h, and a gelatinized semi-solid mass is produced. This is introduced into the mash tun along with water and coarsely ground malt at a temperature of approximately 65°, to give a wort with a specific gravity of about 1·06. After approximately an hour this first wort is drawn off into the "underback" and passed *via* a heat exchanger to reduce its temperature to about 25°, to the "wash-back" in which the fermentation is carried out. (The word "back" is from the same root as the Dutch *bak* and French *bac*, meaning vat.) It is common practice to add the inoculum in the form of a yeast suspension, at this early stage, in order to gain time in the fermentation. This process is known as "pitching". When the yeast is added the fermentation is officially "declared", and is under Excise observation from this time. Several more washings of water are added to the mash to complete the extraction of sugars, starches and other soluble compounds. The temperatures of the waters are commonly increased somewhat at each stage, but by cooling through heat-exchangers on the way to the wash-back the initial fermentation temperature is kept at about 25°, although this temperature may vary somewhat according to requirements. The final water, the third or fourth, is often used as the extractant for the next mash, rather than being transferred to the wash-back.

By these means an aqueous mixture of specific gravity 1·03 to 1·05 is obtained which contains soluble sugars, starch and dextrin complexes, amino acids, growth factors, mineral salts (some small amount often added in the water supply) and amylolytic enzymes. Except for the thermolabile substances (specifically amylolytic enzymes and thiamine), these compounds are extracted in proportionate amounts at each stage. Well over half of the soluble contents are removed in the first "water"; most of that transferred by later washings is already in solution, but must be separated from the bulk of solids remaining in the mash tun. The point at which these extractions are stopped is dependent on the economics of the overall process.

The enzymes made available, mainly from the malt, by this procedure include α- and β-amylases and dextrinase, which continue to attack undegraded starch during the fermentation, and are vital to an efficient grain fermentation process, but are highly influenced by the temperature and pH values reached during the process. The quantity of assimilable sugars available to the yeast can be affected in this way, with the resultant influence on spirit yield.

α-Amylase (α-1,4-glucan 4-glucanohydrolase) acts on the α-1,4 glucan linkages of amylose (straight-chain α-1:4 linked D-glucose) of starch to give a mixture of glucose and maltose, and on the amylopectin (branched α-1:4 linked D-glucose units with 1:6 linkages at the branch points) to produce a mixture of oligosaccharides containing α-1,6-glucosidic bonds (Myrbäck and Neumüller, 1950; Dixon and Webb, 1964). This enzyme is quite stable to heat at 70° in the presence of calcium ions, but is inactivated at pH values below 4·5, at first slowly, but rapidly even at 0° when the pH reaches 4·0 (Luchsinger, 1966).

β-Amylase (α-1,4-glucan maltohydrolase) acts on amylose to form predominantly maltose, and on amylopectin to give a mixture of maltose and highly branched 1:6 linked dextrins (Dixon and Webb, 1964). It does not attack maltotriose. β-Amylase is extremely active for a limited period at 65°; 57° is generally considered to be the most favourable temperature, but this depends on the particular circumstances. Preece and Shadaksharaswamy (1949a) found that in a simple aqueous solution β-amylase activity at 65° does not persist for more than 10–15 min, whereas in malt extract the action is much more prolonged—up to 40 min. The normal pH optimum is 4·6, but in the presence of adequate concentrations of calcium ion the enzyme can be stable at pH values as low as 3·3. The responses of α- and β-amylases to temperature are used to produce preparations with their separate activities. α-Amylase free from active β-amylase can be made by heating malt extract at 70° for 15 min in the presence of 0·1% w/v calcium acetate (Preece and Shadaksharaswamy, 1949b), and β-amylase free from the α-form by maintaining barley extract at pH 3·4 for 24 h (Hopkins *et al.*, 1946).

Dextrinase (dextrin 6-glucanohydrolase) breaks the 1:6 linkages of the oligosaccharides resulting from the activities of other enzymes, to form simpler units which can then be attacked by the amylases. As assimilable sugars are being formed by the enzymes of the grain at the same time as they are being metabolized by the yeast, the pattern of carbohydrates at various stages during the fermentation changes in a complex fashion, but in general the reactions proceed as described earlier (Sections II.A.3.*a–k*, pp. 288–302). When the specific gravity of the medium reaches a more or less constant value, the fermented wash is run into the wash receiver and allowed to settle to remove some solids before being passed to the still (Sections II.A.5.*b* and *c*, p. 304). The size of grain distilleries is such that as much as 10,000 gal wash/h at 4·5% v/v concentration (i.e. almost 500 gal 95% spirit/h) may be produced from a single still. For gin manufacture, the cut is arranged to give as small a concentration as possible of compounds other than ethanol. When the Coffey still is used for the first distillation, further rectification is necessary.

The essential flavour of gin is juniper, but a variety of other organoleptically active substances may be present. The berries of *Juniperus communis*, dried *Angelica officinalis* root and the fruit of *Coriandrum sativum* are most usually added to the spirit before the final distillation, sometimes with the addition of other plant materials such as orange, cinnamon, nutmeg, cardamon and oris. Because botanical crops are seasonal, storage presents its special problems, evaporation of essential oils, moulds growing on damp materials, and insect infestation require attention. Control of temperature and humidity is used successfully to reduce losses (Simpson, 1966).

Manufacturers each have their own secret formulae for the variety and amount of botanicals added to spirit containing 40–50% ethanol; this is distilled and a defined fraction is collected to form London dry gin. The copper stills used for the purpose are of special design: they are usually of pot-still type, heated by steam jackets or coils, and the vapours may be cooled to obtain the correct reflux ratio. The condensate passes down a cooling tube through observation chambers where the specific gravity is measured. First runnings (heads) are mixed later with the last fraction (tails) to form "feints", which may be treated to remove unwanted components and rectified to produce recovered spirit which can then be added to the bulk gin (Simpson, 1966). Gin made in this way is ready for immediate sale, as it is not the practice to mature this type of spirit.

b. Other gins. German gin is made principally from molasses and cereals, the latter being the most important from the point of view of the characteristic flavour of the product (Laatsch and Sattelberg, 1968). In grain distilleries, rye is the main raw material, with occasionally some wheat; 4–10% dry malt is used for saccharification. Three alternative methods are used to prepare the grain mash. The oldest relies on grinding the grain, mixing with water to form a paste, with some malt added to lower the viscosity. After standing at 50° for about 30 min, the mixture is heated in stages to boiling point until the starch is glutinized. Secondly, higher temperatures, up to 160°, may be applied by using pressurized systems: high conversion of the starch is achieved in this way, with resultant high spirit yields. Finally, a more economical so-called cold mashing process may be employed which operates at the optimum temperatures for the action of the grain amylases (see Section II.C.2.*a*, p. 313) without using a dextrinization step; Laatsch and Sattelberg (1968) give details of this procedure.

After saccharification, yeast which has been grown in a special propagation plant is added, and the wort is fermented for a period up to 72 h, after which the fermented wash is distilled. Usually the first stage employs a continuous still, following which rectification is carried out batch-wise with careful separation of the required fractions in order to give the necessary flavour and odour to the finished product.

The fine-grain distilleries manufacture *Korn* and *Doppelkorn*, gin and double gin respectively. The former has an ethanol content of 32% by volume and the latter 40%, and is stored in wooden casks to develop the characteristic grain spirit flavour. These gins bear a close resemblance to grain whiskies (see Section II.E.1., p. 319). Holland gin, originally made in Schiedam in the Netherlands, and *Schnapps* are produced by similar processes, and also derive their character from the original spirit rather than from additives. On the other hand, Genever and Aquavit are made by a different process; they are treated in the same way as London dry gin by the addition of juniper, carraway and other botanicals. Plymouth gin, not necessarily made in the county of Devon, is a further variety. In Germany and Austria *Steinhager*, and in Hungary *Borovicka* are double-distilled directly from crushed fermented juniper berries: this method produces a very strong flavour. Special gins made from a wide variety of fruits and peels are mentioned in Section II.G (p. 322).

D. RUM

Rum, being made from various products related to sugar cane, is traditionally manufactured in the cane-growing countries, particularly the West Indies, although it is also produced in other areas, such as the eastern States of the U.S.A. The most usual raw material for rum fermentations is cane molasses, but cane juice and other by-products of sugar manufacture may be employed. A disadvantage of juice is its seasonal availability.

The first stage of production is mashing, when the solution of sucrose, in whatever form, is treated prior to the fermentation process. The medium may be heated to provide partial or complete sterilization, and clarified by precipitation of non-sugar material. However, in many cases, for instance in Jamaican distilleries, "spontaneous" fermentation, using the natural microbial flora present in the distillery and in the raw materials, is still normal practice. In this way rums of characteristic flavour, or "mark", are produced (MacFarlane, 1947). Budding yeasts now appear to predominate the flora in Jamaican distilleries, whereas a fission yeast, *Schizosaccharomyces mellacei*, was apparently the main organism at the beginning of the century. For these natural fermentations the molasses is not sterilized, and no nutrients are added. Sufficient acetic acid-producing bacteria are usually present in the wash for the production of the amount of acid which is characteristic of the lighter rums. For the preparation of heavier rums, acid supplied from external sources may be added, although there is no fixed practice. Butyric acid, which is known to be present in the product, is said to be formed by the action of *Clostridium* spp. in the wash. Arroyo (1945) has reported that certain butyric acid bacteria (e.g. *Clostridium saccharobutyricum*) can live symbiotically with yeasts, especially those of the fission type. The presence of the bacteria is said to accelerate the rate of fermentation and to improve the quality and ageing properties of the rum. MacFarlane (1947) states that the free acids in Jamaican rum comprise 97–98% acetic and 1% butyric with traces of formic, pelargonic, capric and other aliphatic acids. Higher alcohols are also found in rum (see Section III, p. 323), and these are considered to give "body"; aldehydes are also important from the flavour point of view. Arroyo considers that a substance called "rum oil" is responsible in part for the characteristic odour of rum. The amount of this substance appears to depend on the type of yeast used for the fermentation, fission yeasts being particularly effective in this respect. Since this oil boils at a high temperature, pot-still distillation is more favourable to its presence than the steady lower temperatures maintained in continuous stills. Arroyo (1945) has also suggested that a top-fermenting fission yeast, *Schizosacch. pombe*, might produce a more desirable rum than is obtained with the usual budding type.

Esters are present in rums in similar proportions to the organic acids.

According to MacFarlane (1947) it is probable that the greater part of the esters in light rums is formed by direct esterification during the yeast fermentation, but some contaminants, such as *Torulopsis* spp., are capable of forming esters direct from sugars, and these organisms play an important part in the production of the flavour constituents of rum. The higher esters can be increased in concentration by the "lime salt" process, in which calcium salts of the acids from the retort lees are heated with sulphuric acid in the presence of ethanol to precipitate the calcium as sulphate and so release the organic acids, which are added to the "high wines" before distillation. A low fermentation temperature, although giving a slow fermentation, is also said to produce rum of good quality.

The fusel oil of Jamaican rum has an exceptional composition due to the high proportions of 2-butanol and 1-propanol in relation to 2-methyl-1-propanol. The amount of 3-methyl-1-butanol compared to that of 2-methyl-1-butanol is also markedly greater than in whisky (Suomalainen and Nykänen, 1967).

In many distilleries, especially in countries other than Jamaica, sterilized molasses medium is pretreated in various ways and fermented with selected yeasts under controlled conditions of temperature and pH value in order to obtain the desired characteristics of the distilled rum. Yeast inocula are grown up for this purpose in increasingly large stages from pure culture, under essentially sterile conditions until a sufficiently large amount is available for seeding the main fermentation. In this type of process, fresh inocula are prepared each week to prevent the excessive multiplication of extraneous micro-organisms (Arroyo, 1945). Solids may be added to the wash to increase the rate of fermentation. After completion of this stage the yeast is centrifuged off and used for animal feed or fertilizer, and distillation is carried out either by a series of two or more batch distillations using a rectifying column, or in a continuous still specially designed for the purpose, preferably using a low final spirit concentration. The latter method is used for large-scale production, mainly in the U.S.A., Cuba and Puerto Rico. Objectionable qualities can be imparted by sulphur-containing compounds as the result of the action of contaminating micro-organisms. The distillate is next transferred to oak barrels, where it is stored for several years to mature; this stage is also called "curing".

Arrack, a type of rum made by distilling fermented molasses and rice mashes, or coconut juice, is produced in eastern countries, notably the East Indies, Java, Thailand and Ceylon (Suomalainen *et al.*, 1968). Dates are also used for this purpose in the Middle East. Various moulds may supplement the diastatic activity of the rice. Arrack is matured in teak casks.

E. WHISKY

The term whisky, or the Irish and American equivalent whiskey, covers a group of potable spirits made by the distillation of mashes of cereal grains saccharified by the diastase of malt. Various types of grain can be malted, although barley is often thought of as the source, probably because it is used for the manufacture of Scotch malt whisky (Simpson, 1968) and beer. The malting process, which involves germination of the grain by moistening with water, causes enzymes to develop which are able to convert starches to simple sugars (see Section II.C.2, p. 313) that are assimilable by the yeasts used in the distillery. The importance of efficient mashing, when the grain is extracted with water, is that the maximum quantity of assimilable sugars should be produced (Griffin, 1966; Elks, 1968). Extraction of amino acids is also important at this stage (Pierce, 1966).

The function of the selected yeast (see Section IV, p. 324) is to utilize the highest possible proportion of these sugars, which normally comprise monosaccharides, chiefly glucose, the disaccharide maltose, and tri-, tetra- and higher oligosaccharides, all glucose polymers, which are successively metabolized by yeasts.

One of the most important cellular enzymes required in this connection is maltase (α-D-glucoside glucohydrolase) which removes glucose units from the more complex chains, but at progressively slower rates according to the number of groups present. The mechanism is dependent on the lower affinity of the enzyme for the larger molecules; thus maltose is removed from solution by the yeast before the other sugars.

1. *Scotch Whisky*

A major characteristic of Scottish highland whiskies is that the malt is dried before use by heating in kilns in which peat is burned, so giving the malt the property of imparting to the mash, prepared later for fermentation, certain organoleptically active compounds, particularly of a phenolic nature. These substances later distil with the spirit to give the special flavour of malt whisky. There is no substantial evidence, however, that the compounds produced in this way have any effect on the yeast. Over-heating of malt must be avoided or the amylolytic enzymes are inactivated.

Unpeated kilned malt, and green malt (see Section II.B.1.*d*, ii, p. 309) may also be used, in particular for the manufacture of grain whisky, which may be made from mashes containing a proportion of unmalted cereal in the same way as gin spirit (Pyke, 1965; Section II.C, p. 312). For the production of Scotch grain whisky, maize is the usual source of additional starch (Simpson, 1968). The maize is ground, cooked and mashed with malted barley at a temperature of 60–65°, agitation being provided during the process by revolving rakes in the mash tun. The specific gravity of the liquid is about 1·07 at this stage, but successive extractions are made to produce a final wort with a gravity of 1·04 or thereabouts, and containing approximately 10% assimilable carbohydrates.

For the manufacture of malt whisky, the liquid mash is strained through the layer of solids at the bottom of the mash tun to produce a solution which is almost clear. Grain worts are usually less completely filtered, a fact which affects the fermentation. According to Merritt (1967) the inclusion of solids in the wort results in enhanced yeast growth, and the excretion of more fusel oils and glycerol than is the case with clear worts. The solids, called spent grain or "draff", which remain are used as an animal feed supplement (see Section V, p. 326).

As has been described for gin manufacture (Section II.C.2.a, p. 313), yeast (*Sacch. cerevisiae*) is added before the cooled wort is fully diluted, and the fermentation is declared for excise purposes. In grain distillery fermentations compressed air, in limited amounts, is used to mix the wort and yeast thoroughly. The oxygen supplied in this way is rapidly taken up by the yeast, thus favouring growth (Section II.A.3.a, p. 288). Fermentation then proceeds for a period of time which is dependent on such factors as the concentration of the sugars in the wort, the amount of yeast initially present and the extent of its multiplication, and the temperature. After a period varying from 36 to 100 hours, the fermented wash is passed to a wash charger and thence to the still. In the case of malt whisky this is always a pot still; for grain whisky a continuous type is used (Section II.A.5, p. 303).

After distillation, the spirit for Scotch whisky, both malt and grain types, is matured by storage in oak casks. The maturing period is at least three years, but is frequently much longer. During this time changes take place which mellow the fiery nature of the new spirit, due to chemical reactions and extraction of substances from the wood. Many of the compounds which are detected by analysis of the matured product (Section III.A, p. 323) are different from those originally present, although largely derived from them. Because the strength of the aqueous ethanol at this stage is relatively high, of the order of 65% (v/v), any substances in the material of which the barrels are made which are soluble under these conditions will be extracted. Two classes of compounds are concerned: (i) residual components of sherry which have soaked into the wood during long periods of wine storage; and (ii) constituents of the wood itself and any oxidation products produced by charring. Lignin and its oxidation products are the most important of these.

2. *Irish Whiskey*

The grain used for the manufacture of Irish whiskey may be barley, oats, wheat or rye. Maize is not normally employed. The proportions of the various types is varied in order to obtain particular flavours in the resulting spirit, but usually about 80% of the mixture is malted or unmalted barley, the remaining 20% consisting of oats, wheat and rye (Simmonds, 1919). Whiskey made by this process is distilled in pot stills. The general practice is to carry out three consecutive distillations, and the distillate is collected at a higher ethanol concentration than is the practice for Scotch pot still whisky. Irish pot still whiskey usually contains a larger proportion of secondary constituents, such as esters and higher alcohols, than the Scottish equivalent. Grain spirit is also made in Ireland; the process is much the same as that used in Scotland.

3. *Bourbon and Rye Whiskeys*

In North America the principal types of whiskey are rye and bourbon: the former is the main drink in Canada, although bourbon is popular, while the latter is more common in the U.S.A. Canadian bourbon is lighter than the United States equivalent.

Rye whiskey is prepared from either rye and rye malt, or rye and malted barley. By law the mixture must contain at least 51% rye. Bourbon is made from maize, with either barley malt or wheat malt added to provide the amylolytic enzymes. Rye may also be used in the mash, but there must always be at least 51% of maize present (Prescott and Dunn, 1959). The mashes for these processes are prepared by methods such as are typically used in brewing practice, except that solids are not removed. The yeast employed is *Sacch. cerevisiae*, but bacteria which are often present are said to contribute to the flavour of the finished product (see Section III, p. 323). In some cases the yeast is deliberately propagated along with lactic acid bacteria as a mixed culture. The lactic acid inhibits the development of undesirable micro-organisms, and plays a part in the development of the characteristic aroma, flavour and other characteristics of the whiskey.

Both rye and bourbon fermented mashes are distilled in continuous stills; the type of still determines the quality of the product (Dudding-ton, 1961). The spirit is then matured in white oak or in charred oak barrels. Accelerated ageing procedures are frequently used for these types of whiskey. Methods employed include the use of heat, charred or uncharred wood chips and charred barrels. These treatments produce no significant change in the higher alcohol content of the spirit, and have little effect of the amount of esters, but the solids, furfural content and colour are increased.

F. BRANDY

Brandy is distilled from the fermented juice of fruit, particularly grapes. It is made in many parts of the world, including Spain, Portugal, Greece, South Africa, Australia, Cyprus, Egypt and Algeria, but especially in France, where the finest types come from the district of Cognac, north of Bordeaux, or from Armagnac. Unsulphited grape must is fermented, using a rapid process, by typical yeasts employed in the manufacture of wine (Duddington, 1961; see also Chapter 2, p. 5). The fermented liquor is double distilled, without prior storage, in special pot stills, the first distillate being about 24% (v/v), the second 70–72% (v/v), although higher rectification can be used to remove unwanted flavours. Foreshots and tails are removed and added to the next distillation (Duddington, 1961). As distilled, the spirit is colourless, but true brandies are matured in oak casks for at least three years, and often for very much longer; in this way colour is developed, as with other spirits such as whisky.

Various inferior types are also produced, for example marc brandy, which is distilled from the fermented skins of fresh grapes after the greater part of the juice has been extracted. To obtain a particular tint extra colouring matter is sometimes added. Brandy is bottled at not less than 36% (v/v) (Prescott and Dunn, 1940).

G. LIQUEURS

Liqueurs consist essentially of spirit, flavouring and sugars, often compounded originally by members of religious orders. The base may be brandy, whisky, gin or other forms of potable spirit. Flavourings may be added before the final distillation, by infusion or, less desirably, by the addition of essences. Good liqueurs are matured in wood, like wines and other spirits, and colouring is often added before filtering and bottling. Flavouring substances include: aniseed (to make absinthe and anisette), apricot, cherry and peach (for the corresponding brandies) blackcurrant (cassis), chocolate (crème de cacao), peppermint (crème de menthe), coffee (crème de moka), almond (crème de noyau), bitter-cherry (maraschino), bitter orange peel (curaçao), tangerine (mandarine), wild cherry (kirsch), caraway seed (kummel) and sloe (sloe gin, prunelle). The sugars used as sweetening agents may be substances such as honey or malt rather than sucrose. Various types of bitters are also prepared by blending certain herbs, Seville orange peel, peach kernels and so on, in rum or other spirits.

H. CARBON DIOXIDE

It has been shown (Section II.A.3.a.i, p. 290) that the weight of carbon dioxide produced during anaerobic fermentation is almost equal to the ethanol, and about half of the weight of carbohydrate metabolized. In many large distilleries it has been found economically worthwhile to collect the gas evolved and sell it in compressed or solid form. For this purpose covered vessels are used, and after the first gas produced has flushed out any air from the head space, the carbon dioxide is pumped through scrubbers to remove unwanted impurities, such as traces of alcohols, esters, hydrogen sulphide, nitrous oxide, sulphur dioxide, etc. which are formed by the action of yeast or contaminating micro-organisms. It is dried by passing through a desiccant such as silica gel or activated alumina before being compressed and loaded into pressure vessels or converted into solid blocks (Hodge and Hildebrandt, 1954).

III. Composition of Distillation Products

A. INFLUENCE OF PROCESS CONDITIONS

The organoleptic responses produced by the different types of potable spirit are due to the presence of a complex mixture of substances which originate from several sources. These include volatile compounds present in the raw materials; for example, phenolic products from the kilning of malt, and chemical break-down by heat of grain constituents during distillation. In the present context, however, the compounds of special interest are those resulting from the action of yeast. These can again be subdivided into volatiles produced directly as excretion products by the metabolic reactions of yeast enzymes on substrates in the fermentation medium, and compounds formed indirectly from this first class by interaction with other substances during fermentation or distillation. In the former group are higher alcohols, organic acids, esters and aldehydes, the biochemical mechanism of synthesis of many of which are given in some detail in Chapter 4 (p. 147) in connection with the brewing of beer, and also in Chapter 2 (p. 5) dealing with the production of flavouring compounds in wines. In the second class, esters of alcohols and certain acids, specifically acetic and lactic (Suomalainen et al., 1968), which result from the action of contaminating micro-organisms (Section II.A.3.k, p. 300) or from other extraneous sources.

Technically there is some difficulty, in many cases, in deciding with certainty which compounds are the primary result of yeast metabolism, because the amounts observed are in general extremely small and most of the analytical procedures involve heating, as in distillation, or other treatments which may cause chemical changes and produce artifacts. However, the compounds directly produced by yeasts undoubtedly include higher alcohols such as n-propyl, active amyl and isobutyl, which are present in fermented wash in relatively high concentrations, certain fatty acids, aldehydes and esters. The conditions of fermentation, for example the concentrations of carbohydrates and nitrogen-containing substances, the temperature, pH value and the yeast strain, influence the pattern of congeners (see Chapter 3, p. 73).

For the sake of completeness most of the volatile components of distilled spirits and the corresponding fusel oils that have been recorded in the literature are listed in the Appendix, along with the products in which they have been observed.

B. ORGANOLEPTIC EFFECTS

The relationship between the odour and taste of potable spirits and the small concentrations of congeners is still very incompletely understood. The work of Suomalainen (1968) and co-workers (Suomalainen et al., 1968) has shown by gas chromatographic and similar techniques that the aroma compounds in various potable spirits (brandy and whisky) and the distillate from a nitrogen-free fermentation of sucrose, are qualitatively similar. The ester compositions of whisky and brandy are quantitatively rather alike, as are the higher fatty acids in whisky, brandy and rum. On the other hand, samples of Scotch and bourbon whiskies were shown to differ markedly in their content of some congeners.

It would appear that subtle differences of flavour are produced by small changes in the concentrations or proportions of the minor constituents. The strain of yeast is of importance in this respect as there is considerable evidence, particularly in the case of wine-making (see Chapter 2, p. 5), that there are differences between yeast strains in the patterns of minor products formed, and hence on the sensory responses.

While the chemistry of this subject is now fairly well understood, both as regards the identity and concentration of the compounds and the synthetic routes by which many of them are formed, the problem of relating this knowledge to sensory stimuli is extremely difficult. The reason for this is that no satisfactory method is available for quantifying taste and smell, although differences can be recorded, for example by suitably chosen and organized tasting panels (Gridgeman, 1967). The estimates of the quality of various products still rely, therefore, on subjective assessment by experts with long experience.

IV. Selection of Yeast Strains

For any specific purpose, it is obvious that the ultimate criterion of the suitability of a yeast strain must be the result of its performance under working conditions. Because of the complexity of the raw materials, fermentation, distillation and maturation processes, and the judgement of the properties of the final product, the choice of cultures for the preparation of potable spirits is particularly difficult. On the other hand, for industrial spirit, where only high ethanol content and possibly freedom from particular chemical contaminants is required, the problem can be approached in a straightforward scientific manner. In the latter case, simulated conditions of time, temperature, pH value and so on can be applied in the laboratory, using the raw materials known to be available commercially, and by careful analysis of the fermented liquor the yield and quality of the spirit can be found relatively quickly. Laboratory studies of alcohol and sugar tolerance and growth rate can be carried out. The alcohol tolerance, for instance, can vary from 5·8% to 11·6% for different strains (Stark, 1954; see also Chapter 5, p. 225). If necessary, pilot plant tests can be used to confirm the results on chosen strains. Yeasts can be screened and selected in this way and, if required, new types can be produced by adaptation (Johnson and Harris, 1948), breeding (Lindegren, 1944) or mutation.

The problem with potable spirits is different. In the case of brandy, for instance, where many types of wine made by various fermentation processes can be available for distillation, it is possible to select the most suitable source. This has taken place naturally over the years, and no doubt accounts for the original localization of brandy distilling in certain areas. As discussed in Chapter 2 (p. 5), the yeast population differs widely in different districts, and a very extensive variety of strains of wine yeasts are known.

The well known variations between wines of the same type suggests that many factors influence the congeners in the fermenting liquid, and hence in any distilled product made from the wine. Selection of yeasts for rum manufacture has been carried out, particularly by Arroyo (1945; see Section II.D, p. 316), although many of the strains employed are of natural origin. Little has been published on the effect of yeast on whisky quality. As in the case of the other potable spirits, the large number of separate different makes on the market, with their particular characteristics, and the fact that the production of blended whiskies requires expert control, emphasizes the complexity of the production process, the yeast being only one of the factors concerned.

The special and unusual characteristics of the pot stills used for the distillation of many potable spirits make it impossible to judge the effects of any change in the manufacturing process, including the yeast strain, without repeated full-scale tests. In many cases, in modern practice, selected culture yeasts are specially grown up for use as the inocula for fermentation processes in distilleries; it can therefore be inferred that changes that are observed under these circumstances cannot be due to the genetical make-up of the yeast employed. For these reasons it is difficult to express quantitative views on the effect of yeasts on spirit quality. However, it is possible that as a result of the great efforts being made to understand the biochemical, physical and physiological mechanisms involved in the production and quality of distilled products, it will be possible in the future to select yeast strains by sophisticated laboratory techniques for these purposes. The final development, to tailor-make strains with the desired organoleptic properties, remains for the future.

V. Effluent Disposal

During the course of all distillery processes large amounts of various effluents are produced which must be disposed of by suitable means. Much of this material has a high biochemical oxygen demand (B.O.D.; see Ministry of Housing and Local Government, 1956), and is therefore an embarrassment, as it cannot be discharged, untreated, into rivers, and is costly to dispose of *via* public sewage systems. As explained in Section II.A.3.*a* (p. 288), yeast dry weight is formed which amounts to about 2% of the carbohydrates utilized, along with another 2% of glycerol and 0·5% of organic acids. It will be seen that yeast and its excretion products contribute in a large degree to the effluent problem.

A number of alternative types of effluent treatment are available, the preferred methods being those which produce as end-product a material which has an economic value sufficient to offset the cost of the treatment. The procedure adopted depends largely on the form of the effluent and on the location of the distillery. Although in a certain number of cases materials are eliminated at early stages in the processes, for instance, the draff removed after malt mash has been extracted with hot water in the preparation of wort for grain fermentations, the main effluent consists of solids, including yeast (Wiley, 1954; see Chapter 8, p. 421) which are present in the final wash, and the spent liquor after distillation. The fact that yeast and other substances of biological origin are present makes this effluent of value on account of its content of vitamins, carbohydrates, proteins and so on, which can be used as animal feed supplements. The content of nitrogen and certain minerals such as potassium and phosphate makes the material useful also as a fertilizer.

Although the insoluble solids can be filtered or centrifuged off, dried if necessary and used directly, the solubles present a more difficult problem. Evaporation of water is expensive, but is sometimes considered to be the most practical solution. For this purpose a number of methods are available. The water can be removed in multistage-evaporators, and at a certain moisture content (50–80%) the partially dried product can be used, mixed with other combustible material such as natural gas or fuel oil, for boiler heating, etc. (McAteer, 1968).

Spray-drying can follow the first stages of evaporation of grain spirit spent wash from Coffey stills to produce a useful food additive which contains appreciable amounts of members of the vitamin-B group, organic acids and other substances (Rae, 1966). Many of the important constituents of the spray-dried product originate from the yeast, which, in partly autolysed form, passes through the rough screening applied before entering the still, along with unfermented solubles from the grain. The process consists essentially of evaporating the wash in several stages, after filtering through sloping wedge-wire screens, to about 25% total solids, the coarse solids being sold as "wet grains", or they are dried. A small amount of lime is added, and the concentrate is pumped to large spray-driers through which gases at 260° are passed; these cool to 120° during their residence time in the drier. The product is a brown powder, which is sold mainly to fortify animal feeds.

The simplest method of disposal of effluent is to sediment the insoluble solids and pump the liquid to the sea or into a river, although in many circumstances this is prohibited unless the B.O.D. is low (e.g. 20). For this purpose biological percolating filters or activated sludge tanks can be used to reduce the B.O.D. to acceptable levels for discharge (Ainsworth, 1966). Both these methods depend on the action of various organisms which metabolize the organic constituents of the effluent, often under aerobic conditions. The choice of site for distilleries is to a large extent controlled by the need to use water in relatively large quantities for process and cooling, and to discharge effluent satisfactorily, as large volumes of liquid require to be disposed of in spite of any treatment, other than evaporation, which may be applied. The other main consideration in respect of siting is supply of raw materials, which are required in considerable bulk.

In many cases the disposal of effluent from a distillery is a second stage problem, because the original purpose of using the raw material to make ethanol (or yeast) is to eliminate an effluent problem. This is particularly the case in inland areas where sulphite waste liquor is produced (McCarthy, 1954). Molasses is also, of course, a residue from the manufacturing of sugar. The metabolism by yeast of the residual carbon-containing compounds in these materials substantially reduces the B.O.D., and therefore the effluent disposal problem.

VI. Process and Quality Control

During the manufacturing processes used for the production of all types of spirit, and for checks on the finished products, various analytical and biological tests are carried out. These are largely designed to measure the amounts of nutrients utilized by the yeast during fermentation, inhibitory substances, and the products of the intracellular reactions. Microbiological techniques are used to detect and estimate the presence of harmful contaminating micro-organisms (Section II.A.3. *k*, p. 300), as well as to follow the behaviour of the yeast.

A. SAMPLING

The sampling of the majority of the bulk raw materials used in distillery practice presents problems. Molasses, being very viscous in the undiluted form, is difficult to mix when in large containers, and if too much heat is employed to reduce the viscosity detrimental chemical changes take place. The best method of obtaining a representative sample is probably to collect a continuous small proportionate run-off from a supply line. On long storage of molasses separation of solids may occur. Cereals present an equally difficult problem. Again a continuous sample can often be taken from the feed line of whole or ground grain. For sampling from large stocks in ship's holds or silos accepted methods are available. Distilled spirit, either in bulk or in individual containers, can be dealt with more easily, as also can other free-flowing liquids. The accepted techniques, and the mathematical treatment of sampling, have been described on numerous occasions; a recent example is by Steiner (1967).

Process control during fermentation requires special care, particularly where solids are present, as in unclarified grain mashes. Some of the problems are dealt with in Section II.A.3.*j* (p. 299). Owing to the fact that there is poor agitation during the early and late stages of commercial batch fermentations when little carbon dioxide is being evolved, the solids, including the yeast cells, settle to the lower parts of the vessels. This affects, among other things, the relative amount of liquid in a sample of given volume, and hence modifies the calculation of the total spirit content. For correct evaluation of the process parameters, bulked samples from all depths are required under these special conditions. As examples of the effects observed, the count of bacterial contaminants may be much greater in the vicinity of settled solids, due presumably to adsorptive and nutritional causes as well as to a limited degree of settling. Carbon dioxide concentration is also considerably greater at the bottom of deep fermentation vessels than at the surface; this can have an indirect effect on other phenomena.

B. RAW MATERIALS

1. *Carbohydrate Supply*

Although a wide variety of different starting materials are used for making spirit, the basic constituent is in all cases carbohydrate. In some instances this is in the form of simple sugars, in others starches or other material which requires hydrolytic treatment before becoming available for assimilation by yeast.

An important analysis is therefore for the total content of potentially assimilable carbohydrates. While the most satisfactory determination involves a measurement of the content before and after the raw material has passed through the actual appropriate stages of the manufacturing process, or a procedure which is designed to simulate these in order to obtain a truly realistic value, simpler methods, such as a direct determination of total carbohydrate, may sometimes be used satisfactorily. One of the most widely used analyses is Fehling's titration, or one of its modifications such as the Lane and Eynon method (1923); these may be applied either before or after hydrolysis. The immediately available reducing sugars can be measured directly, without preliminary treatment of an aqueous suspension of the raw material. Total sugar content can be determined after either acid (Lane and Eynon, 1923) or enzymic treatment. When the proportions of individual sugars need to be estimated, chromatography and similar methods are applied.

From the amount of available substrate supplied, calculated as glucose, a close approximation to the expected yield of ethanol can be calculated as:

$$\frac{\text{molecular weight of ethanol} \times 2 \ (92)}{\text{molecular weight of glucose} \ (180)} \times \frac{92}{100} = 0.471$$

The 92% correction is to allow for yeast growth, glycerol formation, etc. (see Section II.A.3.a, p. 288). Any amino acids which are present have a sugar-sparing effect by contributing to yeast growth in place of some of the carbohydrate. A correction can be made for these constituents of the medium if their amount is known.

Detailed methods of analysis have been given for cereals by Lindemann (1962), for sugar-containing materials, including molasses, by Becker and Weber (1962) and for wood hydrolysates by Butschek and Krause (1962a).

Where malted grain is used, the diastatic power may be measured, as in brewing processes, by standard procedures (Institute of Brewing Analysis Committee, 1961). Logically the determination should be carried out under similar conditions to those applied in the particular production process under consideration.

2. Yeast

The proportion of viable cells in the yeast used for seeding fermentations can be determined by a variety of methods. The use of the stain

Lissamine green V has been found to be particularly suitable, at a concentration of 0·1% and at pH 6·6 (Fowell, 1964). Microbial contaminants in the yeast can be counted by the methods mentioned below.

C. PROCESS CONTROL

As the production of spirit by fermentation is a biological process, the most important factors are microbiological. Changes can be followed in a number of ways.

1. Yeast Activity

Because, during most of the process, the rate of fermentation is proportional to the amount of yeast present, a count of the number of cells per unit volume is useful as a means of following the course of events. In a steady-state continuous system the count should remain constant; otherwise there is an exponential increase, followed by a falling off in multiplication rate as nutrients are used up. If the medium has been clarified, a haemocytometer method is adequate, but a more realistic measure may be obtained by determining the rate of carbon dioxide evolution when a known volume of the fermenting liquor is transferred to a suitable apparatus held at a known fixed temperature with shaking.

2. Infecting Organisms

As discussed in Section II.A.3.k (p. 300), certain acid-producing micro-organisms, in particular *Lactobacilli*, are detrimental to saccharification and fermentation. Methods for following the behaviour of such contaminants present technical difficulties. A selective medium for *Lactobacilli* has been described by Rogosa *et al.* (1951) and modified by Costilow *et al.* (1964) who have given techniques for measuring other bacterial, fungal and wild yeast infections in fermentation processes. These selective plating techniques may be precise, but results are often obtained too late to allow of effective remedial action. Microscopic procedures involving fluorescein-labelled specific antisera have been suggested (Kunz and Klaushofer, 1961; Campbell and Brudzynski, 1966), but more research is required before they can be applied routinely for distillery infection problems. Acid production at a higher rate than that to be expected from the yeast alone (see Section II.A.3.a, p. 288) can be taken as an indication of infection by acid-producing bacteria.

3. Physical Measurements

During fermentation, carbon dioxide is evolved at a rate that is proportional to the formation of ethanol (Section II.A.3.a, p. 288). The weight of a given volume of medium decreases and the specific gravity consequently falls. This change can be followed easily by means of hydrometer readings. The final specific gravity, in many processes, is close to that of water. The solution at this stage consists mainly of a mixture of water, ethanol, glycerol, organic acids and residual soluble solids. By experience with a particular type of process, an accurate idea of the behaviour of the yeast can be obtained.

As explained in Section II.A.3.*c* (p. 296), the exothermic reactions within the yeast cells result in the production of heat, which again is proportional to the amount of carbohydrate metabolized. The rise in temperature of the fermenting liquor can therefore be used to follow yeast behaviour, if due allowance is made for heat losses.

Also, in the absence of excessive infection, organic acids, mainly succinic, are produced in relation to the stage of fermentation (Section II.A.3.*a*, p. 288). Changes in pH value are therefore useful as an indication of the progress of fermentation. As the buffering effect of the wash affects the actual pH value reached, this method must be applied in a manner relative to similar fermentations.

4. *Chemical Analysis*

a. Ethanol. The most direct indication of the efficiency of the performance of the yeast is the amount of ethanol present in the wash. Various methods are available, but, because of Customs and Excise regulations, the most common depends on determination of the specific gravity of a distilled sample of known volume (Nicholls, 1947). Alternatively a dichromate titration may be used with suitable precautions (Guymon and Crowell, 1959).

b. Glycerol. The production of glycerol can be followed by analysis, but care must be taken to avoid interference from similar compounds, including sugars. A simple chromatographic separation followed by a periodate titration is suitable (Sporek and Williams, 1954).

c. Acids. Acid formation is measured simply by titration with a standard solution of sodium hydroxide, preferably using a pH meter for the end point; it is convenient to take this as the original pH value of the medium.

D. QUALITY CONTROL

The criteria applied to the finished product depend on the sales requirements. For industrial spirit the ethanol content is needed, and often limits are set to the permissible content of specific compounds. These are straightforward analytical problems. In the case of potable spirits the all important factor is flavour which, being a subjective phenomenon, cannot be measured by chemical or physical means.

Testing is made even more complicated by the lengthy maturing periods often required. The laboratory analyses may therefore be confined to those required for the purpose of Excise control, and for checks on specific requirements of the manufacturer or customer. The selection of suitable batches for sale, blending and so on is then the responsibility of panels of expert judges of the required subjective qualities.

VII. Conclusions

The making of alcohol by distillation of fermented carbohydrate-containing materials must be one of the earliest cases of the preparation by indirect means of a chemical compound in almost the pure state. The function of yeast has been vital to this process, and it is only within recent years that a competitive economic synthetic process has been applied commercially. The relative merits of the biological and chemical systems must rest largely on the cost of the bulk raw material, naturally-occurring carbohydrates and petroleum oils respectively. As long as by-products from other processes, such as molasses, sulphite liquor and wood waste, are available cheaply and in quantity in localities where the comparatively simple necessary technological operations can be undertaken, no doubt industrial spirit will be made competitively in this way. This situation is favoured, particularly in the case of sulphite liquor, by the increasing world-wide official insistence on the reduction of contamination in rivers and inland waters.

The use of yeast in its related capacity, as an agent in the production of potable spirits, is even more certain of remaining a viable proposition for a very long time. The acceptance of the various types and grades of spirituous beverages is so subjective and bound up with tradition, that it is inconceivable that synthetic substitutes could replace the present traditional drinks in the forseeable future.

For many years to come, therefore, it can be expected that the technology outlined briefly in the preceding pages will be further developed, probably gradually becoming more sophisticated technologically with advances in scientific knowledge, while retaining its essential dependence on yeast.

Appendix

COMPONENTS OF DISTILLED SPIRITS AND FUSEL OILS

B = brandy; F = fermented synthetic sugar medium; G = grain and potato spirit; J = fermented fruit juices; M = molasses spirit; R = rum; S = fermented sulphite liquor and wood hydrolysate; V = vodka; W = whisky

Number of carbon atoms	Compound	Synonyms	Formula	Reported presence
Alcohols				
1	Methyl	Methanol	CH_3OH	B G J R V W
2	Ethyl	Ethanol	CH_3CH_2OH	B F G J M R S V W
3	n-Propyl	1-Propanol	$CH_3CH_2CH_2OH$	B G J M R S W
3	Isopropyl	2-Propanol	$CH_3CHOHCH_3$	B J M
4	n-Butyl	1-Butanol	$CH_3CH_2CH_2CH_2OH$	B G J R W
4	sec-Butyl	2-Butanol	$CH_3CH_2CHOHCH_3$	B J R W
4	Isobutyl	2-Methyl-1-propanol	$(CH_3)_2CHCH_2OH$	B G J M R S W
4	tert-Butyl	2-Methyl-2-propanol	$(CH_3)_3COH$	W
5	n-Amyl	1-Pentanol	$CH_3(CH_2)_3CH_2OH$	B S W
5	pri-act-Amyl	2-Methyl-1-butanol; isopentanol; *sec*-butyl carbinol	$CH_3CH_2CH(CH_3)CH_2OH$	B F G J M R S W
5	sec-act-Amyl	2-Pentanol; methyl *n*-propyl carbinol	$CH_3CH_2CH_2CHOHCH_3$	R
5	Isoamyl	3-Methyl-1-butanol; isobutyl carbinol	$(CH_3)_2CHCH_2CH_2OH$	B F G J M R S W
5	sec-Isoamyl	3-Methyl-2-butanol; methyl isopropyl carbinol	$(CH_3)_2CHCHOHCH_3$	B F M
5	3-Pentanol	Diethyl carbinol	$CH_3CH_2CHOHCH_2CH_3$	M
6	n-Hexyl	1-Hexanol; amyl carbinol	$CH_3(CH_2)_4CH_2OH$	B G M R S W
6	Isohexyl	4-Methyl 1-pentanol	$(CH_3)_2CH(CH_2)_2CH_2OH$	W

Number of carbon atoms	Compound	Synonyms	Formula	Reported presence
7	n-Heptyl	1-Heptanol; enanthic alcohol	$CH_3(CH_2)_5CH_2OH$	G M R W
7	Isoheptyl		$(CH_3)_2CH(CH_2)_3CH_2OH$	G
7	Methyl amyl carbinol	5-Methyl 1-hexanol	$CH_3(CH_2)_4CHOHCH_3$	G M R
8	$prim$-n-Octyl	1-Octanol; capryl alcohol	$CH_3(CH_2)_6CH_2OH$	G M
9	n-Nonyl	1-Nonanol	$CH_3(CH_2)_7CH_2OH$	G
9	Methyl heptyl carbinol	2-Nonanol	$CH_3(CH_2)_6CHOHCH_3$	G M R
10	n-Decyl	1-Decanol	$CH_3(CH_2)_8CH_2OH$	G
10	Geraniol	3,7-Dimethyl 2,6-octa-diene-1-ol	$(CH_3)_2C{:}CHCH_2CH_2C(CH_3){:}CH$ CH_2OH	G W
11	n-Undecyl	1-Hendecanol	$CH_3(CH_2)_9CH_2OH$	G M
11	Methyl nonyl carbinol	2-Undecanol	$CH_3(CH_2)_8CHOHCH_3$	G M
12	n-Dodecyl	1-Dodecanol	$CH_3(CH_2)_{10}CH_2OH$	G M
4	2,3-Butylene glycol	2,3-Butanediol	$CH_3CHOHCHOHCH_3$	B G W
5	Furfuryl	2-Hydroxymethyl furan; 2-furan carbinol	$CH{=}C{.}CH_2OH$, $CH{=}CH$ (furan ring, O)	B G R W
7	Benzyl	Phenyl carbinol; α-hydroxy toluene	$C_6H_5CH_2OH$	G
8	α-Methyl benzyl	1-Phenyl ethanol	$C_6H_5CH(CH_3)OH$	G M
8	Phenethyl	2-Phenyl ethanol; benzyl carbinol; β-phenyl ethanol	$C_6H_5CH_2CH_2OH$	B G M R S W
9	α-Methyl-α'-(β-furyl) tetrahydro furan		$C_4H_6OCH_3(C_4H_3O)$	G

Acids				B F G	R S V W
1	Formic	Methanoic	HCOOH		R V
2	Acetic	Ethanoic	CH₃COOH	B F G	R S V W
3	Propionic	Propanoic; methyl acetic	CH₃CH₂COOH	B F G	R S V W
4	Butyric	Butanoic; ethyl acetic	CH₃CH₂CH₂COOH	B F G	R S V W
4	Isobutyric	2-Methyl propanoic; dimethyl acetic	(CH₃)₂CHCOOH	B F G	R S W
5	n-Valeric	Pentanoic; propyl acetic	CH₃CH₂CH₂CH₂COOH	B F G	R S V W
5	Isovaleric	3-Methyl butanoic; isopropyl acetic	(CH₃)₂CHCH₂COOH	B G	R S W
6	n-Caproic	Hexanoic; butyl acetic	CH₃(CH₂)₄COOH	B F G	R S V W
6	Isocaproic	4-Methyl pentanoic; isobutyl acetic	(CH₃)₂CHCH₂CH₂COOH	F	R S W
7	Enanthic	Heptanoic; n-heptylic; oenanthic	CH₃(CH₂)₅COOH	B F G	R S V W
8	Caprylic	Octanoic; n-octylic	CH₃(CH₂)₆COOH	B F G	R S W
9	Pelargonic	Nonanoic; n-exnoic	CH₃(CH₂)₇COOH	B F G	R V W
9	p-Coumaric	p-Hydroxy cinnamic	HOC₆H₄CH=CHCOOH	G	
10	Capric	Decanoic; n-decylic	CH₃(CH₂)₈COOH	B F G	R W
11	n-Undecylic	Undecanoic; hendecanoic	CH₃(CH₂)₉COOH	B	R W
12	Lauric	Dodecanoic	CH₃(CH₂)₁₀COOH	B F G	R W
13	n-Tridecylic	Tridecanoic	CH₃(CH₂)₁₁COOH	B G	R W
14	Myristic	Tetradecanoic	CH₃(CH₂)₁₂COOH	B F G	R W
15	n-Pentadecylic	Pentadecanoic	CH₃(CH₂)₁₃COOH	B	R W
16	Palmitic	Hexadecanoic; n-hexadecylic; cetylic	CH₃(CH₂)₁₄COOH	B F G	R W
16	Palmitolic	7-Hexadecynoic	CH₃(CH₂)₇C≡C(CH₂)₅COOH		W
17	Margaric	Heptadecanoic	CH₃(CH₂)₁₅COOH	B F G	R W
18	Stearic	Octadecanoic; n-octadecylic	CH₃(CH₂)₁₆COOH	B F G	R W
18	Oleic	9-Octadecenoic	CH₃(CH₂)₇CH=CH(CH₂)₇COOH	B F G	R W
18	Linoleic	9,12-Octadecadienoic	CH₃(CH₂)₄CH=CHCH₂CH=CH(CH₂)₇COOH	B F G	R W
18	α-Linolenic	9,12,15-Octadecatrienoic	CH₃[CH₂CH=CH]₃(CH₂)₇COOH	F	

Esters

Number of carbon atoms	Compound	Synonyms	Formula	Reported presence						
				B	F	G	J	R	S	W
3	Methyl acetate		CH_3COOCH_3	B				R	S	
4	Methyl propionate		$CH_3CH_2COOCH_3$					R		W
7	Methyl caproate		$CH_3(CH_2)_4COOCH_3$			G				W
8	Methyl salicylate		$HOC_6H_4COOCH_3$							
3	Ethyl formate		$HCOOCH_2CH_3$	B	F	G		R		W
4	Ethyl acetate		$CH_3COOCH_2CH_3$	B	F	G	J	R	S	W
5	Ethyl propionate		$CH_3CH_2COOCH_2CH_3$	B		G		R		W
5	Ethyl lactate		$CH_3CHOHCOOCH_2CH_3$	B		G		R		
5	Ethyl carbonate		$CO(OCH_2CH_3)_2$				J			
6	Ethyl butyrate		$CH_3CH_2CH_2COOCH_2CH_3$				J	R		W
6	Ethyl isobutyrate		$(CH_3)_2CHCOOCH_2CH_3$					R		W
7	Ethyl isopentanoate		$CH_3CH_2(CH_3)CHCOOCH_2CH_3$					R		
7	Ethyl 3-methyl butyrate		$(CH_3)_2CH_2CH_2COOCH_2CH_3$							
8	Ethyl caproate		$CH_3(CH_2)_4COOCH_2CH_3$	B	F	G		R		W
9	Ethyl enanthate		$CH_3(CH_2)_5COOCH_2CH_3$		F	G		R		
9	Ethyl benzoate		$C_6H_5COOCH_2CH_3$			G		R		
10	Ethyl caprylate		$CH_3(CH_2)_6COOCH_2CH_3$	B	F	G	J	R		W
11	Ethyl pelargonate		$CH_3(CH_2)_7COOCH_2CH_3$	B	F	G	J	R		W
12	Ethyl caprate		$CH_3(CH_2)_8COOCH_2CH_3$	B	F	G	J	R		W
13	Ethyl undecanoate		$CH_3(CH_2)_9COOCH_2CH_3$	B	F			R		W
14	Ethyl laurate		$CH_3(CH_2)_{10}COOCH_2CH_3$	B	F	G		R		W
16	Ethyl myristate		$CH_3(CH_2)_{12}COOCH_2CH_3$	B	F	G	J	R		W
17	Ethyl pentadecanoate		$CH_3(CH_3)_{13}COOCH_2CH_3$	B						
18	Ethyl palmitate		$CH_3(CH_2)_{14}COOCH_2CH_3$	B	F	G		R		W
18	Ethyl palmitolate		$CH_3(CH_2)_7C\equiv C(CH_2)_5COOCH_2CH_3$	B	F	G				W

No.	Compound	Formula	B	F	G	J	R	W
19	Ethyl margarate	$CH_3(CH_2)_{15}COOCH_2CH_3$						W
20	Ethyl stearate	$CH_3(CH_2)_{16}COOCH_2CH_3$	B	F	G			W
20	Ethyl oleate	$CH_3(CH_2)_7CH=CH(CH_2)_7COOCH_2CH_3$						W
20	Ethyl linoleate	$CH_3(CH_2)_4CH=CHCH_2CH=CH(CH_2)_7COOCH_2CH_3$	B	F		J	R	W
21	Ethyl nondecanoate	$CH_3(CH_2)_{17}COOCH_2CH_3$						W
23	Ethyl heneicoanoate	$CH_3(CH_2)_{19}COOCH_2CH_3$						W
24	Ethyl behenate	$CH_3(CH_2)_{20}COOCH_2CH_3$						W
25	Ethyl tricosanate	$CH_3(CH_2)_{21}COOCH_2CH_3$						W
26	Ethyl lignocerate	$CH_3(CH_2)_{22}COOCH_2CH_3$						W
8	Diethyl succinate	$(CH_2COOCH_2CH_3)_2$					R	
4	Propyl formate	$HCOOCH_2CH_2CH_3$					R	
5	Propyl acetate	$CH_3COOCH_2CH_2CH_3$					R	
6	Propyl propionate	$CH_3CH_2COOCH_2CH_2CH_3$					R	
11	Propyl caprylate	$CH_3(CH_2)_6COOCH_2CH_2CH_3$	B					
7	Isopropyl butyrate	$CH_3CH_2CH_2COOCH(CH_3)_2$					R	
5	Butyl formate	$HCOOCH_2CH_2CH_2CH_3$					R	
6	Butyl acetate	$CH_3COOCH_2CH_2CH_2CH_3$	B				R	W
6	Isobutyl acetate	$CH_3COOCH_2CH(CH_3)_2$					R	
9	Isobutyl valerate	$CH_3(CH_2)_3COOCH_2CH(CH_3)_2$	B				R	W
10	Isobutyl caproate	$CH_3(CH_2)_4COOCH_2CH(CH_3)_2$	B		G			W
12	Isobutyl caprylate	$CH_3(CH_2)_6COOCH_2CH(CH_3)_2$						W
14	Isobutyl caprate	$CH_3(CH_2)_8COOCH_2CH(CH_3)_2$						W
16	Isobutyl laurate	$CH_3(CH_2)_{10}COOCH_2CH(CH_3)_2$						W
17	Isobutyl tridecanoate	$CH_3(CH_2)_{11}COOCH_2CH(CH_3)_2$						
18	Isobutyl myristate	$CH_3(CH_2)_{12}COOCH_2CH(CH_3)_2$						
20	Isobutyl palmitate	$CH_3(CH_2)_{14}COOCH_2CH(CH_3)_2$						
24	Isobutyl arachidate	$CH_3(CH_2)_{18}COOCH_2CH(CH_3)_2$						
6	n-Amyl formate	$HCOO(CH_2)_4CH_3$						
8	n-Amyl propionate	$CH_3CH_2COO(CH_2)_4CH_3$					R	
9	n-Amyl butyrate	$CH_3CH_2CH_2COO(CH_2)_4CH_3$					R	

Esters—continued

Number of carbon atoms	Compound	Synonyms	Formula	Reported presence
9	n-Amyl isobutyrate		$(CH_3)_2CHCOO(CH_2)_4CH_3$	R
10	n-Amyl isovalerate		$(CH_3)_2CHCH_2COO(CH_2)_4CH_3$	G
11	n-Amyl caproate		$CH_3(CH_2)_4COO(CH_2)_4CH_3$	G
13	n-Amyl caprylate		$CH_3(CH_2)_6COO(CH_2)_4CH_3$	G
15	n-Amyl caprate		$CH_3(CH_2)_8COO(CH_2)_4CH_3$	G
7	Isoamyl acetate		$CH_3COO(CH_2)_2CH(CH_3)_2$	B F G R W
11	Isoamyl caproate		$CH_3(CH_2)_4COO(CH_2)_2CH(CH_3)_2$	B F R W
13	Isoamyl caprylate		$CH_3(CH_2)_6COO(CH_2)_2CH(CH_3)_2$	B F R W
14	Isoamyl pelargonate		$CH_3(CH_2)_7COO(CH_2)_2CH(CH_3)_2$	W
15	Isoamyl caprate		$CH_3(CH_2)_8COO(CH_2)_2CH(CH_3)_2$	B F W
16	Isoamyl undecanoate		$CH_3(CH_2)_9COO(CH_2)_2CH(CH_3)_2$	W
17	Isoamyl laurate		$CH_3(CH_2)_{10}COO(CH_2)_2CH(CH_3)_2$	B G J W
18	Isoamyl tridecanoate		$CH_3(CH_2)_{11}COO(CH_2)_2CH(CH_3)_2$	G
19	Isoamyl myristate		$CH_3(CH_2)_{12}COO(CH_2)_2CH(CH_3)_2$	B G W
20	Isoamyl pentadecanoate		$CH_3(CH_2)_{13}COO(CH_2)_2CH(CH_3)_2$	W
21	Isoamyl palmitate		$CH_3(CH_2)_{14}COO(CH_2)_2CH(CH_3)_2$	G
23	Isoamyl stearate		$CH_3(CH_2)_{16}COO(CH_2)_2CH(CH_3)_2$	G W
23	Isoamyl oleate		$CH_3(CH_2)_7CH{=}CH(CH_2)_7COO(CH_2)_2CH(CH_3)_2$	R
23	Isoamyl linoleate		$CH_3(CH_2)_4CH{=}CHCH_2CH{=}CH(CH_2)_7COO(CH_2)_2CH(CH_3)_2$	R
25	Isoamyl β-phenyl myristate		$CH_3(CH_2)_{11}CHC_6H_5COO(CH_2)_2CH(CH_3)_2$	J
11	pri-act-Amyl caproate		$CH_3(CH_2)_4COOCH_2(CH_3)CHCH_2CH_3$	B

No.	Compound	Formula	Synonyms	B	F	G	J	M	R	S	W
13	*pri-act*-Amyl caprylate	$CH_3(CH_2)_6COOCH_2(CH_3)CHCH_2CH_3$		B							
17	*pri-act*-Amyl laurate	$CH_3(CH_2)_{10}COOCH_2(CH_3)CHCH_2CH_3$		B							
19	*pri-act*-Amyl myristate	$CH_3(CH_2)_{12}COOCH_2(CH_3)CHCH_2CH_3$		B							
8	Hexyl acetate	$CH_3COO(CH_2)_5CH_3$		B		G			R		
9	Heptyl acetate	$CH_3COO(CH_2)_6CH_3$				G					
10	Octyl acetate	$CH_3COO(CH_2)_7CH_3$				G					
11	Nonyl acetate	$CH_3COO(CH_2)_8CH_3$				G					
11	Methyl heptyl carbinol acetate	$CH_3COOCH(CH_3)(CH_2)_6CH_3$				G					
12	Decyl acetate	$CH_3COO(CH_2)_9CH_3$				G					
13	Undecyl acetate	$CH_3COO(CH_2)_{10}CH_3$				G					
10	Phenethyl acetate	$CH_3COOCH_2CH_2C_6H_5$		B		G					W
11	Phenethyl propionate	$CH_3CH_2COOCH_2CH_2C_6H_5$				G					

Carbonyl compounds

No.	Compound	Formula	Synonyms	B	F	G	J	M	R	S	W
1	Formaldehyde	HCHO	Methanal; oxo-methane; methylene oxide	B			J		R		W
2	Acetaldehyde	CH_3CHO	Ethanal	B	F	G	J	M	R	S	W
2	Glyoxal	CHOCHO	Ethanedial; biformyl	B		G			R		W
3	Acetone	CH_3COCH_3	2-Propanone; dimethyl ketone	B		G			R		W
3	Methyl glyoxal	CH_3COCHO	Pyruvic aldehyde	B		G					W
3	Glyceraldehyde	$CH_2OHCHOHCHO$	2,3-Dihydroxy propanal; α,β-dihydroxy propionaldehyde	B		G			R		W
3	Propionaldehyde	CH_3CH_2CHO	Propanal	B		G			R	S	W
3	Acrolein	$CH_2{=}CHCHO$	Propenal; acrylaldehyde	B		G	J	M	R	S	
4	Acetoin	$CH_3CHOHCOCH_3$	3-Hydroxy-2-butanone; dimethyl ketol; acetyl methyl carbinol	B							W

Carbonyl compounds—continued

Number of carbon atoms	Compound	Synonyms	Formula	Reported presence
4	Diacetyl	2,3-Butane dione; dimethyl glyoxal; biacetyl	$CH_3COCOCH_3$	B G S W
4	Methyl ethyl ketone		$CH_3COCH_2CH_3$	G
4	Aldol	3-Hydroxybutanal; acetaldol; β-hydroxybutyraldehyde	$CH_3CHOHCH_2CHO$	M
4	Butyraldehyde	Butanal	$CH_3CH_2CH_2CHO$	B G M R S
4	Isobutyraldehyde	2-Methyl propanal	$(CH_3)_2CHCHO$	B G R S W
4	Crotonaldehyde	2-Butenal; propylene aldehyde; β-methyl acrolein	$CH_3CH=CHCHO$	R S
5	Methyl isopropyl ketone		$CH_3COCH(CH_3)_2$	G
5	Furfural	2-Furan carbonal; furfuraldehyde	$CH=C.CHO$ with furan ring $CH=CH$, O	B G R S W
5	Isovaleraldehyde	3-Methyl butanal; isoamyl aldehyde; β-methyl butyraldehyde	$(CH_3)_2CHCH_2CHO$	B G S W
5	*act*-Valeraldehyde	α-Methyl butyraldehyde	$CH_3CH_2CH(CH_3)CHO$	B F S
5	Methyl ethyl glyoxal	2,3-Pentanedione; 2,3-diketopentane; methyl ethyl diketone	$CH_3COCOCH_2CH_3$	B G R S W
5	Acetyl acetone	2,4-Pentanedione; diacetyl methane	$CH_3COCH_2COCH_3$	W

No.	Name	Synonyms	Formula	B	F	G	J	R	S	M	W
5	Diethoxy methane	Formaldehyde diethyl acetal	$CH_2(OCH_2CH_3)_2$					R			W
5	Cyclopentanone	Keto-pentamethylene; adipic ketone	$COCH_2CH_2CH_2CH_2$						S		
6	Acetal	1,1-Diethoxyethane; acetaldehyde diethyl acetal; ethylidene diethyl ether	$CH_3CH(OCH_2CH_3)_2$	B		G	J	R	S		W
6	Caproaldehyde	Hexanal	$CH_3(CH_2)_4CHO$	B	F			R	S		W
6	Methyl amyl ketone	2-Heptanone	$CH_3CO(CH_2)_3CH_3$			G		R			
7	Enanthaldehyde	Heptanal; n-heptaldehyde	$CH_3(CH_2)_5CHO$	B		G		R			
7	Methyl hexyl ketone	Octanone	$CH_3CO(CH_2)_4CH_3$			G		R			
7	1,1-Diethoxy propane		$CH_3CH_2CH(OCH_2CH_3)_2$					R			
7	Benzaldehyde		C_6H_5CHO			G		R			
7	p-Hydroxy benzaldehyde		HOC_6H_4CHO						S		
8	Methyl heptyl ketone	2-Nonanone	$CH_3CO(CH_2)_5CH_3$			G		R			
8	1,1-Diethoxy butane		$CH_3CH_2CH_2CH(OCH_2CH_3)_2$			G					
8	Methyl heptenone		$(CH_3)_2C{=}CH(CH_2)_2COCH_3$			G					
8	Acetophenone	Methyl phenyl ketone	$C_6H_5COCH_3$			G					W
8	Vanillin	4-Hydroxy-3-methoxy-benzaldehyde	$OH(CH_2OH)C_6H_3CHO$			G			S		W
9	Syringaldehyde	4-Hydroxy-3,5-dimethoxy-benzaldehyde	$OH(CH_2OH)_2C_6H_2CHO$			G			S		W
10	Methyl nonyl ketone		$CH_3CO(CH_2)_7CH_3$			G					
10	Coniferyl aldehyde	3-(4-Hydroxy-3-methyl phenyl) 2-propen aldehyde	$(CH_3O)(OH)C_6H_3CH{=}CHCHO$			G			S		W
14	Methyl tridecyl ketone		$CH_3CO(CH_2)_{11}CH_3$			G					
Phenols											
6	Phenol		C_6H_5OH			G					W
6	o-Cresol		$C_6H_5(OH)_2$			G					W
7	Gallic acid	3,4,5-Trihydroxy benzoic acid	$C_6H_2(OH)_3COOH$								W
7	Guaiacol	o-Methoxy phenol; methyl catechol	$C_6H_4(OH)CH_2OH$						S	M	W

Number of carbon atoms	Compound	Synonyms	Formula	Reported presence
Phenols—continued				
8	p-Ethyl phenol		$C_6H_4(OH)CH_2CH_3$	G
8	p-Vinyl phenol		$C_6H_4(OH)CH{=}CH_2$	G
8	p-Methyl guaiacol	2-Methoxy-4-methyl phenol; creosol	$CH_2OHC_6H_3(OH)CH_3$	G W
8	Vanillic acid	4-Hydroxy-3-methoxy benzoic acid	$CH_2OH(OH)C_6H_3COOH$	W
9	p-Ethyl guaiacol	2-Methoxy-4-ethyl phenol	$CH_2OHC_6H_3(OH)CH_2CH_3$	G W
9	p-Vinyl guaiacol	2-Methoxy-4-vinyl phenol	$CH_2OHC_6H_3(OH)CH{=}CH_2$	G
10	n-Propyl guaiacol		$CH_2OHC_6H_3(OH)CH_2CH_2CH_3$	
10	Eugenol	4-Allyl-2-methoxy phenol	$CH_2{:}CHCH_2C_6H_3(OCH_3)OH$	R S
10	Syringic acid	4-Hydroxy-3-methoxy cinnamic acid; methyl coniferic acid	$CH_2OH(OH)C_6H_3CH{=}CHCOOH$	W
14	Ellagic acid	4,4',5,5',6,6'-Hexahydroxy-diphenic dilactone	$(OH)_3C_6(CO)(CO)C_6(OH)_3$	W
Bases				
1	Methylamine		CH_3NH_2	M
2	Dimethylamine		$(CH_3)_2NH$	M
3	Trimethylamine		$(CH_3)_3N$	M
2	Ethylamine		$CH_3CH_2NH_2$	M
4	Diethylamine		$(CH_3CH_2)_2NH$	M
6	Triethylamine		$(CH_3CH_2)_3N$	M
5	Pyridine		C_5H_5N	M
Terpenes				
10	Limonene	p-Menthadiene	$C_{10}H_{16}$	G

			B	F	G	J	M	R	S	V	W
10	Borneol	exo-2-Camphanol; bornyl alcohol			G		M		S		
10	Camphor	2-Keto-1,7,7-trimethyl norcamphane			G		M				
10	Citronellol	Dihydro-2,6-dimethyl-2,6-octadiene-9-ol			G		M				
10	Linalool	3,7-Dimethyl-1,6-octadiene 3-ol			G						
10	Fenchol	1,3,3-Trimethyl-2-norcamphanol							S		
10	Menthol	Methyl hydroxy isopropyl cyclohexane					M				
10	α-Terpineol	1-Menthen-8-ol					M				
13	α-Ionone	4-(2,6,6-Trimethyl-2-cyclohexenyl)-3-butene-2-one			G						
15	Sesquiterpene										
15	trans-Nerolidol	3,7,11-Trimethyl-1,6,10-dodecatriene-3-ol			G		M				

Formulas:
- Borneol: $C_{10}H_{17}OH$
- Camphor: $C_{10}H_{16}O$
- Citronellol: $C_{10}H_{20}O$
- Linalool: $C_{10}H_{18}O$
- Fenchol: $C_{10}H_{18}O$
- Menthol: $C_{10}H_{19}OH$
- α-Terpineol: $C_{10}H_{17}OH$
- α-Ionone: $C_{13}H_{20}O$
- Sesquiterpene: $C_{15}H_{24}$
- trans-Nerolidol: $C_{15}H_{26}O$

Miscellaneous

			B	F	G	J	M	R	S	V	W
0	Hydrogen sulphide	H_2S									
10	Naphthalene	$C_{10}H_8$			G						
0	Water	H_2O					M				

Bibliography: Baraud (1961); Baraud and Maurice (1963); Bouthilet and Lowrey (1959); Drews and Specht (1963); Duncan and Philp (1966); Flanzy and Jouret (1963); Guymon and Crowell (1963); Guymon and Nakagiri (1957); Haeseler and Vogl (1953); Hellström (1943); Hirose *et al.* (1962); Jones and Wills (1966); Kepner and Webb (1961); Kröller (1950); Maarse and ten Noever de Brauw (1966); Misselhorn (1963); Nykänen (1963); Nykänen and Suomalainen (1963); Nykänen *et al.* (1968); Oshmyan and Ignatova (1961); Pfenninger (1963); Prillinger and Horwatitsch (1965); Pyke (1965); Ronkainen and Suomalainen (1966); Salo and Suomalainen (1958); Singer (1966); Steinke and Paulson (1964); Suomalainen and Nykänen (1964, 1966); Suomalainen and Ronkainen (1963); Warwicker (1960).

References

Ainsworth, G. (1966). *Process Biochem.* 1 (1), 15–22.

Andreyev, K. P. (1958). *In* "Continuous Cultivation of Micro-organisms" (I. Malek, ed.), pp. 186–197. Publ. House Czechoslovak Acad. Sci., Prague.

Arroyo, R. (1945). "Studies in Rum", Res. Bull. No. 5, Univ. Puerto Rico Agric. Expt. Station. Puerto Rico.

Aschengreen, N. H. (1969). *Process Biochem.* 4 (8), 23–25.

Baldwin, E. (1963). "Dynamic Aspects of Biochemistry", 4th Ed. University Press, Cambridge.

Baraud, J. (1961). *Bull. Soc. chim. France*, pp. 1874–1877.

Baraud, J. and Maurice, A. (1963). *Indust. Aliment. Agric.* **80**, 3–7.

Bauchop, T. and Elsden S. R. (1960). *J. gen. Microbiol.* **23**, 457–469.

Becker, D. and Weber, J. (1962). *In* "Die Hefen" (F. Reiff, R. Kautzmann, H. Lüers and M. Lindemann, eds.) Vol. II, pp. 31–81. Verlag Hans Carl, Nürnberg.

Boinot, F. (1936a). U.S. Patent 2,054,735.

Boinot, F. (1936b). U.S. Patent 2,063,223.

Boinot, F. (1941). U.S. Patent 2,230,318.

Bouthilet, R. J. and Lowrey, W. (1959). *J. Ass. off. agric. Chem.* **42**, 634–637.

Bruch, C. W., Hoffman, A., Gosine, R. M. and Brenner, M. W. (1964). *J. Inst. Brew.* **70**, 242–246.

Butschek, G. and Kautzmann, R. (1962). *In* "Die Hefen" (F. Reiff, R. Kautzmann, H. Lüers and M. Lindemann, eds.) Vol. II, pp. 501–610. Verlag Hans Carl, Nürnberg.

Butschek, G. and Krause, G. (1962a). *In* "Die Hefen" (F. Reiff, R. Kautzmann, H. Lüers and M. Lindemann, eds.) Vol. II, pp. 142–147. Verlag Hans Carl, Nürnberg.

Butschek, G. and Krause, G. (1962b). *In* "Die Hefen" (F. Reiff, R. Kautzmann H. Lüers and M. Lindemann, eds.) Vol. II, pp. 445–478. Verlag Hans Carl, Nürnberg.

Butschek, G. and Neumann, F. (1962). *In* "Die Hefen" (F. Reiff, R. Kautzmann, H. Lüers and M. Lindemann, eds.) Vol. II, pp. 497–500. Verlag Hans Carl, Nürnberg.

Campbell, I. and Brudzynski, A. (1966). *J. Inst. Brew.* **72**, 556–560.

Clerck, J. de (1957). "A Textbook of Brewing" (translated by K. Barton-Wright). Chapman and Hall, London.

Conrad, T. F. (1962). *In* "Die Hefen" (F. Reiff, R. Kautzmann, H. Lüers and M. Lindemann, eds.) Vol. II, pp. 437–444. Verlag Hans Carl, Nürnberg.

Cook, A. H., ed. (1962). "Barley and Malt: Biology, Biochemistry and Technology". Academic Press, London.

Costilow, R. N., Etchells, J. L. and Anderson, T. E. (1964). *Appl. Microbiol.* **12** (6), 539–540.

Dixon, M. and Webb, E. C. (1964). "Enzymes", 2nd Ed. Longmans, London.

Drews, B. and Specht, H. (1963). *Chemiker Ztg* **87**, 696–697.

Duddington, C. L. (1961). "Micro-organisms as Allies: the Industrial Use of Fungi and Bacteria". Faber and Faber, London.

Duncan, R. E. and Philp, J. M. (1966). *J. Sci. Fd Agric.* **17**, 208–214.

Elks, A. H. (1968). *Process Biochem.* **3** (4), 25–28.

Farrell, J. and Rose, A. H. (1967). *In* "Thermobiology" (A. H. Rose, ed.), pp. 147–218. Academic Press, London.

Flanzy, M. and Jouret, C. (1963). *Ann. Technol. agric.* **12**, 39–50.

Fowell, R. R. (1964). "Biology Staining Schedules for First Year Students", 8th Ed., pp. 27 and 31. H. K. Lewis and Co., London.

Gridgeman, N. T. (1967). *In* "Quality Control in the Food Industry" (S. M. Herschdoerfen, ed.) Vol. 1, pp. 235–283. Academic Press, London.

Griffin, O. T. (1966). *Process Biochem.* 1 (4), 241–243.

Guymon, J. F. and Crowell, E. A. (1959). *J. Ass. off. agric. Chem.* 42, 393–398.

Guymon, J. F. and Crowell, E. A. (1963). *J. Ass. off. agric. Chem.* 46, 276–284.

Guymon, J. F. and Nakagiri, J. A. (1957). *J. Ass. off. agric. Chem.* 40, 561–575.

Haeseler, G. and Vogl, J. (1953). *Z. Lebensmittel Untersuch. u.- Forsch.* 97, 460–470.

Hansford, G. S. and Humphrey, A. E. (1966). *Biotechnol. Bioengng* 8, 85–96.

Harris, G. (1958). *In* "Chemistry and Biology of Yeasts" (A. H. Cook, ed.), pp. 437–533. Academic Press, New York.

Harrison, J. S. (1967). *Process Biochem.* 2 (3), 41–45.

Hellström, N. (1943). *Svensk. kem. Tidskr.* 55, 161–168.

Hirose, Y., Ogawa, M. and Kusuda, Y. (1962). *Agric. biol. Chem. Tokyo* 26, 526–531.

Hodge, H. M. and Hildebrandt, F. M. (1954). *In* "Industrial Fermentations" (L. A. Underkofler and R. J. Hickey, eds.), Vol. 1, pp. 73–94. Chemical Publishing Co., New York.

Hopkins, R. H., Murray, R. H. and Lockwood, A. R. (1946). *Biochem. J.* 40, 507–512.

Hospodka, J. (1966). *In* "Theoretical and Methodological Basis of Continuous Culture of Micro-organisms" (I. Makek and Z. Fenzl, eds.), pp. 493–645. Publ. House Czechoslovak Acad. Sci., Prague.

Hough, J. S. (1961). *In* "Continuous Culture of Micro-organisms", Soc. Chem. Indust. Monograph No. 12, pp. 219–232. London.

Hough, J. S. and Ricketts, R. W. (1960). *J. Inst. Brew.* 66, 301–304.

Ingram, M. (1955). "An Introduction to the Biology of Yeasts". Sir Isaac Pitman and Sons, London.

Ingram, M., Mossel, D. A. A. and Lange, P. de (1955). *Chemy. Ind.* pp. 63–64.

Institute of Brewing Analysts Committee (1961). *J. Inst. Brew.* 67, 317–327.

Jencks, W. P. (1968). *In* "Handbook of Biochemistry" (H. A. Sober, ed.), pp. 144–149. Chemical Publ. Co., Cleveland, Ohio.

Johnson, M. C. and Harris, E. E. (1948). *J. Am. chem. Soc.* 70, 2961–2963.

Jones, K. and Wills, R. (1966). *J. Inst. Brew.* 72, 196–201.

Kalyuzhnyi, M. J. (1958). *In* "Continuous Culture of Micro-organisms" (I. Malek, ed.), p. 192. Publ. House Czechoslovak Acad. Sci., Prague.

Kalyuzhnyi, M. J. (1960). *In* "Continuous Fermentation and Cultivation of Micro-organisms" (N. D. Ierusalimskii, ed.), pp. 69–73. Pishchepromizdat, Moscow.

Kepner, R. E. and Webb, A. D. (1961). *Am. J. Enol. Viticult.* 12, 159–174.

Kerr, R. W., ed. (1950). *In* "Chemistry and Industry of Starch", 2nd Ed. Academic Press, New York.

Kervégant, D. (1946). "Rhums et Eaux-de-vie de Canne", p. 185. Éditions du Golfe, Vaunes.

Kneen, E., Sandstedt, R. M. and Hollenbeck, C. M. (1943). *Cereal Chem.* 20, 399–423.

Kormančíková, V., Kováč, L. and Vidová, M. (1969). *Biochim. biophys. Acta* 180, 9–17.

Kreipe, H. (1962). *In* "Die Hefen" (F. Reiff, R. Kautzmann, H. Lüers and M. Lindemann, eds.) Vol. II, pp. 340–383. Verlag Hans Carl, Nürnberg.

Kröller, E. (1950). *Dt. LebensmittRdsch.* 46, 6.

Kunz, C. and Klaushofer, H. (1961). *Appl. Microbiol.* 9, 469.

Laatsch, H. U. and Sattelberg, K. (1968). *Process Biochem.* 3 (10), 28–30, 35.

Lane, J. H. and Eynon, L. (1923). *J. Soc. chem. Ind.* 42, 32T.

Leonard, R. H. and Hajny, G. J. (1945). *Ind. Engng Chem.* **37**, 390–395.

Lindegren, C. C. (1944). *Wallerstein Labs Commun.* **7**, 153–168.

Lindemann, M. (1962). *In* "Die Hefen" (F. Reiff, R. Kautzmann, H. Lüers and M. Lindemann, eds.) Vol. II, pp. 3–30. Verlag Hans Carl, Nürnberg.

Lodder, J. and Kreger-van Rij, N. J. W. (1952). "The Yeasts: A Taxonomic Study". North-Holland Publ. Co., Amsterdam.

Luchsinger, W. W. (1966). *Cereal Chem. Today* **11** (2), 69–75, 82–84.

Maarse, H. and ten Noever de Brauw, M. C. (1966). *J. Fd Sci.* **31**, 951–955.

MacFarlane, J. R. (1947). *Int. Sugar Jl* **69**, 73–96.

Macher, L. (1962). *In* "Die Hefen" (F. Reiff, R. Kautzmann, H. Lüers and M. Lindemann, eds.) Vol. II, pp. 384–436. Verlag Hans Carl, Nürnberg.

Magné, J. H. P. (1917). U.S. Patent 1,212,656.

Malchenko, A. L. and Krishtul, F. B. (1954). U.S.S.R. Patent 103,681.

Malsch, L. (1962). *In* "Die Hefen" (F. Reiff, R. Kauzmann, H. Lüers and M. Lindemann, eds.) Vol. II, pp. 479–496.

Manning, S. A. (1947). "A Handbook of the Wine and Spirit Trade", p. 47. Sir Isaac Pitman and Sons, London.

Maravin, L. N. and Oshmyan, G. L. (1960). *Spirtovaya Prom.* **26** (7), 18–20.

McAteer, D. J. (1968). *Process Biochem.* **3** (4), 60–62.

McCarthy, J. L. (1954). *In* "Industrial Fermentations" (L. A. Underkofler and R. J. Hickey, eds.) Vol. 1, pp. 95–135. Chemical Publishing Co., New York.

Merritt, N. R. (1967). *J. Inst. Brew.* **73**, 484–488.

Ministry of Housing and Local Government (1956). "Methods of Chemical Analysis as Applied to Sewage and Sewage Effluents". H.M. Stationery Office, London.

Misselhorn, K. (1963). *Branntweinwirtsch.* **103**, 401–406.

Mitscherlich, A. (1878). German Patents 4,178 and 4,179.

Mor, J. R. and Fiechter, A. (1968). *Biotechnol. Bioengng* **10**, 159–176.

Myrbäck, K. and Neumüller, G. (1950). *In* "The Enzymes" (J. B. Sumner and K. Myrbäck, eds.) Vol. 1, pp. 653–724. Academic Press, New York.

Nicholls, J. R. (1947). "The Determination of Alcohol", 3rd Tatlock Memorial Lecture. Royal Institute of Chemistry, London.

Nord, F. F. and Weiss, S. (1958). *In* "The Chemistry and Biology of Yeasts" (A. H. Cook, ed.), pp. 323–368. Academic Press, New York.

Nordström, K. (1966). *Acta Chem. Scand.* **20**, 1016–1025.

Nordström, K. (1968). *J. Inst. Brew.* **74**, 429–432.

Nykänen, L. (1963). *Teknillisen Kemian Aikakauslehte* **20**, 129–133.

Nykänen, L. and Suomalainen, H. (1963). *Teknillisen Kemian Aikakauslehte* **20**, 789–795.

Nykänen, L., Puputti, E. and Suomalainen, H. (1968). *J. Fd Sci.* **33** (1), 88–92.

Oshmyan, G. L. and Ignatova, A. V. (1961). *Tr. Tsentr. Nauchn.-Issled. Inst. Spirt. i. Likero-Vodochn. Prom.*, No. 11, pp. 242–250; from *Chem. Abst.*, 1962, **57**, 14291.

Pairault, M. E. A. (1903). "Le Rhum et sa Fabrication". Gauthier Villars et Cie., Paris; quoted by Arroyo, 1945.

Pasteur, L. (1872). *Ann. Chim. Phys.* **25** (4), 145.

Pfenniger, H. (1963). *Z. Lebensmittel-Untersuch. u.-Forsch.* **120**, 100–116.

Pierce, J. S. (1966). *Process Biochem.* **1** (8), 412–418.

Pirt, S. J. (1965). *Proc. R. Soc.* B **44**, 149–157.

Pleeth, S. J. W. (1949). "Alcohol, a Fuel for Internal Combustion Engines". Chapman and Hall, London.

Preece, I. A. and Shadaksharaswamy, M. (1949a). *J. Inst. Brew.* **55**, 373–383.

Preece, I. A. and Shadaksharaswamy, M. (1949b). *Biochem. J.* **44**, 270–274.

Prescott, S. C. and Dunn, C. G. (1940). "Industrial Microbiology", 1st Ed. McGraw Hill, New York.

Prescott, S. C. and Dunn, C. G. (1959). "Industrial Microbiology", 3rd Ed. McGraw Hill, New York.

Prillinger, F. and Horwatitsch, H. (1965). *Rebe Wein* **15**, 72–79.

Prillinger, F. and Horwatitsch, H. (1966). *Rebe Wein* **16**, 115–126.

Pyke, M. (1965). *J. Inst. Brew.* **71**, 209–218.

Rae, I. J. (1966). *Process Biochem.* **1** (8), 407–411.

Rogosa, M., Mitchell, J. A. and Wiseman, R. F. (1951). *J. Bact.* **62**, 132–133.

Ronkainen, P. and Suomalainen, H. (1966). *Suomen Kemistilehti* **39B**, 280–281.

Rose, A. H. (1968). "Chemical Microbiology", 2nd Ed., p. 134. Butterworths, London.

Saeman, J. F. and Andreasen, A. A. (1954). *In* "Industrial Fermentations" (L. A. Underkofler and R. J. Hickey, eds.) Vol. 1, pp. 136–171.

Salo, T. and Suomalainen, H. (1958). *Z. Lebensmittel Untersuch. u.-Forsch.* **10** 421–422.

Simmonds, C. (1919). "Alcohol, its Production, Properties, Chemistry and Industrial Applications". MacMillan, London.

Simpson, A. C. (1966). *Process Biochem.* **1** (7), 355–358, 365.

Simpson, A. C. (1968). *Process Biochem.* **3** (1), 9–12.

Singer, D. D. (1966). *Analyst* **91**, 127–134.

Sporek, K. and Williams, A. F. (1954). *Analyst* **79**, 63–69.

Stakhorskaya, L. K. and Tokgarev, B. I. (1964). *Mikrobiologiya* **33**, 1056–1060.

Stark, W. H. (1954). *In* "Industrial Fermentations" (L. A. Underkofler and R. J. Hickey, eds.). Vol. 1, pp. 17–72. Chemical Publishing Co., New York.

Steinke, R. D. and Paulson, M. C. (1964). *J. agric. Fd Chem.* **12**, 381–387.

Steiner, E. H. (1967). *In* "Quality Control in the Food Industry" (S. M. Herschdoerfer, ed.) Vol. 1, pp. 235–283. Academic Press, London.

Suomalainen, H. (1968). *Suomen Kemistilehti* **41A**, 239–254.

Suomalainen, H. and Nykänen, L. (1964). *Suomen Kemistilehti* **37B**, 230–232.

Suomalainen, H. and Nykänen, L. (1966). *J. Inst. Brew.* **72**, 469–474.

Suomalainen, H. and Nykänen, L. (1967). *Industrie Chim. belge*, **32**, Spec. No. 3, 807.

Suomalainen, H. and Ronkainen, P. (1963). *Teknillisen Kemian Aikakauslehti* **20**, 413–417.

Suomalainen, H., Kauppila, O., Nykänen, L. and Peltonen, R. L. (1968). *In* "Handbuch der Lebensmittelchemie: 7. Alkoholische Genussmittel" (J. Schormüller, ed.), pp. 496–653. Springer-Verlag, Berlin.

Sykes, G. (1965). "Disinfection and Sterilization", 2nd Ed. E. and F. N. Spon, London.

Sykes, G. (1966). *Process Biochem.* **1** (5), 268–272.

Teixeira, C., Andreasen, A. A. and Kolachov, P. (1950). *Ind. Engng Chem.* **42**, 1781–1783.

Thorley, J. F. (1968). *In* "Maintenance of Pure Culture Conditions in Industrial Fermentations" (in press). Soc. Chem. Ind. Symp., London.

Thorne, R. S. W. (1949). *J. Inst. Brew.* **55**, 201–222.

Thorne, R. S. W. (1950). *Wallerstein Labs Commun.* **13**, 319–340.

Trevelyan, W. E. (1958). *In* "Chemistry and Biology of Yeasts" (A. H. Cook, ed.), pp. 369–436. Academic Press, New York.

Underkofler, L. A., McPherson, W. K. and Fulmer, E. I. (1937). *Ind. Engng Chem.* **29**, 1160–1164.

Unger, E. D., Stark, W. H., Scalf, R. E. and Kolachov, P. J. (1942). *Ind. Engng Chem.* **34**, 1402–1405.

Utter, M. F., Duell, E. A. and Bernofsky, C. (1968). *In* "Aspects of Yeast Metabolism" (A. K. Mills, ed.), pp. 197–212. Blackwell Scientific Publications, Oxford and Edinburgh.

Warwicker, L. A. (1960). *J. Sci. Fd Agric.* **11**, 709–716.

Warwicker, L. A. (1963). *J. Sci. Fd Agric.* **14**, 371–376.

Webb, F. C. (1964). "Biochemical Engineering", pp. 249–292. Van Nostrand, London.

White, J. (1954). "Yeast Technology", pp. 275–279. Chapman and Hall, London.

Wiley, A. J. (1954). *In* "Industrial Fermentations" (L. A. Underkofler and R. J. Hickey, eds.) Vol. 1, pp. 307–343. Chemical Publishing Co., New York.

Wilkinson, J. F. and Rose, A. H. (1963). *In* "Biochemistry of Industrial Microorganisms" (C. Rainbow and A. H. Rose, eds.), pp. 379–414. Academic Press, London.

World-wide Survey of Fermentation Industries 1963 (1966). *Pure appl. Chem.* **13**, 405–417.

Chapter 7

Baker's Yeast

S. Burrows

Distillers Company Ltd., Glenochil Technical Centre, Menstrie, Clackmannanshire, Scotland

I. General Principles

A. HISTORICAL

Since early times micro-organisms have been used, often unwittingly, for the fermentation of dough in bread-making. Leaven, mentioned in the Bible, was probably a mixture of yeasts and lactobacilli maintained in a dough medium, and is still used in some remote areas. After each fermentation a portion of the dough was retained for starting the next batch. Since the Middle Ages use has also been made of the excess yeast from brewing and wine-making operations; this is called barm. This material, however, is unsatisfactory owing to its variable quality and unstable nature. It was not until the end of the eighteenth century that the preparation of yeast specifically for bread-making became a distinct manufacturing operation capable of some degree of control (Frey, 1930). In the course of time, the making of baker's yeast has developed into a highly specialized scientifically controlled process which yields a product of consistently high quality and stability.

The first major step in the development of yeast technology came with the introduction of the Vienna process around 1860 (Frey, 1930). By this means the yield of yeast from the traditional process, in which yeast was grown anaerobically on grain mash, was improved by the passage of a gentle stream of air through the mash. As a result of the well known researches of Pasteur, the need for still more intense aeration was recognized, and this resulted in further improvements in yeast yield, accompanied by a reduction in ethanol formation. Shortage of grain during the First World War led to the introduction of molasses, a by-product of sugar manufacture, which was supplemented with phosphate and ammonia to form the growth medium. Later, the classical batch process, in which all the nutrients were added initially, was modified by the use of a *fed batch* or *Zulauf* process (Hayduck, 1919; Sak, 1919). In this method carbohydrate supply was fed incrementally during the growth period, so that sugar concentration and growth rate were under some degree of external control. As a consequence greatly improved yields and yeast quality, particularly stability, were obtained. Although still termed a batch process, this method must be carefully distinguished from the classical batch process of the microbiologists. The term "fed batch process" is therefore sometimes used. In certain cases it is advantageous for commercial purposes to produce both ethanol and yeast substance simultaneously; this is achieved by using methods akin to the old-fashioned Vienna process.

The need for a further improvement in stability for certain purposes, such as use by the armed forces, led to the gradual development of active dried yeast. An inferior quality of dried yeast was indeed known before 1900, and a dried baker's yeast called *Florylin* was available in Germany after 1918 (Benesch, 1954), but concerted efforts to produce a generally acceptable product were not made until after 1920 (Frey, 1957). A good dried yeast was manufactured in Australia during the early 1940s, and rapidly increasing quantities were manufactured in North America during the Second World War (Morse and Fellers, 1949). Further improvements are continually being made, but dried yeast has not yet superseded moist compressed yeast in countries with temperate climates and short distribution lines.

B. BREAD-MAKING

The function of yeast in bread-making is fourfold: (i) to increase dough volume by the evolution of gas during fermentation of the available carbohydrates in the flour; (ii) to develop structure and texture in the dough by the stretching effect of the expansion due to gas production; (iii) to impart a distinctive flavour; and (iv) to enhance the nutritive value of the bread. The last is a minor function, which can generally be disregarded.

1. *Dough Fermentation*

Apart from any sugar which may be added in the bread-making recipe, flour itself contributes about 2% of its dry weight in the form of available carbohydrates. About half of this amount can be utilized immediately, while the remainder is made available by the action of α- and β-amylases in the flour which are activated on the addition of water (Koch *et al.*, 1951; Williams and Bevenue, 1951). In this way a portion of the starch in the flour is degraded to maltose, which is fermentable, and dextrins which are not assimilated.

The immediately available carbohydrates consist of glucose, fructose, sucrose and levosin (a mixture of polysaccharides containing fructose and glucose residues) which are attacked first. As soon as the concentration of these substrates has been reduced to negligible proportions, maltose fermentation commences (Lanning, 1936; Mackenzie, 1958). When a complete fermentation rate/time curve is plotted, a trough between two peaks is usually observed. This is due to the fact that many yeasts have an adaptive maltase system. Such strains, when grown on molasses, do not begin to synthesize maltase until the concentration of easily fermented carbohydrates is reduced to a very low value, a phenomenon known as *diauxie* (Monod, 1942). Strains are, however, known which have a constitutive maltase system (Grylls and Harrison, 1956; Morita *et al.*, 1959). When these are grown in a maltose-free medium such as molasses, they give rate/time curves during dough fermentation which exhibit little if any trough, the two peaks merging into a plateau. This type of curve is also given by all yeast strains when grown on grain wort (which contains maltose, see Chapter 6, p. 283), and is also claimed to be obtained by rich feeding of molasses in the absence of maltose during manufacture (Pyke *et al.*, 1958). The importance of maltose fermentation in bread-making, and the advantageous effect of grain wort in increasing the maltose fermentation rate, was probably first appreciated by Rungaldier (1931).

In bread-making processes employing "bulk fermentation", the maltose fermentation stage occurs during the final "proving" period, during which gas is formed in the dough immediately before baking in the oven, and therefore considerably influences the final loaf volume. In rapid mechanical development processes, or when sugar is added to the dough, glucose, fructose, sucrose and levosin are the most important substrates. During this short fermentation period the rate accelerates, due to: (i) activation of enzyme systems by increase in substrate and coenzyme concentrations; (ii) adaptation (synthesis of new enzyme molecules); and (iii) increase in the amount of yeast present by the growth of new cells (Hoffman *et al.*, 1941). The time taken to reach the maximum fermentation rate is probably as important as the highest level attained, and all these effects are influenced by such factors as the yeast strain, the method of growth and the conditions of storage.

A further important factor in dough fermentation is temperature; indeed, a reproducible bakery process requires close temperature control. In the older procedures the best loaf texture was obtained by using low bulk fermentation temperatures (e.g. 25°) and long fermentation times. Rapid fermentations at higher temperatures (e.g. 35°) are, however, now preferred in mechanical development processes. During baking the temperature rises rapidly, and the loaf volume increases due to expansion of entrained gas; some fermentation by yeast continues during the early stages of the baking period as the temperature of the centre of the loaf rises relatively slowly (Farmiloe *et al.*, 1954). Although the optimum temperature for fermentation is about 40° (White, 1954), increases above this value still result in increase of rate, provided that fermentation was initiated at a lower temperature. The increase is, however, only maintained for a short time.

2. *Dough Structure*

Theories put forward to explain the part played by yeast in producing the rigid dough structure essential for obtaining a satisfactory loaf have been many and varied (Halton, 1959). It is now generally accepted, however, that the main effect is a modification of the protein (gluten) structure produced by mechanical stretching during fermentation. A minor chemical effect, due to excreted metabolic products such as glutathione and cysteine, on the protein disulphide bonds may also be present under some circumstances (Ewart, 1968). Indeed, with some inferior types of dried yeast these chemical effects may produce a weakening of the dough. Such advances in theoretical knowledge were soon followed on the practical plane by the introduction of mechanical methods of dough development, initiated by Baker (1954), whose continuous bread-making process has become commercially successful in

America. In Britain the Chorleywood Bread Process (Axford and Elton, 1960), using intense mechanical mixing, was introduced about 1960. Although this is essentially a batch process, it has led to increased automation in the baking industry. Mechanical dough development is now being challenged by the use of chemical additives, such as cysteine, which produce the same effect with a smaller power consumption (Henika and Rodgers, 1965; Jensen, 1967).

These new techniques eliminate the need for the time-consuming "bulk fermentation", which was relied on to produce the requisite dough structure. They introduce, however, the need for a faster dough fermentation in order to produce the necessary loaf volume sufficiently quickly; this may be accomplished by increasing the proportion of yeast and using a strain of higher specific fermentative activity. A brew process, which comprises a preliminary fermentation in sucrose or other suitable medium, is also sometimes employed to this end. This procedure is reported to have additional benefits in improved loaf volume and structure (Collyer, 1967); there are differences of opinion as to whether it affects bread flavour (Redfern et al., 1968). The yeast manufacturer has benefited from these new developments by increased sales, but the accompanying demand for improved uniformity and quality for use in mechanized processes necessitates stricter control at every stage of production and distribution.

3. *Flavour*

The products of yeast fermentation play an important part in modifying the flavour of baked loaves (Collyer, 1964). The constituents responsible for this effect are probably quite numerous, as shown by several recent studies (Linko et al., 1962; Wiseblatt and Zoumut, 1963). Rapid fermentation methods do not appear materially to alter bread flavour (Collins and Lipton, 1967; but see also Schuldt, 1967). Likewise, changes in yeast strains or manufacturing processes have not, contrary to expectation (Coppock, 1958), resulted in noticeable flavour changes. In the case of active dried yeast, where a higher proportion of yeast dry matter must be used because of its lower activity, and where the drying process itself may increase the number and concentration of flavour constituents, some variations in flavour may be detectable (Burrows and Fowell, 1968). Due regard must therefore be paid to this factor in development work and process control.

4. *Nutritive Value*

Yeast is a rich source of protein, and also of vitamins of the B-group, but because of the small proportion of yeast used in bread the protein content is only very slightly increased by that present in the yeast.

Owing to the low levels in flour of some amino acids, such as lysine and cystine, and their comparative abundance in yeast, some small improvement in the nutritive value of the flour protein is probably obtained by the bread-making process (Wilson, 1944; Bunker, 1947).

The conditions of growth and the composition of the medium have a considerable influence on the vitamin content of baker's yeast; consequently published figures are variable. Thiamine, biotin and nicotinic acid are freely taken up from the medium, while other factors such as riboflavin, p-aminobenzoic acid and pantothenic acid levels are relatively independent of the amount present in solution. Observed ranges of vitamin contents in baker's yeast are given in Table I (Wilson, 1944; Peterson, 1950; Kirchhoff, 1960; Grylls, 1961). At the normal level in bread (0·2–0·8% dry weight) the contribution of yeast to B-group vitamin intake is very small. Yeast with a high content of thiamine may readily be obtained by special propagation methods (see Chapter 10, p. 529), and by this means a significant contribution can be made to thiamine intake (Dawson and Martin, 1941). Today, however, the addition of synthetic vitamins at a suitable stage during manufacture is more economical.

TABLE I. *Vitamin Content of Baker's Yeast. From Wilson, 1944; Peterson, 1950; Kirchhoff, 1960; Grylls, 1961*

Vitamin	Range of content (μg/g dry weight)	
p-Aminobenzoic acid	16	175
Biotin	0·5	36[a]
Choline	2100	5100
Folic acid	19	80
i-Inositol	3000	5000
Nicotinic acid	200	700[a]
Pantothenic acid	69	280
Pyridoxine	16	65
Riboflavin	7·5	85
Thiamine	9	89[a]

[a] Not reinforced by added vitamin.

C. YEAST NUTRITION

The nutritional and cultural requirements of *Saccharomyces cerevisiae*, the organism normally used for bread-making, have been extensively investigated. Much of the published work has been carried out by those interested in the commercial manufacture of baker's yeast, and has been mainly concerned with yield, or yeast dry matter production (e.g. White, 1954). Much less information is available on the effect of nutrients and culture conditions on the quality of the yeast. In this chapter the term *quality* may be interpreted as the balance of properties in the yeast which are of interest to the baker and the manufacturer. No doubt a considerable amount of undivulged know-how has been accumulated by yeast manufacturers, much of which has not been published. Nevertheless, the main principles involved are known; these are discussed below.

1. *Carbon*

A wider range of carbon compounds is available for aerobic metabolism and growth than is the case for anaerobic fermentation. In addition to the sugars glucose, fructose, mannose, sucrose and maltose, baker's yeast will also metabolize lactic, tartaric, succinic, acetic and glycolic acids, and ethanol aerobically. Furthermore, under conditions of low substrate concentration, higher organic acids and other potentially inhibitory substances may be metabolized (Kampf and Behrens, 1966). Amino acids are utilized in the absence of ammonium salts, as a source of both carbon and nitrogen.

One-third of the molecule of the trisaccharide raffinose is normally assimilated, leaving melibiose. Lactose is not metabolized, and the position of pentoses is somewhat obscure. Thus Fanti and Kohan (1964, 1965) report that pentoses are assimilated aerobically but not anaerobically. In Russia and some European countries, sulphite liquor, the residue from paper manufacture, has been used as a substrate both for alcohol and baker's yeast production (Kryuchkova and Korotchenko 1959), but it is probable that when *Sacch. cerevisiae* is used, the hexoses but not the pentoses in this medium are utilized. The recent interest in the growth of yeasts such as *Candida lipolytica* and *C. krusei* on hydrocarbons prompts the question as to whether *Sacch. cerevisiae* can metabolize such unpromising substrates. Some evidence for faint activity has, in fact, been found (Nyns *et al.*, 1968).

The metabolism of carbon compounds is dealt with in another part of this treatise (see Volume 2), and only a few points of special interest need be mentioned here. Glucose and fructose are incorporated directly into the glycolytic cycle, but sucrose and maltose have first to be hydrolysed. Baker's yeast exhibits high levels of invertase activity, with the result that sucrose is very rapidly split and behaves, in effect, as a mixture of glucose and fructose. This high invertase content is turned to good account in the manufacture of invertase concentrate (Cochrane, 1961; Chapter 10, p. 529). The maltase activity, as discussed above (Section I.B.1, p. 351), is variable and is often the limiting factor in maltose fermentations. A number of anomalies have been brought to light in this connection. Thus, the rate of maltose fermentation is often less than that calculated from the maltase activity. Occasionally yeast strains have been found which ferment maltose more rapidly than glucose, or a mixture of maltose and glucose more rapidly than either sugar individually as shown in Table II (Harrison, 1958). Explanations of these effects were sought in alternative pathways, such as direct phosphorylation of maltose with formation of glucose 1-phosphate (Leibowitz and Hestrin, 1945). No direct evidence of this reaction has been found in baker's or brewer's yeast. The known facts can probably be explained by the theory that some sugars, such as maltose and maltotriose, have their own specific permeases which assist transport into the cell. These permeases behave like adaptive enzymes, and may sometimes be the rate-controlling step in the system (Robertson and Halvorson, 1957). On the other hand, glucose and fructose appear to be transported by a facilitated diffusion process (Cirillo, 1961; Heredia et al., 1968).

TABLE II. *Glucose and Maltose Fermentation Rates*

Yeast strain	Method of growth	Fermentation rate (μl CO_2/5 mg moist yeast . h)		
		Glucose, 0·2 M	Maltose, 0·1 M	Mixture of glucose 0·2 M and maltose, 0·1 M
Brewery yeast	Malt wort, anaerobic	146	45	187
Brewery yeast, Haploid A	Malt wort, anaerobic	83	122	159
Brewery yeast, Haploid B	Malt wort, anaerobic	114	122	221
Brewery yeast, Haploid C	Malt wort, anaerobic	236	136	280
Brewery yeast, Haploid D	Molasses, aerobic	246	300	340

Attempts have been made to describe the overall production of yeast cell-substance from sugars by empirical chemical equations. Pasteur found that under strongly aerobic conditions about half the carbon of glucose was oxidized to carbon dioxide to provide energy for growth, and the other half was converted to cell substance. White (1954) gives the Gay-Lussac equation for fermentation of glucose:

$$C_6H_{12}O_6 \longrightarrow 2 C_2H_5OH + 2 CO_2$$

and assumes that all the ethanol is converted quantitatively into yeast substance under suitable conditions of growth. He therefore postulates that the maximum "theoretical" yield of yeast is 56·7 g yeast dry matter from 100 g of glucose. Harrison (1967), on the other hand, takes the empirical figure for yeast dry matter yield of 100 g from 200 g sucrose, and from the chemical composition of the various constituents of the yeast cell derives the overall empirical equation (in molar equivalents) based on 200 g sucrose:

$$0{\cdot}585\ C_{12}H_{22}O_{11} + 0{\cdot}61\ NH_3 + 3{\cdot}205\ O_2 \longrightarrow$$
$$C_{3{\cdot}72}H_{6{\cdot}11}O_{1{\cdot}95}N_{0{\cdot}61} + 4{\cdot}29\ H_2O + 3{\cdot}30\ CO_2$$

The yeast constituents given in the example only add up to 90·5 g; the remaining 9·5 g is made up of potassium, phosphorus, sulphur and other elements.

From this equation the oxygen requirement and heat production during aerobic growth can be calculated. In addition, the effect of a change in the yield factor or in the substrate composition on these parameters can be predicted without recourse to experimentation. Using the equation as it stands as being applicable to commercial practice, with molasses and ammonium salts as substrates, the predicted oxygen requirement (102·5 g O_2) and the predicted heat production (387 kcal/100 g yeast dry matter) are close to the values determined experimentally. These two values have a very important practical bearing in yeast manufacture in connection with aeration and cooling.

A useful general relationship, thought to be applicable to any organic compound or organism, has been suggested by Mayberry et al. (1967). These workers, from a consideration of various possible methods of correlating their own results for various substrates, found that the yield of cell dry matter was equal to about three times the number of equivalents of available electrons in the substrate. The latter value is calculated by multiplying by four the number of moles of oxygen required to balance the chemical equation of complete oxidation of the substrate to carbon dioxide and water.

2. Nitrogen

Thorne (1949) has shown that yeast will take up nitrogen from ammonium ions more rapidly than from single amino acids or simple mixtures of these. A balanced complete amino acid mixture, however, is superior to ammonium ions. It is also known that under aerobic conditions ammonium ions will be absorbed in preference to amino acids when molasses is used as substrate. This material generally contains low levels of just a few amino acids. Urea can be utilized as a source of nitrogen, but it is not normally metabolized very rapidly.

For economic reasons it is used in yeast manufacture in India and South America. Extra biotin may be added to assist assimilation (Atkin, 1950), and the nitrogen content of the yeast may be increased by using excess urea in the medium.

Ammonia nitrogen is rapidly converted within the cell into amino acids, proteins, nucleic acids and numerous minor constituents. The proportions of these are largely determined by the assimilable nitrogen supply and the conditions of growth. The nitrogen content itself may vary within wide limits ($5 \cdot 0$–$10 \cdot 0\%$ of the dry weight) to give a corresponding variation in the yeast properties which are of interest to the baker. Thus, Thorne (1954) has shown that the nitrogen content has an important influence on the fermentative activity. Furthermore, a low nitrogen content is associated with a high carbohydrate content which, as discussed later (Section III.C, p. 395), is of importance in the preparation of dried yeast.

3. Oxygen

Although limited growth of yeast may be obtained under anaerobic conditions, a very rapid supply of oxygen is required for maximum growth and efficient utilization of substrate. Oxygen is, in fact, a major nutrient, about $1 \cdot 0$ g oxygen being required for the production of $1 \cdot 0$ g of yeast dry matter (Maxon and Johnson, 1953). This is most conveniently supplied as air, since pure oxygen, and indeed a large excess of oxygen from air, is inhibitory (White, 1954). The subject of aeration as applied to yeast manufacture is discussed later (Section II.D, p. 374).

4. Phosphorus

Soluble phosphate is rapidly assimilated by yeast during active metabolism, and is essential for growth. As in the case of nitrogen, the amount assimilated depends, among other factors, on the quantity supplied. The phosphate content of the yeast (as P_2O_5) normally varies from $1 \cdot 4\%$ to $5 \cdot 0\%$ of the dry weight, although under special conditions values ranging from $0 \cdot 8\%$ to $10 \cdot 0\%$ dry weight may be obtained according to Markham and Byrne (1968). Values between $1 \cdot 8\%$ and $3 \cdot 5\%$ dry weight are encountered in baker's yeast.

5. Other Elements

Potassium, sulphur and magnesium, but not calcium, are essential, and are generally supplied as inorganic salts. Excesses do not appear to have any detrimental effects unless present in sufficient concentration to cause osmotic damage. Most of the minor elements known to be essential for plants and micro-organisms have, at one time or another, been stated to be necessary for yeast growth. Only three, copper, iron and zinc, are stated by Olson and Johnson (1949) to be essential.

6. Growth Substances

Three growth substances are generally considered to be essential for optimum growth of the strains of baker's yeast in common use. These are the B-group vitamins biotin, pantothenic acid and inositol (White, 1954). Olson and Johnson (1949) also include thiamin and nicotinic acid. The latter two compounds, together with pyridoxine, are synthesized by many strains, but additional exogenous supplies may sometimes be necessary for optimal development of enzyme activities (Schultz *et al.*, 1937). Thiamin, however, has been reported to reduce yield (Matsuda, 1961) and this effect has also been observed by the author.

Under conditions of adequate aeration the biotin requirement is about 0·25 µg/g yeast dry matter (Dawson and Harrison, 1949), although Kautzmann (1961) reported that more is needed at suboptimal oxygen supply rates. The requirement for pantothenic acid is about 12 µg/g, and for inositol 300 µg/g (White, 1954). Many strains of yeast will utilize desthiobiotin as a source of biotin, and β-alanine to replace pantothenic acid although, in certain cases, antagonistic effects can be demonstrated due to the presence of other amino acids (Jakobovits *et al.*, 1951).

D. CULTURAL REQUIREMENTS

1. Growth Rate

Under special conditions, such as high nutrient concentration and low cell density, maximum specific growth rates of 0·6 g/h.g yeast (h^{-1}) may be obtained (Aiyar and Luedcking, 1966). In the manufacture of yeast, where high cell densities and maximum conversion of substrates to cell substance are required for economic reasons and particular yeast properties are required, specific growth rates within the range 0·05 to 0·3 h^{-1} are used. The growth rate required to give optimum yield and balance of yeast properties may be determined by preliminary experiments, and the assimilable carbon (e.g. molasses) feed programmed to limit the yeast growth rate to this value. In a fed-batch process, the exponentially-increasing rate of supply required follows the equation (White, 1954):

$$B_t = B_0 \, e^{\mu t}$$

where B_0 = amount of yeast at zero time,
B_t = amount of yeast at time t
μ = exponential growth factor.

The expression may be restated as:

$$M_t = M_0 \mathrm{e}^{\mu t}$$

where M_t and M_0 are corresponding quantities of substrate, assuming that the yield factor is constant. In commercial practice the hourly growth increment K, often erroneously termed the growth rate, is used where:

$$K = \frac{B_t}{B_{t-1}} = \frac{M_t}{M_{t-1}} = \mathrm{e}^{\mu}$$

This factor is only relevant to a fed-batch process but, because of the misnomer "growth rate", it has sometimes been erroneously applied in theoretical discussion of continuous process mathematics (e.g. Olsen, 1928; Wirtschaftliche Vereinigung der Hefeindustrie, 1935).

It is not essential in fed-batch processes to maintain a rate of feed conforming precisely to an exponential curve over the whole period of the propagation. Thus, constant and even diminishing rates of feed are sometimes used (de Becze, 1940; von Fries, 1956; Drews et al., 1962). Aeration and cooling requirements may necessitate the use of these types of feed pattern, but yeast quality can be affected by their use.

The differential form of the growth equation, namely:

$$\frac{dB}{dt} = B\mu$$

describes the instantaneous conditions in a batch process at time t, and may be used to derive the expression:

$$\frac{\Delta B}{\Delta t} = B\mu$$

which describes the steady-state conditions in a continuous process, where ΔB = constant amount of new yeast formed and removed from the system in time Δt. By setting up a material balance equation, Monod (1950) concluded that the growth rate is equal to the dilution rate, or the volume-fraction overflowing in unit time. Thus, in steady-state propagations of yeast in a single vessel, the growth rate is controlled by the dilution rate, and not by the rate at which substrate is supplied.

The Michaelis-Menten equation, originally put forward to describe the behaviour of single enzyme systems, has often been applied successfully to the behaviour of growing micro-organisms, no doubt on account of the rate-controlling influence of a single step in the chain of reactions (Ierusalimski, 1967). This equation enables the substrate concentrations required to maintain a given growth rate to be calculated.

Subsequent work has suggested that, owing to such influences as physical requirements of additional substrate for maintenance energy, and loss of viability, deviations from the equation may be found (Powell, 1967). For most purposes, however, the Michaelis-Menten equation is adequate. This states that:

$$\mu = \mu_m \frac{S}{K_s + S}$$

where μ = growth rate,
μ_m = maximum growth rate,
S = substrate concentration,
K_s = the "saturation constant", i.e. the substrate concentration at half maximum growth rate.

Before this equation can be used, the constants must be determined. By growing a yeast strain currently in use as a baker's yeast under steady-state conditions in a chemically-defined medium with glucose as the limiting substrate at dilution rates ranging from 0·15 to 0·30 h⁻¹, and determining the residual glucose by the method of Keilin and Hartree (1948), the constants μ_m and K_s were found to be 0·37 ± 0·03 h⁻¹ and 3·6 ± 0·5 × 10⁻⁴ M when the growth temperature was 30° and the pH value 4·0 (S. Burrows, unpublished observations).

The yield of yeast relative to substrate consumed is substantially independent of growth rate within the range 0·08–0·18 h⁻¹. Above this range substrate concentration is sufficiently high to ensure that an increasing proportion of substrate is converted to ethanol irrespective of the fact that excess oxygen is supplied. Below 0·08 h⁻¹ the proportion of the substrate in aqueous media diverted to cell maintenance becomes an appreciable proportion of the whole. Substrate is oxidized completely to carbon dioxide and water to provide energy for maintenance at a rate proportional to the mass of the organism (Pirt, 1965; Hansford and Humphrey, 1966; Mor and Fiechter, 1968). At present, there is insufficient data on yeast yields under steady-state conditions at various growth rates to be able to make a reliable estimate of this maintenance requirement.

Cell size, structure and chemical composition are also related to growth rate. Thus, McMurrough and Rose (1967) observed changes in shape, size, protein and lipid content, and cell wall composition with growth rate. Such profound variations would be expected to result in changes in yeast quality as viewed by the baker. This effect has, in fact, been known for some time (e.g. Braun et al., 1931), and is the reason for the close control of growth rate required in the continuous process for yeast production (Sher, 1960). High growth rate, with the accompanying production of ethanol, in addition to resulting in low yields, is also stated to produce yeast of high fermentative activity but poor stability (White and Munns, 1955).

2. *Temperature*

It is possible to grow yeast reasonably satisfactorily at any temperature in the range 20–40° (White, 1954). The lower the temperature the greater the efficiency of conversion of substrate to cell substance (Monod, 1942), probably because of the lower maintenance energy requirement. At temperatures over 35° losses of yeast yield and fermentative activity occur, but these are offset by improved stability (Pyke, 1958). In commercial practice the range is normally restricted to 25–35°: the upper end of this range is sometimes favoured owing to the greater efficiency of cooling, since very large quantities of heat have to be dissipated. Because of this, some attempts have been made to develop strains of yeast capable of giving good yield and quality at higher operating temperatures. Loginova (1956) cultured yeast at gradually increasing temperatures to obtain a new strain which appeared to be more resistant to higher temperatures than the original. The growth requirements of the new strain were, however, more exacting than those of the original. This effect has also been observed by other workers (Sherman, 1959; Lichstein and Begue, 1960).

In developing a strain which will grow at a higher temperature, the essential qualities of high yield with an acceptable balance of properties must be preserved. No strain of yeast appears to be available which, after growth at 40° for more than an hour or two, behaves satisfactorily as a baker's yeast. This apparent inability to adapt to growth at higher temperatures may be associated with the low calcium requirement compared with that of bacteria. Calcium is reported to protect enzymes from denaturation (Rose, 1962).

3. *Hydrogen Ion Concentration*

Most, if not all, strains of baker's yeast grow well at pH values between 3·5 and 7·0. Some restriction of growth may, however, be noticed at the extremes of this range. Normally values between 3·5 and 4·5 are employed, since under these conditions the growth of many kinds of bacteria which may be present as contaminants is restricted. Adsorption by the yeast cell walls of dark-coloured suspended matter from molasses occurs below pH 5·0 (White, 1954), and for this reason higher pH values are sometimes favoured.

4. *Yeast Concentration*

Providing the necessary conditions of aeration, agitation, growth rate, temperature and pH can be maintained, yeast will grow satisfactorily over a wide range of cell concentrations. Good quality yeast can be grown up to a concentration of 50 g dry matter per litre of medium (Drews *et al.*, 1962); indeed, with special aeration and agitation devices concentrations of 150 g dry matter per litre have been reported (Skiba, 1966). At this concentration about 30% of the culture volume is occupied by cells, and various adverse factors become important. Among these are an excessively high rate of heat production, high demand for oxygen and other nutrients, steep concentration gradients around the cells, and high concentrations of unassimilated or excreted materials. Decrease of growth due to osmotic effects, direct inhibition or to inadequate agitation may therefore occur. Increased agitation, as pointed out by Stoker and Rubin (1967), may not eliminate concentration gradients in the immediate vicinity of the cell, with the result that the "density-dependent inhibition" is impossible to overcome. Thus there is a practical limit to the yeast concentration attainable, and therefore also to the production capacity of a given vessel.

E. CULTURE MAINTENANCE

The problems discussed in this section are normally not the responsibility of the production unit, but of associated supporting laboratories.

1. *Yeast Strains*

Torula yeasts have occasionally been used for baking (Locke *et al.*, 1946) but pure strains of *Saccharomyces cerevisiae* are now almost universally employed. Until about 1950 two main types of baker's yeast were available. One of these was of low activity and high stability during storage; the other had a higher activity, but was somewhat unreliable owing to its poor stability. Various strains of these two types were in use in America, Great Britain and the continent of Europe. Since 1940 methods of hybridization and selection have been greatly improved (see Volume 1, Chapter 7), with the result that a number of hybrids have come into use (Burrows and Fowell, 1961). All these strains appear to be subject to spontaneous variation at a fairly high rate. Special precautions have, therefore, to be taken to prevent changes due to this cause, or the original strain must be maintained by reselection, using an appropriate testing procedure for check purposes (Lincoln, 1960).

2. *Culture Degeneration*

The serial transfer method of culture maintenance often results, after several months or years, in degeneration of the properties of the strain. Thus Thorne (1962), working with brewery yeasts, found a gradual increase in flocculence and decrease in fermentation efficiency which he ascribed to double mutation. According to this theory, among cells which had already mutated to give a lowered response in the property measured, a further mutation for increased growth rate enabled the doubly-mutated strain gradually to outgrow the original. The normally accepted mutation frequency of 10^{-6} to 10^{-7} is, however, much lower than that occurring in dividing yeast cells (Chester, 1963). It is now thought that chromosome rearrangement during vegetative division (mitotic recombination) may be the explanation (Roman, 1962; Emeis, 1965). If this is the case, restriction of growth whilst maintaining viability would constitute the best conditions for preservation. This not only reduces the chance of mitotic recombination, but also delays the growth of any mutants already formed. These high rates of variation, although posing a problem in culture maintenance, have been put to good account in developing particular properties such as temperature resistance (Loginova, 1956), resistance to the inhibitory action of formaldehyde (Rosa et al., 1962) and high dry matter (White, 1954). Emeis (1965) has suggested that the breeding of homozygous strains would reduce the effects of initial recombination, and hence improve strain stability.

3. Maintenance Methods

Several methods of cell culture maintenance, designed to restrict or arrest growth, have been suggested and used.

a. *Freeze-drying*, or lyophilization, is a common method of preserving micro-organisms, but the recovery of viable material is generally low (Heckly, 1963; Brady, 1962; Atkin et al., 1961). Wickerham and Andreasen (1942) were the first to recommend this method, but it is now losing favour owing to the high loss of viability and consequent risk of an increase in the percentage of mutants. In addition, the freezing and drying processes themselves may accelerate mutation (Subramaniam and Rao, 1951; Kirsop, 1955; Wyants, 1962).

b. *Low temperature storage* in liquid nitrogen, or in a dry ice-solvent mixture, is employed by some workers. A protective agent which is hydrophilic, and also penetrates the cell, must be added to prevent the formation of ice crystals. For *Sacch. cerevisiae*, ethanol was found to be most suitable. High recoveries of viable cells were obtained by rapid change of temperature during the freezing and thawing stages, but long-term tests showed that complete loss of viability occurred in three months at $-72°$, and 10% loss in five months at $-195°$ (S. Burrows, unpublished observations).

c. *The oil-culture method*, in which agar-slant cultures are overlaid with mineral oil and stored at 4° (Hartsell, 1956), is used widely in the

U.S.A. and Canada. Using this technique, the cells continue to metabolize at a very slow rate, and viability is maintained at a high level for many months; thus the risk of a mutant multiplying to become a significant proportion of the culture is negligible.

4. Pure Culture Preparation

Starter cultures for factory-scale propagations must be prepared by aseptic methods, using proven culture stocks. A small-scale propagation plant is required if the culture stocks are to be tested at intervals in order to check their freedom from degenerative changes. Starter cultures are usually grown anaerobically in flasks or stainless steel vessels, using media based on malt or molasses and containing about 10% sugar. The greater the amount of culture grown under aseptic conditions that can be supplied to the factory, the lower will be the infection levels in the final product, or the more yeast will it be possible to produce at the accepted level of purity, other factors being equal.

Desirable as it may be to confirm the absence of infecting organisms in the pure starter culture, this is extremely difficult owing to the very high ratio of yeast to possible contaminants. Enrichment culture is the only practical method, but this is time-consuming and not quantitative. Stringent adherence to well-tried aseptic techniques provides the best safeguard.

II. Factory Operations

A. MOLASSES COMPOSITION

1. Constituents of Importance

Since 1914 molasses has increasingly replaced grain wort and other substrates as a source of carbon and other essential nutrients although, as mentioned previously, sulphite liquor is sometimes used. Three types of molasses are employed according to local availability: (i) beet molasses, the concentrated syrup left after the extraction of sucrose from beet juice; (ii) refiner's cane molasses, formed during the purification of raw cane sugar; and (iii) blackstrap cane molasses, the residual by-product after sugar has been separated from cane juice. A less-frequently used material, high-test cane molasses, is a concentrate of the original cane juice; it may be partially inverted to prevent sucrose crystallizing.

The composition and properties of the various forms of molasses show some important differences, and there is also a considerable degree of variation within the types according to the country of origin, processing factory, season and conditions of storage. The availability and prices of the different molasses types fluctuate; hence the yeast factory should be ready to adapt itself to these varying factors: this can be quite a difficult task.

Molasses is generally a viscous dark-coloured liquid, although beet molasses of light colour and low viscosity are occasionally obtainable. It contains variable amounts of suspended particulate and colloidal material, and also micro-organisms. The chemical constituents of interest in yeast manufacture are listed in Table III, together with the range of concentrations to be expected in each type. In one or two cases individual figures only were available, but these are probably typical. The biotin figures include desthiobiotin, which can usually replace biotin. Similarly, the value for calcium pantothenate includes β-alanine. In the case of beet molasses, most of the assimilable sugar is sucrose, whereas in cane molasses appreciable proportions of glucose and fructose are present.

TABLE III. *Partial Analyses of Molasses (Based on information given by: Rogers and Mickelson, 1948; White, 1954; Olbrich, 1956)*

Constituent	Beet	Refiner's cane	Blackstrap cane
Total reducing sugar after inversion, % wt	47–58	50–58	50–65
Total nitrogen, % wt	0·2–2·8	0·1–0·6	0·4–1·5
α-Amino nitrogen, N% wt	0·36	0·03	0·05
Phosphorus, P_2O_5 % wt	0·02–0·07	0·01–0·08	0·2–2·0
Calcium, CaO % wt	0·15–0·7	0·15–1·0	0·1–1·3
Magnesium, MgO % wt	0·01–0·10	0·25–0·8	0·3–1·0
Potassium, K_2O % wt	2·2–5·0	0·8–2·3	2·6–5·0
Zinc, Zn µg/g	30–50	5–20	10–20
Total carbon, C % wt	28–34	28–33	—
Total ash, % wt	4–11	3·5–7·5	7 –11
Sulphur, SO_3 % wt	0·3–0·4	—	—
Biotin, µg/g	0·01–0·13	0·9–1·8	0·6–3·2
Calcium pantothenate, µg/g	40–100	16	20–120
Inositol, µg/g	5,000–8,000	2,500	6,000
Thiamin, µg/g	1–4	—	1·4–8·3
Pyridoxine, µg/g	2·3–5·6	—	6–7
Riboflavin, µg/g	0–0·75	—	2·5
Nicotinamide, µg/g	37–51	—	20–25
Folic acid, µg/g	0·21	—	0·04

A large number of compounds, present in small amounts, have been identified, and some of these are of interest on account of their possible growth-supporting or -inhibiting function. Among these are: (i) nitrogen bases and their derivatives, e.g. adenine, guanine and uridine; (ii) organic acids, e.g. acetic, aconitic, butyric, citric, fumaric, glycolic, lactic, malic, malonic, oxalic, propionic, succinic and valeric; (iii) amino acids, e.g. glutamic acid, asparagine, serine, alanine, valine and γ-aminobutyric acid; (iv) sugars, e.g. arabinose, galactose, kestose and raffinose; (v) alcohols, e.g. D-arabitol, d-erythritol, mannitol and sterols; (vi) aldehydes, e.g. furfuraldehyde and hydroxymethylfurfuraldehyde; (vii) miscellaneous compounds, e.g. sulphurous acid and potassium imidodisulphonate.

2. *Molasses Defects*

Molasses consignments may occasionally be faulty owing to low levels of assimilable sugars, growth substances such as biotin, or magnesium. On the other hand inhibitors for yeast growth may be present. Low sugar content may not be detected by reducing-sugar determination after inversion, since reducing complexes between sugars and amino acids which are not assimilable by yeast may be formed during storage or excessive heating. Reducing-sugar determinations before and after fermentation or growth, using baker's yeast, will however detect this anomaly. Beet molasses is generally deficient in biotin (see Table III), but this can be made up by addition of the synthetic compound or by admixture with cane molasses which is rich in biotin. Magnesium deficiency is easily guarded against by the addition of magnesium salts, since excess is not detrimental.

Several inhibitors have, from time to time, been detected in molasses; these include:

a. *Sulphur dioxide*, which is used in beet processing, and is sometimes present in molasses. The level is reduced during heating for sterilization purposes, especially if cane molasses containing reducing sugars, with which it reacts chemically, is also present. The sulphite ion is not active, hence the inhibitory action becomes rapidly stronger as the pH is reduced (Ingram, 1959). Thus at pH 3·0 inhibition occurs at 10 mg Na_2SO_3/100 ml; at pH 4·0 it begins at 100 mg/100 ml; and at pH 5·0 at 1,000 mg/100 ml.

b. *Hydroxymethylfurfural* is often present in appreciable amounts in cane molasses that has been overheated at the sugar factory, during transport and storage, or at the yeast factory (Ingram *et al.*, 1955). It is known to be inhibitory to yeast metabolism under anaerobic conditions at concentrations as low as 0·05% (Stakhorskaya and Tokgarev, 1964). However, under aerated conditions, where the molasses is fed incrementally to the growing yeast, 0·4% (based on molasses weight) can be tolerated (S. Burrows, unpublished observations). This is due to the fact that the concentration of inhibitor in the medium in contact with the growing yeast is low at the start and increases gradually as the yeast concentration rises. Inhibition is thus minimized, and the yeast presumably has the opportunity to develop resistance to the inhibitor.

c. *Potassium imidodisulphonate* is formed during the processing of beet molasses in the sugar factory from nitrites which have been previously produced by bacterial action; these combine with added sulphites to form a sparingly soluble potassium complex salt (Parker, 1960). This compound has been detected several times in the sandy deposit that settles in beet molasses storage tanks. It inhibits yeast growth in Petri dishes, but is only slightly inhibitory at a concentration of 5% by weight of molasses in incrementally fed propagations (S. Burrows, unpublished observations). Hence, it does not normally cause trouble, but if the sediment in the tank is dissolved by steam heat, in order to remove crystallized sugar, the solution should not be used until it has cooled and redeposited the sparingly soluble inhibitory material.

d. *Fatty acids* detected in molasses include several volatile compounds (Dierssen *et al.*, 1956). Of these, acetic acid (0·5–1·2% w/w) is metabolized by yeast but becomes inhibitory at concentrations over 0·75%. The higher acids, butyric (0–1·0%) and valeric (0–0·1%), are inhibitory at somewhat lower concentrations (0·1–0·5%) depending on the pH value of the medium (Holtegaard, 1956). It is quite possible that these higher acids are, like acetic acid, metabolized when present in concentrations which are not inhibitory (Kampf and Behrens, 1966).

Apart from sulphur dioxide these inhibitors require time-consuming methods for detection and determination. Furthermore, unknown inhibitors may sometimes be present. It is therefore impracticable to obtain complete information on the presence of inhibitors in consignments of molasses entering the yeast factory. The safest procedure is to use molasses from sources previously known to be reliable, and to use incrementally-fed growth processes at every stage in order to keep the ratio of inhibitor to yeast as low as possible.

B. MOLASSES TREATMENT

1. *Clarification*

Refiner's cane and British beet molasses, which are relatively free from suspended particulate matter, have sometimes been used without clarification. Colloids remain in suspension throughout the process, although some may be adsorbed on the yeast, so giving it a creamy colour. Beet molasses from some sources, and most blackstrap molasses, contain considerable amounts of suspended particulate and colloidal matter which must be removed to prevent contamination of the yeast by inert coloured material and blockage of filters during the separation of the yeast. Excessive foaming, with consequent high usage of antifoam agents, is also associated with a high colloid content.

Chemical clarification by co-precipitation with alumina or calcium phosphate may be used (Walter, 1953), but losses of sugar and growth substances may occur (Shukla and Kapoor, 1956). With the advent of the highly efficient polyelectrolyte flocculating agents, such as alginates and polyacrylamides (Scutt, 1967), further developments may be expected in this sphere.

2. Sterilization

Many types of micro-organism are collected by molasses during transport and storage (Bergander and Konigstedt, 1965), but apart from osmophilic yeasts proliferation does not occur until the molasses is diluted in the yeast factory. Most of the bacteria and yeasts are easily destroyed by heating to 100–110° for about one hour; this process of pasteurization is most often used. The pH value should be between 6·0 and 8·0, otherwise caramelization of sugar may result. Any organisms escaping this treatment will multiply rapidly on cooler parts of the plant, making frequent cleaning and steaming of the feed pipes and pumps essential.

The efficiency of pasteurization can be improved by the use of formaldehyde at a concentration of 0·1% in the molasses feed, but not more than 0·03% in the medium in the final stages of growth (Bergander and Konigstedt, 1965). This treatment also limits the growth of surviving bacteria in the feed lines, but not after further dilution in the vessel, where yeast growth is taking place.

Improved commercial sterilization techniques have been introduced in the food industry, notably for milk processing, based on improved understanding of the kinetic and statistical requirements. The temperature and time dependence of the destruction of viable spores is given by the equation (Richards, 1966):

$$\ln \frac{N_0}{N} = At e^{-E/RT} \tag{1}$$

or

$$\ln t = \frac{E}{RT} + \ln C \tag{2}$$

where N_0 = initial concentration of viable spores
N = concentration of viable spores at time t
A = Arrhenius constant
E = activation energy of the rate-limiting reaction
R = gas constant
t = time
T = absolute temperature
and C = constant.

Equation 2 gives the relation between time and temperature for a given fractional reduction in viable spore numbers. It is important to realize, however, that sterility is only relative and, as shown by Equation 1, complete sterilization theoretically requires infinite time. The safe degree of sterility required in practice must therefore be decided upon before the equation can be used. A further complication is that sterility is a statistical concept, with the number of survivors conforming to a Poisson distribution. A more correct term to use would be "probable degree of sterility" (Aiba and Toda, 1967). For example, a required reduction of 100 organisms per cm^3 of medium to one organism per 1,000 m^3 involves a fractional decrease of 10^{11}. Application of the calculated conditions to obtain this result will, on average, give the assumed figure, but this may vary over a range from zero to two or even three organisms per 1,000 m^3.

Equation 2 also shows that sterilization time decreases logarithmically as the temperature is increased. High-temperature short-time conditions are therefore the most efficient. Owing to the lower activation energy of many chemical reactions compared with thermal destruction of bacteria (Richards, 1966; Hospodka, 1966), these conditions also reduce the possibility of chemical degradation of nutrients.

Application of high-temperature short-time treatment may be carried out in practice in a continuous system, using heat-exchangers, and sometimes steam injection for the final heating step (Deindoerfer and Humphrey, 1959). In this case careful control of temperature, pressure and medium flow-rate are required so that the rate of steam condensation and dilution can be controlled and calculated. This method results in a high rate of destruction of the more persistent organisms, with negligible loss of nutrient value. Sterilization by γ-radiation has attractive possibilities, but at present is too expensive in capital equipment and running costs to be practicable.

3. *Mixing and Diluting*

Molasses is stored in large tanks at the yeast factory, and pumped out as required for mixing, diluting and sterilizing. The proportion of the various types used depends on such factors as availability, price and composition. Thus, as already mentioned, beet molasses is deficient in biotin, whereas blackstrap is rich in this growth substance, and generally contains more thiamin and pyridoxine. It is thus advantageous to use as high a proportion of blackstrap as possible. In some countries it is the sole available type, whereas in others beet molasses only can be used. With very high proportions of blackstrap, problems of foaming, colour adsorption by the yeast, and restriction of filtration by colloidal material are encountered. On the other hand, if cane molasses is not available, biotin must be added in order to maintain high yeast yields. Other materials, such as malt and grain extracts, corn-steep liquor and yeast extract, may also be added.

The proportions of the various constituents of the medium are usually adjusted by batch methods using weighing tanks and mixing vessels, but proportioning pumps can be used for continuous operation. Dilution of the medium, which can be carried out either before or after sterilization, must be accurately controlled since this influences the rate of feed of assimilable carbon, and hence the total amount of yeast formed. Since phosphate and nitrogen are most conveniently fed in predetermined amounts, a variable carbon feed results in changes in yeast composition and properties. A fixed rate of dilution may not provide adequate control owing to the variable sugar content of the raw materials. Specific gravity control (Sher, 1961) gives some degree of correction, but different types of molasses have different specific gravity to sugar content relationships, mainly because of varying proportions of inorganic matter. The same difficulty also applies to polarimetric and refractive index measurements.

As pointed out previously, total reducing-sugar determinations after inversion may be misleading; moreover, analysis before and after fermentation or yeast growth is cumbersome and difficult to apply to automatic dilution control. Furthermore, significant amounts of non-sugar carbon may be assimilated. Continuous automatic methods for the determination of total carbon in liquid samples are being developed (Van Hall *et al.*, 1963; Cropper *et al.*, 1967) and could conceivably be used for molasses. By applying such a method to the diluted molasses feed, and also to the spent medium after removal of yeast, the assimilated carbon could be calculated and used for dilution control.

It should be noted that this method measures *assimilated* carbon, not *assimilable* carbon, and is therefore affected by process conditions and the presence of inhibitors or activators. However, if the emphasis is to be placed on yeast growth or yeast yield and not on molasses dilution, control could be exercised by measuring heat production, carbon dioxide evolution, or even yeast yield itself if suitable methods were available.

4. Liquid Feed Systems

The importance of incremental feeding of molasses has already been stressed. Although precise adherence to a predetermined feed rate programme is often practised, it is not essential. Various methods of incremental feed programming are in use, but perhaps the commonest, and until now the most reliable, is the Swedish S.J.A. dosing device (Sher, 1961). This consists of a reservoir of, say, 50 litres capacity, fitted with valves which enable it to be filled to a constant height and then automatically discharged into a propagation vessel. The frequency of this operation is controlled by a metal tape perforated with slots which activate a tipping mercury switch as the tape is made to pass over it. Any desired programme of molasses supply can be arranged by suitable positioning of the slots along the tape.

There are, however, three disadvantages of this system. Firstly, a new tape must be prepared for each new feed programme. Secondly, an intermittent nutrient supply is produced which generally has a low frequency at the start with increasing frequency as the propagation proceeds: this results in considerable variation of such parameters as sugar concentration and oxygen tension which may affect growth and properties of the yeast. Thirdly, the maximum feed rate obtainable may be limited by the size of the measuring reservoir and the time taken for filling and emptying, although these can be designed to suit any specific programme.

Electronic methods of programming, simply requiring manual adjustments of electrical contacts, are now available. These may be used in conjunction with measuring reservoirs or batch flow-meters, but should preferably be used to programme the set points of flow-meters which control valves on the molasses feed lines. Feed-back methods of controlling molasses feed rate are discussed in the sections dealing with aeration (Section II.D, p. 374) and automatic control systems (Section III.D, p. 398).

C. CHEMICAL ADDITIVES

1. *Phosphate*

Phosphorus may be added as calcium superphosphate, ammonium phosphate or phosphoric acid according to availability, purity and convenience. Calcium superphosphate is often used, but has a somewhat variable composition and usually contains an insoluble fraction which causes trouble. If pure ammonium phosphate is available this is probably most suitable. This salt may be metered in as a solution during the propagation, or a known weight can be added at the beginning of a batch process.

2. *Assimilable Nitrogen*

Ammonium sulphate, ammonium chloride, ammonia solution or even anhydrous ammonia are all suitable sources of assimilable nitrogen for industrial use. Urea may also be used but, as mentioned previously, presents some difficulties. These materials are normally added incrementally, mainly because bulk addition results in unacceptably large pH drifts. Olsen (1961) describes equipment for dissolving ammonium sulphate automatically. Ammonium salts sometimes contain appreciable amounts of arsenic, which is rapidly absorbed and accumulated by yeast: arsenic-free salts must therefore be specified.

The amount of assimilable nitrogen fed as the ammonium ion largely determines the nitrogen content of the yeast, although under some circumstances a little nitrogen is absorbed as amino acids from the molasses, thus enhancing the yeast yield. The rate of nitrogen feed may be controlled by means of a *formol* titration carried out on the medium. This determines the concentration of unused ammonium ion or amino-nitrogen, and has therefore been thought to indicate the nitrogen requirements of the yeast. In fact, it measures the balance between supply and demand, and is only a very rough guide to the best rate of feed, particularly since the requirements for growth and for optimum yeast quality may be quite different. It is therefore sometimes preferable to determine the optimum feed pattern by previous experimentation and yeast quality testing, and to use this predetermined programme irrespective of apparent low or high levels as indicated by the formol titration.

3. *Other Nutrients*

Magnesium sulphate is generally added as a safeguard against magnesium deficiency. Any requirement for sulphur is also taken care of by this salt if ammonium sulphate or sulphuric acid are not used. Biotin and thiamin are sometimes added when deficiencies become apparent.

4. *Antifoam Materials*

In highly aerated yeast propagations using molasses very high heads of foam are produced which necessitate the use of vessels that are 30–100% larger than the working volume. The mechanism of stable foam formation is complex, but generally depends on the presence of traces of surface-active colloidal or particulate materials. In the case of beet molasses, high molecular weight peptides may be the active agents (Oldfield and Dutton, 1968).

Mechanical methods of foam breaking are available, and an ultrasonic method has been used experimentally (Dorsey, 1959) but chemical antifoam agents are generally preferred (Solomons, 1967). Fatty materials, sometimes containing mineral oil, are cheap and effective. Synthetic polyglycols (Bristol-Myers Co., 1964) are becoming widely used since they are tasteless, colourless and soluble or miscible with water. Care must be taken that the additives comply with food regulations. Antifoam agents are sterilized by heat before use, and may be injected automatically into the propagation vessel when the foam reaches a certain height and short-circuits a trigger circuit through a probe adjusted to the required level in the vessel.

D. AERATION

1. *Filtration of Air*

The large volumes of air required to supply oxygen, which is essential for yeast growth, must be filtered free from dust and micro-organisms before use for yeast propagation. Passage through beds of oiled rings removes dust, but only a proportion of the micro-organisms. Complete removal of infecting organisms can only be obtained by the use of absolute filters, which ideally have pore dimensions smaller than those of the organisms to be removed. Passage of air through these is generally slow, but reticulated convoluted systems are now available which pass air at sufficiently high rates for commercial purposes (White and Smith, 1964). Deep beds of fibrous filters, when properly designed and used, are very effective in removing particles down to a diameter of 1 μ. The voids in this material are relatively large and allow rapid air flow with only a small pressure differential. Materials used are glass fibre, mineral wool and cellulose–asbestos (Firman, 1965).

In the latter case filtration takes place in the body of the material by impact, gravity, Brownian movement, convection and electrostatic attraction. Although complex, the behaviour may be described by a simple empirical equation (Richards, 1967):

$$\log \frac{F}{F_0} = - kx$$

where F_0 = concentration of particles entering filter
 F = concentration of particles leaving filter
 x = depth of filter
 k = constant

The fractional reduction ratio F/F_0 can be estimated by the use of bacteriological techniques, although this may involve the difficult problem of measuring a very low count down-stream of the filter. If a thin layer of the material is used a chemical method, such as the injection into the air stream of an aerosol containing methylene blue, followed by determination of its concentration by means of an absorptiometer, is preferable. In this way k can be determined, and the thickness required for the necessary fractional reduction can be calculated (Mulcaster and Aukland, 1964). The efficiency of filtration depends in a non-linear manner on the linear air velocity and the particle size (Purchas, 1966). Particles around 0·2 μ penetrate more readily than smaller or larger sizes. Rapid flow rates are, within limits, more efficient than slow rates (Dorman, 1967). The development of channels within the pad, and between the pad and the air-duct, are sources of trouble (Gaden and Humphrey, 1956); these can be detected by the methylene blue test. The temperature of the filter must be above the dew point of the air, otherwise condensation occurs and bacteria are carried through with the water.

2. *Oxygen Solution Rate*

Aeration problems arise mainly because the solubility of oxygen in water is very low and the rate of removal by growing yeast is high. Oxygen must therefore be transferred rapidly from the gaseous phase into solution, and from solution into the yeast cell. The first stage, transfer from gas to liquid, appears to be the rate-limiting step (Calderbank, 1959). The oxygen transfer rate, or solution rate, is given by the equation (Finn, 1954):

$$\frac{dc}{dt} = K_L \, a(C_M - C_L)$$

where
dc/dt = rate of solution of oxygen
K_L = mass transfer coefficient
a = interfacial area
C_M = equilibrium oxygen concentration in liquid phase
C_L = actual oxygen concentration in liquid phase.

Generally, neither K_L nor a can be determined separately, but for a given set of conditions the product $K_L a$ is considered to be constant, and gives an overall measure of the gas-absorbing capacity of the aeration system. On the laboratory scale this can be measured by the sulphite oxidation method (Cooper et al., 1944). For use both on the large and small scale, the oxygen electrode (see Sections II.D.4 and 5, pp. 376 and 377) is an invaluable tool for measuring the balance between oxygen supply and demand, by determining the oxygen concentration in the solution.

Owing to incomplete knowledge, it is not possible to scale up aeration rates accurately, or indeed to predict aeration requirements in one commercial vessel as compared with another of the same size but with different design of aerator or internal geometry. It therefore becomes important to be able to measure oxygen supply and usage rates in order to determine the optimum rate without a considerable amount of empirical experimentation.

3. *Aeration Efficiency*

A number of different types of aerating devices for microbiological processes are available, and the efficiency of these must be considered from three points of view: (i) the oxygen solution rate; (ii) the proportion of oxygen in the air bubbles which dissolves in the medium and is utilized by the yeast; (iii) the power consumption. In a given system factors i and ii are often negatively correlated, since an increased rate of aeration results in a higher solution rate, but in a lower mean proportion of the oxygen in the air bubbles passing into solution. However, in comparing different systems, those with higher solution rates will usually give a greater percentage removal of oxygen from the air. In this case a lower volume of air is used, and consequently less power is needed to operate the compressors. The increased solution rate may, however, require additional agitation, which can be supplied by a turbine within the vessel, but the total power consumption is thereby increased. The economics of oxygen transfer in industrial aeration systems have been discussed by Finn (1969).

The traditional method of aeration employed in yeast factories makes use of a simple sparger, which consists of a system of radiating perforated pipes located at the bottom of the vessel (de Becze and Liebermann, 1944; Walter, 1953). The air outlets comprise a large number of holes of 0·4 to 1·6 mm diameter. Large volumes of air are driven through the distributor by means of compressors, and no additional agitation is employed. This system is not particularly efficient: on the commercial scale, using liquid depths of five metres or more, 5% to 13% of the oxygen in the air is removed. In small-scale vessels as little as 1% may be taken up. Much higher efficiencies are obtained by the use of vessels with stirrers and baffles, but these are more expensive in initial cost and maintenance. Maxon and Johnson (1953) obtained oxygen removal efficiencies of about 20% on the 1·5-litre scale when using a stirring rate of 1,600 r.p.m. Many patents describe stirrers, agitators, mixers, pumps and other devices which are claimed to be more efficient in terms of oxygen solution rate or power consumption than the original air distributor. For instance, Patentauswertung Vogelbusch GmbH. (1961) use a system in which air is dispersed from the trailing edges of hollow rotating blades. This method is claimed to use less power than the conventional system, and probably has an oxygen removal efficiency of between 15% and 20%. Ebner et al. (1967) claim that their air distributor utilizes about 30% of the oxygen in the air, and has been successfully employed in large-scale yeast propagations.

4. Control of Aeration Rate

For a given type of substrate it might be assumed that the oxygen uptake rate would be proportional to the substrate supply rate and to the yeast formation rate. However, it has not yet been demonstrated that this is precisely true, and it is possible that any factor which affects the percentage yield of yeast may also affect the oxygen demand rate (Harrison, 1967). Nevertheless, the direct relationship between substrate feed rate and oxygen demand rate is approximately true, and may be used for calculating oxygen requirements during growth. Thus, aeration rate may be programmed to follow molasses supply rate, remembering that as the aeration rate increases the transfer of oxygen into solution generally becomes less efficient on an air volume basis. Additional increases in air supply to offset this effect must therefore be made.

The rate of aeration may be measured by means of a variable area flow-meter, or more usually by an orifice meter. A manually-operated valve may be used to control the rate according to a programme previously determined experimentally on the basis of optimum yield and yeast quality, together with the known nutrient feed rates. Pneumatic and motorized valves are being increasingly used (Sher, 1961); these are adjusted automatically according to a predetermined programme or by feed-back from an instrument which measures some parameter, such as oxygen concentration in the medium or in the exhaust gases.

Several methods of automatic control of oxygenation in biological systems have been proposed, but it is not certain whether any of these are sufficiently reliable and versatile for use in commercial operations under all conditions. The instruments proposed are: (i) the paramagnetic oxygen analyser, which measures oxygen concentration in the exhaust gas (Shu, 1956); (ii) the hydrocarbon meter, which can be used to determine the concentration of ethanol in the exhaust gas (Patentauswertung Vogelbusch GmbH., 1961b); (iii) the membrane-covered type of oxygen electrode, which measures the oxygen tension in the medium (Clark, 1956; Mackereth, 1964); (iv) the Teflon tube apparatus, which also determines oxygen tension (Phillips and Johnson, 1961; Roberts and Shepherd, 1968); and (v) the infrared gas analyser, which is used to estimate the carbon dioxide concentration in the exhaust gas (Seidel, 1943).

These instruments may be used for controlling the rate of supply of air (Hefe-Patent GmbH., 1963), the agitation rate (Shu, 1956, Svenska Jästfabriks Aktiebolaget, 1966) or the proportion of oxygen in the feed gas (Siegell and Gaden, 1962; MacLennan and Pirt, 1966). As an alternative to varying the air supply, the oxygen demand may be controlled by adjusting the rate of nutrient feed (Ohashi, 1958; Československá Academie Ved, 1966).

5. Practical Difficulties of Proposed Methods

In practice, various difficulties attend the use of most of the proposed methods. The paramagnetic oxygen analyser linked to an agitator is difficult to operate on the large scale owing to the large power surges which result from the varying agitation requirements. Furthermore, this instrument is inconvenient for controlling aeration rates, since changes in the rate of passage of air result in changes in the oxygen concentration for a given oxygen uptake rate.

Certain types of hydrocarbon meter provide a robust and reliable means of measuring ethanol in the exhaust gas from the propagation vessel, from which the concentration in the medium can be calculated. This method has the advantage of not requiring the insertion of a probe into the medium, with the attendant risk of introducing infection, but has several disadvantages. There are a number of delays in the system, such as: (i) the slow rate of re-assimilation of ethanol, once it has been formed; (ii) the diffusion of ethanol out of the medium into the air bubbles; and (iii) the passage of exhaust air through the head space and sampling line to the instrument. These latter delays may amount in total to several minutes. Moreover, ethanol concentration is not an unfailing indication of the oxygen status of the system, since it is also produced in the presence of oxygen when the rate of sugar supply is near to, or in excess of, the maximum yeast growth rate. Furthermore, since control must be based on a positive ethanol concentration, yield will be lost owing to loss of ethanol in the exhaust gas.

Owing to its rapid response, the oxygen electrode appears to show promise, but a number of difficulties have still to be overcome. Because of the extremely rapid uptake of oxygen by yeast, the electrode cannot be situated in a by-pass line as is often the practice for temperature and pH measurements. It must therefore be robust, fully immersible and preferably sterilizable under pressure. A robust temperature-compensated electrode, which is reliable and electrically stable over long periods, has been developed (Ebbutt, 1966). This instrument must be sterilized chemically, and must therefore be withdrawn from the vessel after each propagation. Fully sterilizable electrodes are being developed (Collis and Diggins, 1967), but a reliable model for industrial use was not available at the time of writing, although laboratory types have been described (Johnson et al., 1964; Borkowski and Johnson, 1967).

6. Factors Controlling Oxygen Uptake

Before the oxygen electrode can be used effectively, the optimum oxygen concentration, or partial pressure, in the medium must be known. This has been said to be about 0·2 p.p.m. (Strohm et al., 1959), but it is probable that the value is not constant. Thus, Johnson (1967) discusses three possible situations: (i) where the rate of oxygen utilization by the cell is rate-limiting and independent of oxygen concentration in the medium; (ii) where the rate of entry of oxygen is limited by the activity of the first enzyme in the uptake system, e.g. cytochrome oxidase; (iii) where the rate is diffusion-limited owing to some physical restriction near the surface of the cell.

In the case of enzyme-limited uptake the relation between oxygen tension and yeast growth rate is given by the equation:

$$P = BP_{max} \frac{T_l}{K_0 + T_l} = K_1 B\mu \tag{3}$$

where
P	=	specific rate of oxygen uptake
P_{max}	=	maximum specific rate of oxygen uptake
B	=	amount of yeast in system
μ	=	yeast growth rate
T_l	=	oxygen tension in the medium
K_0	=	Michaelis constant
K_1	=	constant.

The first part of this equation assumes that the oxygen tension in the bulk of the medium is the same as that at the enzyme surface, and the second part that the oxygen uptake rate is proportional to the rate of formation of new yeast.

The diffusion step in the system is described by the equation:

$$P = K_2 B (T_l - T_c) = K_1 B\mu \tag{4}$$

where T_c = oxygen tension inside the cell, or at the site of the initial enzyme reaction

and K_2 = constant.

Under steady-state conditions, using dilution rates corresponding to yeast growth rates between 0·08 and 0·15 h^{-1}, it was found that the oxygen tension was dependent on the growth rate (Table IV). Since the rate of molasses feed and yeast formation were substantially constant, the rate of oxygen absorption, and hence the oxygen tension in the medium, should be constant if oxygen utilization were rate-limiting. Furthermore, use of Equation 4 gives negative values for T_c, but Equation 3 gives reasonable values for P_{max} and K_0 (S. Burrows, unpublished observations), indicating that oxygen uptake was probably enzyme-limited, as has also been suggested by Winzler (1941) and Longmuir (1954). The average saturation (Michaelis) constant (K_0) was 1·9 × 10^{-6} M (1·3 mmHg partial pressure), compared with the value 2·84 × 10^{-6} M obtained by Winzler, and 0·6 × 10^{-6} M by Longmuir. The results do not rule out the possibility that diffusion might, under certain circumstances, have some influence on the oxygen uptake rate. Indeed, Powell (1967) has shown that the diffusion and enzyme reaction mechanisms can be described approximately by a single equation of the Michaelis-Menten form, in which the apparent saturation constant is dependent partly on diffusion.

TABLE IV. *Oxygen Tension and Yeast Growth Rate*
[Temperature, 35°]

Growth rate (h^{-1})	Oxygen tension average over 8 h (mmHg)		
	Expt 1	Expt 2	Expt 3
0·08	0·9	2·4	1·3
0·11	1·9	12·0	—
0·15	—	—	6·5

E. FACTORY HYGIENE

1. *General Cleaning*

Before considering the yeast production processes themselves, the important question of hygiene will be considered briefly. Yeast, being a constituent of food products, must be prepared under hygienic conditions, and contamination with other organisms must be prevented as far as possible. To this end, factory operatives should be fully aware of the need for utmost cleanliness, and be trained in the necessary techniques. Regular cleaning of floors, walls, ceilings and internal structures such as girders and staging is essential. The labour involved in this can be much reduced by improvements in factory design and in the nature of the materials and surface coatings of the equipment. The wide use of non-poisonous non-corrosive materials such as stainless-steel, and the employment of techniques using detergent-disinfectant sprays, assist greatly in improving general hygiene.

Some parts of the factory require special attention. Thus filtration, packing and drying operations tend to create dust, and should if possible be separated from the propagation area in order to avoid recycling of contaminating micro-organisms. These areas should also receive more frequent cleaning, since extraneous organic matter can be accumulated in odd corners with surprising rapidity and create reservoirs of infection.

2. *Sterilization of Equipment*

The interior of pipes, vessels, pumps, etc. must not only be cleaned but sterilized. Until recently the only effective method of cleaning was to dismantle the machinery and pipes and clean manually. Large vessels were cleaned by hand-scrubbing from the inside. After reassembly the equipment was sterilized by passage of steam under slight pressure.

These methods are now being replaced, where possible, by *in-place* cleaning, using efficient chemical detergent-sterilizing agents. Vessels and reservoirs are cleaned by powerful sprays or jets (Olsen, 1961), pipes with moving slugs of resilient material which provide a scrubbing action, and more complicated machinery such as pumps and heat exchangers by frequent circulation of a cleaning agent. Ricketts (1962) outlines several distinct properties required by an efficient detergent-sterilizing agent. Caustic soda and sodium metasilicate, the agents most frequently used, have good solubilizing effect on organic materials, but caustic soda precipitates many inorganic materials, and sodium silicate is not very effective for sterilization purposes. A mixture of several chemicals may prove more efficacious. Thorough rinsing, draining and sometimes steaming follows the chemical treatment.

It is not always realized how large a part the design of equipment can play in the maintenance of clean and sterile conditions. Thorley (1968) has emphasized the need for simplicity and attention to detail in the design of vessels, valves and joints. Polished stainless steel with continuous welding which has been well smoothed, and has no pockets, crevices or hollow sections should be specified.

Mechanical and in-place cleaning techniques are particularly adaptable to automatic methods, and complete systems of this kind are already being manufactured. Dairies and breweries are rapidly adopting these new methods, and yeast factories will undoubtedly quickly follow suit. Such installations require large numbers of automatic valves with associated pipework, electronics, progamming devices and display panels. Existing pipework often needs to be extensively modified, and safety devices must be introduced to prevent contamination of nutrient feeds by cleaning fluids, or the inactivation of chemical detergents and sterilizing by accidental contact with large volumes of nutrient solutions (Stanton, 1965). The automatic cleaning programme can eventually be integrated with an automatic propagation programme, so that all the individual events taking place in a particular growth vessel are completely under automatic control. One advantage of these improved cleaning methods is reduction in the turn-round time between yeast propagations. The significance of this is discussed more fully in Section III.B.7 (p. 393).

3. *Effluent Disposal*

The main effluents from a yeast factory are air, waste cooling water, spent medium and wash water. The waste air from the propagation vessels has a slightly reduced oxygen and a slightly increased carbon dioxide content, but also contains culture yeast and small numbers of other micro-organisms in suspension. These are generally not considered to be harmful, and the waste air is therefore usually allowed to escape untreated into the open air. The waste cooling water is generally disposed of into the river, the sea or the sewers, but disposal of the spent medium presents a difficult problem owing to the increasing controls which are being universally exercised over trade effluents of all kinds.

Sterilization and re-use of spent medium is possible (Gregr *et al.*, 1960) but although the volume of the waste is reduced, the organic matter content is not necessarily affected to any great extent. Numerous organisms are known which are less fastidious than *Saccharomyces cerevisiae*, and these may be grown on the spent wash, thus reducing its carbon content (Soong, 1957). Such methods have not found general acceptance, probably owing to lack of demand for the product (single-cell protein) or to the danger of uncontrolled growth of bacteria or other organisms and subsequent cross-infection of culture yeast propagations. The same dangers also apply to the installation of small sewage disposal units attached to yeast factories. Unless some saleable by-product can be recovered, all such methods are expensive.

A system for evaporation, incineration and recovery of the resulting heat of combustion have been patented (Cederquist, 1962); this has the advantage that all living organisms are destroyed. This method, however, is only economical where the charge for disposal of effluent through the sewers is high (see also Othmer, 1968).

III. Commercial Production Processes

A. BATCH PROCESS METHODS

1. *Seed Production*

Large quantities of yeast are required as inocula for the final commercial propagation stage when batch processes are used. This "seed yeast" must be prepared as aseptically as possible, preferably commencing with a pure culture. Complete asepsis after the laboratory stage is not necessarily practiced, but precautions are taken to reduce contamination to a minimum. The initial seed stages are frequently conducted under conditions of limited aeration, and are true batch processes in the sense that all the nutrients are present in the vessel at the time of inoculation (Butschek and Kautzmann, 1962). This allows the possibility of the vessel and nutrients (molasses and chemicals) to be sterilized together, and reduces the dangers of infection from pipelines and the use of large volumes of air. The quantity of yeast seed may be increased by passing from one vessel to another of increasing size which contains a larger quantity of sterilized nutrient than the previous vessel. This process is continued until enough seed is obtained to start a propagation in a vessel of working capacity 5–10 m³ (Walter, 1953; Butschek and Kautzmann, 1962).

If desired, the first stage can alternatively be carried out directly in a full-sized vessel (100 m³). The initial dilution of the pure culture is greater but, providing sterility can be maintained, this is an advantage since it allows faster growth. With improvements in air filtration, and in techniques of medium and plant sterilization, fully aerated incrementally-fed processes can sometimes be used with advantage. In the preliminary stages a temperature of 29–32° is maintained, and the pH value is held between 3·5 and 4·5 in order to restrict the growth of unwanted bacteria (White, 1954). Typically an initial inoculum of about 200 g of moist yeast, using 1,000 kg of molasses without incremental feed, will produce 190 kg of yeast in 24 h under aerobic conditions. The progress of this propagation is shown in Fig. 1.

The yeast and spent medium, often containing some ethanol, is then pumped to a larger vessel of 50–200 m³ capacity, diluted with chlorinated or otherwise sterilized water, and further growth stages are conducted. Traditionally the next stage is mildly aerated, but the molasses is not fed incrementally. There does not, however, seem to be any fundamental

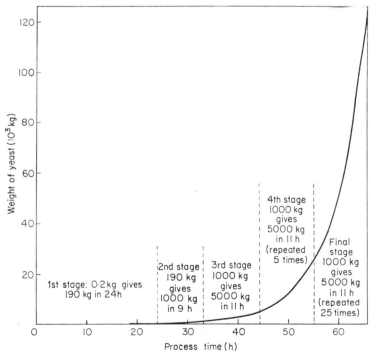

FIG. 1. Progress of yeast growth in commercial production.

reason why full aeration and incremental nutrient feed cannot be applied in this case also. The temperature and pH level are kept within the same range as for the earlier stages. In the example illustrated in Fig. 1, 1,000 kg of yeast are grown from the original 190 kg. This requires about 2,500 kg of molasses by the older process, but only about 1,500 kg if an incremental feed is used.

The third stage, which can be conducted in the same vessel without separation of the yeast from the medium, employs the basic conditions for all the subsequent propagations. For example, 1,000 kg of yeast is grown up to 5,000 kg, using 5,000 kg of molasses which is fed at an exponentially-increasing rate for 11 h, with sufficient aeration at all times to produce maximum yield with no formation of ethanol (Butschek and Kautzmann, 1962). In addition, any ethanol formed in previous stages is re-assimilated (Deloffre procedure, see White, 1954). At this stage the maximum capacity of the vessel with regard to aeration and cooling rates has probably been reached. The yeast crop is concentrated into a cream by means of centrifuges, washed and retained in a special reservoir at low temperature for further use as seed for subsequent propagations. The concentration of yeast in the cream is measured, and the volume required for seeding the next propagation is calculated.

The number of subsequent propagations depends on the total yeast output required, but may be limited by the onset of infection. In the example illustrated in Fig. 1, the yeast from the third stage is divided into five parts, and each is used to seed five propagations of a similar type to that already indicated. The fourth stage produces 25,000 kg of yeast, which after separation and washing provides enough seed yeast for 25 fifth-stage propagations, producing a total of 125,000 kg of pressed yeast for sale.

2. *Final Stage*

The conditions of growth in the early stages outlined above are important from the point of view of producing the maximum amount of yeast, as free as possible from non-viable yeast cells and foreign organisms. In the final production stage the main considerations are high yield and an optimum balance of properties in the yeast. These are obtained by accurate control of nutrient feed rates, temperature, pH level and aeration. The vessels are therefore provided with facilities for controlling these parameters, and in addition are equipped with foam-control and sampling devices.

Large amounts of heat are liberated during aerobic yeast growth; von Fries (1956) gives a figure of 850 kcal/kg molasses. Efficient cooling must therefore be provided by means of internal coils, external jackets or heat exchangers. The passage of cold water through these is usually manually controlled, but automatic means are becoming more general. Resistance thermometers bolted through the side of the vessels provide a rugged and reliable method of measuring and recording temperature. A programmed series of temperature changes is sometimes used in order to provide yeast with the necessary balance of properties (Meyer and Chaffe, 1938).

With experience, good control of the pH level can be attained by a suitable adjustment of the ratio of ammonia to ammonium sulphate in the assimilable nitrogen feed, when it is metered incrementally along with the molasses. However, automatic pH control is preferable. If only ammonia is supplied, the pH value tends to rise, but the upward drift is corrected by the automatic addition of sulphuric acid. The response of this system is rapid, and pH changes due to yeast metabolism are slow, so that a multi-point switching arrangement, using a single meter connected in turn to several vessels, can be used satisfactorily. The pH set point is normally kept low and constant (e.g. 4·0) in the seed stages and early part of the final stage. It may be increased to 5·0 or 6·0 at the end of the final stage in order to reduce adsorption of the coloured constituents of molasses on the surface of the yeast (White, 1954; Rosen, 1968).

Typical fed-batch processes last for 9–15 h, and result in multiplication of yeast by 4–15 times (Butschek and Kautzmann, 1962). The rate of molasses feed during the final period is limited by the maximum aeration available or by the cooling capacity of the plant. A variety of molasses feed patterns may be used, depending on local conditions and preferences (von Freis, 1956; Drews et al., 1962). The duration of the process employed is thus determined by three factors: (i) the amount of seed yeast added; (ii) the growth rate factor employed in programming the molasses feed; (iii) the maximum aeration and cooling available.

At the end of the process the nutrient feeds are stopped, but aeration is allowed to continue for a further 30–60 min in order to "ripen" the yeast. During this period unused nutrients are assimilated, budding cells divide and mature, and the quality of the product is improved (Pyke, 1958; von Fries, 1964). This treatment tends to synchronize the reproductive mechanism of the cells so that they are in a "resting" condition and at the beginning of their budding cycle. Thus, when growth recommences the individual cells are substantially in step. Processes have been suggested which are supposed to accentuate this phasing (Dawson, 1965), but it is doubtful if they would provide any useful improvement in the case of baker's yeast.

3. Separation

When the process is complete, the yeast and spent medium are passed to centrifuges which concentrate the suspension and, after washing the yeast with water, reseparate it. During this procedure some bacteria and finely-divided suspended matter may be removed, in addition to soluble waste material. The yeast cream resulting from this process has a concentration of about 150–200 g dry matter per litre, and is cooled and stored in reservoirs prior to filtering or pressing off the final yeast.

Filtration may be through cloth in presses or, more usually in modern practice, on rotary vacuum filters (Butschek and Kautzmann, 1962). Should the use of the latter result in a soft plastic product, the Mautner (1956) process can be used with advantage. This consists in adding sodium chloride to give a concentration of 0·2–0·6% (w/v) in the yeast cream, thus causing shrinkage of the yeast cells by osmosis. During vacuum-filtration water is sprayed on the layer of filtered yeast on the rotating drum in order to remove excess salt from the immediate vicinity of the cells. The osmotic equilibrium is upset, and the yeast cells begin to swell. This process takes one or two minutes, but filtration only takes a few seconds; thus the cells take up water at the expense of extracellular water in the already filtered yeast. This results in a firm product with a dry matter content between 28% and 30%, which is more acceptable to bakers than yeast with a plastic consistency.

4. *Packaging, Storage and Transport*

The yeast is removed from the presses, or scraped from the rotary filter by adjustable knives, and passed to a hopper where it may be mixed with an emulsifier and water, if desired, in order to facilitate the extrusion and packing processes. The yeast is then either packed into jute bags, compressed and extruded into blocks, or granulated and packed in paper or plastic containers (van Büren, 1964; Schuldt and Seeley, 1966). Schuldt and Seeley state that granulated yeast (called "bulk" yeast by them) is easily resuspended in water, and when packed in 10–20 kg bags is convenient for large-scale automated and continuous baking processes.

The yeast blocks (normally weighing 1 lb or 500 g) are wrapped in waxed paper, packed in cartons and transferred to the cold store for a sufficient time for the yeast to become thoroughly chilled, when it can be safely distributed. Refrigerated vans are being increasingly used for this purpose. The yeast should be at a low temperature when loaded, because at temperatures above 10° yeast will metabolize storage carbohydrate with the production of heat at a significant rate. A large bulk of yeast, because of its self-insulating nature, may produce heat at a rate which causes a continuous rise in temperature, especially towards the centre of the mass. Refrigerated storage ($-7°$ to $0°$) is sometimes practised for the export trade. This allows yeast to be maintained in an active condition for several months (Bailey *et al.*, 1940; Thiessen, 1942). The formation of ice inside the cells must be avoided, since this is likely to cause cell damage (Farrell and Rose, 1967).

B. STEADY-STATE AND CONTINUOUS PROCESSES

1. *Preliminary Considerations*

In the fed-batch processes, such as have been used for the past half-century for the manufacture of yeast, the growing cells can be maintained in virtually a steady-state environment. That is, the concentration of all nutrients and the mean physical parameters such as temperature can be kept constant. The whole system, however, is not strictly in a steady state since either the yeast concentration or the volume of the medium is increasing. In the continuous process, on the other hand, the whole system is in a steady state with medium entering and yeast leaving the system at a constant rate, although there are abrupt changes of environment as the individual cells pass from one stage to the next. It is clear that the change from a fed-batch to a continuous process is not so great as from a classical batch to a continuous process, where a completely non-steady state is replaced by a truly steady-state system.

In spite of the superficial similarity between the fed-batch and continuous systems, and the obvious advantage of increased production rate in the latter case, it has not yet been widely adopted for the manufacture of baker's yeast. Many patents have been taken out (Bührig, 1929; Olsen, 1928), but it is doubtful if any have been operated on a commercial scale except that of Sher (1960).

There are probably several reasons for this apparent lack of interest in continuous processes. Infection problems have frequently troubled yeast factories when using the batch process, but even greater dangers from infecting organisms are inherent in continuous systems (White, 1954). Expensive re-equipment would be necessary to reduce these dangers materially. Further expenditure would also be required on such items as pumps, automatic valves and level controllers for the operation of the process. Incomplete understanding of the theory of steady-state continuous processes has also hindered their application. Until about 1950 the mathematical theory of the process had not been developed, and no published method contained any reference to the need for strict control of growth rate (e.g. Walter, 1953). The term growth rate was sometimes used erroneously, in the sense of the hourly increment of the fed-batch process. It was applied to the building up of the necessary yeast concentration in the system before the onset of steady-state conditions, but there was no clear description or understanding of how conditions in the steady-state phase were controlled (Olsen, 1928; Wirtschaftliche Vereinigung der Hefeindustrie, 1935).

It is now well known that, in a steady-state process at equilibrium, considerations of material balance lead to the conclusion that the growth rate of the organism is the same, to a close approximation, as the dilution rate (Monod, 1950; Maxon, 1955):

$$\mu = \lambda = \frac{\mathrm{d}b}{\mathrm{d}t} \frac{1}{B}$$

where μ = specific growth rate
     λ = dilution rate, or volume fraction of liquid leaving the vessel in unit time

and $\dfrac{\mathrm{d}b}{\mathrm{d}t}$ = rate of formation of new yeast from the constant amount B in the growth vessel.

It follows that the control of dilution rate is essential, and is equivalent to the correct programming of the molasses feed in the batch process. Furthermore, the nutrient feed rate, provided it is independent of the total liquid feed rate, has no effect on growth rate, at least in a single-vessel system, but does affect the steady-state concentration of the organism. In a multi-vessel system, such as that of Sher (1960), the nutrient feed rate may in some circumstances have a minor effect on the growth rate.

In most of the systems described in the literature a dilute nutrient feed of constant concentration is used, and the dilution rate is controlled by varying the rate of feed of this dilute solution. This has the effect of not only altering the growth rate, but also the rate of production of the organism (Herbert et al., 1956). In commercial practice it is desirable to maintain a known constant production rate; this is achieved by feeding concentrated nutrient solution at a constant rate. Any change in growth rate that may be desired, in order to adjust the balance of properties of the yeast, is then produced by altering the feed of dilution water (Olsen, 1961). This has the added advantage that only relatively small volumes of nutrient need to be sterilized by heat, the dilution water being treated by chlorination.

Two methods of maintaining the system in a steady state are available. In the first, dilution water is fed at a predetermined rate, and yeasted medium is pumped out of the vessel at such a rate that the volume of liquid in the vessel remains constant. Transfer of yeast suspension in this way is controlled by a level detector attached to the vessel. By the second method, yeast medium is removed from the vessel at a constant rate and the addition of dilution water is controlled by the level detector.

2. Multivessel Systems

In practice it is found that single-vessel systems give yeast of inferior quality. This stems from the fact that programming of pH level, temperature and growth rate have been found to be advantageous in batch processes. Thus, the ripening period of the batch process must be simulated by the use of an extra vessel which is aerated but not fed. This promotes more complete assimilation of nutrients, and allows the yeast to complete its growth cycle. In theory each change of temperature or pH value in the batch process would require an extra vessel in which these particular growth conditions were reproduced. A satisfactory product can, however, be obtained with quite a small number of vessels; for instance six (Olsen, 1961). An example of the nutrient feeds and dilutions applicable to such a system is given in Table V.

TABLE V. Steady-state Conditions in a Six-vessel System

Vessel	1	2	3	4	5	6
Working volume, l.	20	20	20	20	20	20
Molasses feed rate, g/h	200	200	200	200	200	0
Assimilable nitrogen feed rate, g N/h	4·0	4·0	4·0	4·0	0	0
Assimilable phosphate feed rate, g P_2O_5/h	6·5	0	0	0	0	0
Temperature, deg C	30	30	30	32	33	33
Hydrogen ion concentration, pH units	3·5	4·0	4·5	5·0	5·5	5·5
Transfer rate, l./h	3·0	6·0	9·0	12·0	15·0	15·0
Yeast growth rate, h^{-1}	0·15	0·15	0·15	0·15	0·15	—
Dilution rate, h^{-1}	0·15	0·30	0·45	0·60	0·75	0·75
Retention time, h	6·67	3·33	2·22	1·67	1·33	1·33
Estimated yeast dry matter concentration, g/l.	15	15	15	15	15	15
Total liquid feed rate in nutrient feeds and dilution water, but excluding medium transferred from previous vessel,[a] l./h	3·0	3·0	3·0	3·0	3·0	0

[a] Add to these figures any necessary allowances for evaporation.

A two-vessel system has been suggested by Beran (1958) and Hospodka (1966) on theoretical grounds. However, this system employs such a fast growth rate in the first vessel that it is doubtful if assimilation could have been completed in the unfed second vessel. The use of an unfed vessel appears to be essential for baker's yeast production; this, of course, reduces the efficiency of the system. In a two-vessel arrangement the efficiency is reduced by 50% if the vessels are of equal size, whereas in a six-vessel system the reduction is less than 20%.

In all but the first vessel of a multi-vessel system, yeast is both entering and leaving. The difference between the two rates gives the rate of production of new yeast within the vessel. If now this is divided by the calculated amount of yeast in the vessel, the specific growth rate (μ) is obtained.

3. Build-up

An important practical problem in continuous processes is the rapid efficient build-up of the yeast concentration necessary before steady-state conditions can be imposed. In a multivessel system all vessels must have reached this state before the production of yeast with the correct balance of properties can commence. Two methods are available: in the first, fed-batch processes are carried out in each vessel, so timed as to reach the desired condition simultaneously, when addition of dilution water and transfer of yeast from vessel to vessel commences throughout the system. In the second system, the yeast concentration in the first vessel is built up batchwise from a small inoculum. This vessel is then maintained in a steady-state, and feeds yeast to the other vessels until sequentially they become ready for steady-state operation. The process can be hastened by feeding nutrients to each vessel as it is being filled, so that the amount of yeast is also increasing by growth in every vessel (Olsen, 1961).

4. Steady-state Conditions

In a multivessel system in which several parameters in each vessel can be adjusted, a large number of possible combinations are available. Batch-process conditions may be a guide, but are by no means a certain indication of the conditions required. A simple example has already been given (Table V). With a total molasses feed rate of 1k g/h, the rate of production of yeast dry matter was 225 g/h. The system contained about 1,800 g of yeast dry matter in the steady-state, which was reached in 28 h, the inoculum in the first vessel being 25 g (dry matter). At the end of the run each vessel was drained in turn, the correct transfer rate to the next vessel being maintained with the feed of medium discontinued. This "run-down" phase took 16·55 h, and during this time the rate of yeast leaving the system to the filters was maintained at the same value as during the true steady-state phase.

It will be seen that, although the growth rate in each vessel was the same (0.15 h^{-1}), the transfer rate and the dilution rate increased from vessel to vessel, reaching 0.75 h^{-1} at the end. These high dilution rates may be used to advantage in the control of infection, as indicated below.

5. Infection Control

The maintenance of completely aseptic conditions is at present probably an unrealistic ideal in most yeast propagation plants; consequently the growth of contaminating micro-organisms must be limited, where possible, by control of the process conditions. A study of the growth kinetics of some of the organisms encountered is therefore of interest. In view of this, the yeast *Candida krusei*, a typical fast-growing contaminant of culture yeast propagations, was examined under steady-state growth conditions, using a constant feed rate for the glucose medium but varying the dilution rate from 0.15 to 0.34 h^{-1} by means of an additional water feed. From glucose determinations the saturation constant was estimated to be 5×10^{-5} M, and the maximum growth rate 0.34 h^{-1}. These results compare with values of 3.6×10^{-4} M and 0.37 h^{-1} for culture yeast. Later experiments, using molasses media, suggested that the maximum growth rate may be even higher (Burrows, 1956–67).

The results demonstrate that, when yeast is being maintained under steady-state conditions at a growth rate of 0.15 h^{-1}, the residual sugar concentration is sufficient to support a much higher rate of growth of *C. krusei* owing to its lower saturation constant. This fact is known empirically from practical observation. No adjustment of pH level, temperature or yeast growth rate within the acceptable ranges can materially reduce this advantage; hence the only method of control is by prevention of entry of organisms of this type.

6. *Growth of Bacteria in Multivessel Systems*

With bacteria the case is somewhat different, owing to the relatively large effect of pH on bacterial growth rates. By making a number of simplifying assumptions, equations describing the growth of bacteria in successive vessels of a multivessel system may be calculated. These assumptions are: (i) that no contaminants enter *via* the air or nutrient feeds, all bacterial growth being ascribed to multiplication of the organisms present in the first vessel at zero time; (ii) only one species of micro-organism is present (in practice the most rapidly-growing types would predominate and hence, even in a mixed population, the formulae would give a good approximation to the total infection present); (iii) the rate of growth of the organism is independent of its own population density, and also of the growth rate of the yeast; (iv) the vessels are all of equal volume. The equations increase in complexity from vessel to vessel, so that it was only practical to proceed as far as the fourth vessel in the series. Thus:

if $C_{\bar{n}}$ = concentration of infecting organism in the nth vessel at zero time

    C_n = concentration of infecting organism in the nth vessel at time t

v_n = rate of transfer of medium out of the nth vessel
V_n = working volume of the nth vessel
λ_n = dilution rate in the nth vessel $(= v_n/V_n)$
μ_n = absolute growth rate of infecting organism in the nth vessel
r_n = relative growth rate, i.e. growth rate calculated directly from concentration data without taking the dilution rate into account (whence $r_n = \mu_n - \lambda_n$).

Then, for the first vessel the well-known exponential equation may be applied directly:

$$C_1 = C_{\bar{1}} e^{(\mu_1 - \lambda_1)t} = C_{\bar{1}} e^{r_1 t}$$

For the second vessel, by solving the appropriate differential equation, we obtain, if $V_1 = V_2$

$$C_2 = C_{\bar{2}} e^{r_2 t} + \frac{\lambda_1 C_{\bar{1}}}{r_1 - r_2} (e^{r_1 t} - e^{r_2 t})$$

For the third vessel, in a similar manner, if $V_1 = V_2 = V_3$

$$C_3 = f e^{r_1 t} + g e^{r_2 t} + h e^{r_3 t}$$

where $\quad f = \dfrac{\lambda_1 \lambda_2 C_{\bar{1}}}{(r_1 - r_2)(r_1 - r_3)}$

$$g = \frac{\lambda_2 C_{\bar{2}}}{r_2 - r_3} - \frac{\lambda_1 \lambda_2 C_{\bar{1}}}{(r_1 - r_2)(r_2 - r_3)}$$

and $\quad h = C_{\bar{3}} - \dfrac{\lambda_2 C_{\bar{2}}}{r_2 - r_3} + \dfrac{\lambda_1 \lambda_2 C_{\bar{1}}}{(r_1 - r_2)(r_2 - r_3)}$

Finally, for the fourth vessel, if $V_1 = V_2 = V_3 = V_4$

$$C_4 = C_{\bar{4}} e^{r_4 t} + \frac{\lambda_3 f}{r_1 - r_4} (e^{r_1 t} - e^{r_4 t})$$

$$+ \frac{\lambda_3 g}{r_2 - r_4} (e^{r_2 t} - e^{r_4 t}) + \frac{\lambda_3 h}{r_3 - r_4} (e^{r_3 t} - e^{r_4 t})$$

Although the formulae derived for the second and third vessels represent the sum of two or three exponential expressions, the behaviour of bacteria in these vessels can usually be expressed approximately as a simple exponential when considering the data over a restricted range of time (e.g. 12 h). Hence:

$$C_n = C_{\bar{n}} e^{r_i t}$$

where r_i is an empirical index of growth rate.

It is apparent in all these equations that, if $\mu = \lambda$, r becomes zero and all the exponential expressions reduce to unity; hence

$$C_n = C_{\bar{n}}$$

In other words, if the bacterial growth rate is equal to the dilution rate, no relative increases in infection take place. Since the dilution rate (λ) may be arranged to increase in value from vessel to vessel, so the bacterial growth rate (μ) can be allowed to increase without risking high levels of infection.

This leads to a situation in which it is possible to use relatively high temperatures and pH values in the later vessels, provided conditions of temperature and pH in the earlier vessels can be adjusted to restrict bacterial growth. By applying the above equations to bacterial counts obtained in a laboratory four-vessel system, average bacterial growth rates were obtained for a range of pH and temperature. These were used to design a six-vessel system which was theoretically capable of running indefinitely with low bacterial contamination. The salient features are given in Table VI (Burrows, 1956–67). The weak point of this system is the first vessel, where the dilution rate is lowest. The spontaneous development of a bacterial strain able to grow rapidly at low pH values would, of course, render this system ineffective. Furthermore, it offers no protection against fast-growing micro-organisms which are unaffected by low pH levels, such as certain types of yeast.

TABLE VI. *Bacteriological Control by Dilution in a Six-vessel System*

Vessel	1	2	3	4	5	6 (Ripening vessel)
Temperature, deg C	27	30	31·5	33	33	33
Hydrogen ion concentration, pH units	3·2	3·5	4·0	5·0	5·5	5·5
Yeast growth rate, h^{-1}	0·20	0·20	0·15	0·15	0·15	–
Dilution rate, h^{-1}	0·20	0·40	0·55	0·70	0·85	0·85
Estimated bacterial growth rate, h^{-1}	0·20	0·30	0·45	0·55	0·70	0·70

7. *Batch or Continuous Processes?*

The relative merits of batch and continuous processes for many purposes have been widely discussed, and at first sight the advantages of the continuous system appear to be overwhelming, although various workers have expressed doubts (Hockenhull, 1963; Moss and Saied,

1967). A close examination of the position, particularly as it affects baker's yeast production, reveals several complicating factors. For instance, the economics of the two types of process have been discussed by several authors (Herbert *et al.*, 1956; Deindoerfer and Humphrey, 1959; Miall, 1965; Butterworth, 1967). The conclusion that the running costs of the continuous process are dependent on the dilution rate employed is only applicable where the dilution rate is controlled by the rate of feed of dilute solutions of nutrients. As already mentioned, this is not the case for the continuous process patented by Sher (1960). Furthermore, these comparisons are generally based on the classical batch process, and not the fed-batch system now widely employed for the manufacture of baker's yeast. The feeding of nutrients during the growth period allows higher concentrations of yeast to be reached, and hence in some respects more economical use of vessel space. Another point to be borne in mind is that most multivessel systems envisaged in the literature are unlike those discussed above, in that the second and following vessels are relatively unproductive, being used to take up nutrients not metabolized in earlier vessels.

Conversion to continuous operation requires considerable expenditure, not only on pumps, valves and automatic control equipment, but also on means for improved cleanliness and sterility, such as the treatment of media at high temperatures, and in-place cleaning. Economies in manual labour may be obtained by automation, but additional technicians are needed.

Hepner (1967) has made the valid suggestion that the same amount of effort and automation that has been given to the development of continuous systems, if applied to batch process techniques might have been more rewarding. There is no doubt that the fed-batch process can be made more productive in a number of ways. In a factory where several vessels are employed consecutively for batch propagations, the turn-round time of each can be considerably reduced by in-place cleaning, automatic filling and starting-up, and increased separation capacity, in order to reduce emptying times. Thus, in the example quoted by Olsen (1961), if the turn-round time were reduced from 5·5 h to 3·0 h, between ten and fifteen extra propagations could be carried out in the working week, and production would be increased from 225 to 260 or 270 × 10^3 kg of compressed yeast.

The productivity of the fed-batch process can also be improved by eliminating the relatively unproductive early period when the yeast concentration is low. This can be achieved by increasing the amount of inoculum, so reducing the process time. This change is most effective when the turn-round time is short. Thus, if one unit of seed yeast normally produces eight units in 14 h plus 3 h turn-round time, a ten per cent increase in production is obtained by growing two units up to eight units in 10 h plus 3 h for turn-round. The 270 × 10³ kg quoted above would then become 300 × 10³ kg, after making due allowance for the extra amount of seed yeast that needs to be grown.

Where infection levels are low, further increases can be obtained by adopting a topping-up procedure (Fleischmann Co., 1926). By this method one-third to one-half of the yeast at the end of the batch process is run away to the separators, and the remainder is used to conduct a further propagation of 2 to 6 hours duration. This results in a considerable reduction in the overall turn-round time.

Returning to the continuous process, some of its limitations should be mentioned. In the multivessel process described by Olsen (1961), the comparatively short duration of the propagations produces an awkward discontinuous effect. Thus, 300 × 10³ kg of yeast are produced in 150 h at a rate of 2 × 10³ kg/h, with intervals between runs when no yeast is produced. Therefore, unless there is a reservoir of large capacity between propagation vessels and packing machines, the separators and packing machines are used at less than the maximum efficiency.

The continuous process becomes more attractive as infection is reduced, when runs can consequently be made longer. The fully automated, truly continuous process, running for an indefinite length of time, would undoubtedly offer advantages over batch processes, but is an ideal which still lies in the future. Mutation may, in this case, be the limiting factor, and rapid quality control tests will be required in order to detect mutational changes in yeast quality.

C. DRIED YEAST

1. *Manufacture*

Increasing attention is now being paid to the manufacture of dried yeast owing to its obvious advantages of greater stability and lower moisture content, with resultant reduction in storage and transport costs. These advantages, however, have to be balanced against lower activity, which necessitates a higher rate of usage (on a dry matter basis) in bread-making.

Special propagation methods, differing in some important respects from those used for pressed yeast, are necessary (Oyaas *et al.*, 1948). For example, a special strain of yeast, imparting greater stability during drying, is employed (Merritt, 1957; Pyke, 1958). To the same end the nitrogen feed is restricted to give a nitrogen content between 6·0% and 7·0% on dry matter (Frey, 1957; Thorn and Reed, 1959). Schneider (1954) has recommended the use, towards the end of the growth period, of a higher temperature (36° instead of 30–33°), and a restriction of nutrient feed by adding spent wash containing ethanol instead of fresh molasses. Synchronization of the budding cycle by alternate feed and starvation periods may also be beneficial (Schuldt and Seeley, 1966).

All these special procedures have been arrived at empirically, as the underlying reasons for their use are still not fully understood. Effects on the composition of the cell, or the cell wall, have been described which may throw light on the mechanism. Thus, nitrogen restriction results in an increased total content in the cell of carbohydrates, particularly trehalose, which appears to confer stability during the drying process (Pollock and Holmstrom, 1951). The proportion of lipid in the cell walls is also increased by this means (McMurrough and Rose, 1967), and this may affect cell permeability which is important during reconstitution (see Section III.C.3, p. 397).

After propagation the yeast is separated and filtered by methods that are designed to give the highest possible dry matter. Values between 30% and 38% dry matter are commonly achieved, for instance by the use of filter presses. The yeast mass is then extruded through a screen to give continuous threads, which are chopped finely before being transferred to wire trays (Walter, 1953), to a continuous band (Merritt, 1957; Serwinski et al., 1961) for tunnel drying, or directly into a rotary drum drier (Frey, 1957; Thorn and Reed, 1959). The product from the last process is in the form of hard pellets, which are stated to have excellent keeping quality.

Large volumes of dry air, at temperatures ranging from 35° to 90°, are then passed over the yeast, but the conditions are so arranged that the temperature of the yeast remains low at all stages, a process which is assisted by evaporative cooling. The drying process normally takes from 6 to 24 hours, or until the moisture content has been reduced to about 8%. Fines and agglomerates, if formed, are removed before packing, for which purpose gas-tight tins filled with nitrogen may be used (Oyaas et al., 1948; Thorn and Reed, 1959).

2. Recent Developments in Drying Methods

In addition to the types of driers already in use for the manufacture of active dried baker's yeast, several others have been proposed (see Butschek and Kautzmann, 1962). The direct method, in which heated air is passed over unheated yeast, is invariably used, since the alternative of passing unheated air over heated yeast is much less efficient owing to the low limit which must be imposed on the yeast temperature. In the direct method air at a relatively high temperature can be used, since the latent heat of evaporation supplied by the wet yeast keeps it cool, thus protecting the product from damage.

When the surfaces of crumbled yeast begins to dry the rate of drying decreases since the movement of water towards the surface is limited by diffusion within the particles (Bowmer, 1964). At this stage temperature control becomes important because evaporative cooling is reduced. Other factors controlling the rate of drying, such as air temperature and humidity, also become important, for if the rate of drying is too high the surface layers may become hardened, thus reducing diffusion and hence drying rate.

These conditions can generally be satisfied in conventional tray-, band- and drum-driers, but other types may offer some further advantages. Thus the fluidized bed drier is of simple and cheap construction. Baranowski and Nowicki (1963) have patented a process for low temperature drying in shallow beds, and de la Gandara (1964) has made a detailed investigation of the conditions required. A complete plant utilizing multi-stage fluidized beds for manufacturing active dried yeast is stated to be available (Pressindustria SpA., 1967).

Another attractive method is spray-drying. This system, however, operates most efficiently under short time-high temperature conditions, which give minimum chemical degradation but a high rate of biological damage. It is claimed that certain protective agents reduce the latter effect, so that a fine powdery easily-reconstituted product can be obtained (Toyo Jozo Co. Ltd., 1967). It may also be possible to obtain a viable product by decreasing the temperature and increasing the path length (Birs Brit A.-G., 1962), but a very high tower is required in order to obtain the necessary retention time. As with spray-drying, vacuum- and freeze-drying have not so far proved to be a complete success (Frey, 1957). Other methods, for example mixing moist yeast with a dehydrating agent such as potassium sulphate, silica gel, flour or starch (Bast Hefe- und Spirituswerke GmbH., 1967), aerating a suspension of yeast in corn oil (Johnston, 1959), and drying a layer of yeast foam containing a stabilizing agent (Trevelyan, 1963) have also been patented.

3. *Reconstitution*

It is convenient at this point to discuss the general properties of dried yeast, although this question is not directly relevant to the main section heading. Before dried yeast can be used it must be restored to a hydrated and viable condition. This process of reconstitution is generally carried out by soaking in warm water at temperatures between 38° and 43° (Peppler and Rudert, 1953). If cold water is used, essential constituents leak from the cell and the fermentative activity of the yeast is reduced (Herrera *et al.*, 1956). The extent of leakage may be measured by ultraviolet light absorption of the aqueous extract (Chen and Peppler, 1956), or by measuring the average size of the yeast cells viscometrically (Ebbutt, 1961). These measurements are found to be well correlated with the loss of activity.

Allied to the problem of reconstitution is the question of the relationship between dried yeast properties and moisture content. Using a given method of reconstitution, decreases in moisture content below 10% result in progressively lower fermentative activity (Oldhaver, 1912; Mitchell and Enright, 1957). Moreover, these workers found that, if the moisture content of the over-dried yeast was increased by vapour rehydration, the original higher activity was recovered. Since the drier yeast has a better keeping quality (Merritt, 1957; Thorn and Reed, 1959; Schuldt, 1963), a dry matter content which gives a suitable balance must be chosen.

The above considerations have an important bearing on the marketing of mixtures of flour and dried yeast ready for home-baking. Normally flour has a higher relative humidity than dried yeast at 8% moisture: hence the yeast is rehydrated in the pack and deteriorates rapidly. This difficulty has been overcome by the use of specially dried flour, and by the addition of antioxidants (Universal Foods Corporation, 1966).

Further light on the nature of the reconstitution process is given by the surprising results of Echigo et al. (1966), who found that, not only could the damaging effect of rehydration in cold water be prevented by heating the dry yeast to about 40° just before adding to the water, but that this beneficial effect persisted even when the temperature of the yeast was reduced to 20° before being added to the cold water. The protective effect of the high temperature gradually diminished with time, having completely disappeared in about five hours. A reversible physical effect, reducing cell wall permeability, appears to be involved, rather than a chemical or biochemical process.

4. Comparison with Pressed Yeast

In using dried yeast in the bakery it must be added at between 40% and 50% of the weight normally used for pressed yeast (Thorn and Reed, 1959). Calculated on a dry-matter basis this means that active dried yeast has about 65% of the activity of the pressed yeast. This lower activity is due to a number of causes: (i) the use of special yeast strains; (ii) different method of propagation; (iii) loss of viability during drying; (iv) loss of enzyme activity during drying; (v) loss of enzyme activity due to leakage of co-factors during reconstitution; (vi) excretion during reconstitution of glutathione which adversely affects dough structure and loaf volume.

D. PROBLEMS IN AUTOMATIC CONTROL

1. Plant and Process Control

A large number of operations in yeast manufacture are amenable to automatic control, and many of these have already been mentioned. Some are concerned with ancillary processes, such as in-place cleaning, molasses preparation, yeast separation and yeast drying; others with the yeast growth process itself, and these require special consideration. In common with developments in other industries, automatic process control is being evolved and applied to microbiological systems, including yeast manufacture. Improved methods of measuring temperature, pH level, flow-rate, oxygen concentration, sugar concentration, specific gravity and other parameters are being applied, with the result that more precise control of the process can be exercised and a product of improved and more constant quality obtained.

At this stage of development, which should be strictly termed "plant control" and not "process control" (Pamely-Evans, 1968), two points should be emphasized. Firstly, the values of the controlled variables required to give the optimum yield and quality of product must be known, so that the set points of the various controllers can be fixed or programmed to the best advantage. For some of these, such as temperature and pH value, the information may be already available but for others, such as oxygen tension, more research is required before a satisfactory control can be applied.

Secondly, the process must be sufficiently well understood that the appropriate measurements can be made for the control of the parameter in question. Thus, there is some doubt as to whether the formol titration is a good indication of assimilable nitrogen requirement in a growing yeast culture. Nevertheless, control based on this measurement has been attempted (Sher, 1961). Similarly, oxygen concentration has been taken as a measure of assimilable carbon requirement and used to control molasses feed (Ohashi, 1958). This relationship is indirect, and is probably affected by other factors such as the nutrient balance and the specific growth rate of the organism. The suggestion of Fuld and Dunn (1957) that the refractometric determination of residual sugar in the medium should be used to control the sugar feed appears to be more appropriate. However, practical difficulties such as the non-specificity of the normal method of measurement make the technique unsuitable.

This method, taken to its logical conclusion, requires a complete knowledge of all the factors that affect the quality of the product, and absolute control of these parameters. In yeast manufacture there is now a fairly complete understanding of the effects of most of the independent variables. The main difficulty lies on the control side. This arises because the composition of the nutrient medium (molasses) is difficult to measure and control. Possible methods by which this can be effected have been mentioned previously (Section II.B, p. 368). A further difficulty lies in the variability of yeast strains. This problem is now more fully understood, and improved maintenance and testing methods have brought it under a much greater degree of control.

The next stage of development will be "closing the loop" to provide true process control. This requires methods of measurement of yeast quality which can be used to feed back information to a computer, which would then adjust the appropriate set points in the controllers, or endeavour to discover by trial and error which set points to adjust.

Some automatic methods of measurement of yeast quality are available (Sher, 1961), but require further development. The problem of using these measurements for process control presents difficulties since several process variables can affect a single yeast property. Furthermore, such obscure factors as yeast strain variation or growth inhibitor concentration might be difficult to counteract by normal process control methods. It seems at present that these possibilities must be detected and removed by human intervention before the process commences and the computer takes control.

Increased knowledge of nutritional requirements, enzyme adaptation, methods of preservation of pure strains, and methods of infection control have enabled a gradually increasing degree of control to be exercised on the growth and properties of yeast. For example, in experimental yeast propagations variations of less than 5%, and often as low as 2% including testing errors, can be achieved. Laboratory experimentation using statistical procedures can relatively quickly establish optimum conditions of growth which are applicable to new strains or revised quality requirements. Thus, the concept of unpredictability and high variability in large populations of unicellular micro-organisms is now outmoded. It is concluded that plant control based on predetermined set points will give adequate control until more sophisticated computer-operated feed-back process control methods become possible.

2. Nutrient Feed Control

The factors known to affect yeast yield and properties are temperature, pH value, growth rate, oxygen supply rate, and the nitrogen and phosphorus contents of the yeast. Temperature and pH control are straightforward, yeast growth can be controlled by programming the molasses feed, and oxygen supply rate by oxygen tension and ethanol concentration measurements as discussed previously. This leaves the nitrogen and phosphorus contents of the yeast as major problems: the value of these parameters will depend not only on the feeds of assimilable nitrogen and phosphate, but also on the total amount of yeast formed. This, in its turn, will depend on the percentage yeast yield and on the assimilable carbon content of the molasses. Possible methods of controlling these parameters have already been mentioned (Section II.B, p. 368).

IV. Process and Quality Control

A. PROCESS CONTROL TESTS

Various checks, tests and analyses are carried out, for the purpose of process control, on the factory floor and in the associated works laboratory. Those concerned directly with the growing of the yeast are: analyses of raw materials; determination of pH; temperature; aeration rate; formol titration; yeast concentration; yeast yield; yeast properties, i.e. dry matter, nitrogen and phosphorus contents, fermentative activity, keeping quality, consistency, colour and suspendability; microbial contamination in the air, the nutrient feed solutions and in the separated yeast.

1. Composition of Nutrients

Frequent checks on the concentration of solutions of nutrient chemicals are required, since errors would severely affect the process. Specific gravity determination by hydrometer is generally all that is required. If ammonia solutions are used a simple titration is preferable.

The difficulties of determining available sugar have already been discussed (Section II.B, p. 368). The well known Lane and Eynon method for the determination of reducing sugars after hydrolysis of sucrose is, however, satisfactory for most purposes if its shortcomings are realized. A useful rapid check may be obtained on the degree of dilution during the preparation of molasses by the hydrometer test.

2. Formol Titration

The formol titration is a convenient means of determining the residual assimilable nitrogen content of the medium during yeast growth. A titration is conducted before and after the addition of formalin to the sample (Grant, 1955). Ammonia and α-amino compounds behave as acids under these conditions, and their concentration is easily measured. Although this test is not recommended for the control of the feed of ammonia or ammonium salts, a constant or falling value provides a useful indication of satisfactory yeast growth.

3. Yeast Concentration and Yield

The periodical measurement of yeast concentration provides a method of following growth, and also gives a measure of yeast yield when carried out at the end of the growth period. Turbidimetric and

14—Y. 3

absorptiometric methods (Thorne and Bishop, 1936) are convenient, and may be operated continuously on a sample line (Sher, 1961), but interpretation of the results in terms of yeast dry matter is difficult and subject to inaccuracy, due to the fact that light absorption and light scattering by the yeast suspension depend on the size of the yeast cells, which in turn depends on the yeast strain and the growth conditions. Suitable calibration checks can probably overcome these disadvantages to some extent. Variations in background absorption due to colour and colloid content also present difficulties, although suitable differential techniques might overcome these (Thorne and Beckley, 1958).

Another method often adopted is to take a representative sample of the yeast growth medium, filter this under a standard pressure, weigh the yeast cake so obtained, and determine its dry matter content. In order to calculate the yeast yield, the total volume of medium plus yeast cells in the vessel must be found. This may be done by means of a calibrated dip-stick or by the response of a pressure-measuring device, such as a transducer, situated at the lowest point in the vessel. In the latter case, the specific gravity of the medium in which the yeast is suspended must be known. This method is capable of a precision of about $\pm 2\%$.

4. Dry Matter, Nitrogen and Phosphate

Dry matter, nitrogen and phosphate determinations on yeast samples are generally carried out in the factory control laboratory by conventional techniques. Rapid methods of dry matter determination, such as the specific gravity method (Atkin et al., 1961), or the Karl Fischer procedure, may give an earlier result than the more usual method, but require more skilled operators and more operator time. Oven drying of a 2-g sample at 105–110° for 10–16 h is usually satisfactory, although the type of oven, the internal air circulation and the number of samples dried at the same time must be taken into account.

Owing to the presence of compounds resistant to normal Kjeldahl digestion for the determination of total nitrogen in yeast, it is advantageous when carrying out analyses to use a high temperature and a mercury catalyst (Baker, 1961). A higher result is obtained by the use of the Dumas method (Farmer et al., 1967). The Coleman Nitrogen Analyzer (Sternglanz and Kollig, 1962) is simple to operate and provides a convenient way of applying the Dumas method to routine samples. Its reproducibility is at least as good as that of the Kjeldahl technique, and single results can be obtained within 20 minutes. Nevertheless, for large numbers of samples the Kjeldahl method is probably more suitable. Total phosphorus content is determined by digesting with perchloric acid to destroy organic matter, and measuring the colour formed in the molybdenum blue or vanadomolybdate yellow reactions by an absorptiometric procedure.

The Technicon Autoanalyser, with the incorporation of a digestion unit may be used for automatic total nitrogen determinations, and possibly also for total phosphate (Marten and Catanzaro, 1966). This instrument could feasibly be linked to an on-line analysis system to which cream samples of standard dry matter content would be supplied on a continuous basis. The result of these chemical tests may be used for adjusting and maintaining process variables such as chemical feed additions and filtration conditions.

B. YEAST PROPERTIES AND QUALITY CONTROL TESTS

The quality control tests described in this section measure the properties of the yeast which are of interest to the consumer. They are also of value to the manufacturer, since he must adjust his processes to produce yeast with optimum properties.

1. *Yeast Activity*

The fermentative activity of yeast, and its behaviour in the bread-making process, are determined by methods which may involve test-bakery or small-scale bakery equipment, and normal bakery techniques. When large volumes of dough are handled, however, temperature control is difficult and results are subject to an appreciable error. Owing to the unwieldy and time-consuming nature of such methods, and the need for the employment of highly skilled bakers in order to achieve the best accuracy, they are not applicable to routine determinations on large numbers of samples, but must, however, be used for certain special purposes, notably to correlate laboratory tests with bakery practice.

Many laboratory methods for the routine determination of yeast activity have been proposed, and a critical appraisal of these has been given by Burrows and Harrison (1959). These authors developed a method called the Fermentometer Test, based on gas volume measurement, in which the main source of error, inadequate temperature control, is much reduced by using a small sample of dough which is mixed and fermented in a bottle immersed in a water bath which is temperature controlled to within 0·05°. Up to 24 samples can be tested in a day by a single operator. In order to correct for the effect of ambient temperature on water vapour pressure corrections in an effort to obtain the highest possible accuracy, the published method may be modified by the incorporation of a tube containing anhydrous calcium sulphate between the fermentation jar and the measuring burette in order to dry the gas thoroughly before measuring its volume. This necessitates the replacement of the original aqueous calcium chloride solution (Schultz *et al.*,

1942) in the burette and reservoir by a light oil. Gas-volume measurements give a reliable indication of dough-raising power in spite of the fact that a proportion of the gas produced within the dough escapes. Thus, Mitterhauszerová and Sedlárová (1966) found a strong correlation between dough volume and gas volume in comparative tests.

A major source of variability in any yeast activity test based on dough fermentation is a lack of consistency in the quality of the flour. The use of a flour which has been specially blended to conform to a predetermined specification can do much to reduce this source of error, but a more easily standardized chemically-defined substrate, such as that of Atkin et al. (1945), would be preferable. The use of such a medium, in conjunction with an inert filler (e.g., maize starch or gluten) shows some promise (Matsumoto et al., 1956; Schultz, 1965), but further work on this subject is required.

Automatic activity determination using gas volume methods is quite feasible, and may be incorporated in a process stream if yeast concentration is known and standardized (Sher, 1961). In the instance quoted a liquid medium was used and carbon dioxide evolution was measured by infrared light absorption. The advantages of an on-line continous recording of yeast activity would outweigh any disadvantage of incomplete correlation of the results with those using dough fermentation methods. The special virtue of such a scheme would be that the property being assessed was closely related to that required by the consumer.

2. Osmosensitivity

In the presence of high concentrations of salts and sugars, the yeast cell shrinks owing to osmotic effects. Many ions and large un-ionized molecules do not penetrate the cell wall and hence, unlike some plant and animal cells, no recovery of volume takes place while the cells are in such a solution (Oskov, 1945). Under these conditions the fermentative activity of the yeast is severely reduced. The precise reason for this effect is not, however, known. Possible reasons are: (i) dehydration of enzyme systems in the cell wall; (ii) loss of water from the cell interior; (iii) inhibition of transport of substrates into the cell.

Osmosensitivity is of importance in the preparation of sweet-bun doughs, since an initial sucrose concentration of up to 1.0 M is often reached. In the early stages of dough fermentation a reduction of 70% in the rate of fermentation may sometimes be sustained (Sato and Tanaka, 1961). The fermentometer test (see previous section) may readily be adapted for measuring the response of yeast in sweet doughs by adding a prescribed amount of sugar to the flour before the addition of water. The particle size of the sugar and the degree of mixing must be standardized, owing to the necessity of controlling the rate of sucrose hydrolysis. Factors which control yeast propagation, such as growth rate (White and Munns, 1955), can influence the osmosensitivity of the cells, but genetical factors are also important.

3. *Keeping Quality*

Before tests for keeping quality are discussed, the extent of our knowledge of this property will be reviewed. When yeast is separated from its nutrient medium it quickly assumes a resting state. In this condition its fermentative activity decreases with time at a rate which is dependent on its metabolic state and on the ambient conditions. This loss of activity may be due to: (i) loss of viability; (ii) intrinsic decrease of enzyme activity of the living cells; (iii) inhibition or lysis by its own products of metabolism or those of infecting organisms. It is found in practice that in an uncontaminated and correctly-grown yeast factors (i) and (iii) have a negligible influence (White and Munns, 1955). Little is known of the main cause, which is a decrease of enzyme activity at the cellular level. Keeping quality, as well as being dependent on the strain of yeast, is influenced by the method of preparation. Thus, anaerobically-grown yeast has a much poorer keeping quality than that which is grown aerobically (White and Munns, 1955). Generally speaking keeping quality decreases as fermentative activity increases, if other factors are unchanged (Bergander and Bahrmann, 1957).

The time-course of deterioration in activity is not linear, as indicated

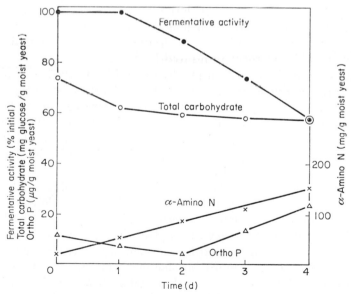

FIG. 2. Changes taking place during yeast storage.

by the curves in Fig. 2. In the experiment illustrated samples of compressed yeast were incubated at 25° in small closed bottles, samples being removed each day for testing. The activity was determined by the fermentometer test (Burrows and Harrison, 1959; see Section IV.B.1, p. 403), and total carbohydrate on the whole yeast by the anthrone method (Trevelyan et al., 1952). Orthophosphate and α-amino nitrogen were determined on the cold water extract by the methods of King (1932) and Yemm and Cocking (1955) respectively. Under the prevailing conditions amino acids were excreted by the yeast at a constant rate, but the extracellular orthophosphate originally present first decreased and then increased. Total carbohydrate decreased sharply at first and then reached a constant value. During these changes the activity remained unchanged for one day, and then decreased rapidly (Burrows, 1956–67). Several workers have observed excretion of amino acids (Sjoblohm, 1965) and metabolism of intracellular carbohydrate (Sjoblohm and Stolpe, 1964; Takakuwa, 1963) during storage of baker's yeast. Furthermore, Takakuwa (1963) has reported a constant and then decreasing time-course for fermentative activity, accompanied by a decreasing and then constant time-course for carbohydrate content. These changes appear to indicate that high activity is maintained whilst certain intracellular carbohydrate fractions are being metabolized. When these are exhausted limited proteolysis sets in (Drews, 1936), with a consequent reduction in enzymic activity. The small but constant rate of excretion of α-amino nitrogen (Fig. 2) suggests that this fraction does not derive from proteolytic break-down of protein. Under these mild conditions decomposition of enzymes in the short time available would scarcely produce sufficient free amino acids or simple peptides to react to ninhydrin.

Using bacteria, Mandelstam (1957) and Willetts (1967) showed that protein turnover during growth was only 0·6% /h, whilst during starvation it was 5%/h. The proteolytic system therefore appears to be latent during growth, and to become active only while resting. The high rate of deterioration associated with very active yeasts may be due to a high general level of enzyme activity, resulting in both high fermentative and proteolytic activities.

Numerous methods for the determination of keeping quality have been proposed. An approximate test sometimes used depends on the fact that excretion products are accompanied by water, and hence a gradual softening of the yeast takes place during storage. This change in consistency can be measured by means of a penetrometer (Choisnev and Auerman, 1963). A more satisfactory method is to measure the fermentative activity by a standard procedure before and after incubation under controlled conditions. Various methods of recording the results are possible: (i) final activity; (ii) loss of activity; (iii) loss of activity as a percentage of the initial activity; (iv) time taken for the activity to reach a predetermined fraction of the initial value. Furthermore, various conditions of time, temperature, humidity and rate of gas exchange may be specified. Rapid tests, say at 35° for one day, or even at 40° for a shorter time, give reasonably satisfactory correlations with more realistic storage temperatures near ambient. Mitterhauszerová (1966) found a good correlation between the results of normal keeping quality tests and the amount of amino acids excreted when a yeast suspension was incubated for four hours at 50°. De Armon *et al.* (1962) have shown that accelerated (high temperature) storage tests give a good indication of behaviour of micro-organisms at lower temperatures.

Oxygen and carbon dioxide concentrations in the immediate vicinity of the resting cells appear to be of considerable importance. Thus, granulated compressed yeast is more difficult to store than consolidated blocks. In each case a certain amount of ventilation appears to be necessary, possibly to allow a reduction in carbon dioxide concentration by diffusion (Schuldt and Seeley, 1966). The effect of the gaseous atmosphere on the deterioration of activity may conveniently be demonstrated by incubating a small amount of yeast in sealed bottles of varying capacities. In a particular experiment the portion of the bottle not occupied by yeast contained normal atmospheric air. Under conditions in which less than 300 ml of air per 0·5 g yeast was present,

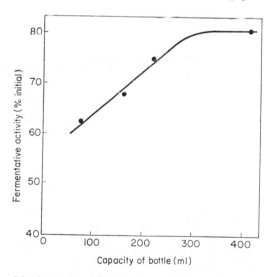

FIG. 3. Relationship between fermentative activity of yeast after storage and volume of sealed air-filled containers.

the rate of deterioration increased as the volume of air decreased (see Fig. 3). This effect may again be associated with endogenous carbohydrate breakdown, since more intracellular carbohydrate is metabolized under aerobic than under anaerobic conditions (Eaton, 1960, 1961). Alternatively, carbon dioxide or other gaseous metabolic products may have an accelerating effect. The reported adverse effect of oxygen (Schuldt and Seeley, 1966) may be due to a rise in temperature, consequent on endogenous metabolism causing more rapid breakdown of intracellular material.

The choice of a suitable test procedure depends on whether the result is to be used for process or quality control. For the former purpose a rapid method, such as 24 h at 35° (Gorokhova, 1962) or even a few hours at 40°, would be quite suitable. However, for quality control, conditions as near as possible to those encountered in practice are preferable. Thus, incubation at 25° for seven days is satisfactory and convenient. Shorter times at rather higher temperatures may require measurements to be made during week-ends, which raises staff problems. Humidity, air circulation rate, size of samples and method of packing must all be stipulated if reproducible results are to be obtained.

4. Consistency

The consistency of the finished product is largely determined by the conditions under which the yeast is finally separated from the wash water, and by the method of packing. As White (1954) pointed out, the determination of total moisture is inadequate as a measure of consistency, since the larger part of the water is held within the cells and therefore does not affect this property. Furthermore, the intracellular water content varies with process conditions and yeast strain. Consistency measurements are, therefore, essentially tests for extracellular water or the ratio of the volume of the cells to the total volume of the compressed yeast sample. Determinations based on the measurement of the proportion of extracellular water utilize the dilution effect obtained when a non-absorbed substance is added to the moist yeast. A recent method (Watson and Levinson, 1967) involves the addition of an "azo-peptone" reagent to the yeast, and the measurement absorptiometrically of the dilution of the coloured compound by the extracellular water. To reduce osmotic effects and subsequent changes in the amount of intracellular water, high molecular weight compounds must be used. Consistency may also be estimated by the use of a penetrometer-type of instrument, which measures, for instance, the depth to which a standard probe sinks when applied under controlled conditions. Alternatively, the time taken for the penetration to a defined point of weighted cutting wires may be used. The viscosity of a standard dilution of the yeast sample indicates the cell-volume ratio, and may be measured by the capillary-flow method (White, 1954) or by a dynamic viscometer (Ebbutt, 1961).

5. *Colour*

A light-coloured yeast is held to be desirable; some form of colour measurement is therefore worthwhile since it indicates the efficiency of the molasses clarification and yeast-washing processes. A commercial reflectometer may conveniently be used, the reflectance from the surface of the yeast being measured photometrically in comparison with that of a standard white reflecting surface. Otherwise, simpler methods by which the sample under test is compared with a standard colour chart may be employed.

6. *Resistance to Suspension*

The property termed here *resistance to suspension* is sometimes referred to as insolubility or, in the reverse sense, "suspendability". It indicates the proportion of the sample which does not disperse completely in water when a suspension is prepared. This property is of importance to the baker, since he requires a yeast which will suspend easily and will not leave a residue of granular material which will separate out rapidly and cause difficulty when handling yeast cream, or when dispersing the cream in the dough. The literature is strangely devoid of references to this property, which does not appear to have any close affinity with flocculence in brewer's yeast, to which there are a large number of references (see review by Rose, 1963). Flocculence occurs whilst the yeast is in suspension, whereas resistance to suspension appears only after the yeast has been separated by filtration. Generally, the greater the pressure applied during separation the greater is the tendency to form granular agglomerates. The effect therefore appears to be due to short-range forces which only come into play when the yeast cells are forcibly pressed together. Other factors known to affect the severity of resistance to suspension, are: (i) genetical make-up; (ii) environmental influences, e.g. the conditions under which the yeast is grown and stored after manufacture; and (iii) the ionic strength of the suspending medium (Hedrick and Feren, 1963). Resistance to suspension often increases during storage of yeast cultures for long periods of time, thus single-cell isolates from old cultures may show a wide range of values; consequently, controlled reselection will usually give an improved strain. Some types of yeast are known to show negligible resistance to suspension however adverse the environmental factor may be. Such strains have tended to be selected for commercial use.

A knowledge of the effect of storage conditions on resistance to suspension is necessary before a reproducible test can be applied. Time, temperature and the moisture content of the sample are important in this respect. Based on practical observations, and particularly on the fact that a 5% sodium chloride solution when used as the suspending medium considerably increases the resistance to suspension, a standard test was devised. The use of sodium chloride tends to minimize differences in ionic strength caused by soluble material in the yeast samples. The crumbled yeast (100 g) is sprinkled on the surface of 400 ml of mechanically-stirred 5% sodium chloride solution in a 600-ml beaker, and stirring is continued for three minutes. After standing for one minute the yeast suspension is decanted, water (50 ml) is added and after swirling, the supernatant is again decanted and the residue is transferred to a measuring cylinder. After two minutes the volume of the compacted yeast residue is measured (S. Burrows, unpublished observations).

The conditions of the test exaggerate the resistance to suspension. An improved understanding of this property has enabled yeast strains with a tendency towards resistance to suspension to be detected.

7. *Infection*

Tests for infecting micro-organisms should be carried out regularly, not only on the finished yeast but at intermediate stages during manufacture, and also on raw materials, such as chemicals, molasses, water and air. The organisms encountered are bacteria of many types, wild yeasts (i.e. any yeasts other then the strain being grown) and moulds. The detection, identification and counting of these organisms in yeast factory samples has been dealt with in some detail by Fowell (1965, 1967). The disadvantage of most of the tests so far devised is that they take 12 to 24 h before an estimate can be obtained, and consequently immediate corrective action is not possible. As a guide to planning subsequent cleaning programmes and plant modifications, however, they are invaluable. As indicated earlier (Section III.B.6, p. 391), a more detailed knowledge of rates and conditions of growth of the commoner micro-organisms is also of value in controlling their proliferation.

V. Associated Laboratory Activities

In addition to a works laboratory for carrying out process and quality control tests, laboratory facilities for culture maintenance, pure culture preparation and experimental propagation are required if yeast quality is to be maintained and improved. Furthermore, well-equipped mechanical, chemical and electrical engineering facilities must be available in order to maintain and improve the efficiency of the factory operations.

Culture maintenance and some aspects of factory operations have already been discussed. Other laboratory activities concerned with strain improvement and optimization of propagation conditions will now be considered.

Because of its easy availability, more is probably known about yeast than almost any other living unicellular organism. In spite of this, however, empirical methods must still be used for much of the development work connected with the improvement of processes and strains for the production of baker's yeast. An important aspect of such research relates to scaling up and down. This involves mainly the consideration of different total numbers of cells and consequently, in contrast to the situation with many chemical reactions, the scale of operation is immaterial. Probably the only exception is in the case of aeration rate and agitation where, as discussed previously (Section II.D, p. 374), many factors come into play, and the effects of the scale are difficult to predict. If, however, oxygen concentration or partial pressure is measured even this difficulty disappears. No aggregates of cells are formed during normal growth, and hence certain difficulties of restricted diffusion encountered in the growth of moulds during the manufacture of antibiotics are not evident.

As a consequence of these considerations, the scale of working in an experimental plant depends only on convenience and the size of samples required for testing. Two experimental plants are used by the author.

1. *Twenty-Litre Plant*

Eight 50-litre vessels with a working capacity of 20 litres, fitted with resistance thermometers and hot and cold water systems for temperature control, spargers for aeration, an external sample-line circulation which is also used for pH measurement and control, and automatic foam control, are provided with a supply of filtered air. The rate at which air enters each vessel is measured by a rotameter. Each vessel is equipped with two feed lines for nutrients, one for molasses or other carbon source, and the other for assimilable nitrogen and phosphorus. Medium is fed from measuring burettes through small positive displacement pumps, the strokes of which are adjusted by cams which revolve once during the complete propagation period. The cams can be cut to give any desired programme of feed rates. At the end of the propagation the yeast is separated on large suction filters.

The arrangements outlined allow individual control in each vessel of temperature, pH value, growth rate and aeration. This provides a convenient plant for strain selection, process development and the study of interactions between strain and process conditions by batch methods. Continuous propagation can be carried out by adding water feed lines to each vessel. Transfer from vessel to vessel, or vessel to separator, is effected by inserting an external weir in the sampling line and returning the sample to the following vessel, or to the filter, as shown in Fig. 4. Any arrangement of single or multivessel systems can be set up in this way; the normal procedure, however, has been to employ two four-vessel systems simultaneously. By this means different yeast strains and process conditions can be compared under closely controlled conditions.

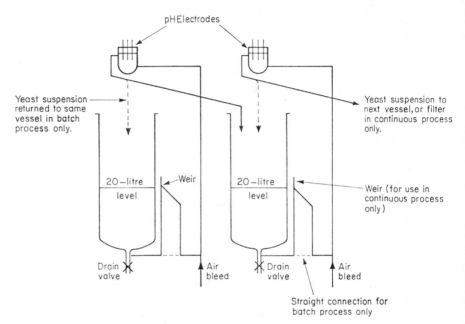

FIG. 4. Diagram of small-scale yeast propagation plant.

2. Two-Hundred-Millilitre Plant

The smaller unit, designed primarily for preliminary screening of yeast strains under a single set of conditions, consists of 16 glass vessels (43 cm × 5 cm) set in a circular thermostated water bath. The vessels are provided with sintered-glass aerating tubes, and are autoclaved with 100 ml of medium which contains phosphate and molasses. Glass electrodes, connected to a manual switching unit and pH meter, provide pH control by addition of sodium carbonate solution. A buffered medium, containing molasses, ammonium sulphate, sodium bicarbonate and an antifoam agent, is fed incrementally by means of stainless-steel syringes. The syringe barrels are fixed and the plungers move upwards at an exponential rate which is controlled by a cam. The system is normally arranged so that 95 ml of medium, containing 9·5 g of molasses, is fed at an exponentially increasing rate for 21 h. This apparatus functions overnight without attention, and provides sufficient yeast for fermentative activity tests and analyses for dry matter and phosphate.

These experimental propagation units are used for the investigation of process modifications, the effect of nutrient composition, process control methods, culture maintenance checks, strain improvement and hybrid selection. A variety of experimental designs are employed, and for the assessment of the results a computer is sometimes required.

VI. Concluding Remarks

In the preceding pages an attempt has been made to indicate how the prevailing atmosphere of rapidly advancing technology has affected the yeast manufacturing industry. Advances are being made continually in plant design, air and nutrient medium sterilization, plant and process control, and testing methods. These, together with strain selection, have led to rapid strides being made in the improvement of yeast quality and purity. Further advances in these and other directions can be expected in the future.

This is not intended to be an exhaustive account of present day practice in yeast technology, but rather a review of the underlying principles, and a critical survey of their practical application. The views expressed are the author's own, and it is hoped that they will prove stimulating and lead to a wider exchange of information, ideas and opinions in this field.

References

Aiba, S. and Toda, K. (1967). *Process Biochem.* **2** (2), 35–40.

Aiyar, A. S. and Luedeking, R. (1966). *Chem. Engng Prog. Symp. Ser.* **62** (69), 55–59.

Armon, I. A. de, Orlande, M. D., Rosenwald, A. J., Klein, F., Fernelius, A. L., Lincoln, R. E. and Middaugh, L. P. R. (1962). *Appl. Microbiol.* **10**, 422–427.

Atkin, L. (1950). *In* "Yeasts in Feeding", (S. Brenner, ed.), pp. 76–78. Milwaukee, Wisconsin.

Atkin, L., Moses, W. and Gray, P. P. (1961). *Wallerstein Adv. Beer Quality* II, pp. 153–155.

Atkin, L., Schultz, A. S. and Frey, C. N. (1945). *Cereal Chem.* **22**, 321–333.

Axford, D. W. E. and Elton, G. A. H. (1960). *Chemy Ind.* 1257–1258.

Bailey, L. H., Bartram, M. T. and Row, S. C. (1940). *Cereal Chem.* **17**, 55–66.

Baker, J. C. (1954). *Bakers' Weekly* **161**, 60.

Baker, P. R. W. (1961). *Talanta* **8**, 57–71.

Baranowski, K. and Nowicki, B. (1963). Polish Pat. 46,567; from *Chem. Abstr.* 1964, **61**, 7666g.

Bast Hefe- und Spirituswerke GmbH. (1967). Brit. Pat. 1,089,686.

Becze, G. de (1940). U.S. Pat. 2,199,722.

Becze, G. de and Liebermann, A. J. (1944). *Ind. engng Chem.* **36**, 882–890.

Benesch, R. A. (1954). *Fd Mf* **29**, 305.

Beran, K. (1958). *In* "Continuous Cultivation of Micro-organisms: a Symposium". Czechoslovak Academy of Sciences, Prague, pp. 122–156.

Bergander, E. and Bahrmann, K. (1957). *Nährung* **1** (1), 74–87.

Bergander, E. and Konigstedt, I. (1965). *Nährung* **9**, 229.

Birs Brit A.-G. (1962). Brit. Pat. 886,533.

Borkowski, J. D. and Johnson, M. J. (1967). *Biotechnol. Bioengng* **9**, 635–639.

Bowmer, G. C. (1964). *Chemy Ind.* pp. 1638–1648.

Brady, B. L. (1962). *In* "Proceedings of Specialist Conference on Culture Collection", (S. M. Martin, ed.), pp. 110–113. Ottawa.

Braun, A., Fische, E. and Rosenburg, F. (1931). Brit. Pat. 346,361.

Bristol-Myers Company, Delaware (1964). Brit. Pat. 960,099.

Bührig, W. A. F. (1929). U.S. Pat. 1,730,876.

Büren, H. J. van (1964). Brit. Pat. 966,954.

Burrows, S. and Fowell, R. R. (1961). Brit. Pats 868,133 and 868,621.

Burrows, S. and Fowell, R. R. (1968). Brit. Pat. 1,135,418.

Burrows, S. and Harrison, J. S. (1959). *J. Inst. Brew.* **65**, 39–45.

Bunker, H. J. (1947). *Chemy Ind.*, pp. 203–205.

Butschek, G. and Kautzmann, R. (1962). *In* "Die Hefen. II. Technologie der Hefen", (F. Reiff, R. Kautzmann, H. Lüers and M. Lindemann, eds.), pp. 501–610. Verlag Hans Carl, Nürnberg.

Butterworth, D. (1967). "Biology and the Manufacturing Industries". Academic Press, London and New York.

Calderbank, P. H. (1959). *Trans. Inst. chem. Engrs* **37**, 173–185.

Cederquist, K. N. (1962). Swedish Pat. 191,762.

Československá Academie Ved (1966). Brit. Pat. 1,045,930.

Chen, S. L. and Peppler, H. J. (1956). *Archs Biochem. Biophys.* **62**, 299–300.

Chester, V. E. (1963). *Proc. R. Soc.* B**157**, 223–233.

Choishnev, K. and Auerman, L. Y. (1963). *Izv. vўssh ucheb. Zaved., Pishchevaya Tekhnol.* (4) 141–143; from *Chem. Abstr.* 1964, **60**, 1077f.

Cirillo, V. P. (1961). *A. Rev. Microbiol.* **15**, 197–218.

Clark, L. C. (1956). *Trans. Am. Soc. artif. internal Organs* **2**, 41.

Cochrane, A. L. (1961). *Soc. chem. Ind. Monograph* No. 11, p. 25.

Collins, T. H. and Lipton, R. A. (1967). *British Baker* **154** (26), 63.

Collis, D. E. and Diggins, A. A. (1967). Paper presented to *Inst. chem. Engng Symp. Biochem. Engng*, London.

Collyer, D. M. (1964). *Bakers' Digest* **38**, 43–54.

Collyer, D. M. (1967). *J. Sci. Fd Agric.* **18**, 428–439.

Cooper, C. M., Fernstrom, G. A. and Miller, S. A. (1944). *Ind. engng Chem.* **36**, 504–509.

Coppock, J. B. M. (1958). *Soc. chem. Ind. Monograph* No. 3, 158–162.

Cropper, F. R., Heinekey, D. M. and Westwell, A. (1967). *Analyst, Lond.* **92**, 436–442.

Dawson, E. R. and Harrison, J. S. (1949). *1st int. Congr. Biochem., Cambridge, Abstr. Commun.*, p. 546.

Dawson, E. R. and Martin, G. W. (1941). *J. Soc. chem. Ind., Lond.* **60**, 241–245.

Dawson, P. S. S. (1965). *Can. J. Microbiol.* **11**, 893–903.

Deindoerfer, F. H. and Humphrey, A. E. (1959). *Ind. engng Chem.* **51**, 809–812.

Dierssen, G. A., Holtegaard, K., Jensen, B. and Rosen, K. (1956). *Int. Sug. J.* **58**, 35–39.

Dorman, R. G. (1967). *Chemy Ind.* pp. 1946–1949.

Dorsey, A. E. (1959). *J. biochem. microbiol. Technol. Engng* **1**, 289–295.

Drews, B. (1936). *Biochem. Z.* **288**, 207–237.

Drews, B., Specht, H. and Herbst, A. M. (1962). *Die Branntweinwirtschaft* **102**, 245–247.

Eaton, N. R. (1960). *Archs Biochem. Biophys.* **88**, 17–25.

Eaton, N. R. (1961). *Archs Biochem. Biophys.* **95**, 464–469.

Ebbutt, L. I. K. (1961). *J. gen. Microbiol.* **25**, 87–95.

Ebbutt, L. I. K. (1966). *Brit. Pat.* 1,033,171.

Ebner, H., Pohl, K. and Enenkel, A. (1967). *Biotech. Bioengng* **9**, 357–364.

Echigo, A., Fujita, T. and Koga, S. (1966). *J. gen. Microbiol.* **12**, 91–99.

Emeis, C. C. (1965). *Proc. Eur. Brew. Conv.* Stockholm, pp. 156–163.

Ewart, J. A. D. (1968). *J. Sci. Fd Agric.* **19**, 617–623.

Fanti, O. D. and Kohan, T. (1964). *Revtd Invest. Agropecuar., Ser.* 2, **1** (1); from *Chem. Abstr.* 1967, **67**, 41251.

Fanti, O. D. and Kohan, T. (1965). *Revtd Invest. Agropecuar., Ser.* 2, **2** (1); from *Chem. Abstr.* 1967, **67**, 41252.

Farmer, S. N., Howard, C. J. and Hughes, L. B. (1967). *Chemy Ind.*, pp. 154–155.

Farmiloe, F. J., Cornford, S. J., Coppock, J. B. M. and Ingram, M. (1954). *J. Sci. Fd Agric.* **5**, 292–304.

Farrell, J. and Rose, A. H. (1967). *In* "Thermobiology", (A. H. Rose, ed.), pp. 147–218. Academic Press, London.

Finn, R. K. (1954). *Bact. Rev.* **18**, 254–274.

Finn, R. K. (1969). *Process Biochem.* **4** (6), 17, 22.

Firman, J. E. (1965). *Filtration and Separation*, March–April, pp. 102–107.

Fleischmann Co. (1926). *Brit. Pat.* 277,476.

Fowell, R. R. (1965). *J. appl. Bact.* **28**, 373–383.

Fowell, R. R. (1967). *Process Biochem.* **2** (12), 11–15.

Fries, H. von (1956). *Chem. Z.* **80**, 411–414.

Fries, H. von (1964). *German Pat.* 1,174,733.

Frey, C. N. (1930). *Ind. engng Chem.* **22**, 1154–1162.

Frey, C. N. (1957). *In* "Yeast its Characteristics, Growth and Function in Baked Products", (C. S. McWilliams and M. S. Peterson, eds.), pp. 7–33. Quartermaster Food and Container Institute for the Armed Forces, Chicago.

Fuld, G. J. and Dunn, G. G. (1957). *Ind. engng Chem.* **49**, 1215–1220.

Gaden, E. L. and Humphrey, A. E. (1956). *Ind. engng Chem.* **48**, 2172–2176.

Gandara, O. de la (1964). *An. R. Soc. esp. Fis. Quim., Sér. B* 60 (4), 345–352; from *Chem. Abstr.* 1965, **62**, 11113a.

Gorokhova, N. V. (1962). *Khlebopek. kondit. Prom.* **6** (8), 16–18; from *Chem. Abstr.* 1962, **57**, 17172h.

Grant, J. (1955). "Sutton's Volumetric Analysis", p. 483. Butterworth, London.

Gregr, V., Dyr, J. and Barta, C. J. (1960). *Czech. Patent* 96,374.

Grylls, F. S. M. (1961). *In* "Biochemists' Handbook", (C. Long, ed.), pp. 1050–1053. E. and F. N. Spon, London.

Grylls, F. S. M. and Harrison, J. S. (1956). *Nature, Lond.* **178**, 1471–1472.

Halton, P. (1959). *Soc. chem. Ind. Monograph* No. 6, pp. 12–25.

Hansford, G. S. and Humphrey, A. E. (1966). *Biotechnol. Bioengng* **8**, 85–96.

Harrison, J. S. (1958). Soc. Chem. Indust. Monograph No. 3, p. 147.
Harrison, J. S. (1967). *Process Biochem.* **2** (3), 41–45.
Hartsell, S. E. (1956). *Appl. Microbiol.* **4**, 350–355.
Hayduck, F. (1919). German Pat. 300,662.
Heckly, R. J. (1963). *Adv. appl. Microbiol.* **3**, 1–76.
Hedrick, L. R. and Feren, C. J. (1963). *J. Bact.* **86**, 1288–1294.
Hefe-Patent GmbH. (1963). British Patent 923,286.
Henika, R. G. and Rodgers, N. E. (1965). *Cereal Chem.* **42**, 397–408.
Hepner, I. L. (1967). *Process Biochem.* **2** (3), 3.
Herbert, D., Elsworth, R. and Telling, R. C. (1956). *J. gen. Microbiol.* **14**, 601–622.
Heredia, C. F., Sols, A. and DelaFuente (1968). *Eur. J. Biochem.* **5**, 321–329.
Herrera, T., Peterson, W. H., Cooper, E. J. and Peppler, H. J. (1956). *Archs Biochem. Biophys.* **63**, 131–143.
Hockenhull, D. J. D. (1963). *In* "Biochemistry of Industrial Micro-organisms", (C. Rainbow and A. H. Rose, eds.), pp. 227–299. Academic Press, London.
Hoffman, C., Schweitzer, T. R. and Dalby, G. (1941). *Cereal Chem.* **18**, 342–349.
Holtegaard, K. (1956). *Int. Sug. Jour.* **58**, 221–223.
Hospodka, J. (1966). *In* "Theoretical and Methodological Basis of Continuous Culture of Micro-organisms", (I. Malek and Z. Fencl, eds.), pp. 595–645. Czechoslovak Academy of Sciences, Prague.
Ierusalimsky, N. D. (1967). *In* "Microbial Physiology and Continuous Culture", (E. O. Powell, C. G. T. Evans, R. E. Strange and D. W. Tempest, eds.). pp. 23–33, H.M.S.O., London.
Ingram, M. (1959). *Chemy Ind.*, pp. 552–557.
Ingram, M., Mossel, D. A. A. and Lange, P. de (1955). *Chemy Ind.*, pp. 63–64.
Jakobovits, J., Wiggins, E. H. and Harrison, J. S. (1951). *J. gen. Microbiol.* **5**, 648–656.
Jensen, C. J. (1967). U.S. Patent 3,309,203.
Johnson, H. J., Borkowski, J. and Engblom, C. (1964). *Biochem. Bioengng* **6**, 457–468.
Johnson, M. J. (1967). *J. Bact.* **94**, 101–108.
Johnston, W. R. (1959). U.S. Pat. 2,919,194.
Kampf, G. and Behrens, U. (1966). *Z. allg. Mikrobiol.* **6**, 237–244.
Kautzmann, R. (1961). *Branntweinwirtschaft* **101**, 253–260.
Keilin, D. and Hartree, E. F. (1948). *Biochem. J.* **42**, 230–238.
King, E. J. (1932). *Biochem. J.* **26**, 292–297.
Kirchhoff, H. (1960). *In* "Die Hefen. I. Die Hefen in der Wissenschaft", (F. Reiff, R. Kautzmann, H. Lüers and M. Lindemann, eds.), pp. 564–589. Verlag Hans Carl, Nürnberg.
Kirsop, B. (1955). *J. Inst. Brew.* **61**, 466–471.
Koch, R. B., Geddes, W. F. and Smith, F. (1951). *Cereal Chem.* **28**, 424–430.
Kryuchkova, A. P. and Korotchenko, N. I. (1959). *Gidroliz. lesokhim. Prom.* **12**, No. 1, 8–10; from *Chem. Abstr.* 1959, **53**, 8465e.
Lanning, J. H. (1936). *Cereal Chem.* **13**, 690–697.
Leibowitz, J. and Hestrin, S. (1945). *Adv. Enzymol.* **5**, 87–127.
Lichstein, H. C. and Begue, W. J. (1960). *Proc. Soc. exp. Biol. Med.* **105**, 500–504.
Lincoln, R. E. (1960). *J. biochem. microbiol. Technol. Engng* **2**, 481–500.
Linko, Y., Johnson, J. H. and Miller, B. S. (1962). *Cereal Chem.* **39**, 468–476.

Locke, E. G., Saeman, J. F. and Dickerman, G. K. (1946). *In* "Wood Yeast for Animal Feed", p. 119. Northeastern Wood Utilization Council, New Haven, Connecticut.

Loginova, L. G. (1956). *Mikrobiologiya* **25**, 415.

Longmuir, I. S. (1954). *Biochem. J.* **57**, 81–87.

Mackenzie, R. M. (1958). Soc. Chem. Indust. Monograph No. 3, pp. 127–136.

Mackereth, F. J. H. (1964). *J. scient. Instrum.* **41**, 38–41.

MacLennan, D. G. and Pirt, S. J. (1966). *J. gen. Microbiol.* **45**, 289–302.

Mandelstam, J. (1957). *Nature, Lond.* **179**, 1179–1181.

Markham, E. and Byrne, W. J. (1968). *J. Inst. Brew.* **74**, 374–378.

Marten, J. F. and Catanzaro, G. (1966). *Analyst, Lond.* **91**, 42–47.

Matsuda, T. (1961). *Vitamins, Kyoto* **23**, 177–182.

Matsumoto, H., Iwa, I. and Minami, H. (1956). *J. Ferment. Technol., Osaka* **34**, 414–419.

Mautner, Markhof, Th. and G. (1956). Brit. Pat. 763,926.

Maxon, W. D. (1955). *Appl. Microbiol.* **3**, 110.

Maxon, W. D. and Johnson, M. J. (1953). *Ind. engng Chem.* **45**, 2554–2560.

Mayberry, W. R., Prochazka, G. J. and Payne, W. J. (1967). *Appl. Microbiol.* **15**, 1332–1338.

McMurrough, I. and Rose, A. H. (1967). *Biochem. J.* **105**, 189–203.

Merritt, P. P. (1957). *In* "Yeast, its Characteristics, Growth and Function in Baked Products", (C. S. McWilliams and M. S. Peterson, eds.), pp. 94–99. Quartermaster Food and Container Institute for the Armed Forces, Chicago.

Meyer, E. A. and Chaffe, P. W. (1938). Brit. Pat. 523,019.

Miall, L. M. (1965). *Chem. Process Engng* **46**, 205–296.

Mitchell, J. H. and Enright, J. J. (1957). *Fd Technol.* **11**, 359–362.

Mitterhauszerová, L. (1966). *Kvasný Prům.* **12** (4), 84–86; from *Chem. Abstr.* 1966, **65**, 2968b.

Mitterhauszerová, L. and Sedlárová, L. (1966). *Kvasný Prům.* **12** (10), 222.

Monod, J. (1942). "Recherches sur la Croissance des Culture Bactériennes". Hermann, Paris.

Monod, J. (1950). *Annls Inst. Pasteur, Paris* **79**, 390–410.

Mor, J. R. and Fiechter, A. (1968). *Biotechnol. Bioengng* **10**, 159–176.

Morita, M., Yamamoto, T. and Tezuka T. (1959). *Hakko Kagaku Zasshi* **37**, 339–343; from *Chem. Abstr.* 1960, **54**, 12476i.

Morse, R. E. and Fellers, C. R. (1949). *Fd Technol., Champaign* **3**, 234.

Moss, F. and Saied, M. (1967). *In* "Progress in Industrial Microbiology", (D. J. D. Hockenhull, ed.), pp. 209–257. Iliffe Books Ltd., London.

Mulcaster, K. D. and Aukland, J. R. (1964). *In* "High Efficiency Air Filtration", (P. A. F. White and S. E. Smith, eds.), p. 157. Butterworths, London.

Nyns, E. J., Fruytier, M. and Wainx, A. L. (1968). *Agricultura* **16** (3), 60–70.

Ohashi, M. (1958). *J. chem. Soc. Japan, Indust. Chem. Sect.* **61**, 1001–1015.

Olbrich, H. (1956). "Die Melasse". Institut für Gärungsgewerbe, Berlin.

Oldfield, J. T. F. and Dutton, J. V. (1968). *Int. Sug. J.* **70**, 40–43.

Oldhaver, P. D. H. (1912). U.S. Pats 1,027,700 and 1,039,999.

Olsen, A. J. C. (1928). Brit. Pat. 299,336.

Olsen, A. J. C. (1961). *Soc. chem. Ind. Monograph* No. 12, pp. 81–93.

Olson, B. H. and Johnson, M. J. (1949). *J. Bact.* **57**, 235–246.

Oskov, S. L. (1945). *Acta path. Microbiol. scand.* **22** (6), 523–559.

Othmer, D. F. (1968). *Process Biochem.* **3** (2), 33–35.

Oyaas, J., Johnson, M. J., Peterson, W. H. and Irvin, R. (1948). *Ind. engng Chem.* **40**, 280–283.

Pamely-Evans, O. G. (1968). "An Introduction to Industrial Cybernetics and Instrumentation", p. 4. Emmott and Co. Ltd., Manchester.

Parker, W. H. (1960). *Int. Sug. J.* **62**, 313–318.

Patentauswertung Vogelbusch GmbH. (1961a). Brit. Pat. 877,652.

Patentauswertung Vogelbusch GmbH. (1961b). Brit. Pat. 897,166.

Peppler, H. J. and Rudert, E. J. (1953). *Cereal Chem.* **30**, 146–152.

Peterson, W. H. (1950). *In* "Yeasts in Feeding", (S. Brenner, ed.), pp. 26–33. Milwaukee, Wisconsin.

Phillips, D. H. and Johnson, M. J. (1961). *J. biochem. microbiol. Technol. Engng* **3**, 261–275.

Pirt, S. J. (1965). *Proc. R. Soc.* **B 163**, 224–231.

Pollock, G. E. and Holmstrom, C. D. (1951). *Cereal Chem.* **28**, 498–505.

Powell, E. O. (1967). *In* "Microbial Physiology and Continuous Culture", (E. O. Powell, C. G. T. Evans, R. E. Strange, and D. W. Tempest, eds.), p. 34. H.M.S.O., London.

Pressindustria SpA., Milan (1967). *Process Biochem.* **2** (9), 49.

Purchas, D. B. (1966). *Process Biochem.* **1**, 177–180.

Pyke, M. (1958). *In* "The Chemistry and Biology of Yeasts", (A. H. Cook, ed.), pp. 535–586. Academic Press, New York.

Pyke, M., Ebbutt, L. I. K., Mackenzie, R. M. and Cunningham, J. (1958). Brit. Pat. 800,030.

Redfern, S., Cross, H., Bell, R. L. and Fischer, F. (1968). *Cereal Sci. Today* **13**, 324–326, 360.

Richards, J. W. (1966). *Process Biochem.* **1**, 41–46.

Richards, J. W. (1967). *Process Biochem.* **2** (9), 21–25.

Ricketts, R. W. (1962). *J. Inst. Brew.* **68**, 391–395.

Roberts, A. N. and Shepherd, P. G. (1968). *Process Biochem.* **3** (2), 23–24.

Robertson, J. J. and Halvorson, H. O. (1957). *J. Bact.* **73**, 186–198.

Rogers, D. and Mickelson, M. N. (1948). *Ind. engng Chem.* **40**, 527–529.

Roman, H. (1962). *In* "Recent Progress in Microbiology", (N. E. Gibbons, ed.), pp. 306–312. University of Toronto Press.

Rosa, M., Vernerova, J. and Stros, F. (1962). *Nährung* **6**, 130–147.

Rose, A. H. (1962). *Wallerstein Labs Commun.* **25**, 5–18.

Rose, A. H. (1963). *Wallerstein Labs Commun.* **26**, 21–37.

Rosen, K. (1968). *Process Biochem.* **3** (7), 45–47.

Rungaldier, K. (1931). *Gambrinus* **58**, 96–100; from *Chem. Abstr.* 1931, **25**, 4941.

Sak, S. (1919). Danish Pat. 28,507.

Sato, T. and Tanaka, Y. (1961). *J. Ferment. Technol.*, *Osaka* **39**, 718–723.

Schneider, K. (1954). U.S. Pat. 2,680,705.

Schuldt, E. H. (1963). Thesis, University Microfilms Inc., Ann Arbor, Michigan, 64–5908.

Schuldt, E. H. (1967). *Bakers' Digest* **41** (11), 90–93.

Schuldt, E. H. and Seeley, R. D. (1966). *Bakers' Digest* **40** (4), 42–44.

Schultz, A. (1965). *Brot Gebäck* **19**, 61.

Schultz, A. S., Atkin, L. and Frey, C. N. (1937). *J. amer. Chem. Soc.* **59**, 2457–2460.

Schultz, A. S., Atkin, L. and Frey, C. N. (1942). *Ind. engng Chem. Anal. Ed.* **14**, 35.

Scutt, J. E. (1967). *Process Biochem.* **2** (9), 29–31.

Seidel, M. (1943). German Pat. 739,021.

Serwinski, M., Amanowicz, J., Kasprzycki, J. and Nowak, B. (1961). *Zesz. nauk. Politech. lódz., Chemia spożywcza* **6**, 141–156; from *Chem. Abstr.* 1964, **61**, 3654e.

Sher, H. N. (1960). Brit. Pat. 845,315.

Sher, H. N. (1961). Soc. Chem. Indust. Monograph No. 12, p. 94–115.

Sherman, F. (1959). J. cell. comp. Physiol. 54, 29–35.

Shu, P. (1956). Ind. engng Chem. 48, 2204–2208.

Shukla, J. P. and Kapoor, B. D. (1956). Proc. Congr. Int. Soc. Sugar-cane Technologists 9th India 2, 498–505.

Siegell, S. D. and Gaden, E. L. (1962). Biotechnol. Bioengng 4, 345–356.

Sjoblohm, L. and Stolpe, E. (1964). Acta Acad. åbo., Mathematica et physica 24 (3), 28 pp.; from Chem. Abstr. 1965, 62, 15107h.

Sjoblohm, L. (1965). Kemistamfundet Medd. 74 (4), 73–79; from Chem. Abstr. 1966 64, 14630d.

Skiba, M. (1966). Abstracts of the 2nd Intern. Congress Food Sci. Technol. Warsaw, p. 53; from Int. Sug. J. 1967, 69, 218.

Solomons, G. C. (1967). Process Biochem. 2 (10), 47–48.

Soong, P. (1957). Taiwan Sug. 4 (9), 16–19.

Stakhorskaya, L. K. and Tokgarev, B. I. (1964). Mikrobiologiya 33, 1056–1060; from Chem. Abstr. 1965, 62, 12189f.

Stanton, J. H. (1965). Tech. Quarterly Master Brewers Assn Amer. 2, 233–237.

Sternglanz, P. D. and Kollig, H. (1962). Analyt. Chem. 34, 544–547.

Stoker, M. G. P. and Rubin, H. (1967). Nature, Lond. 215, 171–172.

Strohm, J. A., Dale, R. F. and Peppler, H. J. (1959). Appl. Microbiol. 7, 235–238.

Subramaniam, M. K. and Rao, P. L. S. (1951). Experientia 7, 98.

Svenska Jästfabriks Aktiebolaget (1966). Brit. Pat. 1,035,552.

Takakuwa, M. (1963). Mem. Ehime Univ. Sect. VI 8 (4), 363–447; from Chem. Abstr. 1963, 59, 10,740a.

Thiessen, E. J. (1942). Cereal Chem. 19, 773–784.

Thorley, J. F. (1968). In "Maintenance of Pure Culture Conditions in Industrial Fermentations", (in press). Soc. Chem. Ind. Symposium, London.

Thorn, J. A. and Reed, G. (1959). Cereal Sci. Today 4 (7), 198–200, 213.

Thorne, R. S. W. (1949). J. Inst. Brew. 55, 201–222.

Thorne, R. S. W. (1954). J. Inst. Brew. 60, 227–237, 238–248.

Thorne, R. S. W. (1962). In "Colloque sur les Levures", pp. 83–102. École de Brasserie de Nancy.

Thorne, R. S. W. and Beckley, R. F. (1958). J. Inst. Brew. 64, 38–46.

Thorne, R. S. W. and Bishop, L. R. (1936). J. Inst. Brew. 42, 15–36.

Toyo Jozo Co. Ltd. (1967). Brit. Pat. 1,062,212.

Trevelyan, W. E. (1963). Brit. Pat. 930,000.

Trevelyan, W. E., Forrest, R. S. and Harrison, J. S. (1952). Nature, Lond. 170, 626–627.

Universal Foods Corporation. (1966). Brit. Pat. 1,052,292.

Van Hall, C. E., Safranko, J. and Stenger, V. A. (1963). Analyt. Chem. 35, 315–319.

Walter, F. G. (1953). "The Manufacture of Compressed Yeast", 2nd ed. Chapman and Hall, London.

Watson, R. W. and Levinson, M. L. (1967). Appl. Microbiol. 15, 398–402.

White, J. (1954). "Yeast Technology". Chapman and Hall, London.

White, J. and Munns, D. J. (1955). J. Inst. Brew. 61, 223–229.

White, P. A. F. and Smith, S. E. (1964). "High Efficiency Air Filtration". Butterworths, London.

Wickerham, L. J. and Andreasen, A. A. (1942). Wallerstein Labs Commun. 5, 165–169.

Willetts, N. S. (1967). Biochem J. 103, 453–461.

Williams, K. T. and Bevenue, A. (1951). Cereal Chem. 28, 416–423.

Wilson, F. A. (1944). *Int. Sug. J.* **46**, 154–156.
Winzler, R. J. (1941). *J. Cell. Comp. Physiol.* **17**, 263–276.
Wirtschaftliche Vereinigung der Hefeindustrie (1935). German Pat. 618,021.
Wiseblatt, L. and Zoumut, H. (1963). *Cereal Chem.* **40**, 162–169.
Wyants, J. (1962). *J. Inst. Brew.* **68**, 350–354.
Yemm, E. W. and Cocking, E. C. (1955). *Analyst, Lond.* **80**, 209–231.

Chapter 8

Food Yeasts

H. J. PEPPLER

Universal Foods Corporation, Milwaukee, Wisconsin, U.S.A.

I. Introduction

A. DEFINITION OF FOOD YEASTS

Food yeasts, classed as "Dried Yeast", comprise non-fermenting stable whole yeast cells, carefully prepared and dehydrated to yield flakes and powders, which are intended for the nutritious and flavorous improvement of foods in human dietary. These nutritional concentrates are obtained either by recovery of spent brewer's yeast, or by harvesting yeast cultivated primarily for this purpose in mineral-supplemented media prepared with various sugar-bearing materials, notably molasses, spent sulphite mill waste, wood hydrolysates and cheese whey.

In the literature food yeasts may also be found designated as *dried yeast, inactive dried yeast, dry yeast, dry inactive yeast, levure-aliment sèche, dried torula yeast, sulphite yeast, wood sugar yeast, xylose yeast, levadura alimenticia, la levure alimentieri, Nahrungshefe* and *Saccharomyces Siccum.*

Food yeasts differ from feed (fodder) yeasts in quality characteristics. Generally food yeasts exhibit lower bacterial and mould counts, higher levels of vitamins, higher protein content, bland flavour, light colour, and the absence of obnoxious micro-organisms, added food ingredients and/or fillers than fodder yeasts. Because food yeasts are intended for human consumption, rigid chemical and microbiological standards have been established by professional groups and governmental agencies. In some instances these specifications are augmented by individual food processors and formulators to meet special processing conditions and desired dietary standards. Principal definitions and minimal standards are published by the United States Pharmacopeia (The National Formulary, XIII Edition), the Food and Drug Administration (U.S. Department of Health, Education and Welfare), and the International Union of Pure and Applied Chemistry (IUPAC).

Several culture strains of yeast are suitable for the production of food yeast: *Saccharomyces carlsbergensis, Sacch. cerevisiae, Sacch. fragilis, Candida lipolytica, C. tropicalis* and *C. utilis*. Neither *C. lipolytica* (favoured in hydrocarbon media; see Section IV.A.4, p. 453) nor *C. tropicalis* are used industrially in North America. Also undeveloped are processes proposed for other yeasts, for example, *Endomycopsis chodati* and *E. fibuliger* (Wickerham and Kuehner, 1956) and *Saccharomyces platensis* (Bertullo and Hettich, 1966).

B. EVOLUTION OF INTEREST

Man's involvement with yeast extends from prehistoric times through several periods of special interest. Today yeast is grown as a major, primary end-product on every continent. In the earliest ages, when life depended upon Nature's provisions, primitive people encountered spontaneous fermentation of honey and slurries of fruits and cereals. Eventually household methods evolved for bread, beer, cheese and other fermented foods (Jacobs, 1944). When the leavened coarse cereal porridges, batters or doughs were baked, yeast entered Man's diet in its most useful form, as a concentrate of dead easily-extracted cells.

Ancient Aryans, Egyptians, Assyrians and Babylonians exhibited skills in bread-making and brewing. Egyptian paintings show bakers and brewers working in adjoining rooms. They produced two kinds of bread and two types of beer: (i) refined white bread and clarified beer for the nobility; and (ii) coarse, dark bread and turbid beer for the workmen, who gained thereby the better nutrient value provided by the retained yeast (von Gontard, 1948). A comparable crude yeast-retained beverage food, Kaffir beer, is still brewed today by the Bantu (Novellie, 1968).

Although the brewing art is a centuries-old practice, the usefulness of its left-over yeast went unrecognized until the late nineteenth century when spent brewer's yeast was added to livestock feeds (Poff, 1899). The possibility of growing yeast as a food for direct human consumption occurred to Delbrück (1910) and his coworkers at the Institut für Gärungsgewerbe in Berlin, nearly 40 years after Pasteur related the mystery of fermentation to living yeast cells. The ideas of Poff (1908), Delbrück, Völtz (1910) and others stimulated extensive feeding experiments with brewer's yeast on all classes of domestic animals (Braude, 1942). Greater attention to the value of yeast for human nutrition followed the discovery of nutritive factors in yeast by Wildiers (1901) and Funk (1911, 1912), who originated the term vitamin, and the demonstration that yeast protein can replace protein of vegetable and animal origin (Osborne and Mendel, 1919).

Concurrently, industrial systems of yeast propagation were developed by Delbrück (1915), Hayduck (1914, 1915) and Sak (1921). They successfully combined Hansen's (1896) pure-culture concept with Pasteur's (1874, 1876) discoveries that yeast thrives on ammonia, and cell growth is stimulated by aeration which represses alcohol formation. Delbrück and Hayduck aerated molasses solutions fortified with phosphate and magnesium salts, and synthetic ammonia. Hayduck (1919) contributed the important idea of adding nutrients in step with the rate of cell increase. This principle of incremental feeding, or *zulauf*, inaugurated a new era in yeast technology (see Chapter 7, p. 349).

In the 30 years following 1915, interest in yeast as a dependable food source rose steadily as reports of beneficial effects in numerous feeding experiments and nutritional studies accumulated (Braude, 1942; Hock, 1960). In human nutrition as well, yeast was soon recognized as a valuable source of protein and a major natural reservoir of the vitamin-B complex which eventually helped eradicate deficiency diseases, notably beriberi and pellagra (King, 1950). Yeast also became a test organism for assaying some of the new essential nutrients, namely, biotin (Wildiers, 1901), thiamin (Schultz *et al.*, 1937), inositol (Wooley, 1941) and pyridoxine (Atkin *et al.*, 1943). It was in this period of peak interest in dried yeast that food yeast got its name. According to Thaysen (1957), a committee constituted by the Royal Society in 1941 coined the term *food yeast* to centre attention upon primary yeast grown expressly for human consumption, aiming to set it apart from dried yeasts obtained as by-products.

Mounting research efforts in the 1930s and 1940s explored production possibilities of high-fat yeasts (Enebo *et al.*, 1946; Lundin, 1950; Woodbine, 1959) and food yeast beyond the traditional grain and molasses substrates and selected cultures of *Saccharomyces*. Of greatest significance were the studies with asporogenous yeasts grown in pulp-mill waste at the Institut für Gärungsgewerbe by Lindner and associates (Fink and Lechner, 1941; Schmidt, 1947). Details of their industrial processes based on *Candida utilis* growth in wood sugar substrates were revealed by post-war inspection teams of operating units (Locke *et al.*, 1945; Locke, 1946). These ventures and the concurrent molasses-based studies in England brought fresh waves of commercial interest in food yeast production in Europe, the United States, Jamaica, Taiwan, Japan and elsewhere (Thaysen, 1957; Wiley, 1954). Despite the contribution made by food yeasts to the protein famine in carbohydrate-based national diets, operations have ceased at two of the new locations: at the Frome plant in Jamaica, and the Green Bay facility in Wisconsin (U.S.A.).

Today, new challenges face us as we contemplate this planet's people crisis (Anon., 1968b). One of the complex consequences of overpopulation, a shortage of food, has renewed the search for protein sources, creating another surge of interest in yeast and other microbial protein. Under the attractive name of single-cell proteins, or SCP (Mateles and Tannenbaum, 1968), several pilot-plant projects are exploring the potentials of SCP obtained with yeast and bacterial conversion of petroleum fractions (Champagnat *et al.*, 1963; Johnson, 1967; Miller and Johnson, 1966; see Section IV.A.4, p. 453). Other efforts to close the protein gap involve novel schemes for transforming or upgrading a wide variety of agricultural and industrial debris: cellulose accumulations, process left-overs, crop and livestock residues, municipal wastes and sewage. A most unusual project, at the University of New South Wales, would produce pure protein from grasslands by imitating the fermentation mechanism of ruminants (Anon., 1968a). This old technique—the use of mixed cultures of micro-organisms—deserves more attention.

C. PRODUCTION STATISTICS

Dried yeast is produced in at least 20 countries, whose total annual output now is greater than the estimated 180,000 metric tons dry weight reported in World Wide Survey of Fermentation Industries 1963 (1966), the only published compilation of statistics covering the global production of fermentation products. These earlier estimates of yeast production are compared in Table I. The relative abundance of yeast protein is greatest in Europe, concentrated chiefly as feed yeast in the countries of eastern Europe. Russia, East Germany, Poland, Hungary, Czechoslovakia and Rumania contributed about 60% of the world total reported for 1963. Molasses and spent sulphite liquor continue to be the principal carbon sources used. According to Bunker (1964), Russia is increasing yeast production so rapidly that by 1970 it hopes to make two million tons of dried yeast per annum, mainly for use in animal feeds. Half of the anticipated production will come from conversion of spent sulphite liquor, wood hydrolysates and other plant debris; and the other half is to be derived from petroleum fractions and natural gas.

TABLE I. *Annual World Production of Yeast* (*In part from Bunker*, 1964; *and World Wide Survey of Fermentation Industries*, 1966)

Source	Annual production (metric tons of dry matter)	
	Dried yeast	Baker's yeast
Europe	125,500	67,600
North America[a]	37,500	61,000
The Orient	21,300	12,900
South America	1,200	6,300
Africa	2,200	2,350
Total	187,700	150,150

[a] 1968 estimated production.

Detailed information obtained by Bunker (1964) places the total food yeast output at a mere 1,800 tons, or 1% of the annual world dried yeast tonnage estimate. This includes 1,000 tons of molasses-grown food yeast out of Taiwan's yearly production of 13,000 tons, all of which at

TABLE II. *Yeast Production in the United States: 1968 Estimate* (*Based in part on Census of Manufactures*, 1963)

Product	Annual production (metric tons dry matter)	Sugar sources
Baker's yeast		
Compressed	53,500	Molasses
Active dry	2,300	Molasses
Food yeast		
Saccharomyces sp.	14,100	Grain, molasses, whey
Candida utilis	1,000	Spent sulphite liquor
Feed yeast		
Saccharomyces sp.	9,500	Grain, molasses, whey
Candida utilis	1,800	Spent sulphite liquor
Autolysates, extracts	3,500	Grain, molasses
Total	85,700	

the outset of the project was intended for human consumption (Chien, 1960). Not included is the United States food yeast tonnage from primary manufacture, approximately 5,000 tons.

In the United States the yeast industry comprises seven manufacturers operating at 16 locations. Fourteen factories (five companies) in 1968 produced about 55,800 metric tons of yeast dry matter as some form of baker's yeast. As shown in Table II, this represents nearly 65% of the domestic harvest. The remainder of the annual production comprises 18% food yeast, 13% feed yeast and 0·5% miscellaneous extracts and autolysates.

Although a wide variety of carbohydrates are available in great abundance in the United States, yeast is called on to work over only a

TABLE III. *Utilization of Sugar Sources for Yeast Propagation and Fermentation during 1965 in the United States (From: Agricultural Statistics, 1966; Larkin, 1966; Internal Revenue Service, 1966)*

Raw material	Total supply	Amount utilized	
		Yeast	Fermentation
Molasses (10^6 gal)	615	50	117
Sulphite liquor (10^6 gal)	12,000	75	298
Whey (10^6 gal)	3,530	5	10
Fruit products[a] (10^3 tons)	4,624	Nil	2,454
Grain[b] (10^3 tons)	154,700	c	3,569

[a] Grapes and raisins.

[b] Major ingredients for beer and spirits production.

[c] Secondary yeast recovered as by-product of brewing and distilling operations; no estimate available.

Table IV. *World Molasses Production: 1964 (From Larkin, 1966)*

Source	Amount produced (10^6 gal)
Europe	1,162
North America	1,013
Asia	575
Africa	184
Oceania	84
Others	466
Total	3,484

small fraction of the supply (Table III). Yeast-making consumes 8% of the U.S.A. domestic molasses supply (615 million gal), the principal source of energy and nutrients.

Fermentations yielding organic acids, vinegar, industrial ethanol, etc. use up 19%. The remainder (73%) is expended as an ingredient for cattle feeds. Huge quantities of cheese whey and spent sulphite-mill waste are available in the United States and around the world, but less than 1% of the domestic supply is used in media for food yeast production. Elsewhere, especially in Europe and Japan, spent sulphite liquor is a major raw material, primarily for feed yeast.

Molasses occurs universally (Table IV), but no estimate of its utilization for yeast propagation has been reported.

II. Characteristics of Food Yeasts

In their industrial evolution yeasts have been chosen for chemical performance rather than functional composition. At first cultures were screened for their fermentation capabilities, practical performance, and cultural stability to meet the needs of bread-making, brewing and alcohol production. Later, attention was given to yield and efficiency of converting substrate carbon to cell substance. More recently, the search is centred on strains capable of assimilating components of unusual carbon sources.

A. BIOLOGICAL PROPERTIES

Food yeast in today's market is represented by three species of the genus *Saccharomyces* and two species of the genus *Candida*. Four of these yeasts are merchandised in the United States: *Saccharomyces carlsbergensis*, recovered from beer; *Saccharomyces cerevisiae*, molasses-grown strains of baker's yeast; *Saccharomyces fragilis* cultured on cheese whey; and *Candida utilis*, propagated on spent sulphite liquor. In Germany *C. tropicalis* replaced *C. utilis* in some yeast factories (Windisch, 1948; Butschek, 1962). *Candida tropicalis* is mentioned often in the Russian and Japanese literature. Butschek (1962) believes it may be identical with *Monilia murmanica* appearing in Russian reports.

Differentiating growth properties of cultures used for food yeast production are compared in Table V. All of these yeasts exhibit in common the ability to utilize ammonia, urea and some amino acids, synthesize B-group vitamins, absorb many of these vitamins from the medium, and proliferate well aerobically at 30° to yield approximately 50 g dry cells per 100 g glucose. They grow as well in molasses media as in the simple solution tested by Olson and Johnson (1949), which contains the following components (per litre): 10 g glucose, 6 g ammonium

TABLE V. *Some Biological and Biochemical Characteristics of Cultures Used for Food Yeast Production (From Lodder and Kreger-Van Rij, 1952; Wickerham, 1957)* [+ denotes assimilation (growth); ±, a few strains weakly positive; −, no growth]

Nutrient	Saccharomyces cerevisiae	Saccharomyces carlsbergensis	Saccharomyces fragilis	Candida utilis	Candida tropicalis
Glucose	+	+	+	+	+
Galactose	+	+	+	−	+
Maltose	+	+	+	+	+
Sucrose	+	+	−	+	+
Lactose	−	−	+	−	−
Xylose	−	−	−	+	+
KNO$_3$	−	−	−	+	−
Ethanol	±	−	+	+	−
Average size (μ)	5 × 9	7 × 9	4 × 6	4 × 7	7 × 9

dihydrogen phosphate, 2·5 g L-asparagine, 1 g sodium citrate, 0·25 g magnesium sulphate, 0·2 g potassium dihydrogen phosphate, 10 mg inositol, 4 mg thiamin hydrochloride, 1 mg pyridoxine hydrochloride, 0·5 mg calcium pantothenate, 400 μg zinc (as sulphate), 150 μg iron (as ammonium ferrous sulphate), 25 μg copper (as sulphate), 20 μg biotin.

Individually food yeast cultures exhibit some unique differences which account for their exploitation. The most versatile member of the group is *Candida utilis* (Henneberg) Lodder and Kreger-van Rij, formerly *Torulopsis utilis* (Henneberg) Lodder. This yeast was found by Henneberg (1926) as a contaminant in several German yeast factories producing feed and food yeast from wood sugars. He named it *Torula utilis*. At that time every small, almost round yeast cell with an oil droplet inside was considered a "typical *Torula*", a genus Turpin had established erroneously in 1838, although Persoon had given this name in 1796 to a group of moulds (Lodder and Kreger-van Rij, 1952). Despite clarification of its taxonomy, the designation of food yeast comprised of *C. utilis* as torula yeast persists in the trade and technical literature.

Because it attacks more carbon and nitrogen compounds than other common yeasts, *C. utilis* is favoured industrially to modify molasses, wood hydrolysates, spent sulphite liquors, distillery wastes and residual liquors from food and chemical processing. In addition to the nutrients listed in Table V, *C. utilis* assimilates fatty acids such as acetic, propionic, butyric and caproic (Rieche *et al.*, 1964), lactic acid, acetoacetic acid, pyruvic acid, glycerol, mannose, acetaldehyde, urea, asparagine, peptones, ammonium and nitrate nitrogen (Dunn, 1952). Unlike most yeasts, *C. utilis* synthesizes biotin (Chang and Peterson, 1949).

Candida tropicalis (Cast.) Berkhout also grows well on xylose and *n*-alkanes but it does not utilize nitrate. Originally isolated by Castellani in 1910, *C. tropicalis* was adapted in Germany to feed yeast production from wood sugars (Windisch, 1948). In some factories cultures labelled *Candida arborea* were used (Thaysen, 1957; Butschek, 1962). However, Lodder and Kreger-van Rij (1952) identified one of two submitted cultures as *C. tropicalis*; the other proved to be *C. utilis*. Since the identity of *C. arborea* remains in doubt, it is without status (*nomen nudum*).

Candida pseudotropicalis (Cast.) Basgal ferments and assimilates lactose. Some strains elaborate extracellular proteinase (Ahearn *et al.*, 1968). Windisch (1948) reported its use for food yeast production from whey. But current production in the United States employs *Sacch. fragilis*, the perfect (ascosporogenous) form of *C. pseudotropicalis*. This preference is consequent to the intensive pioneering studies conducted by the U.S. Department of Agriculture (Wasserman *et al.*, 1958; Wasserman, 1960).

Candida japonica Diddens and Lodder, a large yeast (6–10·5 × 6–20 μ), is propagated chiefly on spent sulphite liquor in Japan.

Candida lipolytica (Harrison) Diddens and Lodder has great potential as a food yeast from hydrocarbons. Among its unusual properties are the absence of fermentative ability, assimilation of only one sugar (glucose), elaboration of extracellular lipase and protease (Ahearn *et al.*, 1968), and conversion of a variety of *n*-alkanes and gas oils to cell substance (Filosa, 1960; Champagnat *et al.*, 1963; Champagnat and Lainé, 1966; Johnson, 1967; Evans, 1968).

Domestic food yeast manufacture depends chiefly upon strains of traditional brewer's and baker's yeasts, *Sacch. carlsbergensis* Hansen and *Sacch. cerevisiae* Hansen. For production of food yeast by primary culture, usually in molasses media, *Sacch. cerevisiae* is preferred, since many of its strains exhibit stronger oxidative dissimilation of sugars and grow at higher temperatures than strains of *Sacch. carlsbergensis*.

B. COMPOSITION

No other microbe matches the universal attention accorded yeast in biology, chemistry and technology. Though more than a century of literature has accumulated, the composition of yeast is a relatively modern concern, closely linked to the discovery and search for vitamins and amino acids, and the evaluation of their nutritional value. Since this voluminous documentation of yeast properties has merited review elsewhere (Eddy, 1958; Kirchhoff, 1960; Hock, 1960; Baird, 1963), treatment here centres on data germane to five types of commercial food yeast grown on four different substrates: molasses, wood sugars, whey and wort (recovered brewer's yeast).

1. Gross Composition

Variations in raw materials, propagation conditions and treatments of the harvested slurry affect yeast composition. With the best control of commercial operating conditions, yeast production lots given nominally the same regimen usually show small differences in chemical constitution. The nature of the growth medium and degree of aeration are major factors influencing yeast content of carbohydrates, proteins, fats and vitamins (Singh et al., 1948; Chen, 1959; Suomalainen, 1968).

Frey (1930) reported the following composition for baker's yeast dry matter: 52·4% protein, 37·1% polysaccharides, 1·7% fat, and 8·8% ash. Later Frey et al. (1936) determined the elemental constituents to be 45% C, 32% O, 9% N, 6% H and 8% inorganic matter. Sperber (1945) reported the average composition of yeast cells to be 47% C, 31% O, 7·5% N, 6·5% H, 8% ash. Modern data cited by Harrison (1967)

TABLE VI. *Gross Composition of Food Yeasts* [Analysis on dried samples immediately after propagation, before enrichment with vitamins. A and B = two different strains of *Saccharomyces cerevisiae* grown on molasses (Seeley, 1951); C = *Candida utilis* grown on spent sulphite liquor (Wiley et al., 1951; Seeley, 1951). Brewer's yeast, debittered (Seeley, 1951); distiller's yeast, by-product of alcohol production (Seeley, 1951)]

Analysis	Content in dried yeast				
	Primary grown			Brewer's yeast	Distiller's yeast
	A	B	C		
Moisture (%)	9·2	4·4	5·8	4·7	7·0
Nitrogen (%)	6·5	8·1	8·0	7·9	6·6
Purine nitrogen (%)	0·3	0·4	0·4	0·4	0·4
Polysaccharides (%)	22·5	30·3	29·6	30·7	34·0
Fat[a] (%)	4·8	4·2	4·8	2·6	7·5
Fibre (%)	0·9	0·4	0·8	0·4	1·5
Ash (%)	9·7	5·8	9·0	7·3	6·1
Thiamin (µg/g)	28	165	5	136	65
Riboflavin (µg/g)	62	100	45	38	52
Niacin (µg/g)	283	585	415	525	195
Pyridoxine (µg/g)	34	20	29	40	26

[a] Ether extract of acid-hydrolyzed sample.

indicate similar values for carbon, hydrogen, oxygen and nitrogen; and Harrison calculated the inorganic elements as 0·6% S, 1·1% P, 2·2% K, and 5·7% other minerals.

Chemical characteristics of different types of yeast produced as food yeast were compared by Seeley (1951). Part of his data, shown in Table VI, reveal the marked effect of differences in yeast strains and composition of the medium. Food yeast in the market, Table VII, also varies widely in composition due in part to the types of media and yeast involved, and in part to vitamin enrichment of the harvested yeast. Methods of analysis are recommended by the International Union of Pure and Applied Chemistry (Commission on Characterization and Evaluation of Dried Yeast, 1963).

TABLE VII. *Approximate Composition of Trade Food Yeasts* (*Data from manufacturers' technical bulletins*) [A = molasses-grown *Saccharomyces cerevisiae*; B = sulphite-grown *Candida utilis*; C = cane molasses-grown *Candida utilis*; D = debittered brewer's yeast, roller-dried; E = debittered brewer's yeast, spray-dried; F = whey-grown *Saccharomyces fragilis*; G = average of collaborative analyses (Commission on Characterization and Evaluation of Dried Yeast, 1963)]

Analysis	Content in dried yeast						
	A	B	C	D	E	F	G
Moisture (%)	5	<7	7	6	6	7	5
Protein (N × 6·25) (%)	50	>50	51	45	45	54	49
Fat (%)	6	5	3	6	1·5	1	3
Ash (%)	7	<8	10	8	7	9	5
Thiamin (µg/g)	150	>120	25	150	125	20	131
Riboflavin (µg/g)	70	>40	50	45	35	50	77
Niacin (µg/g)	500	>300	335	400	500	135	610
Pyridoxine (µg/g)	30	30	—	40	50	40	15

2. Nitrogenous Compounds

About one-half of the dry weight of yeast is crude protein (N × 6·25) consisting of 80% amino acids, 12% nucleic acids, and 8% ammonia (Frey, 1930; von Soden and Dirr, 1942; Harrison, 1968). Around 7% of the total nitrogen occurs as free amino acids (Roine, 1946). The presence of large amounts of purine and pyrimidine bases lowers the true protein of yeast to 40% of the dry weight (von Soden and Dirr, 1942; Bunker 1963). Other constituents of yeast whose nitrogen is only a small part of the total include glutathione, lecithin, adenylic acid, vitamins, enzymes and coenzymes (Dunn, 1952; Brunner, 1960).

The nutritive value of dried yeast hinges upon the quality of its protein and vitamin content. With a relatively high digestibility and biological value, both at 87%, as compared with whole hen's egg at 96% and 97% respectively (Murlin *et al.*, 1944), yeast protein has proven a superior protein supplement of cereals (Sure, 1946, 1948). The abundance of lysine and tryptophan in yeast protein accounts for the improvement of cereal diets. When yeast protein is supplemented with methionine, its protein utilization efficiency nearly equals that of casein (Klose and Fevold, 1947; Sure, 1948; Harris *et al.*, 1951). Table VIII compares the content of nutritionally essential amino acids of food yeasts marketed today.

TABLE VIII. *Essential Amino Acid Composition of Commercial Food Yeasts* [Yeasts A, B, C, D from Technical Bulletins (1967). A = *Saccharomyces cerevisiae* from molasses; B = *Candida utilis* from spent sulphite liquor (Peppler, 1965); C = *Saccharomyces fragilis* from whey; D = debittered brewer's yeast; E = baker's yeast (Harrison, 1968); F = *Candida utilis* from molasses]

Amino acid	Content in yeast (g/16 g N)					
	A	B	C	D	E	F
Lysine	8·2	6·7	8·8	7·3	9·7	10·7
Valine	5·5	6·3	6·6	5·2	5·9	5·7
Leucine	7·9	7·0	9·9	6·3	7·7	8·1
Isoleucine	5·5	5·3	5·5	5·7	7·3	7·3
Threonine	4·8	5·5	5·5	4·8	7·0	4·8
Methionine	2·5	1·2	1·5	1·2	3·5	1·4
Phenylalanine	4·5	4·3	3·9	4·4	5·6	4·1
Tryptophan	1·2	1·2	1·5	1·1	1·7	0·5
Cystine	1·6	0·7	—	0·9	1·2	0·3
Histidine	4·0	1·9	2·5	1·5	3·6	2·8
Tyrosine	5·0	3·3	—	—	4·5	1·4
Arginine	5·0	5·4	4·9	4·7	4·3	4·7

Kwolek and Van Etten (1968) devised an effective method of reducing amino acid data to two numbers: R, the amount of essential amino acids in a protein; and $V(r)$, the pattern, as compared with the FAO/WHO reference protein of whole hen's egg (Van Etten *et al.*, 1967). For their yeast data (Kwolek and Van Etten, 1968), given in Table IX, the amounts of essential amino acid (R values) compare well with the reference protein; however, better agreement in pattern [lower value for $V(r)$] would be expected from the benefits observed in comparative feeding trials (Seeley, 1951; Baird, 1963). Murlin *et al.* (1946) found that the nutritive value of food yeast is greater than the biological value of a mixture of amino acids combined in the same proportion occurring in yeast.

TABLE IX. *Nutritionally Essential Amino Acid Patterns of Food Yeasts and Common Proteins (Based on criteria established by Kwolek and Van Etten, 1968) [R measures total quantity of essential amino acids in protein; V(r) measures agreement of patterns]*

Material	R	$V(r) . 10^3$
Hen's egg[a]	0·513	0·00
Saccharomyces cerevisiae dried yeast[b]	0·457	14·55
Candida utilis dried yeast[b]	0·496	16·25
Dried brewer's yeast[b]	0·434	8·14
Baker's yeast[b]	0·459	7·43
Cow's milk	0·513	7·75
Soybeans	0·424	6·68

[a] Reference pattern adopted by FAO/WHO report (1965); 0·513 indicates 0·513 g essential amino acid per g protein.

[b] Calculated by W. F. Kwolek (personal communication).

The possibility of improving the nutritional value of yeast by raising its methionine and cystine content was investigated by Chiao and Peterson (1953). Of 20 yeasts, representing 11 genera, propagated in fortified beet molasses medium, the methionine content of the best producers averaged 0·74% of the dry yeast, nearly the same as *Sacch. cerevisiae* commercial yeasts. The addition of theoretical precursors of methionine (choline, cystine) to the medium did not affect the methionine content, but cystine markedly increased the yield. The fat synthesizer, *Rhodotorula gracilis*, had the highest total sulphur amino acid content among the yeasts tested, 1·0% methionine and 0·27% cystine (dry basis). A later survey of 271 strains of yeast by Nelson *et al.* (1960) uncovered no high producers of methionine, lysine and tryptophan. However, Jensen and Shu (1961) increased the lysine content of yeast three- to four-fold by the addition of 2-oxo-adipic acid during the growth period. Such enrichment of yeast becomes less attractive as the cost of amino acids from non-biological sources decreases. At the present time (early 1969) L-lysine is available at $5·50/kilo. DL-Methionine is priced at $1·60/kilo of 99% material, and $4.50/kilo of pharmaceutical-grade product. At $90/kilo DL-tryptophan is the highest priced of all essential amino acids available in commercial quantity.

3. *Vitamins*

Although yeasts contain more than ten water-soluble vitamins collectively designated the vitamin-B complex, only three compounds— thiamin, riboflavin, and niacin—are specified in commercial food yeasts.

Considered with these, however, are pyridoxine and folic acid (folacin), both of which are included in the revised Recommended Dietary Allowances (1968) promulgated by the National Academy of Sciences for the maintenance of good nutrition.

Analyses for the principal B-group factors occurring in commercial food yeasts are compared in Table X. The higher thiamin values, as compared with out-of-the-fermenter values (Table VI), result from enrichment of the growth medium, or the harvested yeast slurries, before drying (Van Lanen, 1954). Other B-group vitamins, notably niacin and biotin, are similarly incorporated by cellular uptake to predetermined levels (Van Lanen *et al.*, 1942; Van Lanen, 1946; Chang and Peterson, 1949).

TABLE X. *Vitamin Contents of Trade Food Yeasts* (In part from Seeley, 1951; Wiley *et al.*, 1951; Wasserman, 1961; Powell and Robe, 1964; and Technical Bulletins, 1967) [A = *Saccharomyces cerevisiae*, molasses-grown; B = *Candida utilis*, spent sulphite liquor substrate; C = Brewer's yeast, debittered, roller-dried; D = Brewer's yeast, debittered, spray-dried; E = *Saccharomyces fragilis*, whey-grown; F = *Candida utilis*, cane-molasses grown]

Vitamin	Content in dry product (μg/g)					
	A	B	C	D	E	F
Thiamin HCl	165	130	150	125	20	25
Riboflavin	100	45	45	35	50	50
Niacin	585	400	400	500	330	335
Pyridoxine HCl	20	30	40	50	40	—
Folacin	13	21	5	49	14	20
Calcium *d*-pantothenate	100	40	100	120	115	120
Biotin	0·6	0·8	1	1	2	2
p-Aminobenzoic acid	160	11	5	—	24	—
Choline chloride	2,710	2,860	3,800	4,850	4,550	5,500
Inositol	3,000	4,500	3,900	5,000	3,000	—

The folic acid values shown in Table X, as taken from manufacturers' bulletins and earlier reports, are too high for some yeasts. Recent assays of a large series of commercial samples (Peppler, 1965) indicate average folic acid contents of 9·1–12·7 μg/g dry weight of *C. utilis*, and 9·8 to 12·1 μg/g *Sacch. cerevisiae*.

Vitamins of the B-complex occur in yeast mainly in bound form as components of enzymes and coenzymes (Kirchhoff, 1960; Williams *et al.*, 1950). Thiamin (vitamin B_1, aneurin) is linked with phosphoric acid, forming thiamin pyrophosphate (TPP, cocarboxylase), the prosthetic group of carboxylase (Green *et al.*, 1940; Jansen, 1954).

Riboflavin (vitamin B_2, riboflavine, lactoflavin, vitamin G) occurs in yeast flavoproteins as riboflavin-5'-phosphoric acid and flavin adenine dinucleotide, both tied to protein. In the "old yellow enzyme" of Warburg and Christian (1933), the prosthetic group is a mononucleotide (Horwitt, 1954).

Pantothenic acid (vitamin B_3, Bios IIa), one of the vitamins first isolated from yeast (Williams *et al.*, 1933), is a constituent of coenzyme A.

Niacin (nicotinic acid, niacinamide, PP factor) is present in yeast as the amide of nicotinic acid, which is a constituent of various coenzymes, especially diphosphopyridine nucleotide (DPN, NAD, Coenzyme I, cozymase) and triphosphopyridine nucleotide (TPN, NADP, Coenzyme II). Funk (1913) found niacin in yeast while attempting to concentrate the anti-beriberi factor (vitamin B_1). Hundley (1954) reviewed the detailed chemistry and biochemistry of niacin.

Pyridoxine (vitamin B_6, adermin) occurs as pyridoxal and pyridoxamine esterified with phosphoric acid in coenzymes (transaminase, codecarboxylase) bound to protein (Kereztesy and Umbreit, 1954). It was first isolated from yeast cell-free juice (Kuhn and Wendt, 1938).

Biotin (Bios II, Bios IIb, vitamin H) is found in yeast largely bound with protein (György, 1954). As the "bios" of Wildiers (1901), biotin was the first yeast growth stimulant studied, and it marked the beginning of the vitamin age. Later, after many B-group factors had been isolated and characterized, it was named biotin (Kögl, 1937).

Other B-group vitamins and their derivatives, for which yeast is a good source material, include *p*-aminobenzoic acid, a growth factor for many bacteria (Rubbo and Gillespie, 1940); inositol, a growth factor for yeast (Eastcott, 1928) and some bacteria; choline, the basic component of lecithin; and folic acid (folacin), found conjugated in yeast as pteroylhexaglutamic acid (Pfiffner *et al.*, 1945, 1946). Miscellaneous new and unidentified compounds of the vitamin-B complex have been reviewed by Cheldelin (1954).

4. *Inorganic Elements*

The ash content of commercial food yeasts is usually within the range 6%–8% of dry cell weight. As shown in Table XI, most of the ash consists of potassium and phosphorus, with lesser amounts of calcium, magnesium and sulphur. Primary-grown food yeasts are dependable sources of phosphorus, magnesium and calcium. Yeast grown on wood sugar substrates are consistently lowest in sodium. Selenium is present in commercial food yeast in biologically active form (Kelleher

TABLE XI. *Major Inorganic Elements in Trade Food Yeasts* [A and B = different strains of *Saccharomyces cerevisiae*, molasses-grown (Seeley, 1951); C = brewer's yeast, debittered roller-dried (Seeley, 1951); D = brewer's yeast, debittered spray-dried (Technical Bulletins, 1967); E = distiller's food yeast (Seeley, 1951); F = *Candida utilis*, sulphite-grown (Technical Bulletins, 1967). *Italics* indicate five elements for which dietary allowances are recommended (Recommended Dietary Allowances, 1968)]

Element	Content in yeast dry matter					
	A	B	C	D	E	F
Potassium (%)	2·0	2·7	1·6	0·9	2·1	2·1
Phosphorus (%)	1·1	1·1	1·8	1·5	1·2	2·0
Calcium (%)	0·4	0·6	0·8	0·1	0·1	0·9
Magnesium (%)	0·2	0·3	0·2	—	0·2	0·2
Sulphur (%)	0·4	0·5	0·4	—	0·3	0·4
Sodium (%)	0·2	0·1	0·2	—	0·06	0·02
Zinc (µg/g)	42	89	107	39	280	125
Iron (µg/g)	92	1,010	71	200	157	175
Copper (µg/g)	21	122	53	35	98	17
Lead (µg/g)	2·5	5·9	3	—	6·9	2
Manganese (µg/g)	4	28	4	5·3	35	35
Iodine (µg/g)	1·6	0·5	2·9	—	1·4	3·8
Total ash (%)	6·6	5·8	6·0	7·0	6·0	7·5

et al., 1958). Other trace metals detected in yeast have been reviewed by Peterson (1950), White (1954) and Eddy (1958).

Minute amounts of arsenic and lead are present in commercial food yeasts. IUPAC (International Union of Pure and Applied Chemistry) standards of dried yeast specify maxima of 5 µg arsenic and 5 µg lead per gram of air-dried sample. Results obtained in the collaborative analysis of two different samples of dried yeast indicate mean values of 0·38 µg arsenic, 1·17 µg lead and 34·2 µg iron per gram air-dried sample, and 1·31% phosphorus (Commission on Characterization and Evaluation of Dried Yeast, 1963).

5. *Carbohydrates and Lipids*

The carbohydrate content of food yeasts varies from 22% to 34% of the dry matter (Table VI). It fluctuates with the composition of the growth medium and the level of aerobiosis maintained in it. For yeast grown at aeration rates practised in primary food yeast production, Chen (1959) recovered 25% of the cell dry weight as carbohydrates and isolated four fractions which accounted for nearly 94% of the total: 33% trehalose, 27% glucan (yeast cellulose), 21% mannan (yeast gum) and 12% glycogen.

Trehalose, a non-reducing reserve disaccharide (α-D-glucopyranosyl-α-D-glucopyranoside), occurs in yeast as the free sugar. Food yeast derived from brewer's yeast contains little, if any, trehalose (Trevelyan, 1958).

As major components of the yeast cell wall, yeast glucan and mannan are associated with proteins and lipids (Eddy, 1958). Cell wall material comprises about 15% of the dry yeast. Usually brewer's yeast contains less mannan than baker's yeast (Trevelyan, 1958). This polysaccharide is also associated structurally with yeast invertase (Eddy, 1958). Both glucan and mannan are well-defined, non-reducing, highly-branched polysaccharides. Enzymic and acid hydrolysis of isolated yeast glucan liberates only D-glucose, characterizing the complex mainly as $\beta(1 \rightarrow 3)$ -glucosan. Acid hydrolysis of yeast mannan yields only D-mannose. Structural studies of this polysaccharide indicate that it is predominantly $\alpha(1 \rightarrow 3)$-linked mannosan (Eddy, 1958).

Yeast glycogen, principally branched $\alpha(1 \rightarrow 4)$-glucosan, is a soluble polydisperse reserve carbohydrate considered identical with glycogens of animal origin. The quantity present, some loosely bound to protein, fluctuates in yeast with cultural changes in available sugar and oxygen.

Yeast lipids, usually expressed as crude ether extract, comprise only 2–3% of the cell dry matter (Frey, 1930). Current analytical procedure favours extraction after acid hydrolysis of the dried yeast. This treatment generally doubles the crude fat value of most food yeasts (see Tables VI and VII).

Triglycerides, lecithin and ergosterol are the major constituents of yeast lipids. In true yeast fat, triglycerides of oleic and palmitic acids predominate, resembling the composition of common vegetable fats (Hoogerheide, 1950; Eddy, 1958). Food yeast produced from brewer's yeast is lowest in total lipids, and its triglycerides contain less oleic acid, and more palmitic acid, than baker's yeast (Suomalainen, 1968; Suomalainen and Keränen, 1968).

Lecithin (phosphatidylcholine) accounts for 75% of the phospholipids in yeast (1–2% of dry weight). Cephalins (kephalin, phosphatidylethanolamine) make up the remaining 25% (Hoogerheide, 1950; Eddy, 1958).

The sterol content of dried yeast, mainly ergosterol, varies from 1% to 3% of the dry matter in food yeasts to 10% in superior strains of *Sacch. cerevisiae* (Dulaney et al., 1954; Dunn, 1958). In Latvian dried yeast Zirins (1967) found 0·65–0·75% ergosterol in the dry matter of baker's yeast, 0·25–0·34% in *C. tropicalis*, and 0·11–0·15% in *C. utilis*. Slurries of high-yielding ergosterol strains of *Sacch. cerevisiae* are irradiated with ultraviolet light (235–315 nm), converting ergosterol to calciferol (vitamin D_2). Assays of the dried irradiated yeast range from

120,000 to 180,000 I.U. per gram (one I.U. = 0·25 µg crystalline D_3, or activated 7-dehydrocholesterol).

Minor constituents of yeast lipids include zymosterol, cerevisterol, cerebrin and squalene. From the ether extract of food yeasts, Forbes *et al.* (1958) isolated in crystalline form a compound exhibiting antioxidant properties. J. Amsz and A. M. Moustafa (personal communication) demonstrated stabilization of corn oil and tallow with food yeast (*C. utilis*). Autolysis of the yeast cells before drying increased the antioxidant effect.

C. QUALITY SPECIFICATIONS

Dried yeast for human use is produced in accordance with quality guide-lines set by governmental agencies, professional associations and food processors. Food yeast must conform to specifications of yeast type, colour, flavour, microbial content, chemical composition and vitamin potency. The prevailing definitions of dried yeast are published by the International Union of Pure and Applied Chemistry (IUPAC), The British Pharmaceutical Codex, the National Formulary (N.F. XIII) of the American Pharmaceutical Association, and the food additive regulations of the Food and Drug Administration (FDA), U.S. Department of Health, Education and Welfare. In addition some food manufacturers impose additional chemical and bacteriological specifications for dried yeast added to products for thermal processing. Their limitations pertaining to bacterial spores are often more exacting than those recommended by the National Canners' Association.

Food yeast in domestic commerce generally conforms to the standards prescribed by the American Pharmaceutical Association and the food additive orders of FDA. Dried yeast was first described in the Pharmacopoeia of the United States (1947), XIII Revision. It was transferred to The National Formulary in 1960 (N.F. XI).

Dried Yeast, N.F. XIII, is defined, in part, as "the dried cells of any suitable strain of *Saccharomyces cerevisiae* ... or *Candida utilis* ... grown in media other than those required for beer production. ... Such yeasts, properly designated as to species, are commonly known as Primary Dried Yeast ... and Torula Dried Yeast". "Brewer's Dried Yeast" is defined as "a by-product from the brewing of beer ... washed free of beer prior to drying". When the washing step includes one or more alkaline treatments to remove hop resins adsorbed on the yeast cells, the product is designated "Debittered Brewer's Dried Yeast". Each type of "Dried Yeast", N.F. XIII, must be free of fillers (starch, corn meal, etc.) and contain a minimum of 45% protein (N × 6·25), not more than 7% moisture and 8% ash; and, in each gram, the equivalent of not less than 120 µg thiamin hydrochloride, 40 µg riboflavin and 300 µg niacin. Live bacteria and mould count maxima are 7,500 and 50/g respectively. In addition, the U.S. Food and Drug Administration (FDA) has established a zero tolerance for *Salmonella*.

Dietary usage of food yeast is further restricted (since 1963) by FDA on the basis of its folic acid content. When included in foods as a flavourant, dried yeast (*Sacch. cerevisiae*, *Sacch. fragilis*, *C. utilis*) may be added to the extent that the folic acid content it contributes does not exceed 40 µg/g yeast [about 8 µg of free folic acid (pteroylglutamic acid)/g]. Additions to foods in special dietary usage are limited to 400 µg total folic acid per day (100 µg free folic acid/adult). Analyses of primary dried yeast of recent production show both total and free folic acid levels to be well within the limits prescribed by FDA. For each type of food yeast, the average total folic acid content is about 12 µg/g. The free folic acid content of both dried yeasts is consistently less than 1 µg/g yeast (Peppler, 1965).

Standards and methods specified by IUPAC in 1963, define Dried Yeast as "the whole organism of one individual yeast or a mixture of several yeasts belonging to the family Saccharomycetaceae . . . and to the family Cryptococcaceae . . . obtained either as a by-product of fermentation processes or by special culture and conforming to such standards as may be laid down". Specifically excluded are yeasts which have been extracted, those containing more than 20% fat, and yeasts which carry inert fillers or substances that are not incorporated components of normal yeast cells. Nine standards are laid down for "Dried Yeast": upper limits are specified for moisture (10%), ash (10%), lead (5 µg/g), arsenic (5 µg/g), live bacteria (7,500/g) and moulds (50/g). Minimum levels are set for nitrogen (7·2%, equivalent to 45% protein), for thiamin (10 µg/g), riboflavin (30 µg/g) and niacin (300 µg/g). Dried Yeast must also be free of starch and bacteria of the genus *Salmonella*.

The British Pharmaceutical Codex in 1949 stated that food yeast be dried by "a process which avoids decomposition of the vitamins present", and in 1954 required, in each gram, not less than 100 µg thiamin, 40 µg riboflavin, 300 µg niacin (Pyke, 1958). The Swedish Pharmacopoeia in 1948, according to Pyke (1958), specified only the minimum content for thiamin (150 µg/g).

Colour, flavour, granulation and packaging of food yeast are not specified in official standards. In the trade, however, these properties assume economic and aesthetic importance according to the requirements of dietetic preparations, pharmaceutical products and food processing applications. The National Formulary (N.F. XIII) merely states, "Dried Yeast occurs as yellowish white to weak yellowish orange flakes, granules or powder, with an odour and taste characteristic of the type". Dried Torula Yeast is described in identical terms except for the final phrase, which reads "with a characteristic odour and bland taste".

III. Uses of Food Yeast

World-wide interest in food yeast is centred about two nutritional attributes: high quality concentrated protein, and abundance of B-complex vitamins. Further, food yeast usage is also favoured because of its mild, often bland, flavour, its non-hygroscopic and stable form (Garber et al., 1949), its compatibility with a variety of diets and ease of incorporation into familiar foods (Brenner et al., 1948, 1949; McCay, 1950). In the broadest sense food yeast is a major raw material. As dried yeast it can be smoked or otherwise treated to improve its flavour. Before drying, the concentrated suspension of yeast cells may be elevated in vitamin potency, or autolysed, or plasmolysed, or extracted. In the following pages a catalogue of food yeast uses, and the nature of transformed food yeast products, derivatives and isolated compounds is given.

A. FOOD ADDITIVE

Powdered dried yeast is an ingredient of foods, whether processed and packaged, prepared in households and institutions, or formulated to supplement and improve flavour and nutritive values of food mixtures at relatively low cost.

1. As Flavourant and Nutrient

The usual powdered form of food yeasts may be added to standardized cereal products defined by the U.S. Department of Health, Education and Welfare (Code of Federal Regulations, 1968; referred to as 21 CFR). Three types of food yeast (Sacch. cerevisiae, Sacch. fragilis, C. utilis) are optional ingredients in all breads, buns and rolls in amounts up to 2% of the weight of the flour used (21 CFR 17·1–17·5). Seeley et al. (1950) determined the nutritional value of white bread with added food yeast. In rye bread, a non-standard product, bakers could use as much as 5% C. utilis food yeast to enhance the flavour. Pretreatment of yeast intended as an ingredient of bread doughs is desirable to inactivate free glutathione and other reducing compounds which affect gluten elasticity, loaf texture and volume. Mild oxidizing agents (bromates, iodates, peroxides) are suitable for treatment of the liquid yeast before drying. A process for complexing glutathione in C. utilis food yeast from spent sulphite liquor is described by Ferrara and Dalby (1967).

Enriched macaroni products and enriched noodle products may be fortified with food yeast to supply all or part of the vitamin and mineral requirements. In each pound of product the limits are 4–5 mg thiamin, 1·7–2·2 mg riboflavin, 27–34 mg niacin, 250–1,000 U.S.P. units vitamin D, 13·0–16·5 mg iron (as Fe), and 500–625 mg calcium (21 CFR 16·9–16·12 and 16·14).

For enrichment of breakfast cereals and canned baby and geriatric foods, food yeast levels of 0·5–2% are commonly used. Supplements of 1–3% food yeast are acceptable in many foods and prepared dishes: biscuits, muffins, crackers, chocolate cake, baked macaroni, meat dishes, peanut butter (up to 20% dried yeast), desserts and puddings (Sure, 1946). Doughnuts of satisfactory palatability can be made with dried yeast replacing 10% of the flour (McCay, 1950). Syrups used on pancakes and waffles, cookies, soups, gravies, confections, salad dressings, snacks, sausages, specialty products and pet foods are excellent vehicles for incorporation of the flavour and nutrient qualities of different food yeasts (Lyall, 1964).

INCAPARINA, the successful cereal mixture for the people of Central America, is fortified with 3% food yeast (Béhar, 1963). A similar high-protein cereal blend, ARGENTARIA, contains 2% dried yeast (Bressani and Elias, 1968).

Grated cheese dry-blended with 2–5% food yeasts exhibits improved flavour and storage stability (Traisman and Kurtzhalts, 1957). Seasoning blends may contain as much as 10–15% food yeast.

Some ancillary benefits not usually apparent in food yeast applications include the antioxidant properties of yeast (J. Amsz and A. M. Moustafa, personal communication; Forbes et al., 1958; Pinkos, 1967) and its moisture-holding and bulking capacities (Lyall, 1964).

2. As Dietary Supplement

Nutritionally, food yeast is used in the form of powder, capsules and compressed tablets. It is a convenient way for hospitals, institutional kitchens and households to furnish protein and the vitamin-B complex in available form (Parsons et al., 1945; Price et al., 1947) in the nourishment of patients, health food enthusiasts and persons otherwise on restricted diets or in confinement. Klapka et al. (1958), experimenting with hospital diets, reported food yeast to be a practical way of increasing protein, vitamin and mineral intake of 300 mental patients. Yeast-supplemented dishes especially well received were soups, meat casseroles, chili, gravies, tomato sauces, salads and cooked vegetables.

In America the usual adult therapeutic dose of dried yeast is 40 g per day (N.F. XIII); but in Britain 8 g per day is recommended (British Pharmaceutical Codex). One heaped teaspoonful of food yeast weighs about 5 g. A daily dietary supplement of 10 g of food yeast powder or fine flakes (N.F. grade) contributes substantially to the recommended dietary allowances for adults. It supplies about 100% of the recommended daily allowance of thiamin, 25% riboflavin, 20% niacin, 15% pyridoxine and 7% protein.

Yeast tablets, despite their higher cost, are a popular food complement. The adult dosage usually prescribed is four tablets, three times daily. Based on a tablet weight of 6·8 gr (0·44 g), the common size, the ordinary daily dosage of 12 tablets (5·28 g) of a popular proprietary blend of food yeasts provides the following recommended dietary allowances: 84% thiamin, 40% riboflavin, 24% niacin, 45% pyridoxine and 4% protein. These levels can be increased substantially with primary-grown yeasts and brewer's yeast fortified with vitamins and minerals (see Section III.B.2.a, p. 444). A novel dietetic yeast-starch tablet, developed by Griffon and Tixier (1968), is made directly from fresh yeast without heat.

B. MODIFIED PRODUCTS

In its harvested liquid or cream state, yeast of food quality is also a raw material which is readily converted, in a variety of ways, to flavorous nutritious concentrates in paste or dry form. Some yeast is also treated to enable isolation of enzymes, nucleic acids and other cellular components.

Before, as well as after dehydration, food grade yeast can be altered in composition to ennoble yeast with unusual food flavours, and increase its dietary content of vitamins, minerals and amino acids.

1. Flavour-Enhanced Products

a. *Yeast flavour blends.* Dried yeast powders are excellent carriers of food flavours. Smoked yeast is an especially popular food ingredient. It imparts a bacon-like taste and aroma to products formulated with it. Other flavour-bearing yeast powders appear from time to time. One British manufacturer offers a wide range of yeast-based flavour blends which impart to foods such flavours as cheese, ham, tomato, onion, celery and paprika (Lyall, 1964).

b. *Autolysates.* In autolysis, viable yeast is induced to digest itself by its cellular enzymes (Joslyn, 1955; Joslyn and Vosti, 1955). A high protein, primary-grown strain of *Sacch. cerevisiae* which autolyses under mild conditions is preferred for quality-assured extracts (Albrecht and Deindoerfer, 1966). Because of its lower cost, debittered brewer's yeast is commonly used for yeast extract production.

Self-digestion of viable yeast occurs rapidly in an agitated slurry (15–18% yeast solids) brought slowly to 45–50°. To speed up autolysis small amounts (3–5%) of sodium chloride, ethyl acetate or chloroform are added. The reaction is continued for 12–24 h, or until the desired concentration of soluble nitrogen is attained. The digested mixture may be pasteurized and dried directly. In the usual practice, however, a clear extract is concentrated to a thick syrup. It may be used in this form or dehydrated in conventional driers (East *et al.*, 1966).

One spray-dried extract available in the United States, derived from autolysed primary-grown yeast, contains 96·5% solids (53·5% organic matter and 43·0% ash), 7·3% nitrogen, 6·1% monosodium glutamate and 36·5% sodium chloride (Albrecht and Deindoerfer, 1966). A vacuum-dried yeast extract in granular form, derived from autolysed brewer's yeast in Great Britain, contains 95·5% solids, 26% ash (includes 15% sodium chloride) and 8·5% nitrogen (Anon., 1964a). Both extracts are hygroscopic and form clear light-coloured water solutions. They contribute meaty flavour and aroma to soups, meats, sausages, gravies, sauces, sea-foods and vegetables. When dried without separation of undigested yeast matter, principally cell walls, the non-hygroscopic powder contains about 92% solids, 7–9% nitrogen, 2–3% free amino acids, 6–8% ash and 50% solubles.

Yeast autolysates with sweet, tart, fruity or malty taste were developed by Lendvai (1963). His invention embodies a three-stage process: plasmolysis and liquefaction of pressed yeast with fermentable carbohydrates, controlled fermentation of the mixture, and autolysis at 55° in the presence of a high concentration of sugar. The process and products obtained thereby are in commercial development.

c. *Plasmolysates.* Intimate contact of viable yeast cells with high concentrations of salt, sugar or certain acetate esters extracts cellular materials without significant involvement of yeast catabolic enzymes. Meaty-flavoured extracts obtained by salt plasmolysis are unusually high in salt content, and thus are limited in their application to food formulations.

A novel yeast product of enhanced nutritional value and diminished yeast flavour was patented by Schwarz BioResearch, Inc. (1962). It is the result of yeast plasmolysis with sugars, followed by phosphorylation of sugars and B-group vitamins (*viz.*, thiamin, riboflavin, and pyridoxine) by the enzymes released. The dried end-product contains 85–95% solids, 3–7% nitrogen, 10–20% organic phosphate and 40–60% carbohydrate.

d. *Hydrolysates.* The third, and most efficient, means of making yeast extract is accomplished by controlled cooking of yeast in acid solution. A typical hydrolytic process begins with dried yeast reslurried in water to a solids concentration of 65–80%. After acidification with hydrochloric acid, hydrolysis is conducted in a wiped film evaporator fitted with a reflux condenser. Reaction time varies from six to twelve hours, depending on the level of amino nitrogen desired (Ziemba, 1967).

Neutralization of the hydrolysate is followed by filtration, decoloriza-tion and concentration to a syrup (45% solids) or paste (85% solids), or the liquid is spray-dried (95% solids). Yeast hydrolysate syrups contain about 42% solids, 18% sodium chloride, 2·5–3·0% nitrogen, and 3·5% free glutamic acid. Liquid hydrolysates diluted with two parts of yeast cream can be dried successfully on drum driers. While hydrolysis affords the highest yields of yeast extract, some loss of vitamins, protein and flavour occurs.

Rosenthal and Pinkalla (1960) produced a novel food flavouring composition by reacting a mixture of three parts of a vegetable protein hydrolysate with one part of dried torula yeast at 40° for 20 h. The final product, a buff-coloured stable homogeneous paste, is recommended for the improvement of the flavour and texture of processed cheddar cheese spreads.

2. Products Increased in Nutritive Value

a. *Fortified food yeast*. Therapeutic formulations of food yeast effecting high vitamin potency, improved mineral balance and increased amino acid content are prepared to meet a wide range of specifications. Four methods of enrichment may be practised: (i) fortification of the fer-mentation broth with crystalline vitamins and amino acids, or their precursors (see Chapter 10, p. 529); (ii) addition of crystalline vitamins, minerals and amino acids to the yeast suspension before drying; (iii) irradiation of yeast cream with ultraviolet light; (iv) blending dried yeast with crystalline vitamins and minerals.

By means of methods i or iv, the more common routes, food yeast can be fortified to vitamin potencies about 50 times greater than the levels normally occurring in yeast. Enrichment with thiamin, riboflavin and niacin is the usual practice: but in view of the revised Recommended Dietary Allowances (1968), which include for the first time vitamin B_6 and vitamin B_{12}, these too will be increased to higher concentrations in food yeasts. At the cost of pure vitamins today, enrichment of food yeast is relatively inexpensive. USP- or NF-quality vitamins are at present (early 1969) available in quantity at the following prices (per kg): thiamin hydrochloride, \$14; riboflavin, \$32; niacin, \$3; pyridoxine hydrochloride, \$26; and cyanocobalamin, \$8/g activity.

Enrichment of food yeast with thiamin pyrophosphate, the bio-logically-active form of vitamin B_1, is also practised. One health food formulator fortifies dried torula yeast, which is naturally low in sodium, with ten B-group vitamins and the oxides of calcium and magnesium to balance the phosphorus content.

Upon irradiation with ultraviolet light, ergosterol in yeast is converted to calciferol (vitamin D_2) (see Chapter 10, p. 529). One domestic yeast-maker employs a strain of *Sacch. cerevisiae* with a very high ergosterol content in preparing a special food yeast containing 120,000 I.U. vitamin D_2 (i.e. 3 mg) per gram dry product.

Various combinations of dried yeast, autolysates of yeast and extracts of protein hydrolysates are made to interest food processors and health food enthusiasts. One product of dietary significance combines, prior to drying, equal parts of yeast autolysate and enzyme-hydrolysed lactalbumin. In this high nitrogen (9·5–10·5%) powder, the free amino nitrogen content ranges from 35% to 40% of the total nitrogen.

The lysine content of yeast is increased 300–400% when adipic acid precursors are added to the propagating medium (Jensen and Shu, 1961; Rieche *et al.*, 1966). Although commercially feasible, this route to lysine-rich yeast is limited by the high cost of the precursors. Availability of cheap L-lysine hydrochloride favours direct blending of yeast as well as cereal grains and other foods deficient in lysine.

b. *Low purine yeast.* A "depurinized" dried yeast in the trade contains about 1·5% nucleic acid, as compared with the normal yeast content of 7–8%. Cold acid treatment of yeast cream, followed by neutralization and hot salt extraction, effectively depletes yeast of its nucleic acids.

3. *Concentrated and Isolated Constituents*

Since yeast is available in ton quantities, it is a principal source of biocatalysts, nucleic acids and their derivatives. Harrison (1968) reviewed the occurrence and recovery of yeast biochemicals. Only concentrates and purified compounds in volume production are dealt with here (see also Chapter 10, p. 529).

a. *Enzymes and coenzymes.* Invertase (β-D-fructofuranoside: fructohydrolase; E.C. 3.2.1.26; β-fructosidase, saccharase, sucrase) is obtained primarily from *Sacch. cerevisiae.* Two commercial forms are produced: an invertase-rich dry yeast used to prepare high test molasses from sugar cane juice (Peppler and Thorn, 1960), and a clear liquid concentrate extracted from autolysed yeast for the candy industry (Reed, 1966). Product activity in the trade is expressed as the velocity constant k, calculated according to A.O.A.C. Method 29.024 (1965). Current dry product yields k values between 3 and 4, equivalent to about 1,400 to 1,800 invertons/g (Johnston *et al.*, 1935). Liquid invertase preparations in commerce are generally standardized at a k value of 0·3 for the single strength solution. Double strength solutions (k value of 0·6) are also available. Established methods of invertase recovery from yeast were reviewed by Neuberg and Roberts (1946). Suomalainen *et al.* (1967) found invertase activity of baker's yeast was nearly tripled when grown with mannose as the carbon source instead of glucose. Highly purified invertase is obtained by processing *C. utilis* cells. This yeast is also the preferred source for uricase and 6-phosphogluconate dehydrogenase.

Lactase (E.C. 3.2.1.23, β-galactosidase), isolated from *Sacch. fragilis*, is available in purified form and as spray-dried cells (Myers and Stimpson, 1956). Limited applications have created a narrow market (Pomeranz, 1964; Reed, 1966).

Other purified biocatalysts isolated from *Sacch. cerevisiae* in substantial quantity include alcohol dehydrogenase, hexokinase, L-lactate dehydrogenase, glucose 6-phosphate dehydrogenase, glyceraldehyde 3-phosphate dehydrogenase, inorganic pyrophosphatase, coenzyme A, the oxidized and reduced diphosphopyridine nucleotides (NAD, $NADH_2$), and the respiratory coenzymes, the phosphates of adenosine (ADP, ATP), cytidine, guanosine and uridine. Their preparation from yeast follows similar extraction and purification procedures (Reece *et al.*, 1959; Harrison, 1968). Of these compounds, ATP and NAD are produced in greatest volume.

While yeast is also a good source of catalase, the enzyme of animal origin dominates the market because it is recovered more economically.

b. Nucleic acids and derivatives. Recovery and processing of large amounts of ribonucleic acid (RNA) serves two purposes: (i) as raw material for the enzymatic conversion to flavour-potentiating nucleotides; and (ii) as starting material for the purification and derivation of nucleoside and nucleotide analogues favoured in biochemical research. Production of the potent flavouring agents, guanosine 5'-monophosphate (GMP) and inosine 5'-monophosphate (IMP), stems from two large yeast operations in Japan. Both extract RNA from carbohydrate-grown high-yielding (12% RNA) strains of *C. utilis*. Mutants of *C. tropicalis* cultivated on paraffinic substrates also produce high RNA yields (Takeda Chemical Industries, 1967). For use in food applications these flavour potentiators are supplied in two blends: a mixture (1:1) of disodium inosinate and disodium guanylate, and 5% of this blend diluted in monosodium glutamate.

c. Ergosterol. Extraction of yeast ergosterol for conversion to vitamin D_2 by irradiation (see Chapter 10, p. 529) is practised only in a small way in West Germany (World Wide Survey of Fermentation Industries, 1966). Domestic interest in extracting ergosterol from brewer's yeast has been noted (Anon., 1968c), but no product has been marketed.

IV. Technology

Food yeast production developed as an extension of processes yielding yeast crops intended for use in baking, brewing, distilling or livestock feeding. Consequently American varieties of food yeast emanate from four types of production: (i) baker's yeast propagated by fed-batch processes in molasses media to specified nitrogen and vitamin compositions (see Chapter 7, p. 349); (ii) *Candida utilis* grown continuously in spent sulphite liquor to deplete it of pollutants before discharge to natural waters; (iii) *Saccharomyces fragilis* grown batchwise in cheese whey for the purpose of lowering its biochemical oxygen demand; and (iv) debittered brewer's yeast derived from spent beer yeast.

A few exceptions to such incidental approach may be cited. The German programmes to develop technology for converting wood sugars were directed primarily to food yeast and its nutritional adequacy (Schmidt, 1947; Butschek 1962). The propagating ventures in Jamaica, Taiwan, South Africa, Wisconsin and the Philippines were planned for food yeast production (Thaysen, 1957; Kaiser and Jacobs, 1957; Chien, 1960). The widespread investigations of the industrial potential of fat synthesis in yeasts and other micro-organisms focused on food applications, although no large-scale process developed (Woodbine, 1959; Prescott and Dunn, 1959).

In the United States today yeast is a primary product in 16 plants, but only four of these manufacture dried food yeast, two from molasses (*Sacch. cerevisiae*), one from wood sugars (*C. utilis*) and one from cheese whey (*Sacch. fragilis*). In addition, three manufacturers provide the major portion of debittered brewer's yeast.

A. PRIMARY CULTIVATION

Technologically the manufacture of food yeast follows, with appropriate modifications and refinements, the principles practised in baker's yeast production (Chapter 7, p. 349). The essential steps in a conventional system are illustrated in Fig. 1. Industrial practices reported in the literature indicate the extent of process variations around the world. Thus, in Taiwan, six Waldhof fermenters convert molasses to yeast (Chien, 1960). This type of fermenter was designed to grow yeast from pulp mill wastes. Since only general treatment of yeast technology is presented here, the following works are recommended: Woods (1951), White (1954), Wiley (1954), Pyke (1958), Drews *et al.* (1962), Butschek (1962), Suomalainen (1964), Freund (1964), Peppler (1967, 1968), Harrison (1967).

1. *Molasses-Grown Yeast*

Biosynthesis of yeast is accomplished with proven cultures grown aerobically through successively larger culture generations in diluted molasses media fortified with ammonia and phosphorus (see also Chapter 7, p. 349). In the early stages, asepsis is rigorously followed. As fermenter volumes increase, however, control of microbial contaminants,

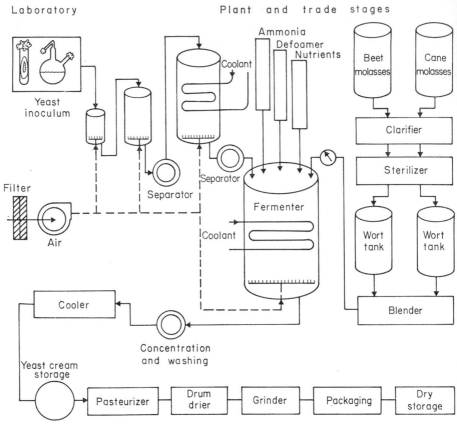

FIG. 1. Food yeast process stages for molasses media.

even in semi-closed systems, is difficult. The challenge is met with superior sanitation practices, sterilized raw materials (including air), efficient crop recovery procedures and prompt processing.

In the United States, food yeast production from molasses employs *Sacch. cerevisiae.* Elsewhere, notably in Taiwan and South Africa, *C. utilis* is cultivated. While beet molasses is preferred for food yeast production in the U.S.A., its higher price dictates blending it with the cheaper cane molasses. Average analytical values (Table XII) provide a general comparison. Detailed comparisons were made by Rogers and Michelson (1948) and Becker and Webster (1962). Both substrates supply fermentable sugars, essential minerals (potassium, magnesium, phosphorus, zinc, iron, copper), vitamins (biotin, pantothenic acid, pyridoxine, thiamin) and amino nitrogen (mainly asparagine, aspartic acid, glutamic acid). Beet molasses is richer in total organic nitrogenous compounds than blackstrap, but half is betaine which is not assimilated by *Sacch. cerevisiae*. Both types of molasses are nearly equal in amounts of fermentable sugars, potassium, trace minerals, niacin and inositol. Cane molasses is substantially richer in magnesium, calcium, biotin, pyridoxine, pantothenic acid and thiamin. To produce food yeast, properly clarified molasses (Peppler, 1967) needs added amounts of nitrogen and phosphorus. American manufacturers employ phosphoric acid, its ammonium salts, anhydrous ammonia and ammonium sulphate. Wherever ammonia is lacking, urea is used economically in cane molasses fermentations (Chien, 1960).

TABLE XII. *Molasses Composition*

Analysis	Average value	
	Beet molasses	Cane molasses
Invert sugar (%)	57	59
Nonfermentables (%)	2·1	3·5
Ash (%)	6·3	5·9
P_2O_5	0·02	0·1
CaO	0·5	0·8
MgO	0·1	0·7
K_2O	3·7	2·2
Vitamins (µg/g)		
Biotin	0·08	0·7
Thiamin	0·6	1·0
Pyridoxine	5·5	35
Nitrogen (%)	1·6	0·4
Betaine	0·8	0
Amino nitrogen	0·4	0·15

The last stage of food yeast propagation is conducted in a high-capacity fermenter fabricated of corrosion-resistant materials and fitted with a cooling coil, aeration grid, defoaming agent controller, molasses metering system and sampling device. Ebner *et al.* (1967) described a yeast fermenter fitted with a mechanical defoamer and self-priming (without compression) aeration system. After charging the sterilized fermenter with potable water, sterile clarified molasses, seed yeast and the required chemicals, a propagation proceeds according to predetermined schedules for air input, molasses increments and allocated ammonia and phosphate salts (see Chapter 7, p. 349).

Efficient biosynthesis depends on careful management of two critical factors: air input and sugar concentration. Over-feeding and under-aeration result in formation of ethanol, which may be lost to the atmosphere. Losses in available carbon are avoided by maintaining the sugar level below 0·1% and the oxygen absorption rate above 2 mM O_2/litre min (Chen, 1959; Strohm and Dale, 1961). According to Maxon and Johnson (1953) and Johnson (1959), synthesis of yeast matter of average composition requires an equal weight of oxygen. Harrison's (1967) calculations confirm this. He further determined the overall formula for yeast synthesis (on a weight basis) to be:

$$200·0 \text{ g sucrose } + 10·4 \text{ g NH}_3 + 102·5 \text{ g O}_2 =$$
$$100·0 \text{ g yeast dry matter (including 9·5 g inorganic matter) } +$$
$$145·2 \text{ g CO}_2 + 77·2 \text{ g H}_2\text{O.}$$

In practical terms the production of 100 lb of yeast dry matter containing about 50% protein (N × 6·25) requires 400 lb molasses, 25 lb aqua ammonia, 15 lb ammonium sulphate, 7 lb monoammonium phosphate and 75,000 ft^3 air.

Upon completion of the propagation schedule, the fermenter contents are cooled while the cell crop is concentrated and washed by recycling with potable water through centrifugal separators. The light-coloured suspension of cells, commonly called yeast cream, is chilled and stored, or processed directly. After enrichment with vitamins, if needed, to meet established standards or customer specifications, the cream is pasteurized, dried, milled to a powder and packed in plastic-lined fibre drums.

Each lot is sampled for chemical, bacteriological and physical analyses. Besides meeting the quality and purity standards discussed above, dried yeast of proper colour and particle size must be produced. Powders of lightest colour are the result of superior molasses clarification (Vámos and Pozsár, 1968, 1969) and elimination of metal-ion contaminants, especially copper ions. High quality powders frequently show reflectometer readings of 60–65%, and at least 70% of the product passes through U.S. Standard Sieve No. 120.

The conventional system outlined above can be adapted to semi-continuous yeast manufacture. However, the relatively low demand for food yeast, the high cost of redesign, limitations in co-ordinating baker's yeast production, and greater vulnerability of continuous propagations to unwanted micro-organisms have, in the United States, terminated interest at the pilot stage, although according to Hospodka (1966) continuous processes have been used in various other countries. Modern systems designed around the Waldhof fermenter (Wiley et al., 1951; Kaiser and Jacobs, 1957; Chien, 1960) are unique engineering achievements which permit essentially automatic continuous yeast synthesis. A competing propagator, the tower fermenter, is gaining favour in the socialist republics (Lefrançois, 1963; Lefrançois and Revuz, 1964).

2. *Yeast from Sulphite Liquor*

Owing to its composition (see Chapter 6, p. 283), spent sulphite liquor must be processed by methods unlike those practised with molasses substrates. The Waldhof fermenter system, shown in Fig. 2, was developed to maintain the aerobiosis needed for efficient sugar removal by the yeast of choice, *C. utilis*. Waldhof-type plants originated in Germany and have been adopted throughout Europe, western Asia,

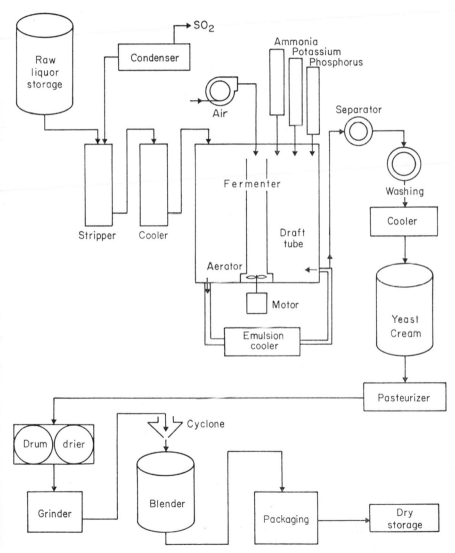

FIG. 2. Food yeast process stages for wood sugar substrates.

Japan and Taiwan. Two sulphite yeast plants were built in the United States. They were described by Wiley (1954), Dyck (1957) and Kaiser and Jacobs (1957). One is operating at the present time.

For food yeast production, blended sulphite liquor is stripped of sulphur dioxide, adjusted to pH 4·5 with ammonia, and cooled. This liquor is fed continuously to the draft tube of a 60,000-gal modified Waldhof fermenter at a rate of approximately 100 gal/min, depending upon the sugar content. Sugar and phosphorus levels in the liquor are monitored automatically, providing accurate proportioning of phosphate additions according to the sugar load to the fermenter. Potassium additions, as potassium chloride, are also based on sugar input. Propagation temperature is held near 32° by continuously recycling fermenter emulsion through an external refrigerated cooler. Yeast is harvested by withdrawing emulsion at a rate equal to the liquor input, defoamed mechanically, separated, washed repeatedly, concentrated to about 15% solids, enriched with vitamins, dehydrated on steam-heated double-drum driers, pulverized and packed in plastic-lined fibre drums or multiwall bags.

In Table XIII are data illustrating the complexity of sulphite liquor composition, and the range of carbon compounds utilized by *C. utilis* from two dissimilar liquor blends. The degree to which non-sugar carbon is assimilated is further reflected in the yeast yield, calculated on the sugar supplied (Table XIV).

TABLE XIII. *Analysis of Liquor Feed and Fermentation Beer*

Analysis	Content (g/litre)			
	Blend I		Blend II	
	Feed	Beer	Feed	Beer
Hexose	11·7	0	9·6	0
Pentose	7·1	0	14·0	0
Non-fermentable (as glucose)	5·7	5·7	3·9	5·0
Ethanol	1·1	0·6	1·5	0·6
Ether soluble (as acetic acid)	7·2	3·3	17·5	6·6
Steam distillate (as acetic acid)	15·9	6·1	16·9	15·8
Lignin	46·0	46·2	47·6	45·5
Formic acid	1·7	1·6	5·5	2·4

In spite of the protein needs and wants of others, the abundance of pulp mill waste, and the acknowledged superior performance of yeast in it, no new fermenter units are at present in prospect in the United

TABLE XIV. *Yeast Performance on Ammonium Base Liquor (Adapted from Peppler, 1965)* [Range over 10-day period at 32°, 2,300 ft³ air/min, rotor speed 390 r.p.m.; discontinuous operation (weekend shut down)]

Parameter	Range
Sugar in feed (g/litre)	15–20
Sugar loading (lb/h)	750–1076
Sugar utilization (%)	77–89
Retention time (min)	182–217
Yeast yield (% on sugar)	54–67

States. Where new yeast plant expansion is occurring, as in Europe and Asia, the dried yeast is intended for non-food uses.

3. Whey-Grown Yeast

By means of the conventional batchwise method with sparger-type fermenters, the lactose-assimilating yeast *Sacch. fragilis* is grown in sterilized cottage cheese whey (Powell and Robe, 1964). Ammonium salts, phosphate and sulphate are added to supplement the small amount of nitrogenous material available in whey during yeast growth. The pH value of the fermentation usually fluctuates within the range 5·0 to 5·7, but it may move higher. Automatic control monitors the sulphuric acid needed for pH adjustment. In a propagation cycle of six to eight hours, the daily yield is about 5,000 lb of dried yeast from 180,000 lb whey.

With vast quantities of whey available ($1·4 \times 10^{10}$ pounds annually), efforts to upgrade it through yeast fermentation are continuing. Amundson (1967) is experimenting with a continuous propagator. Chapman (1966) reported plans for a commercial plant in New Zealand. Its rated capacity is a ton of dried yeast per day. To encourage the efforts being made in the cheese industry towards pollution abatement, the U.S. Department of Agriculture supported extensive pioneering studies with *Sacch. fragilis* (Wasserman *et al.*, 1958; Wasserman, 1960, 1961). Numerous other yeasts have been evaluated for efficiency of protein synthesis in whey-based media (Stimpson and Young, 1957). In a continuous culture system, Atkin *et al.* (1967) attained higher cell yields and lactose conversion efficiencies for *Trichosporon cutaneum* than with *Sacch. fragilis*.

4. Hydrocarbon-Grown Yeast

World-wide interest in protein, and successful pilot demonstration at

Lavera, France (Anon., 1964b) that microbial protein can be produced from petroleum fractions, have greatly accelerated research efforts toward large-scale applications (Anon., 1966; Anon., 1968e; Champagnat and Lainé, 1966; Filosa, 1966; Evans, 1968; Guenther, 1965; Llewelyn, 1968; Miller and Johnson, 1966; Mateles and Tannenbaum, 1968; Perkins and Furlong, 1967). In the forefront of these studies is yeast, usually adapted strains of *C. lipolytica* (Dostálek *et al.*, 1968; Johnson, 1967; Klug and Markovetz, 1967). *Candida tropicalis* (Otsuka *et al.*, 1966) and *Trichosporon japonicum* (Toyo Koatsu Industries, 1967) also grow well on paraffinic compounds. However, as Darlington (1964) and others (Guenther, 1965) point out, the oxygen requirement on hydrocarbon substrates is exceptionally high.

At present no food yeast is produced commercially from hydrocarbons. However, the current widespread experimental interest and commercial planning, though directed primarily to biomass for livestock feeding, holds in the offing the hope of ultimate production of food quality yeast.

At a recent symposium on the biological uses of petroleum products, Llewelyn (1968) summarized the processes employed for the production of yeast protein. These include both batch (Lavera) and continuous (Grangemouth) systems. In the Lavera operations, *C. lipolytica* is grown in an aqueous emulsion containing minerals, yeast extract and gas oil. During a two-day batchwise propagation, yeast utilizes the normal alkanes in preference to the branched compounds. Recovery of yeast from the emulsion employs conventional methods. In the washing steps, however, scrubbing with surfactants and solvent extraction is necessary, as Shacklady (1969) notes, to remove unmetabolized gas oil adsorbed on the cell mass. This step also removes most of the natural lipids in the yeast (Table XV). Experience at Lavera has encouraged new construction at Grangemouth, Scotland (Evans, 1968). Instead of gas oils, it will feed normal paraffins to the bioreactor. This has the advantage that a higher proportion of the substrate is metabolized, and consequently the preparation of the final product is simplified. Japan has announced plans to produce yeast from normal paraffins in 1970 (Anon., 1968d). The Russians have an experimental unit making daily 15 tons of microbial protein from natural gas as well as from paraffins (Anon., 1968e). Joint projects have also been proposed (Anon., 1967): these are based on the British process (British Petroleum Co., 1966) described by Evans (1968). Gulf Oil has begun the first American large-scale experimental production of yeast protein from n-paraffins at Wasco, California (Anon., 1969).

According to Llewelyn (1968) the typical elementary composition of yeast grown on hydrocarbons is $(C_{23}H_{40}O_{12}N_4)$ + ash, a formula re-

TABLE XV. *Composition of Yeast Grown on Hydrocarbons* (In part from Llewelyn, 1968)

Analysis	Substrate		
	Gas oil	*n*-Paraffin	Sulphite liquor[a]
Nitrogen (%)	11·3	10·4	8
Protein (N × 6·25) (%)	70·5	65·0	50
Lipid (%)	0·5	8·1	5
Ash (%)	7·9	6·0	< 8
Moisture (%)	5·0	4·2	< 7
Amino acids (g/16 g N)			
Lysine	7·8	7·0	6·7
Valine	5·8	5·4	6·3
Leucine	7·8	7·0	7·0
Isoleucine	5·3	4·5	5·3
Threonine	5·4	4·9	5·5
Methionine	1·6	1·8	1·2
Phenylalanine	4·8	4·4	4·3
Tryptophan	1·3	1·4	1·2
Cystine	0·9	1·1	0·7
Histidine	2·1	2·0	1·9
Tyrosine	4·0	3·5	3·3
Arginine	5·0	4·8	5·4

[a] *Candida utilis*, see Tables VII and VIII.

markably close to that found for yeasts grown on carbohydrate sources, which approximates to $(C_{24}H_{40}O_{12}N_4)$ + ash (see Section II.B.1, p. 430; Chapter 6, p. 283; Chapter 7, p. 349). The individual amino acid contents fall within the ranges given in Table VIII for carbohydrate-grown yeasts. Nutritional tests on pigs and chickens showed that, when used at commercial levels, hydrocarbon-produced protein performed very satisfactorily. Extended toxicity tests, which are essential for a product of this type, have been in progress for some years in order to establish the complete absence of long-term ill effects (Llewelyn, 1968; Shacklady, 1969).

The yeasts selected for propagation on hydrocarbons are discussed in Section II.A (p. 427).

B. BY-PRODUCT RECOVERY

Surplus yeast obtained during beer manufacturing is readily freed of adsorbed hop resins by repeated washings with dilute alkali (Singruen, 1953; Pyke, 1958). Nay (1954) described a satisfactory process. Into 1,500 lb of brewer's yeast slurry (about 15% solids) are mixed one pound each of caustic soda and tripolyphosphate: the supernatant is then decanted, or the yeast is recovered by centrifugation. More than one alkaline wash may be needed. The final suspension of debittered yeast is made in potable water and mildly acidified (pH 5·5–5·7). Before drying, the yeast slurry may be fortified with vitamins and mineral nutrients (Dunn, 1952). Alternately, products of autolysed, plasmolysed or hydrolysed debittered brewer's yeast can be prepared for nutritional and flavouring purposes (see Section III.B.1, p. 442).

References

A.O.A.C. (1965). "Official Methods of Analysis", 10th Ed., pp. 490–491. Association of Official Agricultural Chemists, Washington, D.C.

Agricultural Statistics (1966). U.S. Dep. Agric., U.S. Government Printing Office, Washington, D.C.

Ahearn, D. G., Meyers, S. P. and Nichols, R. A. (1968). *Appl. Microbiol.* **16**, 1370–1374.

Albrecht, J. J. and Deindoerfer, F. H. (1966). *Fd Engng* **38** (10), 92–95.

Amundson, C. H. (1967). *Am. Dairy Prod. Mfg Rev.* **29** (7), 22, 23, 96–99.

Anon. (1964a). *Fd Engng* **36** (3), 134.

Anon. (1964b). *Chem. Engng News* **42** (2), 42.

Anon. (1966). *Chem. Wkly* **99** (34), 91–94; **100** (2), 61–62.

Anon. (1967). *Chem. Engng News* **45** (2), 46–48.

Anon. (1968a). *Fd Engng* **40** (5), 15.

Anon. (1968b). *Chem. Engng News* **46** (43), 5, 118–144; (44), 90–107.

Anon. (1968c). *Chem. Wkly* **103** (4), 57–58.

Anon. (1968d). *Chem. Wkly* **103** (22), 47.

Anon. (1968e). *Chem. Wkly* **102** (9), 49.

Anon. (1969). *Chem. Engng News* **47** (21), 17.

Atkin, C., Witter, L. D. and Ordal, Z. J. (1967). *Appl. Microbiol.* **15**, 1339–1344.

Atkin, L., Schultz, A. S., Williams, W. L. and Frey, C. N. (1943). *Ind. Engng Chem. analyt. Edn* **15**, 141.

Baird, F. D. (1963). "The Food Value and Use of Dried Yeast", pp. 1–36. Cerevisiae Yeast Institute, Chicago, Ill.

Becker, D. and Webster, J. (1962). *In* "Die Hefen", (F. Reiff, R. Kautzmann, H. Lüers and M. Lindermann, eds.) Vol. II, pp. 31–78. Verlag Hans Carl, Nürnberg.

Béhar, M. (1963). *J. Am. med. Wom. Ass.* **18**, 384–388.

Bertullo, V. H. and Hettich, F. P. (1966). German Patent 1,209,090.

Braude, R. (1942). *J. Inst. Brew.* **48**, 206–212.

Brenner, S., Dunlop, S. G. and Wodicka, V. O. (1948). *Cereal Chem.* **25**, 367.

Brenner, S., Wodicka, V. O. and Dunlop, S. G. (1949). *J. Am. diet. Ass.* **25**, 409.

Bressani, R. and Elias, L. G. (1968). *Adv. Fd Res.* **16**, 1–103.

British Petroleum Co., Ltd. (1966). Brit. Patent 1,049,067.

Brunner, R. (1960). *In* "Die Hefen", (F. Reiff, R. Kautzmann, H. Lüers and M. Lindemann, eds.) Vol. 1, pp. 368–402. Verlag Hans Carl, Nürnberg.

Bunker, H. J. (1963). *In* "Biochemistry of Industrial Micro-organisms", (C. Rainbow and A. H. Rose, eds.), pp. 34–63. Academic Press, London.

Bunker, H. J. (1964). Report to the Royal Society's National Committee for the International Biological Programme (unpublished).

Butschek, G. (1962). *In* "Die Hefen", (F. Reiff, R. Kautzmann, H. Lüers and M. Lindemann, eds.) Vol. II, pp. 611–664. Verlag Hans Carl, Nürnberg.

Census of Manufacturers (1963). U.S. Bureau of the Census, Vol. II, Industry Statistics, Part I. U.S. Government Printing Office, Washington, D.C.

Champagnat, A. and Lainé, B. (1966). Brit. Patent 1,049,065; 914,567 (1963).

Champagnat, A., Vernet, B., Lainé, B. and Filosa, J. (1963). *Nature, Lond.* **197**, 13–14.

Chang, W. S. and Peterson, W. H. (1949). *J. Bact.* **58**, 33–34.

Chapman, L. P. J. (1966). *N.Z. J. dairy Technol.* **1** (3), 78–81.

Cheldelin, V. H. (1954). *In* "The Vitamins", (W. H. Sebrell and R. S. Harris, eds.) Vol. III, pp. 575–600. Academic Press, New York.

Chen, S. L. (1959). *Biochim. biophys. Acta* **32**, 470–479, 480–484.

Chiao, J. S. and Peterson, W. H. (1953). *J. agric. Fd Chem.* **1**, 1005–1008.

Chien, H. C. (1960). *Fd Engng*, Feb. 1960, pp. 92–95.

Code of Federal Regulations (1968). Title 21, Parts 1 to 119. U.S. Government Printing Office, Washington, D.C.

Commission on Characterization and Evaluation of Dried Yeast (1963). *Pure appl. Chem.* **7**, 147–153.

Darlington, W. A. (1964). *Biotechnol. Bioengng* **6**, 241–242.

Delbrück, M. (1910). *Wschr. Brau.* **27**, 31, 375.

Delbrück, M. (1915). *Z. Spiritusind.* **38**, 121, 137, 161, 235.

Dostálek, M., Munk, V. and Volfová, O. (1968). *Biotechnol. Bioengng* **10**, 33–43.

Drews, B., Specht, H. and Herbst, A. M. (1962). *Branntweinwirtschaft* **102**, 245–247.

Dulaney, E. L., Stapley, E. O. and Simpf, K. (1954). *Appl. Microbiol.* **2**, 371–378.

Dunn, C. G. (1952). *Wallerstein Labs Commun.* **15**, 61–79.

Dunn, C. G. (1958). *Econ. Bot.* **12**, 145–163.

Dyck, A. W. J. (1957). *Paper Ind. Paper Wld* No. 4, 26–28.

East, E., Smith, B. J. and Borsden, D. G. (1966). *Fd Mf.* Sept. 1966.

Eastcott, E. V. (1928). *J. phys. Chem., Ithaca* **32**, 109–111.

Ebner, H., Pohl, K. and Enenkel, A. (1967). *Biotechnol. Bioengng* **9**, 357–364.

Eddy, A. A. (1958). *In* "The Chemistry and Biology of Yeasts", (A. H. Cook, ed.), pp. 157–249. Academic Press, New York.

Enobo, L., Anderson, L. G. and Lundin, H. (1946). *Archs Biochem.* **11**, 383–390.

Evans, G. H. (1968). *In* "Single-Cell Protein", (R. I. Mateles and S. R. Tannenbaum, eds.). M.I.T. Press, Cambridge, Mass.

FAO/WHO Expert Report (1965). "Protein Requirements", WHO Tech. Report, Ser. 301, Rome.

Ferrara, P. J. and Dalby, G. (1967). U.S. Patent 3,295,990.

Filosa, J. A. (1960). Brit. Patent 914,568.

Filosa, J. A. (1966). U.S. Patent 3,264,196.

Fink, H. and Lechner, R. (1941). *Angew. Chem.* **54**, 281.

Forbes, M., Zilliken, F., Roberts, G. and György, P. (1958). *J. Am. chem. Soc.* **80**, 385–389.

Freund, O. (1964). *Proc. Am. Soc. Brew. Chem.*, pp. 70–73.

Frey, C. N. (1930). *Ind. Engng Chem.* **22**, 1154–1162.

Frey, C. N., Kirby, G. W. and Schultz, A. (1936). *Ind. Engng Chem.* **28**, 879–884.

Funk, C. (1911). *J. Physiol., Lond.* **43**, 395; *Brit. J. State Med.* (1912).

Funk, C. (1913). *J. Physiol., Lond.* **46**, 12–20; 173–179.

Garber, M., Marquette, M. M. and Parsons, H. T. (1949). *J. Nutr.* **38**, 225.

Gontard, A. von (1948). *Yeast* **1** (1), pp. 4–5. Anheuser Busch Inc., St. Louis, Mo.

Green, D. E., Herbert, D. and Subrahmanyan, V. (1940). *J. biol. Chem.* **135**, 795–796.

Griffon, H. and Tixier, G. (1968). U.S. Patent 3,378,377, April 16.

Guenther, K. R. (1965). *Biotechnol. Bioengng* **7**, 445–446.

György, P. (1954). *In* "The Vitamins", (W. H. Sebrell and R. S. Harris, eds.) Vol. I, pp. 527–618. Academic Press, New York.

Hansen, E. C. (1896). "Practical Studies in Fermentation". C. Griffin and Co., London.

Harris, E. E., Hajny, G. J. and Johnson, M. C. (1951). *Ind. Engng Chem.* **43**, 1593–1596.

Harrison, J. S. (1967). *Process Biochem.* **2** (3), 41–45.

Harrison, J. S. (1968). *Process Biochem.* **3** (8), 59–62.

Hayduck, F. (1914). *Z. Spiritusind.* **37**, 69.

Hayduck, F. (1915). *Z. Spiritusind.* **38**, 161.

Hayduck, F. (1919). German Patent 300,662 and 303,221; U.S. Patents 1,449,105 and 1,449,107 (1923).

Henneberg, W. (1926). "Handbuch der Gärungsbakteriologie" Vol. II, 2nd Ed. Berlin.

Hock, A. (1960). *In* "Die Hefen", (F. Reiff, R. Kautzmann, H. Lüers and M. Lindemann, eds.) Vol. I, pp. 928–953. Verlag Hans Carl, Nürnberg.

Hoogerheide, J. C. (1950). *In* "Yeasts in Feeding", Proc. of Symposium, Nov. 1948, pp. 20–25. Garrard Press, Champaign, Ill.

Horwitt, M. K. (1954). *In* "The Vitamins", (W. H. Sebrell and R. S. Harris, eds.) Vol. III, pp. 300–394. Academic Press, New York.

Hospodka, J. (1966). *In* "Theoretical and Methodological Basis of Continuous Culture of Micro-organisms", (I. Malek and Z. Fencl, eds.), pp. 493–645. Publishing House of the Czechoslovak Academy of Sciences, Prague.

Hundley, J. M. (1954). *In* "The Vitamins", (W. H. Sebrell and R. S. Harris, eds.) Vol. II, pp. 452–578. Academic Press, New York.

Internal Revenue Service (1966). Publication 67, U.S. Treasury Dept., U.S. Government Printing Office, Washington, D.C.

Jacobs, H. E. (1944). "Six Thousand Years of Bread", pp. 31–35. Doubleday, Doran and Co. Inc., Garden City, New York.

Jansen, B. C. P. (1954). *In* "The Vitamins", (W. H. Sebrell and R. S. Harris, eds.) Vol. III, pp. 404–460. Academic Press, New York.

Jensen, A. L. and Shu, P. (1961). *Appl. Microbiol.* **9**, 12–15.

Johnson, M. J. (1959). *In* "Recent Progress in Microbiology", (G. Turnevall, ed.), pp. 397–402. Almqvist and Wiksell, Stockholm.

Johnson, M. J. (1967). *Science, N.Y.* **155**, 1515–1519.

Johnston, W. R., Redfern, S. and Miller, G. E. (1935). *Ind. Engng Chem. analyt. Edn* **7**, 82–84.

Joslyn, M. A. (1955). *Wallerstein Labs Commun.* **18** (61), 107–122.

Joslyn, M. A. and Vosti, D. C. (1955). *Wallerstein Labs Commun.* **18** (62), 191–205.

Kaiser, W. and Jacobs, W. (1957). *Fd Process.* No. 1, pp. 22, 23, 58.

Kelleher, W. J., Gitler, C., Sunde, M. L., Johnson, M. J. and Bauman, C. A. (1958). *J. Nutr.* **67**, 433–444.

Kereztesy, J. C. and Umbreit, W. W. (1954). *In* "The Vitamins", (W. H. Sebrell and R. S. Harris, eds.) Vol. III, pp. 222–242. Academic Press, New York.

King, C. G. (1950). *In* "Yeasts in Feeding", Proc. Symposium 1948, pp. 1–5. Garrard Press, Champaign, Ill.

Kirchhoff, H. (1960). *In* "Die Hefen", (F. Reiff, R. Kautzmann, H. Lüers and M. Lindemann, eds.) Vol. I, pp. 564–589. Verlag Hans Carl, Nürnberg.

Klapka, M. R., Duby, G. A. and Pavcek, P. L. (1958). *J. Am. diet. Ass.* **34**, 1317–1321.

Klose, A. A. and Fevold, H. L. (1947). *Archs Biochem.* **13**, 349–358.

Klug, M. J. and Markovetz, A. J. *Appl. Microbiol.* **15** (4), 690–693.

Kögl, F. (1937). *Proc. R. Soc.* **B124**, 1–9.

Kuhn, R. and Wendt, G. (1938). *Ber. dt. chem. Ges.* **71**, 780.

Kwolek, W. F. and Van Etten, C. H. (1968). *J. agric. Fd Chem.* **16**, 496–499.

Larkin, L. C. (1966). Econ. Res. Service, ERS—327, U.S. Dept. Agr.

Lefrançois, L. (1963). *Inds agric. aliment.* **80** (9.10), 927–928.

Lefrançois, L. and Revuz, B. (1964). *Inds agric. aliment.* **81** (12), 1175–1181.

Lendvai, A. (1963). U.S. Pat. 3,051,576.

Llewelyn, D. A. B. (1968). *In* "Microbiology", (P. Hepple, ed.), pp. 63–84. Elsevier Publ. Co., London.

Locke, E. G. (1946). *Pulp Pap. Int.* **20** (1), 20–26.

Locke, E. G., Saeman, J. F. and Dickerman, G. K. (1945). "FIAT Final Report No. 499".

Lodder, J. and Kreger-van Rij, N. J. W. (1952). "The Yeasts—a Taxonomic Study". North Holland Publishing Co., Amsterdam.

Lundin, H. (1950). *J. Inst. Brew.* **6** (1), **56**, 17–28.

Lyall, N. (1964). *Fd Engng* **36** (5), 98–99.

Mateles, R. I. and Tannenbaum, S. R. (1968). *Econ. Bot.* **22**, 42–50.

Maxon, W. D. and Johnson, M. J. (1953). *Ind. engng Chem.* **45**, 2554–2560.

McCay, C. M. (1950). *In* "Yeasts in Feeding", Proc. of Symposium, Nov. 1948, pp. 59–63. Garrard Press, Champaign, Ill.

Miller, T. L. and Johnson, M. J. (1966). *Biotechnol. Bioengng* **8**, 549–565; 567–580.

Murlin, J. R., Edwards, L. E., Fried, S. and Szymanski, T. A. (1946). *J. Nutr.* **31**, 715–736.

Murlin, J. R., Edwards, L. E. and Hawley, E. E. (1944). *J. biol. Chem.* **156**, 785–786.

Myers, R. P. and Stimpson, E. G. (1956). U.S. Patent 2,762,749.

"The National Formulary", 13th Edition—NF XIII (1970), pp. 764–765. American Pharmaceutical Association, Washington, D.C.

Nay, E. V. (1954). *Brewers' Dig.* **29** (5), 52T–54T, 61T.

Nelson, G. E. N., Anderson, R. F., Rhodes, R. A., Shakelton, M. C. and Hall, H. H. (1960). *Appl. Microbiol.* **8**, 179–182.

Neuberg, C. and Roberts, I. S. (1946). "Invertase", Monograph No. 4, pp. 1–62. Sugar Research Foundation Inc., New York.

Novellie, L. (1968). *Wallerstein Labs Commun.* **31** (104), 17–29.

Olson, B. H. and Johnson, M. J. (1949). *J. Bact.* **57**, 235–246.

Osborne, T. B. and Mendel, L. B. (1919). *J. biol. Chem.* **38**, 223–227.

Otsuka, S. I., Ishii, R. and Katsuya, N. (1966). *J. gen. appl. Microbiol., Tokyo* **12**, 1–11.

Parsons, H. T., Williamson, A. and Johnson, M. L. (1945). *J. Nutr.* **29**, 373.

Pasteur, L. (1874). *C. r. hebd. Scéanc. Acad. Sci., Paris* **78**, 213–216.

Pasteur, L. (1876). "Études sur la Bière", Gauthier-Villars, Paris. Translated by F. Faulkner and D. C. Robb, "Studies on Fermentation", (1879). MacMillan and Co., London.

Peppler, H. J. (1965). *J. agric. Fd Chem.* **13**, 34–36.

Peppler, H. J. (1967). *In* "Microbial Technology", (H. J. Peppler, ed.), pp. 145–171. Reinhold, New York.

Peppler, H. J. (1968). *In* "Single-Cell Protein", (R. I. Mateles and S. R. Tannenbaum, eds.). M.I.T. Press, Cambridge, Mass.

Peppler, H. J. and Thorn, J. A. (1960). U.S. Patent 2,922,748.

Perkins, M. B. and Furlong, L. E. (1967). U.S. Patent 3,355,296.

Peterson, W. H. (1950). *In* "Yeasts in Feeding", Proc. of Symposium, Nov. 1948, pp. 26–33. Garrard Press, Champaign, Ill.

Pfiffner, J. J., Calkins, D. G., O'Dell, B. L., Bloom, E. S., Brown, R. A., Campbell, C. J. and Bird, O. D. (1945). *Science, N.Y.* **102**, 228.

Pfiffner, J. J., Calkins, D. G., O'Dell, B. L., Bloom, E. S., Brown, R. A., Campbell, C. J. and Bird, O. D. (1946). *J. Am. chem. Soc.* **68**, 1392.

"Pharmacopoeia of the United States", 13th Revision—U.S.P. XIII (1947). Pp. 606–607. American Pharmaceutical Association, Washington, D.C.

Pinkos, J. A. (1967). *Fd Prod. Dvel.* **1**, 35–36.

Poff, E. (1899). "Die Landwirtschaftlichen Futtermittel", cited by A. C. Thaysen (1957), *in* "Yeasts", (W. Roman, ed.), p. 156. Academic Press, New York.

Poff, E. (1908). *Illte landw.. Ztg.* **28**, 33, 295.

Pomeranz, Y. (1964). *Fd Technol., Champaign* **18**, 682–690.

Powell, M. E. and Robe, K. (1964). *Fd Process.* **25** (2), 80–81, 95.

Prescott, S. C. and Dunn, C. G. (1959). "Industrial Microbiology", 3rd Edition, McGraw-Hill, New York.

Price, E. L., Marquette, M. M. and Parsons, H. T. (1947). *J. Nutr.* **34**, 311–319.

Pyke, M. (1958). *In* "The Chemistry and Biology of Yeasts", (A. H. Cook, ed.), pp. 535–586. Academic Press, New York.

Recommended Dietary Allowances (1968). 7th Rev. Ed. National Academy of Science, Washington, D.C.

Reece, M. C., Donald, M. B. and Crook, E. M. (1959). *J. biochem. microbiol. Technol. Engng* **1**, 217–228.

Reed, G. (1966). "Enzymes in Food Processing". Academic Press, New York.

Rieche, A., Hilgetag, G., Martini, A. Thonke, M. and Lorenz, M. (1964). *Zentbl. Bakt. ParasitKde, Abt. II* **118** (1), 53–65.

Rieche, A., Krueger, S., Martini, A. and Lorenz, M. (1966). *Mber. dt. Akad. Wiss. Berl.* **8** (12), 891–897.

Rogers, D. and Mickelson, M. N. (1948). *Ind. Engng Chem.* **40**, 527–529.

Roine, P. (1946). *Suomen Kemistilehti* **B 19**, 37–49.

Rosenthal, W. A. and Pinkalla, H. A. (1960). U.S. Patent 2,928,740.

Rubbo, S. D. and Gillespie, J. M. (1940). *Nature, Lond.* **146**, 838–839.

Sak, S. (1921). Danish Patent 28,507; U.S. Patent 1,884,272 (1932).

Schmidt, E. (1947). *Angew. Chem.* **59**. 16.

Schultz, A. S., Atkin, L. and Frey, C. N. (1937). *J. Am. chem. Soc.* **59**, 948–949; *J. biol. Chem.* **129**, 471 (1939); *Ind. Engng Chem. analyt. Edn* **14**, 35–39 (1942).

Schwarz Bio Research, Inc. (1962). Brit. Patent 952,713.

Seeley, R. D. (1951). *In* "Yeast", (A. von Gontard, ed.) **1** (5), 1–19. Anheuser-Busch Inc., St. Louis.

Seeley, R. D., Ziegler, H. F. and Sumner, R. J. (1950). *Cereal Chem.* **27**, 50–60.

Shacklady, C. A. (1969). *Fd Mf.* **44** (4), 36–37, 40.

Singh, K., Agarwal, P. N. and Peterson, W. H. (1948). *Archs Biochem.* **18**, 181–193.

Singruen, E. (1953). *Brewers' J.* **108** (2), 66, 69–70.

Soder, O. von and Dirr, K. (1942). *Biochem. Z.* **312**, 252, 263.

Sperber, E. (1945). *Arkiv. Kemi, Mineral. Geol.* **21A** (3), 1.

Stimpson, E. G. and Young, H. (1957). U.S. Patent 2,809,113.

Strohm, J. A. and Dale, R. F. (1961). *Ind. Engng Chem.* **53**, 760–764.

Suomalainen, H. (1964). *Branntweinwirtschaft* **104** (16), 402–405.

Suomalainen, H. (1968). *In* "Aspects of Yeast Metabolism", (A. K. Mills and H. Krebs, eds.), pp. 1–31. Blackwell Scientific Publications, Oxford.

Suomalainen, H. and Keränen, A. J. A. (1968). *Chem. Phys. Lipids* **2** (3), 296–315.

Suomalainen, H., Christiansen, V. and Oura, E. (1967). *Suomen Kemistilehti* **B 40**, 286–287.

Sure, B. (1946). *J. Am. diet. Ass.* **22**, 114–118, 494–502, 766–769; **23**, 113–119.

Sure, B. (1948). *J. Nutr.* **36**, 59–63, 65–73.

Takeda Chemical Industries, Ltd. (1967). Neth. Appl. Patent 6,612,422.

Technical Bulletins (1967). Knudsen Creamery Co. ("Wheast"); Standard Brands Inc. ("Dried Fragilis Yeast" and "Debittered Brewer's Dried Yeast"); Universal Foods Corp. ("Red Star Primary Grown Dried Yeasts" and "Torula Dried Yeast N.F. XII"); Butland Industries Ltd. ("Dried Torula Yeast").

Thaysen, A. C. (1957). *In* "Yeasts", (W. Roman, ed.), pp. 155–210. Academic Press, New York.

Toyo Koatsu Industries Inc. (1967). Fr. Patent 1,493,259.

Traisman, E. and Kurtzhalts, W. (1957). U.S. Patent 2,817,590.

Trevelyan, W. E. (1958). *In* "The Chemistry and Biology of Yeasts", (A. H. Cook, ed.), pp. 369–436. Academic Press, New York.

Vámos, L. and Pozsár, K. (1968). *Branntweinwirtschaft* **108** (20), 452–462.

Vámos, L. and Pozsár, K. (1969). *Branntweinwirtschaft* **109** (1), 1–3.

Van Etten, C. H., Kwolek, W. F., Peters, J. E. and Barclay, A. S. (1967). *J. agric. Fd Chem.* **15**, 1077–1089.

Van Lanen, J. M. (1946). *Archs Biochem.* **12**, 101–111.

Van Lanen, J. M. (1954). *In* "Industrial Fermentations", (L. A. Underkofler and R. J. Hickey, eds.) Vol. II, pp. 191–216. Chemical Publishing Co. Inc., New York.

Van Lanen, J. M., Broquist, H. P., Johnson, M. J., Baldwin, I. L. and Peterson, W. H. (1942). *Ind. Engng Chem.* **34**, 1244–1247.

Völtz, W. (1910). *Z. Spiritusind.* **33**, 47.

Warburg, O. and Christian, W. (1933). *Biochem. Z.* **266**, 377.

Wasserman, A. E. (1960). *Dairy Engng* **79**, 12–17.

Wasserman, A. E. (1961). *J. Dairy Sci.* **44**, 379–386.

Wasserman, A. E., Hopkins, W. J. and Porges, N. (1958). *Sewage ind. Wastes* **30**, 913–920.

White, J. (1954). "Yeast Technology". Chapman and Hall, London.

Wickerham, L. J. (1957). *J. Am. med. Ass.* **165**, 47–48.

Wickerham, L. J. and Kuehner, C. C. (1956). U.S. Patent 2,764,487.

Wildiers, E. (1901). *Cellule* **18**, 313.

Wiley, A. J. (1954). *In* "Industrial Fermentations", (L. A. Underkofler and R. J. Hickey, eds.) Vol. I, pp. 307–343. Chemical Publishing Co., New York.

Wiley, A. J., Inskeep, G. C., Holderby, J. M. and Hughes, L. P. (1951). *Ind. Engng Chem.* **43**, 1702–1711.

Williams, R. J., Eakin, R. E., Beerstecher, E. and Shive, W. (1950). "The Biochemistry of B. Vitamins". Reinhold, New York.

Williams, R. J., Lyman, C. M., Codyear, G. H., Truesdail, J. H. and Holaday, D. (1933). *J. Am. chem. Soc.* **55**, 2912.

Windisch, S. (1948). *Brauwelt* **3** (10), 203–207.

Woodbine, M. (1959). *In* "Progress in Industrial Microbiology", (D. J. D. Hockenhull, ed.) Vol. I, pp. 181–245. Interscience Publishers Inc., New York.

Woods, R. (1951). *In* "Yeast", (J. D. Veron, ed.), **2** (1), 1–24. Anheuser-Busch Inc., St. Louis.

Wooley, D. W. (1941). *J. biol. Chem.* **140**, 453–459.

World Wide Survey of Fermentation Industries 1963 (1966). *Pure appl. Chem.* **13**, 405–417.

Young, H. and Healey, R. P. (1957). U.S. Patent 2,776,928.

Ziemba, J. V. (1967). *Fd Engng* **39** (1), 82–85.

Zirins, F. (1967). *Latv. PSR Zināt. Akad. Vest.* **1**, 140–141; from *Chem. Abstr.* 1967, **67**, 990.

Chapter 9

Yeasts as Spoilage Organisms

H. W. Walker and J. C. Ayres

Department of Dairy and Food Industry, Iowa State University,
Ames, Iowa, U.S.A. and Department of Food Science,
The University of Georgia, Athens, Georgia, U.S.A.

I. Introduction

Microbial spoilage of foods is commonly the result of the combined activities of yeasts, moulds and bacteria; however, depending upon the environment, one of these groups of microbes may prevail over the others. Yeasts ordinarily do not compete well in mixed populations and, therefore, cause spoilage under conditions which are favourable for their growth but unfavourable for growth of most bacteria. Ingram (1958), in his review on yeasts as spoilage agents, has emphasized the conditions in a food or in the environment that predispose towards spoilage by yeasts. Factors of paramount importance in determining the ability of yeasts to compete with other organisms are numbers and types of yeasts infecting the food initially; nutrients available in the food; pH value; redox potential of the food; temperature during storage; and relative humidity or water activity (a_w) of the surroundings or food product.

In mixed populations, bacteria will usually outgrow yeasts in substrates having a neutral or slightly acid pH value. As the pH value decreases below 5·5, yeasts are generally favoured and outgrow bacteria at this level of acidity. Recca and Mrak (1952) have reported growth of various yeasts at pH values as low as 1·5–2·0. Certain oxidative yeasts utilize organic acids and alcohols, leading to deterioration of a food product in which these metabolic by-products are desirable. In addition, depletion of these organic acids raises the pH value and favours growth of bacteria that can cause serious spoilage problems.

Most yeasts grow best under aerobic conditions but many can also grow slowly under conditions of low oxygen tension. Thus, yeasts sometimes may be responsible for fermentative spoilage of certain products from which air has been excluded. Ingram (1958) has pointed out that the redox potential is highest at the surface of a food, for which reason the more aerobic yeasts develop on the surface of the food product; examples of this phenomenon are film yeasts on brines and mycodermas on wines. Acid tolerance and ability to grow under low redox potential account, in part, for the ability of yeasts to prevail in the fermentation of fruit juices and pickling solutions.

The optimum temperature for growth of most yeasts is 25–30° while the maximum temperature can vary between 30° and 46–47° (Lund, 1958). The minimum temperature for growth is frequently around 0°;

but growth below zero has been recorded on a number of occasions. Berry (1934) found a species of *Saccharomyces* capable of growth at − 2·2° in cider. Berry and Magoon (1934) attributed spoilage of strawberries packed in 50% sucrose to a *Torula* sp. which grew at − 4°; spoilage occurred after four weeks of storage. Shrader and Johnson (1934) and McCormack (1950) reported growth of yeasts at − 17·8° in orange juice concentrate and oysters, respectively. Michener and Elliott (1964) have reviewed extensively the minimum growth temperatures for non-psychrophilic and psychrophilic organisms; the lowest growth temperature they recorded was for the pink yeast isolated by McCormack (1950) which grew on a microbiological medium at − 34°.

Yeasts occurring in substrates with a high sugar content such as honey, jams, jellies, syrups, dried fruits and fruit concentrates are usually referred to as osmophilic yeasts or osmoduric yeasts. For purposes of this discussion, osmophilic yeasts are those which proliferate in an environment of high osmotic pressure, and osmoduric yeasts are those which tolerate but do not multiply in environments of high osmotic pressure. Several extensive reviews have been published which cover various aspects of osmophilic yeasts (Mrak and Phaff, 1948; von Schelhorn, 1950a; Scarr, 1953; Scott, 1957; Ingram, 1955, 1957, 1958; Ōnishi, 1963; Phaff *et al.*, 1966).

In the literature previous to 1952, most of the yeasts capable of growing in high sugar concentrations have been referred to as species of *Zygosaccharomyces*. Kroemer and Krumbholz (1931) in their studies on osmophilic yeasts divided the *Zygosaccharomyces* into two groups. The first group was related to *Zygosacch. priorianus* and, of the disaccharides, only maltose was fermented. The second group, related to *Zygosacch. nadsonii*, was less sugar-tolerant and did not ferment maltose, but did ferment sucrose and raffinose with the production of much volatile acid. Ingram (1959) suggested dividing yeasts resistant to sugars into two groups; namely, semi-osmophiles fermenting sucrose, represented by *Saccharomyces rosei*, and osmophiles, represented by *Sacch. rouxii* and *Sacch. mellis*, which tolerate the highest concentrations and which do not ferment sucrose.

Lodder and Kreger-van Rij (1952) have abandoned the use of the sub-genus *Zygosaccharomyces*. They believed that no fundamental difference existed between the *Zygosaccharomyces* and the *Saccharomyces* based on sexual reproduction and have classified this group of yeasts as *Saccharomyces* spp. Nearly all the osmophilic yeasts are now included in the following species: *Sacch. rouxii*, *Sacch. rouxii* var. *polymorphus* and *Sacch. mellis*.

Yeasts that multiply in enviroments of salt concentrations so high that "ordinary" species of yeasts cannot grow may also be termed osmophilic. Little information is available on significant differences which may exist between salt-tolerant and sugar-tolerant yeasts, and on the mechanisms involved in the adaptability to this generally unfavourable environment. Yeasts that flourish in substrates containing high concentrations of sugar do not develop well in an environment of high osmotic pressure resulting from high concentrations of sodium chloride (Phaff et al., 1966). Kroemer and Krumbholz (1932), after their work on the comparison of the effects of different concentrations of several salts and glycerol on various osmophiles, decided that the effect of all these solutes was mainly osmotic. However, Ingram (1957, 1959) is of the opinion that factors other than osmotic pressure are involved; he has pointed out that in the data of Kroemer and Krumbholz, equal molarity of sodium chloride was much more inhibitory than that of other salts. Ōnishi (1963) also is of the opinion that salt tolerance and sugar tolerance of the osmophilic yeasts are quite different in nature. He proposes that the mechanism of salt tolerance may depend on the environmental influence of the ions or the undissociated molecule of sodium chloride. Interestingly enough, Ōnishi has presented evidence that the main metabolic activities of the salt-tolerant yeasts are inhibited almost completely by a high concentration of salt in the cell-free state but not in the intact cells.

Scott (1957), in his excellent review and commentary on water requirements of spoilage micro-organisms, discusses the relationships between osmotic pressure, vapour pressure, equilibrium relative humidity and water activity (a_w). The equilibrium relative humidity of a solution is the humidity at which the rates of evaporation and of condensation are equal and, obviously, will vary with the amount and type of material in the solution, which in turn affects the water activity. The value of a_w is numerically equal to the corresponding relative humidity (H_R) expressed as a fraction, i.e., $H_R/100$. Relative humidity, as Scott points out, more strictly applies to the surrounding atmosphere, but under equilibrium conditions the two terms are interchangeable.

On the basis of water requirements, yeasts can withstand drier conditions than most bacteria but usually require more moisture than moulds. According to Scott (1957), bacteria as a class grow at an a_w as low as about 0·75 (saturated sodium chloride), whereas some yeasts and moulds grow very slowly at 0·62 a_w. The a_w values at which organisms can grow will vary with temperature, pH value, availability of nutrients and availability of oxygen. More meaningful comparisons of limitations of growth of micro-organisms in concentrated solutions could be made with the use of a_w values rather than with the use of per cent concentration of solids, which is frequently used in literature on the growth of osmophilic yeasts.

II. Syrups

A. HONEY

Spoilage of foods of high sugar content by alcoholic fermentation is generally caused by the action of sugar-tolerant yeasts. These yeasts are capable of developing in concentrated sugar solutions which suppress non-osmophilic yeasts such as *Saccharomyces cerevisiae* and *Sacch. ellipsoideus*. Honey, being a food of high sugar content, has received much attention as to causes of its spoilage.

Species of *Zygosaccharomyces* have been found particularly active in causing spoilage of honey even when outnumbered by other sugar-tolerant yeasts. Lochhead and Heron (1929) pointed out that Nussbaumer in 1910 should be given credit for demonstrating that sugar-tolerant yeasts are an important factor in the fermentation of honey. Nussbaumer apparently isolated and described two types of *Zygosaccharomyces* from fermented honey. He also showed that the heating of honey inoculated with spores of these yeasts for 30 min at 70° prevented fermentation. Since that time, several workers have isolated species of *Zygosaccharomyces* from fermented honey; namely, Richter (1912), Fabian and Quinet (1928), Lochhead and Heron (1929), Lochhead and Farrell (1931b).

Richter (1912) identified *Zygosacch. mellis-acidi* as the organism responsible for an alcoholic fermentation in extracted honey, including samples which had already crystallized. This yeast also caused a gassy fermentation in honeycombs which resulted in the contents streaming over the hives. *Zygosaccharomyces japonicus*, *Zygosacch. barkeri*, *Zygosacch. mellis*, *Zygosacch. priorianus*, *Zygosacch. nussbaumeri* and *Zygosacch. richteri* have been isolated from various samples of normal and fermented honeys (Fabian and Quinet, 1928; Lochhead and Heron, 1929). Later, Lochhead and Farrell (1931b) reported that they encountered *Zygosacch. richteri* most frequently in both normal and fermented honeys. Although species of *Saccharomyces*, *Schizosaccharomyces*, *Torula* and *Mycotorula* are capable of growing in solutions of high sugar concentration, Lochhead and Landerkin (1942) concluded that the most prominent forms of sugar-tolerant yeasts in honey are members of the genus *Zygosaccharomyces*.

Honey is hygroscopic; Fabian and Quinet (1928) reported that 100-g samples of honey will absorb a minimum of 25·90% and a maximum of 32.99% moisture at 20°. The critical moisture point at which fermentation occurred was apparently 21%. Fabian and Quinet advanced the theory that honey, being hygroscopic, absorbs sufficient water at the surface to lower the concentration of sugars to a point that the yeasts and bacteria present can grow; yeasts usually predominate under these circumstances. Thus, micro-organisms capable of surviving in concentrated sugar solutions grow and cause fermentation of the product.

While extent of yeast infection and moisture content are important predisposing causes of fermentation of honey, certain inexplicable cases of fermentation with comparatively low loads of yeasts, and non-fermentation with relatively high moisture and yeast contents, suggest that other factors are involved. Honey has been found to contain an active principle which stimulates fermentation by species of *Zygosaccharomyces* in synthetic media (Lochhead and Farrell, 1931a; Farrell and Lochhead, 1931). This activating effect of honey is impaired by heating in alkaline solution and by exposure to moderate heat. Inositol is present in this material but is not necessarily the active principle; growth is apparently dependent upon the presence of another unidentified substance or substances.

A study throughout the season for honey production by Lochhead and Heron (1929) showed the presence of sugar-tolerant yeasts in floral nectar and nectar from the beehives. Many of these yeasts were identical with yeasts from honey tanks, air of the honey-extracting house and in fermented honey, which emphasized the need for good handling and sanitizing practices during the processing of honey. Spores of these yeasts have been recovered from the fresh pollen taken from the legs of bees, in pollen from the beehive and in the honey sac of bees (Wilson and Marvin, 1929). Lochhead and McMaster (1930–31) found no outstanding differences among yeast infections of honeys with respect to floral origin; however, honeys from nectar of buckwheat origin showed a tendency toward higher yeast content than honeys of other floral origins, and an increased susceptibility to fermentation.

Examination of soil from different locations for the presence of sugar-tolerant or osmophilic yeasts capable of fermenting honey showed the soil around the apiary to be regularly infected (Lochhead and Farrell, 1930a). Evidently these yeasts remained viable through the winter in frozen soil; Lochhead and Farrell reported that, at the time of taking samples (December through March), the average daily minimal temperatures for these months were: $-31 \cdot 6°$, $-15 \cdot 4°$, $-15 \cdot 3°$ and $-7 \cdot 4°$, respectively. These yeasts were considered psychroduric rather than psychrophilic since they were incapable of growing below $10°$. The organisms isolated under these circumstances were *Zygosaccharomyces richteri*, *Zygosacch. nussbaumeri* and *Torula* (*Mycotorula*) *mellis*.

Recommended methods for prevention of fermentation in honey include storage at low temperature and the application of heat. In addition, Lochhead and Farrell (1930b, 1936) have reported on the use of chemical preservatives to prevent fermentation by osmophilic yeasts. Sodium hypochlorite and sodium *p*-toluene sulphonchloramide (Chloramine-T) were ineffective at concentrations of 0·5% and exerted a marked bleaching action in addition to affecting the taste. Hexylresorcinol completely inhibited fermentation at a concentration of 0·01%; sodium salicylate was effective at a concentration of 0·06%. Studies on freshly extracted honey inoculated with *Zygosacch. nussbaumeri* showed that 0·025% sodium benzoate, 0·01% sodium sulphite and 0·01% sodium bisulphite were all effective in preventing fermentation during storage, but the latter two compounds affected both colour and flavour. Sodium benzoate showed no detectable influence on flavour, colour or general appearance.

B. MAPLE SYRUP

Maple syrup and maple sugar are made from the sap of the sugar maple (*Acer saccharum*). The sap, as it occurs in the vascular bundles of the tree, is relatively free of micro-organisms; however, it is easily contaminated by the collecting equipment and during processing of the product. Usually a composite of micro-organisms is found in the collected maple sap, and spoilage may be caused by one particular group of organisms or an association of several groups. Such an association of unidentified grey and red yeasts with bacteria apparently is responsible for so-called "red sap" (Edson, 1912; Jones, 1912). Edson, in one survey on the occurrence of micro-organisms in maple sap, found that a sequence of species occurred in which the predominating organisms during the early days of the maple sugar season belonged to yeast-like organisms, while bacteria were relatively few in number; as the season advanced these conditions reversed, with the bacteria predominating. However, in subsequent seasons, he observed just the opposite sequence.

Maple syrups produced during the early periods of the maple season are characteristically light amber in colour and delicately flavoured; this type of syrup is highly valued (Naghski *et al.*, 1957; Naghski and Willits, 1957). As the season advances, the syrup becomes progressively darker and acquires a stronger flavour. These changes have been associated mainly with the growth of micro-organisms in the sap from which the syrup is made. The association of colour of maple products with quality has been stressed to the point that colour has become the chief criterion for grading maple syrup. Syrup made from sap fermented by yeasts shows development of a dark colour and a strong caramel flavour accompanied by a large increase in free reducing sugars.

According to Frank and Willits (1961), the high osmotic pressure of standard-density maple syrup prevents bacterial spoilage. However, a few moulds and yeasts can survive but will not grow unless the syrup becomes diluted. Syrups in bulk containers or in small packages usually have a free head space above the syrup. Frank and Willits noted that stored syrup is subjected to sufficient changes in temperature so that, during the warm cycle, moisture from the syrup distils into the head space and then condenses as water droplets during the cold cycle. These droplets collect on the syrup surface and dilute it. Microbial contaminants will then cause spoilage; yeasts may be involved in this type of spoilage but moulds are the most frequent offenders.

The moisture content of freshly made maple syrup, as measured by Fabian and Hall (1933), ranged from 26·3% to 36·5%. Samples of maple syrup showing visible signs of fermentation contained from 32·7% to 34·6% moisture. The amount of alcohol present in samples of fermented maple syrup from which yeasts were isolated ranged from 0·83 to 2·66% by weight. A minimum of 21 days and a maximum of 29 days were necessary for visible spoilage to occur in maple syrup inoculated with spoilage yeasts and stored at room temperature in a desiccator containing a moist atmosphere.

Yeasts isolated from samples of fermented maple syrup made in Michigan and Vermont were identified by Fabian and Hall (1933) as *Saccharomyces aceris-sacchari*, *Sacch. behrensianus*, *Sacch. monacensis*, *Zygosaccharomyces mellis*, *Zygosacch. barkeri*, *Zygosacch. japonicus* and *Zygosacch. nussbaumeri*. Vegetative cells of these organisms were killed when heated in nutrient broth at a temperature of 55° for 5 min; however, in normal, unfermented maple syrup, 60° for 5 min was required for thermal destruction except for two cultures which were killed after 10 min at 60°. The ascospores when heated in nutrient broth were killed in most instances after 5 min at 65°; a temperature of 75° for 5 min was required to kill ascospores in maple syrup.

Many of the objectionable flavours in maple syrup are due to the action of micro-organisms in the maple sap; thus, any means of controlling the microbial load of the sap influences the quality of the final product. Edson (1912) and Jones (1912) showed a close correlation between the number of organisms present in the sap, the care taken in tapping the trees and attention given to the buckets in which the sap was collected, and the souring of sap. Also, when the sap is collected in plastic bags (transparent under ultraviolet exposure) bright sunlight during cool weather results in an almost sterile product (Naghski and Willits, 1953). Schneider *et al.* (1960) have shown that bacterial strains isolated from maple syrup are more sensitive to ultraviolet rays than yeasts from the same source. Zinc from galvanized containers will sharply reduce numbers of *Cryptococcus albidus* in association with bacteria in maple sap (Frank *et al.*, 1959).

Microbial contaminants can be introduced into syrup during packaging in the sugar house or during home use of small-sized containers.

Frank and Willits (1961) observed that the growth of such yeasts and moulds was inhibited better by low concentrations of esters of *p*-hydroxybenzoic acid than by conventional mould inhibitors (propionate, sorbate, benzoate). Sodium propyl *p*-hydroxybenzoic acid at a concentration of less than 0·02% was also inhibitory under these circumstances.

C. SUGAR CANE SYRUPS, MOLASSES, RAW SUGAR CANE AND FRUIT SYRUPS

Various yeasts have been found in sugar, syrups and molasses; and they frequently can be traced to the original cane (Hall *et al.*, 1937; Owen, 1949; Scarr, 1951; Shehata, 1960). Many of the yeasts observed by these authors are osmophilic, or at least are tolerant of high concentrations of sugar. According to Shehata, eight species of yeasts predominated in fresh cane juice; namely, *Saccharomyces carlsbergensis* var. *alcoholphila*, *Sacch. cerevisiae*, *Pichia membranaefaciens*, *Pi. fermentans*, *Candida krusei*, *C. guilliermondii*, *C. intermedia* var. *ethanophila*. Only three species predominated in fermenting juice: *Saccharomyces carlsbergensis* var. *alcoholophila*, *Sacch. cerevisiae* and *Schizosaccharomyces pombe*. Some other species of yeasts which have been recovered by the authors cited above include *Zygosaccharomyces nussbaumeri*, *Zygosacch. major*, *Zygosacch. globiformis*, *Torula communis*, *Sacch. zopfii* and *Schizosaccharomyces mellacei*. Barbados molasses has as its predominant flora two groups of yeasts—*Zygosacch. nussbaumeri*, *Zygosacch. major* (Hall *et al.*, 1937).

Theoretically both cane and beet syrup leaving the processing plant have only bacterial spores, and possibly a small number of very resistant yeast cells which have survived the temperature used for evaporation of clarified juices. According to Owen (1949), in the raw sugar factory or the refinery, these surviving micro-organisms are of no immediate concern, since the syrups are rarely kept for any length of time and pass continuously to the vacuum-pan for production of crystallized sugar. Nevertheless, the presence of viable yeast cells may be of great concern if the syrup is to be canned and distributed for consumption.

Owen (1913) established that *Saccharomyces zopfii* can be responsible for the fermentation of canned cane syrup. The gas formed during the fermentation may produce enough pressure to cause bursting of the cans. The cane syrups involved in this study had a density between 70° and 75° Brix, which was considered unfavourable for bacteria and yeasts. In addition, the temperature (78°) at which the syrups were processed and sealed in the cans was considered capable of destroying any yeast cells. However, this organism grew and fermented a cane syrup of 71·5° Brix containing approximately 60% total sugars. Also, this yeast was capable of developing a feeble fermentation even after being heated at 90° for 10 min. Syrups below 70° Brix density are susceptible to deterioration by a number of the yeasts listed above.

Scarr (1953) occasionally recovered osmophiles from sugar cane, which carries many other types of yeast; while in cane molasses (an osmophilic habitat) osmophilic yeasts were predominant. She also found that the majority of yeasts from unrefined sugar could be assigned to the species *Sacch. bisporus* but that *Sacch. rouxii* was also of common occurrence. Owen, in 1918, had observed that torulae are of constant occurrence in sugars. He reported that they were active in destroying reducing sugars, under widely varying conditions, but their ability to invert sucrose was relatively weak and only occurred under especially favourable conditions. The fermentation of molasses and syrups by these torulae often resulted in the development of various esters which imparted a characteristic aroma to the deteriorating products.

In raw sugar, sucrose exists as crystals surrounded by a thin film of molasses in which residual sucrose and non-sucrose constituents are dissolved or suspended. This film of molasses is generally too concentrated to permit microbial growth; but, if it is sufficiently dilute, it will support growth of osmophilic micro-organisms. The chemical and physical properties of raw sugar frequently provide an environment which favours the growth of osmophilic yeasts over many other micro-organisms; these properties are: low a_w, high acidity, low redox potential and a high carbon/nitrogen ratio (Tilbury, 1966).

Growth of those micro-organisms which can cause deterioration of raw cane sugar is controlled primarily by the water activity of the sugar. In tropical climates, high atmospheric temperature along with high relative humidity may permit absorption of moisture by the molasses film to a level sufficiently high to permit microbial growth. Direct measurement of the osmotic pressure of this film has not been possible, but a_w values for raw sugars have been estimated by Tilbury (1966) to lie between 0·60 and 0·75, only rarely reaching 0·80. Xerophilic moulds grow at a_w levels of 0·65 and osmophilic yeasts at levels of 0·60 (Mossel and Ingram, 1955). These organisms grow very slowly near the minimal a_w levels, and in practical situations sugar of equilibrium relative humidity of up to 65% to 70% is considered safe for storage, but above this value microbiological deterioration is more likely to occur. As much as 2% of the sucrose content may be lost due to microbial activity (Tilbury, 1966).

Scarr, as reported by Tilbury (1966), has found *Saccharomyces rouxii* and *Sacch. mellis* to be the most commonly occurring yeasts in raw sugar. Additional yeasts which have been isolated from raw sugar are: *Sacch. bisporus, Sacch. cerevisiae, Sacch. elegans, Sacch. rosei, Hansenula anomala, H. subpelliculosa, Pichia fermentans, Torulopsis globosa, T. glabrata, T. etchellsii, T. versatilis, T. dattila, Schizosaccharomyces pombe, Endomycopsis ohmeri, Candida tropicalis* (Tilbury, 1966). The role of many of these species in the deterioration of raw sugar is unknown.

Tilbury (1966) has suggested that storage of raw sugar above 30° to 35° favours deterioration by osmophilic yeasts rather than by moulds. Few of the osmophilic moulds grow well above 30°, whereas the optimal growth temperature for osmophilic yeasts is frequently between 25° and 30°. In addition, Ingram (1959) has observed that the optimal temperature for growth of osmophilic yeasts increases as the concentration of solids in the medium increases.

During the initial stages of deterioration of raw sugar by yeasts a slight rise in the polarization of the sugar occurs because certain yeasts, such as *Sacch. rouxii*, utilize fructose selectively in invert sugar with a subsequent loss in laevo-rotatory activity (Browne, 1918; Peynaud and Domercq, 1955). The ability of such yeasts as *Sacch. rouxii*, which do not produce invertase, to invert sucrose is probably attributable to acid inversion, the acid being formed during the initial fermentation of residual monosaccharides (Scarr, 1951).

Tilbury (1966) has enumerated as many as 10^7 osmophilic yeasts per gram of raw sugar, but has stated that it is not possible to define safe limits for the numbers of micro-organisms in raw sugar because these would be meaningless without reference to the equilibrium relative humidity of the sugar. His conclusion was that the isolation of osmophilic moulds and yeasts in any sugar of an equilibrium relative humidity of 70% or more should be regarded as unsafe.

Fermentation by yeasts and growth of moulds are the most frequent causes of spoilage of stored fruit juice-syrup mixtures (Morgan, 1938, Tarkow *et al.*, 1942; Morse *et al.*, 1948). Sodium benzoate or a combination of sodium benzoate and citric acid have been recommended to prevent this spoilage; however, sodium benzoate may change the flavour and citric acid alone has little preservative action. Morse *et al.* (1948) showed that 0·33% acetic acid, 6·29% citric acid or 6·25% lactic acid was necessary to prevent growth of *Sacch. cerevisiae* in strawberry juice- and raspberry juice-syrup mixtures. Tarkow *et al.* (1942) observed that, in apple-, pineapple- and grapefruit-syrups, 40% glucose inhibited the growth of *Sacch. cerevisiae* to a greater extent than did 40% sucrose. In almost all cases an equal mixture of both sugars yielded results that were intermediate to those of either sugar alone used in the same concentrations. Erickson and Fabian (1942) found fructose and glucose were the only sugars that had a germicidal action on the yeasts they studied; namely, *Sacch. cerevisiae, Sacch. ellipsoideus, Zygosacch. mellis, Torula lactis-condensi* and an acid-tolerant yeast isolated from sweet pickles. Preserving concentrations of these sugars were identical, but a 5–15% greater concentration was required for sucrose. Lactose had no preserving action. Presence of organic acids enhanced the preservative and germicidal activity of glucose and fructose; acetic acid was the most effective, followed by lactic and citric acids.

III. Confectionery Products

A. CANDY

Certain confectionery products are susceptible to spoilage by osmophilic yeasts. These yeasts can flourish in sugar solutions containing as much as 80% solids and are frequently responsible for the "bursting" of chocolate centres (Paine *et al.*, 1927; Pouncey and Summers, 1939). Weinzirl (1922) regarded anaerobic bacteria of the *Clostridium sporogenes* type as the chief cause of exploding chocolates, but stated that yeasts were also capable of causing this type of spoilage. He reported that fondants covered with chocolate and inoculated with *Saccharomyces cerevisiae* exploded after about ten days.

An instance in which a large manufacturer of chocolates experienced considerable loss from such spoilage was investigated by Shutt (1925). The chocolates were cracked or broken and some of the filling was forced out of the candy. The interior was spongy with numerous small holes throughout the mass; however, there was little or no odour and the taste was described as pleasant. The pH value of the samples ranged from 5·0 to 6·0. Pink and white yeasts were frequently isolated, although the white yeasts were the only ones to produce gas when inoculated into chocolates. These yeasts were not identified, but Church *et al.* (1927) observed spore formation in sugar media and in fondant, or at least a tendency toward yoke formation, in all but one yeast they had isolated from such outbreaks.

The addition of invert sugar or invertase has been used as a means of preventing the bursting of chocolates (Paine *et al.*, 1927; Pouncey and Summers, 1939). Paine *et al.* (1927) added invertase to the fondant before the fondant centres were moulded. The invertase inverted the sucrose, increasing the total sugar solubility which, in turn, caused sufficient increase in the density and osmotic pressure of the syrup phase to prevent fermentation. Fondants containing fruits present a special case because of a decrease in density of the syrup as a result of diffusion of fruit juice.

Pouncey and Summers (1939) devised a crystal technique for predicting the susceptibility of a confectionery product to fermentation. This technique is based on the following assumptions: (i) there is a limiting level of osmotic pressure beyond which osmophilic yeasts cannot grow; (ii) the osmotic pressure is directly related to the vapour pressure of a solution; and (iii) the vapour pressure of a solution can be determined by observing the behaviour of crystals of certain salts when in the atmosphere of such sugar products.

Control of osmophilic yeasts in candy can be maintained by good sanitation practices and by heat treatment. For example, Shutt (1925) traced an outbreak of yeast spoilage to a wooden barrel used to store excess filler. The spoilage problem was eradicated by installing vessels capable of being cleaned and sterilized between batches. According to Pouncey and Summers, if a chocolate centre is heated to 60° for 20 min, and is not later contaminated from sugar dust or other sources of osmophilic yeasts, spoilage should not occur even if the water activity (a_w) is such that it permits growth of yeasts. Control by adjustment of acidity is not practical because of modifications of the flavour.

B. CRYSTALLIZED FRUIT

Crystallized fruit is made by dipping the material successively into increasingly concentrated solutions until the fruit is thoroughly impregnated with sugar. Storage of this product under humid conditions can produce a surface film of syrup which may become infected with yeasts. Pouncey and Summers (1939) stated that they always found osmophilic yeasts on glacé cherries and peel. Mossel (1951) reported a case of fermentation in sugared fruits caused by *Saccharomyces rouxii*. The manufacturer believed his product to be safe from fermentation because the fruits (pears and cherries) contained at least 70% solids. This yeast caused fermentation in concentrations of up to 70% fructose in 0·5% yeast extract solution. This same yeast was isolated also from beet molasses and from bees found in the factory where the fruit was made.

Yeasts may cause trouble by growing inside glacé cherries and producing gas which becomes trapped, and causes opaque patches (Oliver, 1962). Such growth has been attributed to the uneven penetration of sugar into the fruit tissue so that areas of relatively low sugar concentration are formed where yeast growth can occur.

IV. Fruits

A. DATES

Dates are of two types, namely the soft and the dry varieties. According to Zein *et al.* (1956), the two types of dates differ greatly in the types of sugar present and in their moisture and sugar content. Soft varieties at their *rutab* or "soft" stage have a moisture content of about 60% and a sugar content of about 32%, consisting mostly of reducing sugars. The dry varieties, on the other hand, have a moisture content of approximately 20% and a sugar content of about 65%, mostly sucrose. Dates are normally sold as dried fruit containing from 12% to 25% moisture. Souring of dates by yeasts generally occurs when the moisture content exceeds 23–25% (Fellers and Clague, 1932, 1942).

Yeast spoilage of dates has been observed at all stages of processing from harvesting to packing and distribution. According to Mrak *et al.* (1942a), yeast growth and fermentation of dates are usually slow and frequently overlooked, although undesirable odours may be present. Occasionally this type of fermentation is accompanied by the formation of subcutaneous gas pockets or rupturing of the skin. Fellers and Clague (1942) characterized the spoiled fruit as soft dark-coloured dates, often closely matted together, which have a peculiar aromatic odour at times resembling ethyl acetate and acetic acid. The flavour of such dates was described as yeasty, sour or simply abnormal.

Rodio (1924), as reported by Mrak *et al.* (1942a), observed the formation of a white crust in the wrinkles of the pericarp, and particularly in the cracks of the epidermis of dates; yeasts isolated from this material were described as *Zygosaccharomyces cavarae*. Isolation of yeasts by Melliger (1931) from two types of Egyptian dates is also discussed by Mrak *et al.* (1942a). One type of date contained sufficient sugar to prevent fermentation and was termed *amhat*; the other type contained insufficient sugar to prevent fermentation and was termed *hayami*. *Zygosaccharomyces cavarae* and several unidentified species of *Zygosaccharomyces* and *Saccharomyces* were obtained from the *amhat* dates. *Hanseniaspora melligeri, Torulopsis pulcherrima* and unidentified species of *Torula* and *Mycoderma* were isolated from *hayami* dates. Esau and Cruess (1933), in turn, isolated two general groups of yeasts from fermented dates. One group was termed *Saccharomyces* because of high alcohol production, and the second group was called "torula yeast" because of moderate alcohol production.

Mrak *et al.* (1942a) reported that on California dates, which are mostly of the dry and semi-dry varieties, species of *Zygosaccharomyces* occurred most commonly followed by *Ha. melligeri* (*Ha. valbyensis*). In order of occurrence the species of *Zygosaccharomyces* were *Zygosacch. japonica* var. *soya, Zygosacch. barkeri, Zygosacch. globiformis* and *Zygosacch. nadsonii*. Other perfect yeasts isolated in small numbers were *Sacch. cerevisiae, Sacch. carlsbergensis* var. *polymorphus, Pi. chodati* var. *fermentans* and *Ha. subpelliculosa*. The genus *Candida* was the most frequently occurring imperfect type; species isolated included *C. chalmersi, C. tropicalis, C. krusei* and a strain of *T. dactylifera*.

From soft varieties of dates Zein *et al.* (1956) isolated mostly *Ha. valbyensis* and *C. guilliermondii*. In addition, they found a few cultures of *Candida tropicalis*, *C. krusei*, *C. mycoderma* and *T. inconspicua*. They explained the absence of species of *Saccharomyces* as possibly due to the use of a short incubation time or to the relatively low sugar content of the recovery medium. *Hanseniaspora valbyensis* cannot grow in a sugar solution of as high a concentration as can species of *Zygosaccharomyces* (Recca and Mrak, 1952).

Yeasts isolated in pure culture from soured dates produced souring in pasteurized sound dates of moisture content in excess of 25% (Fellers and Clague, 1932, 1942). These yeasts were sugar-tolerant and grew well on dates containing 30% moisture or on date syrups containing 65% soluble solids, mainly invert sugar. All of the yeasts grew abundantly at temperatures ranging from 20° to 37·5°, the optimum being about 30°.

Souring of dates was controlled by drying to 23% moisture or by pasteurization (Fellers and Clague, 1932; Clague and Fellers, 1933). Pasteurization of dates caused a slight increase in moisture content, but this did not exceed 22% in the final product (Clague and Fellers, 1933). This increase in moisture lessened the tendency for sugar crystals to form. For control of souring, a pasteurization treatment at 71° and 71% relative humidity for one hour was recommended for obtaining the most desirable results.

B. FIGS AND PRUNES

Figs may also undergo an alcoholic fermentation with production of acetic acid, a process that is referred to as "souring". Rand and Pierce (1920) reported that the causal agent of this spoilage was a yeast. Caldis (1930) pointed out that the term "souring" had been used indiscriminately to cover all kinds of spoilage of figs. However, he restricted the term to specific symptoms. According to his description, souring occurs when the colour of the pulp, which in the healthy tissue is pink, becomes colourless and later the pulpy material becomes watery. A pink liquid may exude through the eye of the fig and may drop on the leaves, or a jelly may form at the eye. Gas bubbles can be seen in the pulp and skin. The pulp disintegrates and smells strongly of alcohol, and is often covered by a white scum. In this condition, the figs begin to shrivel and the dry fruit either drops to the ground or continues to hang on the twig.

When Caldis (1930) examined fruits showing these characteristics, he always found two types of yeast to be the responsible agents; a third was also present but was found incapable of producing the disease. All three were asporogenous and described as wild yeasts. The type most frequently found was a top yeast that produced a scum in liquid media and in the fig cavity; it was referred to as of the Mycoderma type. The second form was described as a bottom yeast, growing poorly on artificial media; it was apiculate. The third form was characterized as a round yeast of the torula type.

Mrak *et al.* (1942b) have pointed out that figs undergo this internal fermentation because botanically the fruit of the fig is a syconium; i.e. it is a more or less hollow receptacle the inner walls of which are lined with flowers when immature and by seed-like fruit when ripe. The fig has an opening at the flattened end which is sealed by overlapping scales until the fruit begins to ripen. At this time, these scales loosen and an opening called the "eye" is formed. When the fruit is mature, the flesh of each floret becomes juicy and an ideal substrate for micro-organisms.

Micro-organisms can be introduced into the fruit *via* different agencies, of which insects are probably the most important. Beneficial as well as harmful insects may play a role; for example, the fig "wasp" (*Blastophaga psenes*) is necessary for pollination of the Calimyrna variety of fig. Mrak *et al.* (1942b) have implicated this insect as a carrier of yeasts to the fig. Other insects which have been incriminated as agents in the transmission of yeasts into the fig are the dried fruit beetle, *Carpophilus hemipterus* (Caldis, 1930; Smith and Hansen, 1931; Davey and Smith, 1933; Miller and Mrak, 1953) and *Drosophila* spp., mainly *Drosophila melanogaster* (Miller and Phaff, 1962).

Yeasts which have been most frequently isolated from the gut and exterior of the dried fruit beetle are *Candida krusei* and *Hanseniaspora valbyensis* (Miller and Mrak, 1953). Additional yeasts that were recovered included *Kloeckera apiculata*, *Pichia membranaefaciens*, *Rhodotorula mucilaginosa*, *Saccharomyces rosei*, *Torulopsis albida*, *T. lactis-condensi* and *T. stellata*. Miller and Mrak showed that this beetle was most strongly attracted to cells of *C. krusei*, then to *Ha. valbyensis*, to fig tissue and to *T. carpophila*, in that order. The average sugar tolerance for these yeasts was 50–55° Brix syrup. The temperature range of growth for these potential spoilage organisms was 5·0–39·5°; however, *C. krusei* was able to grow at 46° and *T. albida* at 0°.

A study of soured figs by Mrak *et al.* (1942b) revealed that of 115 yeasts isolated from the interior of infected figs prior to drying, most were species of *Saccharomyces* and *Candida*; however, a few cultures of species of *Pichia*, *Hanseniaspora*, *Kloeckera*, *Torulopsis*, *Hansenula* and *Debaryomyces* were also isolated. In order of frequency of recovery the species of *Saccharomyces* were *Sacch. cerevisiae*, *Sacch. tubiformis*, *Sacch. fragilis*, *Sacch. cerevisiae* var. *ellipsoideus*, *Sacch. carlsbergensis* varieties *monacensis* and *polymorphus*. The *Candida* spp. were *C. krusei* and *C. chalmersi*. The sugar tolerance of the yeasts listed above was generally low; most of them grew in 40° but not 50° Balling fig syrup. The production of volatile and fixed acids was low and not sufficient to cause the typical spoilage called "souring". Mrak *et al.* (1942b) concluded that typical fig souring results from the associative action of yeasts and acetic bacteria.

Natarajan *et al.* (1948) noted that spoilage of figs during drying occurred most frequently in the cooler climates or during periods when the weather is unfavourable. Yeast populations increased greatly during sun-drying under unfavourable weather conditions; however, the yeast count declined as the fruit became dehydrated.

Dried figs and prunes often become covered with a white "sugar-like" coating during storage and develop gas pockets (Baker and Mrak, 1938; Mrak and Baker, 1940; Mrak, 1941). This coating consists of a mixture of sugar and yeasts, and forms more commonly on bin-stored fruit than on packaged fruit. The sugary substance on packaged fruit consists mainly of sugar crystals because most fruit is heat-treated prior to packaging.

Most of the yeasts isolated from dried prunes and figs were *Zygosaccharomyces* spp. (Baker and Mrak, 1938; Mrak and Baker, 1940). About one-fourth of the isolates were similar to *Zygosacch. mandshuricus*; a second large group resembled *Zygosacch. cavarae*. Other cultures isolated, in order of importance, were species of *Torulopsis*, *Hansenula*, *Saccharomyces*, *Hanseniaspora*, *Debaryomyces*, *Mycoderma*, *Schizosaccharomyces*, *Pichia* and *Zygopichia*. Phaff *et al.* (1966) have repeatedly isolated *Schizosacch. octosporus* from both dried figs and prunes. According to the latter authors this is a less commonly occurring osmophile which was first isolated by Beijerinck in 1894 from spoiled currants.

Mrak and Baker (1940) found that when sun-dried prunes were sealed in cans and stored at 26·7° fermentation and gas formation occurred after a few months. Freshly dehydrated prunes stored under similar conditions did not ferment unless inoculated with a *Zygosaccharomyces* sp. isolated from prunes. This type of spoilage occurred when the prunes contained over 22% moisture. Control of this type of deterioration was attained by dehydration or rapid sun-drying to a low moisture content, fumigation with ethylene oxide and proper storage.

C. MISCELLANEOUS FRUITS AND BERRIES

The fruit fly, particularly *Drosophila melanogaster*, is a serious insect pest in tomato fields, in boxed fruit and in processing plants. These flies, in addition to depositing numerous eggs in damaged portions of the fruit, contribute to the microbial contamination and, consequently, to the fermentative spoilage of tomatoes (De Camargo and Phaff, 1957). For optimal development, *Drosophila* flies require yeasts or other microorganisms in their diet. Yeast flora of the flies and of the fermenting tomatoes were essentially identical. *Hanseniaspora uvarum*, *Kloeckera apiculata*, *Pichia kluyveri* and *Candida krusei* were the predominant yeasts in both flies and tomatoes.

Lowings (1956) found *Kl. apiculata* to develop abundantly on injured strawberry fruits. Moulds would attack both wounded and sound fruits, while the yeast attacked the wounded fruit only. Abundant fluid, softening and pale colour was typical of berries infected with *Kloeckera*; however, yeasts rarely were a major cause of rotting because of the more rapid development of other rot-producing organisms. As mentioned in the Introduction, Berry and Magoon (1934) attributed spoilage of frozen strawberries packed in 50% sucrose to a *Torula* sp. (probably a *Rhodotorula* sp.). This organism grew at $-4°$.

Chilled citrus salad is a mixture of intact grapefruit and orange sections which, in some commercial products, may also contain pineapple chunks and maraschino cherries. The fruit is hand-packed in jars and covered with sucrose syrup which may or may not contain sodium benzoate as a preservative. The unheated product is held at about $-2·2°$ until marketed (Rushing and Senn, 1962). In this product, if it is held at $-1·1°$ to $+4·4°$, yeasts are the predominant flora; at higher temperatures bacteria prevail and are the most prevalent organisms.

Attempts to produce on a commercial scale pickled cherries or "cherry olives" resulted in considerable yeast fermentation even in the presence of 0·03% sodium benzoate (Bowen *et al.*, 1953). The formula of vinegar, sugar and salt usually gave a concentration in the finished product of approximately 1·8% acetic acid, 1·5% sodium chloride and 20% sugar, which was considered satisfactory for preservation. The yeast responsible for this fermentation was of the genus *Zygosaccharomyces*; this particular strain was resistant to 0·1% sodium benzoate. A combination of 2% sodium chloride and 30% sugar with 0·1% sodium benzoate was recommended for control. Pasteurization at 60–65° for 10 min destroyed the yeasts but lowered the quality of the product.

Steele and Yang (1960) implicated yeasts as being the ultimate source of polygalacturonase which causes softening of brined cherries. Apparently the enzyme was produced by yeasts in infected cherries before the cherries were brined; however, the enzyme remained active and in sufficient concentration to cause rapid disintegration of much of the fruit.

D. JAMS, JELLIES AND PRESERVES

Jellies, jams, preserves and marmalades are products prepared from fruits and depend upon a high soluble-solids content and high acidity to prevent spoilage. The chance of spoilage of cooked jams and jellies by osmophilic yeasts is not great because these organisms are killed during heating, and any contaminants which are able to grow at all grow slowly. Osmophilic yeasts in substrates with sufficiently high solids content to prevent growth do not necessarily die, but may remain viable for several weeks or months. Thus, the concentration of soluble solids in sugar-preserved products must be maintained at a level that prevents growth of yeasts and moulds. The required legal minimum concentration of soluble solids in jams and jellies is 65% (Oliver, 1962; Desrosier, 1963). However, Kroemer and Krumbholz (1931) have stated that the growth of certain yeasts is prevented by about 68% soluble solids. Von Schelhorn (1950a, b) has found that *Zygosaccharomyces barkeri*, when cultivated at its optimal pH range of 4–5 and at 30°, is inhibited only by a sugar concentration equivalent to an equilibrium relative humidity of 62%; according to her, this corresponds to a pear concentrate containing approximately 82% solids. Conditions that favour growth of yeasts, such as a favourable pH value and temperature, also favour the possibility of spoilage; thus slightly acid jams and jellies may ferment more readily than definitely acid products.

Subba Rao *et al.* (1965) have described spoilage of an Indian preserve called *murrabba* by osmophilic yeasts. *Murrabba* is a preserve which depends upon a high concentration of sucrose (approximately a 70° Brix syrup) to retard spoilage and to retain the shape of the original fruit or vegetable. However, osmophilic yeasts such as *Saccharomyces rouxii*, *Sacch. mellis* and *Sacch. fermentati* have caused fermentation of this food. Addition of acetic acid and sodium benzoate to the syrup was recommended to control growth. Acetic acid at a concentration of 0·2% with 50 p.p.m. of sodium benzoate effectively inhibited growth of *Sacch. rouxii* and apparently was lethal to *Sacch. mellis*.

V. Fish, Poultry and Red Meats

Normally, yeasts do not have a major role in the spoilage of unfrozen fish, poultry and red meats, because they are only a small part of the initial microbial population. The yeasts grow slowly in comparison with most bacteria, especially on refrigerated meats which are rapidly spoiled by psychrophilic bacteria. Therefore, when yeasts are a problem in the spoilage of these products, bacterial numbers have been restricted in some manner. Yeasts may reach large populations when preservative methods such as pasteurizing levels of ionizing radiation, or when preservatives such as chlortetracyclines, have been used as bacteriostatic agents (Tarr *et al.*, 1952; Eklund *et al.*, 1965).

Yeasts usually constitute only a small proportion of the microbial population on marine products, although numerous varieties have been recovered. Snow and Beard (1939) found that only 0·5–1·0% of the organisms on fresh salmon from the North Pacific area were unidentified yeasts. Ross and Morris (1965) and Morris and Ross (1965), in an extensive survey of yeasts on marine fish, isolated species of *Debaryomyces*, *Torulopsis*, *Candida*, *Rhodotorula*, *Pichia* and *Cryptococcus*. *Debaryomyces kloeckeri* comprised almost half of the isolates; *T. inconspicua* and *C. parapsilosis* were the next most common.

Many of the yeasts associated with marine products grow well at low temperatures. Phaff *et al.* (1952) isolated yeasts from shrimp (*Penlaus setiferus*), most of which grew at 5°. The isolates included species of *Rhodotorula*, *Trichosporon*, *Torulopsis*, *Pullularia*, *Candida* and *Hansenula*. The *Trichosporon* species were not well adapted to growth at low temperatures, but several species of *Rhodotorula*, *C. guilliermondii* and *T. aeria* showed appreciable growth at 2·5° when given sufficient time.

A pink yeast which grows at low temperatures has caused spoilage of oysters in the New England area (Hunter, 1920). Oyster growers of New England previous to 1920 had experienced difficulty in shipping oysters for long distances because of the development of a pink colour in the liquor or on the oyster itself. Oysters which appeared to be in good condition when shipped were often pink upon reaching their destination. Hunter (1920) reported that a pink yeast belonging to the torulae was responsible for this discolouration. The principal source of contamination of oysters by this yeast was during handling in the oyster house. The shell pile was also a favourable location for proliferation of the yeast; these shells were later deposited on the oyster beds and acted as an inoculum for the water. Phaff *et al.* (1952) assumed that this organism was a *Rhodotorula* sp. or *Sporobolomyces* sp.

Later, McCormack (1950) discovered that several packages of oysters stored at −17·8° for a period of one month showed, upon thawing, a pink-coloured liquor and pink and red spots on the oysters. A pink yeast was isolated which was capable of growing in Sabouraud's broth and in inoculated oysters at temperatures of −17·8° to −34°. The pink yeast isolated by Hunter (1920) grew at room temperature (21–25°) but not at 37°.

Eklund and coworkers (1965) surveyed the yeasts associated with the meat of the king crab (*Paralithodes comischatica*) and Dungeness crab (*Cancer magister*). A yeast-like organism, resembling *Aureobasidium pullulans*, and 15 different species distributed among the genera *Rhodotorula*, *Cryptococcus*, *Torulopsis*, *Candida* and *Trichosporon* were represented. Members of the genera *Rhodotorula* and *Trichosporon* were isolated most frequently.

Yeast counts ranged from 200 to 1,000 per gram on freshly processed Dungeness crab meat, and from 30 to 500 per gram on king crab meat (Eklund *et al.*, 1965). In the processing of crab meat, the butchered crabs are placed in a boiling tank and cooked prior to picking of the meat from the shells. Contamination of the product with yeasts and bacteria, therefore, probably occurs during picking and processing.

All these isolates except *A. pullulans* grew in the range 0·5—5·8. *Trichosporon* spp. and *Cr. diffluens* grew well even at 0·5°. All these organisms grew well between 11° and 22° except *A. pullulans* which did not grow well at 11°. Only six of the isolates exhibited any growth at 37°; and only two of these, *Trich. pullulans* and *Rh. glutinis*, grew well.

Yeasts are generally regarded as saccharolytic rather than proteolytic. However, Eklund *et al.* (1965) detected proteolytic activity in several isolates from crab meat. The organism most active in attacking casein and crab proteins was a *Trichosporon* sp. An intracellular proteinase system of a *Trichosporon* sp. has been shown to be an active agent in the ripening of Trappist-type cheese (Vorbeck and Cone, 1963). Three of the isolates, a *Trichosporon* sp., *Trich. pullulans* and *C. scottii*, were very active in the production of lipase.

Dried, salted and smoked fish, because of their low moisture content, are relatively free from bacterial growth; yet, they are susceptible to mould and yeast growth. Boyd and Tarr (1955) and Geminder (1959) delayed growth of moulds and yeasts by dipping fish before smoking in brine solutions containing 0·5% and 1·0% sorbic acid, or the sodium or potassium salts of sorbic acid. They did not consider loss of quality from activity of yeasts and moulds to be as serious as that from bacterial action.

Yeasts are not of major importance on untreated fishery products, because they are rapidly overgrown by bacteria at refrigeration temperatures. However, since yeasts are apparently more resistant to radiation than bacteria, yeast and mould counts have been observed to exceed 10^5 per gram of crab meat irradiated at 0·2 and 0·4 megarad and stored at 0·5° and 5·6°. In contrast, yeast and mould counts seldom exceeded 10^3 per gram on untreated crab meat (Eklund *et al.*, 1966). The predominance of yeasts is only temporary, and they are overgrown by bacteria in the final spoilage flora.

As in fish, spoilage of processed poultry is usually the result of bacterial rather than yeast and mould activity. Nevertheless, under conditions which eliminate or retard bacteria, yeasts may be an important part of the spoilage pattern. Identification of types of micro-organisms occurring on eviscerated poultry has shown that yeasts comprise only a small percentage of the total population. Numbers of yeasts have been appreciably larger on carcasses treated with antibiotics of the tetracycline group (Ziegler and Stadelman, 1955; Ayres *et al.*, 1956; Njoku-Obi *et al.*, 1957; Yacowitz *et al.*, 1957; Barnes and Shrimpton, 1958; Silvestrini *et al.*, 1958; Walker and Ayres, 1956; Wells and Stadelman, 1958). Ayres *et al.* (1956) and Wells and Stadelman (1958) proposed that, if the bacterial population is maintained at a minimum, substrate is then available for growth of yeasts. Additional factors probably are involved, since Whitehill (1957) and Walker and Ayres (1959) have presented evidence that the bacterial flora occurring on meat and poultry produced substances or conditions which limit the growth of yeasts. Simpson *et al.* (1959) never observed yeasts in sufficient numbers to be considered significant for commercially processed poultry treated with chlortetracycline.

Various species of yeasts were recovered from processed poultry (treated with antibiotics and untreated) by Njoku-Obi *et al.* (1957), Wells and Stadelman (1958) and Walker and Ayres (1959). The yeast flora recovered by different workers varied little and included *Saccharomyces cerevisiae, Sacch. dairensis, Rhodotorula minuta, Rh. glutinis, Rh. aurantiaca, Rh. mucilaginosa, Torulopsis holmii, T. albida, T. candida, T. famata, Candida parapsilosis, C. rugosa, C. scottii, C. krusei, C. intermedia* and *C. pelliculosa. Candida parapsilosis,* a potential human pathogen, was recovered by only one group of workers (Njoku-Obi *et al.*, 1957). During their studies on the public health aspects of antibiotic-treated poultry, Thatcher and Loit (1961) failed to find *C. parapsilosis* and believed that this organism was not a common contaminant of poultry products.

Yeasts are of minor importance in the spoilage of fresh red meats. Large numbers and many kinds of micro-organisms can be carried into the packing house on the animal (Empey and Scott, 1939a; Ayres, 1955). The chief sources of such organisms, as pointed out by these authors, are the hide, hair and viscera of the animal. Ayres (1960) recovered yeasts only occasionally from choice steaks stored at 0°, 5° and 10°. Yeasts from three asporogenous genera, *Torulopsis, Candida* and *Rhodotorula*, were recovered. Empey and Scott (1939b) have stated that spoilage of chilled beef held at temperatures close to −1° was caused by growth on the beef surface of a mixture of bacteria, yeasts and moulds. The initial contamination consisted of more than 99% bacteria that grow at 20° and less than 1% of these were viable at −1°. Yeasts were more prevalent at −1° than at 20°, but were not the predominant spoilage organisms.

Lea (1931b) and Vickery (1936a) have reported the growth of asporogenous yeasts at −1° to −1·6° on beef. Vickery (1936a, b) observed that some of these yeasts were lipolytic and could be responsible for appreciable lipolysis of beef fat. Ingram (1958) has pointed out that many yeasts attack fats readily and that this activity may contribute greatly to the spoilage of food by yeasts.

Growth of moulds and yeasts has been noted on lamb and mutton carcasses stored at $-5°$ by Haines (1931) and Lea (1931a). First a decrease in total numbers of micro-organisms occurred, followed by an increase due mainly to the growth of yeasts and moulds. Colonies of these organisms were first observed by Lea (1931a) after about seven weeks of storage, and became fairly numerous on the inside of the flanks after twelve weeks. The fat of the lamb and mutton did not appear to be attacked by these micro-organisms when held for three days at a mean temperature of 12° subsequent to storage. Haines (1931) suggested $-10°$ as a satisfactory storage temperature for controlling microbial growth on lamb and mutton carcasses.

According to Scott (1936, 1937), the critical water contents of muscle, expressed as percentages of the dry weight, are between 45 and 55 for growth of asporogenous yeasts belonging to the genera *Candida*, *Geotrichoides* and *Mycotorula*. Yeasts, in general, are capable of growth on much drier substrates than are bacteria. A decrease in the relative humidity in the atmosphere from $99·3\%$ to $98·0\%$ during storage of meat caused little or no change in the growth characteristics of these yeasts. At 97% relative humidity, retardation of growth was only slight, but at 96% both the lag period and generation time increased, particularly for *Candida* spp. At 91% relative humidity, *Candida* did not grow and growth of *Geotrichoides* became weaker than that of *Mycotorula*. When the relative humidity was 90%, there was no growth of *Geotrichoides* and the growth of *Mycotorula* was so weak that it was apparently close to the limiting water content for its growth. Growth of all the yeasts became irregular at vapour pressures below 94% saturation. For yeast growth, slime was present at 99% relative humidity when the population was between 2×10^6 and 1×10^7 cells per cm^2 depending on the size of the individual yeast cells. At 97% and 98% relative humidity, growth was manifested as small transparent discrete nodules, while at 96% relative humidity these became opaque white. The characteristic yeasty odour was most pronounced on the moist muscle.

A pasty yeast-like slime, which is of microbial origin, may occur on sausage-type meats. Kühl (1910) found a white yeast to be the cause of slimy sausage. He described the fermenting power of this yeast as being unusually slight. Mrak and Bonar (1938) refer to reports by Cesari in 1919 and by Cesari and Guilliermond in 1920 in which several species of unnamed yeasts were encountered on certain types of dried sausages produced in France. Mrak and Bonar (1938) isolated several cultures capable of forming a dull slime when inoculated on sterile sausages; however, these organisms occurred naturally as a mixture of yeasts and various bacteria. Their isolates resembled *D. guilliermondii* var. *nova zeelandicus* (identified as *D. hansenii* by Lodder and Kreger-van Rij, 1952). These isolates could tolerate concentrations of 20% NaCl.

The microbial flora of the surface slime of refrigerated frankfurters consists primarily of lactic acid bacteria, yeasts and micrococci. Drake and coworkers (1958, 1959) isolated members of the genus *Debaryomyces* most commonly from the surface of packaged frankfurters. They isolated the following species: *D. kloeckeri*, *D. hansenii*, *D. subglobosus*, *D. nicotianae*, *C. lipolytica*, *C. zeylanoides*, *C. catenulata*, *T. candida*, *T. gropengiesseri* and *Trich. pullulans*. Wickerham (1957) has reported also that luncheon meats are a good source of nitrite-assimilating *Debaryomyces*.

When frankfurters are irradiated, the surviving flora consists mainly of yeasts and members of the genus *Bacillus*. *Debaryomyces subglobosus* and *T. candida* were the yeasts commonly isolated by Drake *et al.* (1959) from irradiated frankfurters. Growth of yeasts on the surface of frankfurters is increasingly retarded as the concentration of carbon dioxide is increased in the atmosphere used for storage (Ogilvy and Ayres, 1953).

Dyett and Shelley (1965) encountered odours that they associated with yeast spoilage on samples of British fresh sausage stored at room temperature (22°). This sausage consisted of a mixture of raw chopped meat, cereal, spices, salt and water packed in a natural casing. Microscopic examination of the microflora revealed numerous yeasts but a preponderant number of bacteria. Yeast counts made on samples of sulphited sausages stored at room temperature and in the refrigerator (3–5°) substantiated the fact that yeast contamination was much lower than that for bacteria. Sausages incubated at 22° without sulphite had an estimated 10^8 yeasts per gram; with sulphite, 10^7 per gram; with sulphite and incubation at 3·5°, 10^4 yeasts per gram after seven days. The yeasts were not identified.

Little information is available on the yeast flora of meat brines; however, species of *Debaryomyces* are commonly found in brines used for the preservation of various foods. *Debaryomyces* spp. exhibit high tolerances to salt and organic acids; these characteristics, in conjunction with their ability to grow well at low temperatures, are important factors responsible for the prevalence of these yeasts in foods that are preserved by salting and brining. Sturges (1923) isolated yeasts from ham brines which he called "Torula". Mrak and Bonar (1939) in their studies of various brined and pickled products isolated *Debaryomyces* spp. from the single sample of ham brine examined. They were of the opinion that *Debaryomyces* spp. had the greatest salt tolerance and were the most common yeasts present in brines. Costilow *et al.* (1954) found

D. membranaefaciens var. *hollandicus* to be responsible for film formation on brines used for hams, beef tongue, bacon sides and Canadian bacon; they also found *D. kloeckeri*, a non-film-forming yeast, in subsurface brine samples. Ingram (1962) has also reported the presence of yeasts in curing brines in small numbers. He found them commonly occurring in hams and on bacon in comparatively small numbers, and considered them to be of little consequence in spoilage. Yet, he did note one instance of hams spoiled by internal growth of yeasts. According to Ingram (1962), the bacon from sugar-fed pigs is usually more acid than other bacons and favours development of yeasts on vacuum-packed bacon slices from this type of animal; the ultimate source of these yeasts was unknown. Gibbons and Rose (1950) had previously demonstrated in their studies on Wiltshire bacon that bacterial growth was considerably decreased if the pH value of the meat was below 5·6. The development of yeasts did not necessarily lead to rapid spoilage, because the yeasts were not strongly proteolytic and did not give rise to characteristic spoilage odours.

VI. Dairy Products

Bacteria usually cause fermentation of milk and cream, but yeasts do have the ability to grow under conditions unfavourable to many bacteria and occasionally are of importance in dairy products. Yeasts, capable of growth in the presence of large amounts of acid, may actually be favoured and grow better in sour milk than in sweet milk. Hammer (1926) has stated that with certain strains of yeasts the production of objectionable odours occurs much more quickly and is more pronounced in milk which contains lactic acid, either added as such or produced by starter cultures, than in milk of the same lot without acid. In addition, some yeasts can grow in high concentrations of sodium chloride, or may be responsible for production of gas in sweetened condensed milk in which sugar supposedly suppresses growth of organisms.

Gas formation by yeasts is favoured in dairy products containing some acid; however, with improved methods of handling and processing, many of these problems have been brought under control. Hunter (1918) reported that large amounts of cream were lost during the hot summer months from an undesirable fermentation known as "foamy cream". This fermentation was characterized by a yeasty or fruity odour, and was caused by a lactose-fermenting yeast with an optimal temperature for growth near 37° accounting for the prevalence of foamy cream during hot weather. In the receiving rooms of creameries, can covers were frequently thrown violently when the seals were broken on the cans, and the cream would foam out of the can (Hammer and Cordes, 1920).

In addition to the actual loss of cream, the cream could not be used for the best quality butter because the characteristic odour and flavour carried over into the finished product.

Hammer and Cordes (1920) isolated lactose-fermenting yeasts from all the samples of yeasty or foamy cream they examined. The isolates could be divided into two distinct types, and the names *Torula cremoris* and *Torula sphaerica* were proposed for them. According to the classification of Lodder and Kreger-van Rij (1952), these species are *Candida pseudotropicalis* and *Torulopsis sphaerica* respectively. Because of their wide distribution, these yeasts are almost impossible to keep out of cream. Hammer and Cordes (1920) recommended good sanitation and effective pasteurization to keep them under control.

The organisms can grow in materials in which the acid prevents growth of other organisms. *Candida pseudotropicalis* is ellipsoidal in shape and grows optimally at 37°; *T. sphaerica* is spherical in shape and grows poorly, if at all, at 37°. Both organisms produce carbon dioxide vigorously and a yeasty odour in milk at favourable temperatures. Both organisms are commonly present in sweet or sour cream; however, a typical foamy condition is not produced in cream when the organisms are used in pure culture. These yeasts produce little, if any, coagulation of the cream and the gas escapes easily; with a coagulated product, a foamy mass is produced. Hammer and Babel (1957) state that the typical foaminess in cream is the result of associative action; the yeasts produce the gas and some other organism causes coagulation. Allen and Thornley (1929) have referred to a number of early papers that report the prevalence of lactose-fermenting yeasts in milk and milk products. These papers are not discussed here because it would be mainly further cataloguing of dairy products such as yoghurt, cream, starter milk and cheese in which lactose-fermenting yeasts were observed.

Non-lactose-fermenting yeasts can cause a gassy fermentation or "blowing" in cans of sweetened condensed milk. Pethybridge in 1906 reported yeasts as the cause of blowing in tins of Irish condensed milk, and recommended prevention by ensuring that the supply of milk was clean. Hiscox (1923) and Savage and Hunwicke (1923) further substantiated this by their observations that the swelling or blowing of cans of sweetened condensed milk was almost invariably caused by yeasts. Hiscox (1923) found a tin of sweetened condensed milk to be so badly blown that it burst on handling. Microscopic examination of the contents showed the presence of many yeasts. Two types of asporogenous yeasts were isolated which did not ferment lactose or maltose but did ferment sucrose. The first yeast, spherical in shape, caused gas production and coagulation of milk when sucrose was added to milk, but produced only slight blowing at the end of five weeks in milk containing 60% sucrose. A considerable amount of gas dissolved in the milk, which frothed on opening, and the contents had a slight alcoholic taste and odour. The second yeast, oval in shape, produced a vigorous fermentation in milk containing 70% sucrose.

Hammer (1919) also isolated a yeast which he considered responsible for gas formation in sweetened condensed milk. The organism grew in broth saturated with sucrose and produced gas in cans of sweetened condensed milk within two to six days at 37°. Hammer (1919) named this organism *Torula lactis-condensi*; Lodder and Kreger-van Rij (1952) have named it *Torulopsis lactis-condensi*.

Olson and Hammer (1935) investigated a similar outbreak of gas formation in sweetened condensed milk marketed in barrels. Gas development was more prevalent during the summer months than at other times of the year. The heads of the barrels frequently were bulged. When these barrels were opened, gas rushed out; milk was often thrown for some distance. The spoiled milk had a yeasty odour and yeasts were observed in large numbers when microscopic examination was made. Two types of yeasts were isolated; one was identified as *T. lactis-condensi* and the other appeared to be similar to the unnamed spherical yeast isolated by Hiscox (1923). Hammer (1919) named the spherical yeast *Torula globosa*, now called *Torulopsis globosa* by Lodder and Kreger-van Rij (1952).

On the basis of their findings and those of others, Olson and Hammer (1935) proposed that gas formation in sweetened condensed milk is commonly caused by *T. lactis-condensi*, or by *T. lactis-condensi* and *T. globosa* growing together, and that when the two species are growing together *T. lactis-condensi* is the more numerous of the two.

Certain yeasts are acid-tolerant and can grow in cheese and produce flavour defects. Harding *et al.* (1900) concluded that a class of flavour defects in Cheddar cheese designated as "sweet" was caused by yeasts. Cheeses of typical flavour contained few yeasts, whereas cheeses having a sweet flavour contained large numbers of yeasts, at least during the first stages of ripening. Harrison (1902) reported that, in Canadian cheddar cheese, yeasts increased in number from a count of 10^4 per gram in a two-day old cheese to $1 \cdot 5 \times 10^6$ in 23 days and then decreased to 5×10^5 in 45 days. These yeasts were capable of growth in acid whey and of production of bad flavours and mottles in coloured cheddar cheese when in association with starter bacteria. Hood *et al.* (1952) have established that the extent to which the flavour defects developed was closely related to the number of yeasts present in the milk at setting time. According to Edwards (1913) the prevalent flavour defects due to yeasts are fruity, bitter, yeasty, dirty and rancid.

Fruity off-flavours described as pineapple, apricot or strawberry were apparently due to the formation of esters during the progress of the fermentation; however, in many cases the flavour became masked, giving a definite sweetish flavour. Others who have associated certain off-flavours in cheese with yeasts are Van Slyke and Price (1938), Hood and White (1931) and Johns and Katznelson (1941). Pasteurization is effective in controlling this defect.

The presence of lactose-fermenting yeasts in cheese, in addition to producing esters which impart a distinctive flavour, can produce a more open texture than usual, decrease the acidity and cause increased decomposition of the casein. Allen and Thornley (1929) found that two lactose-fermenting yeasts, a *Torula* sp. and *Zygosaccharomyces lactis*, caused such changes in Cheddar cheese. A severe outbreak of "gassy" fermentation in Swiss cheese was traced to a lactose-fermenting yeast by Russell and Hastings (1905). They described the flavour as being disagreeably sweet; very gassy cheeses were badly bleached. Weiser (1942) reported *Torula cremoris* in association with an anaerobic bacterium to be the causative agents of gassy Swiss cheese. These authors related the outbreaks of gassy cheese to unsanitary practices on the farm and in the plant.

Yeasts may produce unpleasant odours and a yeasty or bitter taste in yoghurt. Soulides (1955) isolated three *Torulopsis* spp. from yoghurt, one of which was identified as *T. molischiana*. Lactose-fermenting yeasts can produce a gassy, alcoholic fermentation and a fruity odour in yoghurt. Soulides (1956) found these lactose-fermenting yeasts to be closely related to *Torula cremoris*.

Certain yeasts may develop on the surface of cottage cheese in the presence of an abundant air supply; other yeasts, however, can grow throughout the cheese mass. Considerable surface growth of non-pigmented species can occur without being particularly noticeable; colonies of pink yeasts, of course, are very conspicuous. In general, these organisms have little effect on flavour of cottage cheese, but a yeasty flavour may be imparted by certain lactose-fermenting species (Hammer and Babel, 1957). Other flavours in cottage cheese which have been attributed to yeasts have been described as "acidy", bitter and fruity (Foter *et al.*, 1941; Morgan *et al.*, 1952; Deane *et al.*, 1954). Wales and Harmon (1957) found that acetylmethylcarbinol, a desirable flavour and aroma compound in cottage cheese, rapidly disappeared when *Rhodotorula flava* and *Geotrichum candidum* were present.

Pink yeasts are commonly present in milk and cream in small numbers and produce red spots only occasionally because of their relatively slow growth. In fact, most yeasts have a characteristically slow activity in milk as noted by many workers, but probably first recorded by Hastings in 1906. According to Cordes and Hammer (1927) sweet cream produced under conditions prevalent at that time contained a few hundred and occasionally a few thousand pink yeasts per ml. Sour cream nearly always contained these organisms, sometimes in relatively high numbers. These pink yeasts can tolerate considerable amounts of acid, and accordingly sometimes appear on the surface of sour milk or cream or on soft cheese as distinct colonies. Nissen (1930) was of the opinion that red yeasts are especially harmful in dairy products because of their ability to peptonize milk and decompose butterfat. According to Hammer and Babel (1957), *Rh. glutinis* is the most commonly occurring pink yeast on the surface of dairy products. Yeasts of the genera *Sporobolomyces* and *Bullera* are apparently unimportant in dairy products from the standpoint of numbers present and of the production of defects (Olson and Hammer, 1937).

Pink yeasts were repeatedly isolated from butter by Cordes and Hammer (1927) but colonies were never observed on the product. They isolated *Torula glutinis* most frequently and encountered *Torula rubicunda* and *Torula paraglutinis* only in small numbers. These organisms usually grew better at 21° than at 27°; some failed to grow at the latter temperature. Orla-Jensen (1931) observed that torulae which do not ferment sugar developed freely in butter over a period of time, and hydrolysed butter fat more or less vigorously provided the butter was sufficiently acid. He observed one torula which hydrolysed fat and coloured the butter red; however, the growth of this organism was inhibited by common salt. Miklik (1953) also has mentioned a case of rancidity in butter traced to the action of a *Rhodotorula* which he called *Torula rhodense*. The splitting of the butterfat was enhanced by the presence of lactic acid bacteria.

High numbers of yeasts in butter are objectionable, and are an indication of inadequate sanitary practices (Tanner, 1944). The presence of yeasts usually indicates one of several things: mistreated cream, unsanitary utensils and equipment, lack of refrigeration or improper pasteurization. High yeast counts are usually traceable to improperly cleaned and sanitized churns (Macy *et al.*, 1932).

The importance of yeasts in the deterioration of butter has not been conclusively established, but it would be expected that they could cause deterioration if present in excessive numbers. A yeasty flavour in butter has been attributed by Hammer (1948) to the development of lactose-fermenting yeasts, *T. cremoris* and *T. sphaerica*, in the cream before it reaches the plant. Grimes and Doherty (1929) maintained that the growth of lactose-fermenting yeasts in cream held for several days interfered with normal churning.

Nelson (1928) found the action of non-lactose-fermenting yeasts to be slow in milk, and concluded that they did not have much influence on the changes brought about in dairy products by micro-organisms. However, Chinn and Nelson (1946) thought that non-lactose-fermenting yeasts may cause significant deterioration of dairy products under some circumstances. From cream and butter they isolated species of *Rhodotorula, Pullularia, Torulopsis, Trichosporon* and *Candida,* all of which could be destroyed by pasteurization. Defects produced by representative cultures when inoculated into cream and unsalted butter ranged from no off-flavours to pronounced off-flavours. Bitterness and proteolytic and lipolytic defects were most common and were enhanced by the presence of *Streptococcus lactis.*

Skim milk or dissolved skim milk powder is frequently used as the aqueous phase in the manufacture of margarine. Ripening of the margarine milk by bacterial starter cultures can result in the accumulation of lactose-fermenting yeasts. A yeasty or sour taste and smell has been associated with the presence of these organisms. A soapy sharp taste and excessively high acidity have resulted from the growth of lipolytic yeasts and bacteria. Various types of yeasts, particularly the torulaceae, utilize lactic acid and other organic acids present in margarine, thereby reducing the acidity and producing conditions favourable for the growth of spoilage bacteria (Anderson and Williams, 1965). However, in modern well-managed plants the risk of spoilage from these organisms is considered to be relatively small.

VII. Fermented Foods

A. PICKLES

Salt-stock pickles are the principal type of stored fermented cucumbers. Salt stock is used in making such products as sour pickles, sweet-sour pickles, mixed pickles and relish. There are two types of fermentation which take place; namely, lactic acid and yeast fermentations. The lactic acid fermentation is desired, while the yeast fermentation leads to spoilage. Completion of this fermentation, according to Frazier (1967), requires six to nine weeks, depending on temperature and salt concentration. In salting pickles for dills or for salt stock, the chemical and physical changes are similar but may vary in magnitude, depending on the product to be formed.

Yeasts are present throughout the fermentation and storage of salt-stock pickles. The prevalence of yeasts is influenced by temperature, relative amount of air, salt concentration and amount of sunlight. Vaughn (1954) and Costilow *et al.* (1955) believe that most yeasts are detrimental to cucumber fermentations, because they consume lactic acid which, together with salt, preserves the pickles for future processing.

Film-forming yeasts and yeast-like organisms occur very commonly on brines used for the storage of various fruit, vegetable and meat products. These organisms often cause undesirable changes in flavour, odour and composition of food products on which they grow, and have been commonly known as *Mycoderma vini* or "wine flowers" when they occur on fruit products, as *Mycoderma cerevisiae* on cereal products, and as "film yeasts" or "scum" on pickle brines (Joslyn and Cruess, 1929). Etchells and Bell (1950b) stress the need for a distinction to be made between surface yeasts and sub-surface yeasts in relation to cucumber brines, a point frequently overlooked by some authors. Mrak and Bonar (1939) isolated from various food brines strains of *Debaryomyces membranaefaciens*, *D. membranaefaciens* var. *hollandicus*, *D. guilliermondii* var. *nova zeelandicus*, *Pichia membranaefaciens* and *Mycoderma decolorans*. The film-forming species of *Debaryomyces* are probably the most widely distributed yeasts associated with food brines (Mrak and Bonar, 1939; Etchells and Bell, 1950b; Etchells *et al.*, 1953). The high salt tolerance and ability to assimilate a large number of compounds as a source of carbon make the *Debaryomyces* particularly adaptable to brines. In addition to *Debaryomyces* spp., Etchells and Bell (1950b) recovered *Endomycopsis ohmeri*, *Zygosaccharomyces halomembranis* and *Candida krusei* from films on commercial brines. Etchells *et al.* (1953) are of the opinion that, after the *Debaryomyces* spp., the predominant film-forming yeasts are (listed in order of approximate importance) *Zygosacch. halomembranis*, *E. ohmeri*, *C. krusei*, *Hansenula anomala* and *Pi. alcoholophila*.

Brines containing 4% or 10% sodium chloride do not inhibit the development of acid-consuming film-forming yeasts (Jones and Harper, 1952). They recommended approximately 22% salt to inhibit development of these yeasts completely. However, Mrak and Bonar (1939) observed that some *Debaryomyces* spp. grew in cucumber brines containing 24% salt; *Pichia* and *Mycoderma* spp., on the other hand, grew poorly in brines containing 15·1% salt. Joslyn and Cruess (1929) have stated that the presence of 10% salt greatly decreases the tolerance of film-forming yeasts for organic acids. According to them, few of these yeasts could grow in the presence of 15% salt; yet some yeasts were able to grow in concentrations of 20% or more when organic acids were absent. Mrak and Bonar (1939) also made the observation that species of *Debaryomyces* and of *Mycoderma* maintained in culture collection had a much lower salt tolerance than those obtained from pickle brines.

The surface of cucumber brines sheltered from direct sunlight will usually be covered by luxuriant films attributable to the growth of film-forming yeasts. Control of these films may be achieved by various means.

According to Etchells and Bell (1950a), some operators remove the film by skimming; some keep the brine surface stirred to prevent film formation; and others prefer to leave the film undisturbed. Removal of the surface layer of yeasts, followed by ultraviolet irradiation for two hours each day, also prevents surface growth (Jones and Harper, 1952). However, since ultraviolet radiation does not penetrate to any appreciable depth, many of these yeasts grow below the surface beyond the effective range of the rays and still cause undesirable changes in the cucumber stock.

Sub-surface yeasts have been associated with a gaseous fermentation and a type of spoilage known as "bloater" or hollow cucumber formation (Jones *et al.*, 1941; Etchells *et al.*, 1953). Bloating can be in the form of lens-shaped gas pockets in the tissue, or the gas pressure may press the entire seed portion of the cucumber toward the skin, leaving a large gas-filled cavity. Etchells and coworkers (1953) related *Torulopsis caroliniana, Hansenula subpelliculosa, Zygosaccharomyces* spp. and *Brettanomyces* spp. with bloater formation. Bell and Etchells (1952) also recovered a yeast closely related to *Zygosacch. globiformis* which contributed to a gaseous-type spoilage in commercially prepared sweet cucumber pickles.

According to Jones and coworkers (1941) relatively large percentages of bloaters are formed when brines of high salinity are used during the early portion of the curing period in salt stock production. Etchells (1941) found active yeast fermentation occurring in brines of 20%, 30% and 40% saturation with respect to salt. Addition of lactic acid and of sugar in quantities as small as one per cent of the cucumbers at the beginning of curing or during active fermentation resulted in increased bloater production.

In an isolated case of gaseous fermentation of sweet stock which occurred during the finishing process, Etchells and Jones (1941) observed that bloaters or hollow cucumbers comprised about one-fourth of the stock from a 40-bushel tank. Since fermentation had ceased some time previous to the initial observations, bacteria and yeasts were not recovered by routine plate counts. Microscopic observation of the liquor however, revealed a preponderance of yeast cells and few bacterial cells. In addition, chemical analyses showed the production of carbon dioxide and alcohol and the absence of non-volatiles, which tended to support the contention that yeasts were responsible for the gaseous fermentation and resulted in the production of bloaters. In this instance, bloater formation was attributed to finishing the pickles at too low an acidity. A final acidity of approximately 20 grains (2·0%) of acetic acid or higher is generally recommended unless benzoate or pasteurization is used (Etchells and Jones, 1941; Vaughn, 1954).

Although the acidity of finished sweet pickles should not be less than 2·0% expressed as acetic acid, the sugar content may vary between 20% and 40% according to the quality (Vaughn, 1954). Thus, sweet pickles are susceptible to fermentation by species of osmophilic or osmoduric yeasts, many of which are included in the genus *Zygosaccharomyces*, now called *Saccharomyces* by Lodder and Kreger-van Rij (1952). Fermentation of the sugar can occur during the finishing process or in the finished pickles packed in glass, kegs or barrels. A strictly enforced programme of sanitation is necessary to control them; sodium benzoate or pasteurization may be used as an additional control in the final product.

Dakin and Day (1958) believe that a limited range of species of yeasts are responsible for the spoilage of British acetic-acid preserves, and have identified *Saccharomyces acidifaciens* as the primary agent and *Pichia membranaefaciens* as an occasional cause of spoilage. They described yeast spoilage as usually characterized by gas formation, with musty, sulphurous or yeast-like off-odours. In clear pickles they observed a creamy or brown sediment and cream-coloured colonies on vegetables protruding above the surface of the cover liquid. In more viscous products, the surface was partially or completely covered by a layer of cream- to brown-coloured yeasts.

The distinctive aroma and flavour of genuine dill pickles is the result of a fermentation by lactic acid bacteria and the blending of dill and spices which are added to the brine. If yeasts are present in great numbers during this fermentation, utilization of fermentable carbohydrate and lactic acid decreases the total acidity and increases the pH value of the brine. These conditions in turn favour the growth of undesirable organisms and lessen the quality of the final product.

Yeast populations in commercial fermentations generally decline in numbers during the first three to five days before initiation of rapid growth (Costilow and Fabian, 1953). Peak populations are attained between ten and twenty days after brining, with a steady decline thereafter. *Torulopsis holmii* is frequently the predominant species in cucumber fermentations in northern areas of the U.S.A., and is usually responsible for the early vigorous yeast fermentation. *Torulopsis rosei* occasionally is more prevalent than *T. holmii* (Etchells *et al.*, 1952; Costilow and Fabian, 1953). In southern regions of the U.S.A. *T. caroliniana* frequently dominates the early stages, followed by *Brettanomyces versatilis* during the later stages of the fermentation (Etchells and Bell, 1950a). Both groups of workers regularly found *Hansenula subpellicu-losa* in these fermentations. The types of yeasts occurring in the fermentations from the two regions were similar; the principal differences pertained to the species of yeasts which prevailed during the various stages of production. On the basis of their extensive studies on occurrence and sequence of yeasts in cucumber fermentations, Etchells *et al.* (1952, 1961) concluded that yeasts occur in the following approximate order of frequency: *Br. versatilis, H. subpelliculosa, T. caroliniana, T. holmii, Sacch. rosei, Sacch. halomembranis, Sacch. elegans, Sacch. delbrueckii, Br. sphaericus* and *H. anomala.* These yeasts were found in brines containing from 4% to 18% salt by weight and ranging in pH value from 3·1 to 4·8. The principal sub-surface yeast activity in fermentations containing 4·5% salt was attributed to *C. krusei* and *C. tropicalis.* At the higher brine concentrations (9·2–15·8% salt) *Br. versatilis, H. subpelluculosa, T. caroliniana* and *Sacch. rosei* were considered to be most prevalent.

Brine yeasts generally are not regarded as the causative agents of the softening of cucumber salt-stock. During the brine fermentation of cucumbers for pickles, Bell *et al.* (1950) detected an enzyme similar to polygalacturonase, which they correlated with softening of the salt-stock. The principal sources for this enzyme are possibly micro-organisms occurring during and subsequent to storage in brine and in the fruit that is being brined. Later, the same workers (Bell *et al.,* 1951) reported that the seeds, leaves, petioles, stems, flowers and fruit of the pickling cucumber contained the de-esterifying pectic enzyme, pectinesterase. However, Luh and Phaff (1951) found six cultures of yeasts that could cause noticeable change in pectin broth. These six cultures were all identifiable with *Sacch. fragilis* and a variety of this species, together with its imperfect form *C. pseudotropicalis* and certain varieties of the latter. They named the enzyme yeast polygalacturonase because it differed from that of mould origin in that pectic acid substrate was only partially hydrolyzed. Also, Etchells *et al.* (1953) isolated a strain of *Sacch. cerevisiae* from soft dill pickles which appeared to be identical taxonomically with strains isolated from spoiled citrus concentrate that produced pectinase. Roelofsen (1953) reported polygalacturonase activity in yeasts from such genera as *Candida, Pichia, Saccharomyces* and *Zygosaccharomyces.* All gradations occurred from very little activity to a marked polygalacturonase activity. Later, Bell and Etchells (1956) found that only six of 139 isolates from brines gave definite tests for glycosidic hydrolysis of pectin; namely two strains of *Sacch. fragilis* and four of *Sacch. cerevisiae.* All came from sources other than cucumber brines. They concluded that insufficient evidence was available to incriminate yeasts from cucumber brines as a potential source of the salt stock-softening enzyme, polygalacturonase (see Chapter 10, p. 529).

Sorbic acid is an active inhibitor of catalase-positive organisms in general, and the lactic acid fermentation can occur in the presence of this antibiotic substance (Emard and Vaughn, 1952). For this reason,

Phillips and Mundt (1950) and Jones and Harper (1952) advocated the use of 0·1% sorbic acid to control film yeasts in cucumber fermentations. They reported that sorbic acid at this concentration had no adverse effect on the flavour of the finished pickles; yet, Borg et al. (1955) reported an inhibition of lactic acid fermentation and claimed that the use of sorbic acid impaired the curing and colour of the fermented cucumbers.

Evidently the toxic action of sorbic acid is directly related to the undissociated acid. Several workers have demonstrated the increased effectiveness of sorbic acid at pH values of 5·0 or lower (Costilow et al., 1955; Sheneman and Costilow, 1955; Bell et al., 1959). The efficient use of sorbic acid as an antimicrobial agent would be at pH values yielding the greatest concentrations of undissociated acid. At pH 4·8, 50% of the acid is undissociated; thus, at this pH value or below, one-half of the total sorbic acid or more would be effective.

Sheneman and Costilow (1955) reported that sorbic acid at a concentration of 0·1%, in combination with 0·5% acetic acid, prevented growth of Zygosacch. globiformis, a yeast which had been isolated from spoiled pickles. Complete inhibition occurred with this combination in sucrose solutions ranging from 2% to 40%. When no sorbic acid was present, 2·0% acetic acid was required to inhibit the yeast in sucrose solutions of 20% and above, and 3% acetic acid was necessary in media containing 2% and 10% sucrose. Fabian and Wadsworth (1939) previously had shown that, on an equal acidity basis, the preserving capacity of acetic acid was greater than that of lactic acid.

Evidence presented by Costilow et al. (1955) has demonstrated that yeasts found to be most prevalent in cucumber fermentations are completely inhibited by 0·01% sorbic acid in an 8% salt medium at pH 4·6; but as the pH value is increased and/or the salt concentration decreased, more sorbic acid is required for complete inhibition. During three different brining seasons, Etchells et al. (1961) isolated yeasts from experimental fermentations containing brine concentrations of 4·5%, 9·2%, 11·9% and 15·8%. The addition of sorbic acid at concentrations of 0·025%, 0·050% and 0·10% drastically suppressed the yeast populations but did not eliminate them. Furthermore, the patterns of species occurring with or without sorbic acid were similar qualitatively. Costilow (1957) reported that treatments with concentrations of 0·01 –0·10% sorbic acid of salt stock fermentations greatly reduced occurrence of bloater-type spoilage. The occurrence of severe bloaters and of poor seed cavities was reduced approximately 75%; the incidence of small separations was unchanged. However, some yeast activity, both sub-surface and film-forming, developed in fermentations treated with the 0·01% concentration, and occasionally with the 0·02% treatment (Costilow et al., 1957).

B. OLIVES

Microbiological spoilage of brined olives is similar in many ways to that of brined cucumbers. Bacteria, yeasts and moulds may cause deterioration during the pickling process, during subsequent storage in barrels, or in the package used for final distribution. According to Vaughn et al. (1943) and Vaughn (1954), bacteria are probably responsible for the most serious losses, but yeasts and moulds may be troublesome at times. They are of the opinion that deterioration by yeasts and moulds generally reflects gross neglect during the final phases of fermentation and subsequent storage in barrels.

Mrak et al. (1956) have outlined briefly the various phases in the preparation of Spanish-type green olives as practised in California. According to them, the fruits are harvested and placed in tanks in which they are treated with 1·25–2·0% lye, depending on the variety and maturity of the fruit. Exposure to lye is a debittering process in which a glucoside, oleourapein, is hydrolysed and removed. After this bitter component is removed by leaching with water, the fruit is covered with salt brine in the same tank or in barrels. If the fruit is transferred to barrels, brine (about 6·5%) is added, and during the next ten to twelve days more salt is added to correct for the dilution caused by the aqueous phase of the olive. However, if the olives are brined in the vat, salt is added in relatively small amounts each day for a week or more; thus, attainment of a full strength brine occurs over a period of time. When the brines reach full strength, the barrels are closed and stored in the open; vats are usually covered with a microcrystalline wax, sprayed on the surface in a melted condition. The vats are usually in buildings in which temperature control is possible.

During fermentation and subsequent storage of olives, yeasts develop even in completely filled barrels and in vats with a surface covering. Mrak and Bonar (1939) found that the films on fermenting green olives contained Mycoderma decolorans (C. mycoderma) and Pi. membranaefaciens. Both species grew well in the presence of 9·6% salt but poorly in 15·2% salt, and both utilized lactic, citric, tartaric, malic and acetic acids. Mrak et al. (1956) observed that most of the yeasts isolated from olive brines belonged to two genera, Candida and Pichia. Of the species of Candida, C. krusei was found most frequently; and of the Pichia, Pi. membranaefaciens was the most frequently isolated. Candida mycoderma, which is considered to be the imperfect form of Pi. membranaefaciens, frequently was isolated also.

During the first period of the fermentation, which may last from two to seven weeks, most of the sugar is converted to organic acids (Mrak *et al.*, 1956). The predominant species during this period are *C. krusei*, *C. tenuis*, *C. solani*, *T. sphaerica*, *T. holmii*, *T. versatilis*, *H. subpelliculosa* and *D. nicotianae*. All these species, except *D. nicotianae* which ferments weakly, ferment one or more sugars. These species disappear, and during the last period of fermentation and storage *Pi. membranaefaciens*, *C. mycoderma*, *C. rugosa*, *T. inconspicua*, *T. magnoliae* and *T. etchellsii* predominate. Most of these yeasts, particularly the first four species listed, are primarily oxidative. Under aerobic conditions, these species can utilize the organic acids important to the production of olives, which results in a decreased acidity which, in turn, favours the development of spoilage organisms. In addition, heavy yeast growth on the surface of brines in improperly filled barrels of olives imparts off-flavours and odours (Vaughn *et al.*, 1943).

"Stuck" fermentation frequently occurs with Manzanillo olives (Vaughn *et al.*, 1943). Under these conditions sugar is depleted without increased acid production. "Stuck" olive fermentations result when yeasts predominate to the exclusion of the desirable lactic acid bacteria, and addition of sugar serves only to aggravate the problem. Control of these fermentations can be obtained by replacing the undesirable yeast population with desirable lactic acid bacteria, or by addition of normal actively-fermenting brines in concentrations of 10–20%, or in extreme cases by removal and replacement of the brine with fresh actively-fermenting brine. In laboratory-induced "stuck" fermentations, Vaughn and his coworkers never isolated or observed bacteria, but only yeasts.

A common defect associated with green olives in California has been known as "yeast spots" (Vaughn *et al.*, 1943). This abnormality is characterized by the formation of raised white spots or pimples between the inner surface of the epidermis and flesh of the olives. Although called yeast spots, most of these pimples have been found to contain strains of *Lactobacillus plantarum*.

Sediments and gas formation sometimes occur in glass-packed fermented green olives. This fermentation is usually attributable to the presence of fermentable sugars left in the olives or added with the spices. Lactic acid bacteria are usually responsible for this refermentation; but, if the containers are not filled with brine or vacuum sealed, yeasts may produce much of the sediment and even form a pellicle in the headspace (Vaughn, 1954). Ayers *et al.* (1930) attributed an outbreak of clouding of green olive brine to the growth of a *Mycoderma* yeast. They described the clouding as being due to relatively large flakes appearing in the brine, some of which floated while some settled to the bottom causing a white deposit. Certain bacteria can also cause this condition. Refermentation can be controlled by pasteurization.

C. SAUERKRAUT

The preservation of cabbage as sauerkraut is the result of a lactic acid fermentation of shredded cabbage in the presence of not less than 2%, nor more than 3%, of salt. Although many kinds of micro-organisms may be found in the juice of kraut, the lactic acid bacteria are the most important. Yeasts do not develop to any great extent unless air is present. However, yeasts will practically always develop on free juice upon the surface of sauerkraut, forming a white scum. Occasionally, perhaps because of failure to weigh the kraut properly so that juice does not come to the surface, certain aerobic bacteria and yeasts grow near the surface causing a faulty fermentation and dark kraut near the surface (Pederson, 1931).

A number of workers have reported on the occurrence of pink sauerkraut, caused by the growth of aerobic yeasts, frequently in the presence of high concentrations of salt (Butjagin, 1904; Wehmer, 1905; Brunkow et al., 1921; Fred and Peterson, 1922; Pederson, 1931; Pederson and Kelly, 1938). Yeasts may be found in great numbers in kraut, but pigment is not necessarily evident. High salt concentration was mentioned frequently as the factor favouring growth of these organisms, and high temperature and high acid content favoured growth and pigment production. Fred and Peterson (1922) and Marten et al. (1929) have stated that, when cabbage is allowed to ferment at a temperature of 20° or above, or to ferment in the presence of 3% or more of sodium chloride, a decided pink develops. In fact, any inhibition of normal fermentation can favour the development of pink yeasts.

The yeasts produce a pigment that varies from a light pink to an intense red. Since these yeasts grow only aerobically, they usually develop on the surface of the vats of sauerkraut, in air pockets or on exposed shreds of cabbage. In addition to producing an objectionable colour, these yeasts also utilize sugar and acid which are essential to preserving and producing an acceptable product.

VIII. Fruit Juices, Fruit Drinks and Carbonated Beverages

A. APPLE JUICE

Fresh apple juice contains an assortment of micro-organisms including yeasts, moulds and bacteria. Observations on the yeast flora of apples and in apple juice have been reported by a number of workers (Pearce and Parker, 1908; Marshall and Walkley, 1951; Clark et al., 1954; Challinor, 1955; Williams et al., 1956; Beech, 1958a, b; and Lüthi, 1959). Representatives of the genera Candida, Cryptococcus, Rhodotorula and Torulopsis have been recovered from healthy skin and cores, and from freshly pressed juice of the apples.

During growth and maturation of apples on the tree, yeasts and bacteria on the surface of the fruit continually undergo change in population density, even when variation in climatic conditions is at a minimum (Marshall and Walkley, 1951). In sound fruit, micro-organisms are not necessarily confined to the cuticle but may also be present, but to a lesser extent, in the core. Yeasts are not found in the flesh unless internal or external breakdown has occurred due to fungal activity; then yeasts may be present in areas still free from mould contamination. Marshall and Walkley (1951) are of the opinion that apples with yeast counts in excess of 2×10^6 may be assumed to be damaged; and with counts in excess of 5×10^6, the fruit is unsuitable for apple juice manufacture. Marshall and Walkley (1952) also made the observation that, during commercial processing of apple juice, the incidence of yeasts relative to moulds showed a progressive increase whereas the incidence of bacteria diminished. Beech (1958b), during microbiological studies on the different stages of cider-making, found that the juice was frequently contaminated with yeasts occurring on various pieces of processing equipment; however, treatment of the juice with sulphur dioxide improved the yeast flora of the juice for cider fermentation (see Chapter 3, 73).

Strains of yeast found in the fermented juice are chiefly sporogenous, and those found on the fruits or in freshly pressed juice are asporogenous. Species of *Candida* are usually numerically preponderant on apples (Clark *et al.*, 1954; Beech, 1958a, b). According to these authors and Williams and his coworkers (1956), *C. pulcherrima* and *C. malicola* are most prevalent on apples; Williams *et al.* also isolated numerous *Rh. glutinis*. In fact, all the yeasts isolated by Clark *et al.* from Canadian apples were reported as members of the family Cryptococcaceae and included the genera *Candida*, *Cryptococcus*, *Rhodotorula* and *Torulopsis*. However, yeasts isolated from cider included species of *Debaryomyces*, *Pichia* and *Saccharomyces*.

Unclarified apple juice contains 40–50 times more micro-organisms than clarified juice (Fabian and Bloom, 1942). It was also observed that unclarified juice was usually not saleable as fresh apple juice after approximately seven days due to alcoholic fermentation, and that clarified apple juice was spoiled for sale because of acetic acid fermentation.

At low temperatures, deterioration of apple juice due to the metabolic activity of yeast cells is minimized. Poe and Field (1947) detected no quality changes in juice stored at 4° for 60 days. However, storage of juices at low temperatures is expensive and not always practical; therefore, chemical additives have been studied for their inhibitory effect on yeasts in apple juice. Sodium benzoate at a level of 0·1% will inhibit the growth of yeasts (Rice and Markley, 1921; Tanner and Strauch, 1926;

Fellers, 1929; Poe and Field, 1947), but benzoic acid at this concentration imparts an undesirable flavour to the juice (Fabian and Marshall, 1935; Poe and Field, 1947; Do and Salunkhe, 1964). On the other hand, sorbic acid in low concentrations causes no noticeable flavour changes and effectively prevents yeast fermentation in fresh unpasteurized apple juice (Ferguson and Powrie, 1957; Weaver et al., 1957; Do and Salunkhe, 1964). Sorbic acid may not completely prevent spoilage of apple juice because control of yeasts creates conditions favourable for growth of acetic acid bacteria. Therefore, Ferguson and Powrie (1957) recommended the use of 0·035% sorbic acid and 50 mg ascorbic acid per 100 ml of juice for preservation of juice without refrigeration for a minimum period of two weeks. Robinson and Hills (1959) recommended a combination of sodium sorbate (0·06–0·12%) and mild heat (48·9° for 5 min) for increasing the storage life of fresh apple cider.

Other chemicals that have been examined for inhibition of fermentation in apple juice include vitamin K_5. Vitamin K_5 is effective at 50 p.p.m. but lightens the colour of the apple juice (Yang et al., 1962). Hope (1963) has suggested the use of sodium fluoride; this compound, at concentrations of 50–70 p.p.m., will preserve apple juice during a two-day holding period at 15·6–22·2° and still permit subsequent fermentation by wine yeasts. Hope (1963) points out that the toxicity of sodium fluoride to humans may be a problem; but it is unlikely that a lethal dose could be consumed at a concentration of 50–70 p.p.m. since it would require ingestion of about 20 gal of juice.

Challinor et al. (1948) found that apple juices given an ion exchange treatment became very much less liable to spoilage by wine yeasts. Apparently this treatment effected the removal or partial removal of nutrients necessary for the growth of these organisms. The most pronounced reductions occurred in total phosphorus, calcium and magnesium.

Cruess and Irish (1923) recommended pasteurization of a carbonated apple juice, which was a popular beverage at that time, to prevent spoilage from yeasts. The fruit syrup used for making this beverage was of such low Balling value (45–65°) that treatment was necessary. Pasteurization of the beverage in bottles or cans at 65·6° for 30 min was adequate for destroying all yeast cells in the carbonated juice.

Beech et al. (1963), in their studies on the microbiological stability of juices and concentrates, found that few of the commonly occurring members of the juice flora survived concentration and storage at 5° if the specific gravity of the product exceeded 1·240. However, osmophilic yeasts not only survived but also fermented concentrates with specific gravities of 1·300. One yeast isolated from a cider factory vigorously fermented such concentrates at 5° and 15° over a pH range of 3·1–3·8. Two strains of *Saccharomyces rouxii* were slightly less vigorous. Osmo-tolerant yeasts showed no growth but survived for three months in reduced numbers in concentrates. These yeasts included *Debaryomyces kloeckeri*, *Candida pulcherrima*, *Kloeckera apiculata* and a *Torulopsis* sp. Osmophobic yeasts isolated from ciders showed no survival after one week even at a specific gravity no higher than 1·200. These yeasts in-cluded such species as *Sacch. uvarum*, *H. valbyensis*, *Saccharomycodes ludwigii* and *Br. claussenii*. These authors recommended sterile storage for concentrates which have a specific gravity below 1·300.

Fields (1962, 1964) found that yeasts, moulds and bacteria produced diacetyl and acetylmethylcarbinol in apple juice. The main producers of diacetyl were yeasts, namely *Saccharomyces* spp. He considered the presence of diacetyl in apple juice as presumptive evidence of poor sanitary practice in the processing plant; however, the use of diacetyl was considered to be of limited value as an indicator of microbial deterioration because it decreased in quantity upon storage (Murdock, 1964; Holck and Fields, 1965). Nevertheless, Fields (1962) has proposed standards for acetylmethylcarbinol content of apple juice. He proposed that 1·0 p.p.m. of acetymethylcarbinol or below was acceptable, and generally indicative of sound raw materials and good processing condi-tions. Values of 1·1–1·6 p.p.m. for acetymethylcarbinol were indicative of questionable practices and materials; and values above 1·6 p.p.m. indicated poor raw materials and poor sanitary conditions.

B. CITRUS JUICES AND DRINKS

Fermentation of citrus juices is due to yeasts which may have found their way into the juice during the interval between the halving of the fruit and the time the juice is packed. According to Braverman (1949), sound fruit before being halved contains no yeasts, and any yeasts occurring in the juice came from the outside. He indicated that there may be as many as $1·6 \times 10^8$ micro-organisms on the surface of an average orange, but did not give the relative numbers of yeasts, moulds and bacteria.

Faville and Hill (1951) found that citrus juices (orange, grapefruit and tangerine) extracted under the most ideal conditions may contain a wide range of micro-organisms. However, the high moisture, sugar and organic acid contents of citrus juices particularly favour the growth of yeasts. They maintained that it is essential that the juice be kept as free as possible from micro-organisms during extraction and subsequent processing, particularly if the juice is to be used for frozen concentrate. The fewer the number of micro-organisms in a frozen concentrate, the better will be the possibility of resisting spoilage from improper handling between the time it is produced and that when it is consumed. Freezing kills many, but not all, micro-organisms; in fact, some organisms may be preserved in this manner. Therefore, when frozen juices are thawed and the temperature rises, these organisms begin renewed activity. Citrus juices prepared for freezing should be as free as possible of yeasts before being frozen.

Nolte and von Loesecke (1940) found that all the unpasteurized citrus juices they examined contained yeasts, moulds and bacteria in considerable numbers. Yeasts isolated from raw grapefruit juice were identified as *Saccharomyces cerevisiae, Sacch. apiculatus* (probably *Hanseniaspora*), *Sacch. ellipsoideus, Sacch. pastorianus, Torula* (probably *Rhodotorula*) spp. and *Mycoderma cerevisiae*. None of these yeasts survived pasteurization. Recca and Mrak (1952) conducted a broader survey in which they observed the types of yeasts occurring on the fruit, in single strength juices and in concentrates. From fresh oranges, lemons and grapefruit they isolated *C. pulcherrima, H. anomala, Trich. cutaneum* and *Pull. pullulans*. From single strength orange and lemon juice and beverages, they recovered *C. krusei, C. parapsilosis, H. melligeri, Pi. fermentans* as well as *Zygosacch. vini, Zygosacch. priorianus, Zygosacch. globiformis, Sacch. cerevisiae, Sacch. carlsbergensis, Kl. lindneri* and *Trich. fermentans*. Most of the organisms recovered from concentrates were species of *Candida*, including *C. melibiosi, C. melinii, C. tropicalis* var. *lambica* and *C. pelliculosa*. Other yeasts isolated less frequently were *Zygosacch. major, Zygosacch. japonicus, Zygosacch. mellis-aceti, Sacch. cerevisiae, Sacch. elongosporus, Kl. jensenii, Rh. mucilaginosa* and *T. glabrata*.

Recca and Mrak (1952) found all the cultures of *Zygosaccharomyces* except *Zygosacch. globiformis* and *Zygosacch. mellis-aceti* grew well in the presence of 65% sugar. The two exceptions grew more slowly and sparsely than did the others. The following additional isolates grew slowly in 65° Brix medium: *C. guilliermondii, Sacch. elongosporus, C. melibiosi, C. parapsilosis, Sacch. cerevisiae, C. tropicalis, Kl. lindneri, T. glabrata* and *Trich. fermentans*. Several species failed to grow in 55° Brix medium, including all species of *Pichia, Rh. glutinis, Sp. odorus, Zygosacch. mellis-aceti* and strains of *C. guilliermondii* and *C. krusei*. *Trichosporon cutaneum* failed to grow in a 35° Brix medium.

A large number of these organisms grew at a temperature of 8·5°, but relatively few at 2·5° (Recca and Mrak, 1952). *Saccharomyces oviformis, Sacch. carlsbergensis* var. *mandshuricus, Sacch. elongosporus, C. melibiosi, C. pulcherrima, Zygosacch. fermentati, Rh. glutinis* and *Sp. odorus* were the only organisms that grew at 2·5°; none grew at 0°. Over half the isolates grew at pH 1·5 and practically all at pH 2·0. Although *Trich. cutaneum* failed to grow at pH 3·0, the results substantiate the general assumption that yeasts can tolerate and grow in media having very low pH values.

Berry *et al.* (1956) studied the growth characteristics of *Zygosacch. vini*, an osmophilic yeast known to cause spoilage of orange juice. At 30°, the yeast showed generation times of 2·6 and 2·7 h in media of 12° and 20° Brix respectively, and reached a final cell concentration of about 5×10^7 ml in two days. Growth rates for the test organism became progressively slower as the temperature decreased from 30°, and also as the Brix value increased from 12° to 20°. The yeast grew in 58° Brix at a temperature of 15·6° but not at 10°. At 4·4° growth occurred slowly in all concentrations through 42° Brix.

During World War II, canned orange juice concentrate was frequently not refrigerated; under these conditions the appearance of swelled cans of orange juice concentrate became a major problem. Viable yeasts obtained from such swelled cans of concentrate, and which could be grown on media dissimilar to their previous environment, did not always provide sufficient evidence that yeasts caused the spoilage (Continental Can Co., 1945). The assertion was made that yeasts which are present in the concentrate in a more or less dormant state may be revived to an active condition in a favourable medium. In this revitalized condition they may be sufficiently active to produce carbon dioxide and alcohol when again placed in orange juice concentrate.

Personnel in the laboratories of the Continental Can Co. (1945) found four such yeasts in orange concentrates which could not be isolated on a concentrated medium but could be isolated on acidified dextrose agar. All cans of the product inoculated with revived cells swelled and gave a positive alcohol test after incubation for seven days at room temperature. One of these yeasts at a level of $3·05 \times 10^4$ cells per ml of orange concentrate could survive for five minutes at 65·6°, but not for 10 min. These yeasts were not identified.

During World War II considerable citrus concentrate was shipped to Britain from the U.S.A. Spoilage as a result of yeast fermentation was a frequent problem. The product was shipped at ambient temperatures, which occasionally involved several weeks of warm weather, according to Ingram (1949a). Spoilage was observed sometimes when the cans were unloaded from the ships into cold storage; in severe cases the cans burst and juice dripped from the shipping cartons. In other instances the juice did not ferment until after it had been bottled. On occasion, entire batches of bottles were lost before the product left the distribution centre; on other occasions, however, fermentation and even explosion of the bottles occurred after the product had been received by the consumer.

While studying this spoilage, Ingram (1949a, b) isolated two organisms; one was a pink yeast with the general properties of a *Rhodotorula* species and an osmophilic species of *Zygosaccharomyces* capable of growing in 65° Brix orange concentrate. In pure culture, the osmophilic species could grow in sugar concentrations of up to 70% and at pH 3·0 or even more acid (Ingram 1949a, 1950). The optimum pH value was 4·5 and the optimum temperature varied from 25° to 35° with increasing sugar concentration. The red yeasts proved to be unimportant in so far as spoilage was concerned.

Infection of sterile cans of juice with small numbers of the *Zygosaccharomyces* sp. produced fermentation under warm conditions. The numbers of yeasts apparently had to reach about 10^6 per ml before fermentation became obvious. Therefore, Ingram (1949a) proposed that sound material may contain relatively large numbers of yeasts without evincing any obvious change. At 25° an infection of a few yeasts could cause a can of juice to ferment in approximately 50 days. In commercial channels the fermentation occurred mainly in the summer, and was particularly serious with temperatures of 21° or above. Ingram (1949) found 350 p.p.m. sulphur dioxide (the legal maximum) to be an effective control against low levels of infection, but not against levels of 100 or more cells per ml.

Several strains of yeasts isolated by Ingram (1960) from benzoate-preserved citrus beverages resembled *Sacch. acidifaciens* (*Zygosacch. mandshuricus*). In fact, several serious outbreaks of fermentation caused by this organism occurred in Britain in commercial citrus fruit squashes. These squashes contained about 500 mg per litre of benzoic acid (the statutory limit is 600 p.p.m.), and some of these products had a pH value near 2·5.

A yeast closely resembling a strain of *Sacch. acidifaciens* or *Sacch. elegans* was found to grow readily in non-carbonated orange drink. These organisms grew over a wide range of temperatures and were frequently not inhibited by preservatives at the permitted levels in Australia (McDonald, 1963). Australian health regulations limit the use of preservatives in such drinks to 0·006% sulphur dioxide or 0·02% benzoic acid, or a proportionate mixture of the two compounds. These preservatives, and sorbic acid at 0·02%, showed no inhibiting effect for this yeast. The suggestion was made that under the circumstances the processor must rely on efficient pasteurization of the orange drink for maintenance of shelf-life.

Microbial aspects for processing of concentrates for lemonade are analogous to those of conventional orange juice processing except that a sparser microbial population is encountered in the juice than is normal for orange juice (Cole, 1955). A breakdown of differential counts from quality control work for several years indicated that yeasts predominated and accounted for about 95% of total counts, with moulds accounting for about 4% and bacteria averaging not above 1%. Total counts of the finished product direct from the freezing unit averaged less than 300 organisms per ml. Cole (1955) maintained that the hermetically-sealed product is not likely to ferment even at normal room temperatures; in fact, he observed a reduction in count during storage for 30 days at room temperature.

Prevot and Thouvenot (1960) found *Pseudomonas liquefaciens* associated with numerous cells of a *Saccharomyces* sp. in lemonade which had undergone fermentation. They concluded that the sugar became contaminated while being held in storage by the manufacturer and was the source of the yeast. These workers also analysed fruit drinks which had in the upper part of the bottle several centimetres of a whitish magma which dissociated on shaking into whitish, very light flocs. The majority of the organisms were yeasts which appeared to be *Sacch. fructuum*. This yeast is frequently associated with unpasteurized juice and they concluded that the soda-fruit juice was contaminated by non-pasteurized juice.

Citrus fruits may be invaded or contaminated by yeasts and moulds even before harvest. To prevent entry of mouldy, rotten or otherwise damaged fruit into the concentrating plant, careful sorting and elimination of defective fruit should be the last step before washing and sanitizing the fruit (Vaughn *et al.*, 1957). Routine microscopic examination of the extracted juice can serve as a control in determining the efficiency of sorting. McDonald (1963) has developed a direct microscopic technique for detecting viable yeasts in pasteurized orange drink. Plate counts can also supply additional evidence that spoilage has been caused by yeasts. The presence of alcohol in measurable amounts is a good indication that spoilage has been caused by yeasts (Continental Can Co., 1945). Murdock (1964) investigated the feasibility of measuring diacetyl as an index to contamination by yeasts. The yeast that he used produced diacetyl, but also, subsequently, metabolized it.

C. GRAPE JUICE

A variety of yeasts occur on grapes and in grape products. Yeasts isolated during a survey of such products were identified by Mrak and McClung (1940) as species of *Saccharomyces, Zygosaccharomyces, Pichia, Zygopichia, Debaryomyces, Hansenula, Torulospora, Hanseniaspora, Kloeckeraspora, Torulopsis, Mycoderma, Kloeckera, Rhodotorula* and *Candida*. Pederson (1936a), in his studies on the preservation of grape juice, found the counts on grapes following stemming varied from 6×10^4 to $1 \cdot 2 \times 10^6$ per cm^3. These organisms were primarily yeasts.

They reported that the majority of these yeasts and bacteria were killed at temperatures well below 62·8°.

Pederson *et al.* (1959a, 1961) have reported that the grape juice industry has shifted almost entirely from the old method of storing juice in five-gallon carboys to cool storage in tanks at −5·5° to −2·2°. Excess tartrates are precipitated during this storage. When juice was stored in carboys, the carboys were filled at processing temperature, and either spray-cooled or allowed to cool slowly at the cellar storage temperature. For modern cool storage, the juice is flash-heated at 70–85° and flash-cooled to about 0° before pumping to storage tanks. After the tartrates and other argols are precipitated, the juice is repasteurized and bottled for retail sale. This method of storage apparently minimizes the heating effects; nevertheless, problems of yeast and mould contamination and growth have been serious at times.

Since quality of grape juice is maintained better when stored at a cool temperature, processors would like to hold juice in tanks until bottled juice is needed, as well as to distribute bottling throughout the year. At times this may require holding juice into the following season to equalize differences in supply resulting from variations in production of grapes. The original assumption was that, if yeasts happen to be present, they would not grow at low temperatures and cause spoilage and economic loss. However, Pederson (1936b) and coworkers (1959b) found that some samples of grape juice fermented readily during storage at −2·2°. Pederson, at first, attributed this fermentation to excessive growth of yeasts prior to cooling, resulting in fermentation after cooling and before death of yeasts and bacteria began to occur at this low temperature. Yet, examination of a few samples revealed a high count of yeasts, and he concluded that certain strains became acclimated, grew and fermented the product. Later, Lawrence *et al.* (1959) demonstrated that some varieties of yeasts will grow better at 1° than at room temperature (21°).

According to Lawrence *et al.* (1959), contamination of grape juice by yeasts occurs during the time the juice is transferred to the tanks, or upon contact with surfaces of the tank or from the atmosphere. Grape juice in cool storage does not offer a favourable medium for yeast growth; the low temperature suppresses growth or even causes the death of many strains. Lawrence and his coworkers suggested that a selection of a few strains of yeasts best adapted to such growth conditions occurs; in fact, they found strains of the genus *Candida* to be most adaptable to these conditions. They found the strains of *Candida* to be true psychrophiles, with optimal growth temperatures of approximately 11° and incapable of growth in grape juice at 21° or above.

Other strains of yeasts that they found in grape juice belonged to the genera *Torulopsis*, *Hanseniaspora* and *Saccharomyces*.

The conditions of a tank of juice in storage is judged by its alcohol content (Pederson *et al.*, 1959a). Fermentative yeasts of the genera *Saccharomyces*, *Hanseniaspora* and *Torulopsis* produce alcohol; the genus *Saccharomyces* is the most active and most likely to be responsible for fermentation in so-called "wild-tanks". Since yeasts of the genus *Candida* are not alcohol-producers, their presence would not be indicated by alcohol content. Nevertheless, like the other yeasts, they affect flavour, and also remain in the juice when it is siphoned from a tank. Upon pasteurization, the yeasts are killed and settle to form a fine precipitate in the bottoms of bottled juice. Ordinary filtration removes a high percentage but not necessarily all yeasts.

In bulk storage of grape juice at $-5\cdot5°$ to $-2\cdot2°$, the quality of the product may remain materially unchanged for a year or more. Even with the most careful operation, however, yeast contamination may occur and may lead to considerable loss (Pederson *et al.*, 1959a, b, 1961). According to Pederson and his coworkers, contamination of juices arises from residual yeasts harboured in the pores of wood or the coating of tanks, the air in the room, foam on the surface of containers, intermediate holding tanks, pipe-lines with crevices where juice accumulates, improperly-made gaskets, valves and other places where juice containing a few yeast cells can accumulate and foster growth.

Wickerham and Duprat (1945) found some home-canned grape juice undergoing fermentation with the production of considerable gas. They described the contents as considerably altered and alcoholic, but as pleasantly palatable. They identified the yeast as *Schizosaccharomyces versatilis*. The isolation of this yeast from home-canned grape juice in Michigan was regarded as somewhat unusual since this organism is regarded as being of tropical or subtropical origin. Also, yeasts are generally killed by most pasteurizing treatments, and their presence in a canned food is indicative of deficient heat-processing or leakage. Another instance of a yeast, *Candida tropicalis*, in a home-canned product, apricots, was reported by Bouthilet *et al.* (1949).

The storage of grape juice inoculated with species of *Saccharomyces*, *Candida*, *Hanseniaspora* and *Torulopsis* and held at $-5\cdot5°$ to $-2\cdot2°$ may be extended for several months by addition of small quantities (0·01% to 0·15%) of several fungistats such as sorbic, benzoic, capric and caprylic acids (Pederson *et al.*, 1961). These concentrations of benzoic or sorbic acid are apparently below the taste threshold of the majority of people. They suggested further that the effectiveness of these additives, with low temperature, compared favourably with additives used in heat-processing, or to that of several chemical substances used in conjunction with process temperatures of various foods.

D. CARBONATED BEVERAGES

Carbonated beverages are essentially sparkling water with added flavouring syrups containing edible acid. According to Sharf (1960) the prepared flavoured syrups are usually in the range of 28° to 34° Baumé (50·6° to 62·0° Brix) and may be blended with carbonated water before filling, or dispensed separately into the bottles, closed and later mixed.

In studies of deterioration of carbonated beverages, yeasts have been found to be the most frequent causative biological agents (McKelvey, 1926; Levine and Toulouse, 1934; Levine, 1940; Witter et al., 1958; Sharf, 1960). According to Witter et al. (1958) the soft drink industry depends mainly on the inherent growth-inhibiting properties of the product and good plant sanitation to prevent spoilage. Sucrose concentration, acidity and degree of carbonation are the major characteristics which prevent microbial spoilage. Witter and his collaborators found no lethality for a yeast isolated from spoiled carbonated orange drink without carbonation regardless of the degree Brix or pH value. In fact, in the uncarbonated beverage, pH had little effect on the growth of the yeast; but in the presence of carbonation it was very important, the lower the pH value under these conditions the greater was the lethality. McKelvey (1926) worked with a mixture of spores from 27 yeasts isolated from carbonated beverages and reported that the higher the concentration of sugar in syrups used for carbonated beverages, the longer was the time to kill the yeasts at 70°.

Levine (1940) and Levine and Toulouse (1934) have stated that, at that time, 85–90% of the cases of spoilage in carbonated beverages was attributable to yeasts. Yeasts formed cheesy, flaky, stringy and powdery sediments. Levine and Toulouse (1934) found that few of the varieties of yeasts grew in heavy syrups (36° to 42° Baumé); however, many species can grow in the weaker simple syrups of 28° to 32° Baumé. Moderate quantities of citric acid stimulate yeast growth, as does the presence of air; therefore, these organisms will grow in bottled beverages, except the extremely acid and highly carbonated types.

Payne and Perigo (1960) have reported also that spoilage of carbonated soft drinks in cans by yeasts is a hazard. They found some canners were pasteurizing the product to prevent spoilage. They examined the effects of benzoic acid and sulphur dioxide on *Saccharomyces chevalieri*, an isolate from a can of carbonated soft drink, and found neither inhibitor to be satisfactory. Sharf (1960) has commented that benzoates do not control high-level infections, and cannot be used to compensate for poor ingredients or processing. In the United

Kingdom the use of benzoic acid or sulphur dioxide for carbonated beverages is permitted.

Yeasts generally enter the bottling plant on such items as dust, sugar, water, crowns and empty bottles. It is recommended that yeasts be kept out of the plant, or destroyed in the final beverage, to avoid all possibility of spoilage. Papadakis (1960) considered the isolation of *Saccharomyces* from the product as an indication of low standards of hygiene during production.

Citrus fruit juices containing 6% or higher of fruit juice are not carbonated, and must be filled into the bottles at 80° and closed. If this temperature is not maintained, there may be survival of yeasts and, to a lesser extent, moulds and bacteria. Citrus beverages containing less than 6% juice are prepared as carbonated drinks of low pH value. These do not show microbial growth if properly prepared; yet according to Sharf (1960) added growth stimulants in the juice may permit yeast growth, and 0·1% sodium benzoate, the maximum allowed in the U.S.A., may be added as a further safeguard. Sulphur dioxide may be used also, but beyond certain levels may cause flavour changes. Sorbic acid has not been particularly effective against yeasts in carbonated beverages.

IX. Alcoholic Beverages

A. WINE

Fermentation, the process by which grape juice becomes wine, is caused by yeasts (see Chapter 2, p. 5). Several reviews and books are available in which organisms associated with spoilage of wine are discussed (Bioletti, 1911; Cruess, 1943; Amerine and Cruess, 1960).

According to Cruess (1943), wine grapes ordinarily arrive at the cellar heavily contaminated with moulds and wild yeasts, but only slightly seeded with *Saccharomyces ellipsoideus*, the true wine yeast. According to his review, the addition of 100–200 p.p.m. of sulphur dioxide to the crushed grapes or must will effectively inhibit the growth of mould, wild yeasts and bacteria, but allow growth and fermentation by *Sacch. ellipsoideus*. He states that the use of sulphur dioxide is a common practice in all wine-making regions of the world.

Film yeasts on wine can decrease the total alcohol and acid by oxidation. Bioletti (1911) reported that if a normal wine, especially one strong in alcohol, is left with its surface exposed to the air, it will become covered with a whitish film consisting of yeast-like cells. This film was known as "wine flowers" and consisted of many varieties of yeasts called collectively *Mycoderma vini*. Bioletti described these organisms as a serious enemy to wine, rendering it insipid and cloudy. Amerine and Cruess (1960) described *Candida mycoderma* (sometimes erroneously called *Mycoderma vini*) as a non-sporulating film yeast that grows on wines of low alcohol content, and ferments fruit juices where it forms a chalky-white film. It is frequently found on distilled material and on pickle brines. *Candida tropicalis* is frequently found on fruits and forms a pellicle on fruit juices. In addition, most species of *Hansenula* form films on liquid culture media and on wines of low alcohol content. According to Cruess (1943), spoilage of wine by film yeasts is evidence of negligence on the part of the wine-maker. If the casks and bottles are full at all times and sealed, these micro-organisms will not develop and cause spoilage.

Ordinarily, *Acetobacter* spp. are considered to be responsible for the souring of wine to vinegar, although Nickerson (1943) isolated a yeast which he named *Zygosaccharomyces acidifaciens* from a bottle of soured domestic red wine. The pH value of the wine was 2·2, but only yeasts were observed microscopically and by plating on nutrient agar.

Clouding or sediment formation due to yeasts occurs in bottled wines. Sweet Sauternes and Sauterne-type wines, for example, will ferment with accompanying clouding by the wine yeast, *Sacch. ellipsoideus*, unless the wines are treated with 300–400 p.p.m. of sulphur dioxide to prevent yeast growth, or are pasteurized, or "germ-proof" filtered to render them sterile (Cruess, 1943). Some of these yeasts will tolerate more than 400 p.p.m. of sulphur dioxide.

Clouding of dry table wines and fortified dessert wines has occurred with varying frequency (Amerine and Cruess, 1960). Phaff and Douglas (1944) obtained *Zygosacch. mellis* from cloudy dessert wines. Scheffer and Mrak (1951) isolated yeasts from cloudy wines described as brilliantly clear when bottled. These yeasts were identified as *Sacch. chevalieri*, *Sacch. carlsbergensis* var. *monoacensis*, *Sacch. oviformis*, *Pi. alcoholophila* and *C. rugosa*. According to Scheffer and Mrak, these yeasts do not impart any off-flavours or odours to the wine, but the colour change and/or sediment reduce consumer acceptance. In South Africa, Van der Walt and Van Kerken, as reported by Amerine and Cruess (1960), recovered species of *Brettanomyces* from about 50% of the samples of cloudy wines. Other yeasts recovered to a lesser degree were *Sacch. acidifaciens*, *Sacch. oviformis*, *Sacch. cerevisiae* and *Pi. membranaefaciens*.

Control of spoilage by clouding can be accomplished by germ-proof filtration, use of sulphurous acid and by pasteurization (Amerine and Cruess (1960). Of the three methods, germ-proof filtration probably affects the flavour and bouquet of the wine least. The aged clear wine is filtered through germ-proof sterile filter pads into sterile bottles that are then closed with sterile caps or corks.

Also, according to Amerine and Cruess (1960), unfortified fruit wines are pasteurized in California and Oregon wineries by flash-heating to 62·8° to 73·9° and bottled hot. They suggest that the death temperature of yeasts in wine is lower than in fruit juices because of the alcohol content of the wine. Sweet Sauterne-type wines have been preserved with addition of sulphurous acid, but this method has been found to be unreliable, even when high concentrations are used. Federal regulations set a maximum of 350 p.p.m. More than this amount can be added to the wine before bottling if the bottled wine is allowed to rest for several months before marketing. The sulphur dioxide content gradually decreases after bottling, but these high concentrations of sulphur dioxide reduce the quality of the product otherwise.

B. BEER

Wild yeasts may interfere with normal fermentation (see Chapter 4, p. 147) and cause spoilage of beer by imparting off-flavours and by interfering with clarification (Cosbie et al., 1942; Bunker, 1961; Gilliland and Harrison, 1966). No attempt has been made to cover the entire literature on flavour deterioration of beer caused by yeast, but several citations are used to point out some of the common problems that may occur.

A yeast which closely resembles *Mycoderma cerevisiae* grows well in various types of beer if sufficient oxygen is present (Cosbie et al., 1942). A pellicle develops on the surface of the beer at a rapid rate even at low temperatures (13–15°). Slight agitation results in the formation of a powdery or flocculent sediment. This yeast apparently is not restricted appreciably by pH level or by high alcohol content; exclusion of air has been suggested as a means of controlling yeast growth in an infected beer.

Certain wild yeasts can cause fining difficulties. Clarification of finished beer is achieved by the addition of finings, such as isinglass. The colloidal fining agent tends to coagulate and to carry down suspended particles such as protein, bacteria and yeasts. Some wild yeasts cannot be removed in this manner; the occurrence of such yeasts is serious since the consumer expects a clear brilliant beer. Such types of yeasts are *Saccharomyces cerevisiae* var. *turbidans* and *Sacch. carlsbergensis* (Bunker, 1961). *Pichia membranaefaciens* is another wild yeast which may form a film and may be responsible for a non-fining haziness (Bunker, 1961).

Undesirable yeasts may also impart off-flavours to beer (Bunker, 1961; Gilliland and Harrison, 1966). A wild yeast of the *Sacch. cerevisiae* type imparted a "fruity" flavour to the beer; this flavour was correlated with a high ester content. A soapy flavour in beer has been attributed partly to ethyl hexoate produced by a yeast, and a buttery flavour to the formation of diacetyl. Some strains of *Sacch. carlsbergensis* under certain circumstances will produce a sulphurous odour in beer. Additional odours and flavours have been traced to wild yeasts in beer (Gilliland and Harrison, 1966), but no attempt will be made to catalogue them here.

Heating beer at 46° to 48° for 10 min will kill or strongly impair cells of brewer's yeast, but cells of wild yeasts apparently can endure this treatment (Lund, 1951). Lund, observed that heating at 60° for 20 min stabilized bottom-fermented beer infected with strains of *Sacch. ellipsoideus, Sacch. turbidans* and *Sacch. odessa*. Most of these yeasts were killed between 50° and 56° but vegetative cells of *Sacch. odessa* withstood 58°, and its spores survived 60° for this time period.

X. Malt Extract

Malt extract containing 50% sugars can be fermented by *Saccharomyces rouxii* (*Zygosaccharomyces japonicus*) (English, 1951, 1953). However, the organism was capable of fermenting almost saturated solutions of maltose and glucose solutions of as high a concentration as 90% (w/v). The factor limiting fermentation was the relative humidity of the malt extract; the critical level was about 73%. If the relative humidity was raised above this level, either by increasing the moisture content or decreasing the proportion of sugars in the total solids, the extract became susceptible to fermentation by this osmophilic yeast.

Proszt and Vas (1962) observed a linear correlation between the logarithm of initial cell counts and the time necessary to initiate gas production in malt solutions. Gas production was observed when counts reached 3.2×10^6 to 1.0×10^7 per ml. In solutions of different osmotic pressures, *Sacch. cerevisiae* produced gas more actively than *Sacch. rouxii* provided the sucrose concentration did not exceed 55%. At 65% sucrose concentration, only *Sacch. rouxii* produced gas.

XI. Salad Dressings and Catsup

Mayonnaise is sufficiently stable to be kept at normal room temperatures for reasonable lengths of time. According to Worrell (1951), the expected shelf life of such a product is three to six months. Biological agents are probably not the principal causes of spoilage, but fermentation by organisms can occur if they are introduced in ingredients or from contaminated equipment. Use of high quality ingredients and proper sanitation of the plant can eliminate fermentation. Micro-organisms are usually inhibited by the natural acidity of the product.

Several outbreaks of spoilage of salad dressing by yeasts have been described (Williams and Mrak, 1949; Fabian and Wethington, 1950a, b; Zuccaro *et al.*, 1951). Fabian and Wethington reported that yeasts were not ordinarily found in large numbers in salad dressings, mayonnaise and related products. In those instances in which outbreaks of spoilage due to yeasts occurred, improper cleaning of processing equipment was usually responsible.

Williams and Mrak (1949) described spoilage in a starch-base salad dressing in which considerable gas was generated but little change in pH level occurred. The causative organism had morphological and physiological characteristics similar to those of *Zygosaccharomyces globiformis*. In this particular case, the source of contamination was traced to a centrifugal pump used to transfer the dressing to the filling machine. The problem of spoilage was solved if the pump was dismantled regularly and thoroughly cleaned. Fabian and Wethington (1950a, b) also cited a species of *Zygosaccharomyces* as the cause of spoilage of salad dressing in four different outbreaks. One case of spoilage was traced to improperly cleaned and unsanitized machinery.

In these outbreaks gassy fermentation was the only evidence of spoilage. The fermentation did not noticeably affect the odour or taste of the products. In fact, chemical analysis showed the amount of acid, salt, oil and the pH value to be within the range of that for unfermented products. Analysis of a number of samples of different types of salad dressings by Fabian and Wethington (1950b) showed a pH range of 2·9 to 4·4. These acid conditions would be detrimental to bacterial growth and would favour yeast growth.

Zuccaro *et al.* (1951) hypothesized that spoilage of this type of food could result from organisms becoming coated with oil, with the result that heat processes ordinarily sufficient to destroy spoilage organisms would be inadequate; and, subsequently, these same organisms might migrate from the oil phase into the water phase and cause spoilage. Experiments with *Sacch. cerevisiae, Sacch. ellipsoideus*, a *Zygosaccharomyces* sp. isolated from spoiled French dressing and a *Saccharomyces* sp. isolated from spoiled salad dressing, showed that all were killed in approximately 0·3 min at 71·1° in phosphate buffer, pH 5·6. At pH 4·0 the thermal death time of *Saccharomyces* sp. was reduced to 0·13 min. In oil, the organisms could withstand 2·0 to 3·5 min at this temperature. They concluded that a pasteurizing treatment of French dressing adequate to kill yeast in the aqueous phase was inadequate for killing yeast in the oil phase.

Spoilage of tomato products is usually caused by bacteria belonging to the lactic acid types; however, in the past, yeasts have been observed to cause spoilage of certain tomato products. Ayers (1926) isolated a yeast from gassy catsup; the yeast was not identified, but was described as producing four spores by isogamic conjugation on gypsum blocks (Pederson and Breed, 1926). Pederson and Breed (1926, 1929) observed that spoilage of this type can be controlled by the presence of acetic acid, sugar and salt in the product, and that many manufacturers use this knowledge in the production of catsups that do not spoil readily under normal conditions either in the bottle or after being opened.

XII. Soy Sauce and Miso Paste

According to Ōnishi (1963) spoilage of soy sauce and miso paste may be caused by salt-tolerant yeasts. Soy sauce mashes will contain about 18% sodium chloride, while miso paste is prepared from soya bean paste containing solutions of 7·20% sodium chloride. Raw soy sauce, the liquid part of the mash separated by a hydraulic press, may develop a white dry pellicle on the surface of the liquid during storage; a gaseous fermentation may also occur. These symptoms are caused by the growth of film-forming yeasts such as *Saccharomyces rouxii* var. *halomembranis* (formerly *Zygosaccharomyces salsus* and *Zygosacch. japonicus*) and species of *Pichia*, although Ōnishi has pointed out that salt-tolerant yeasts of the *Sacch. rouxii* type are also important in the development of the characteristic flavour of soy sauce and miso paste.

This type of spoilage ordinarily does not occur in good quality well-matured soy sauce, perhaps because of the presence of yeast-inhibiting compounds in the final product (Yokutsuka, 1954). The spoilage of miso paste by yeasts depends mainly on the water content. Mogi and Naka-jima, as quoted by Ōnishi (1963), claim that deterioration from yeasts will not occur during long storage if the miso paste contains less than 45% water.

XIII. Frozen Foods

Yeasts seem to occur more frequently than bacteria in spoiled frozen foods. Ingram (1958) attributed this to the fact that freezing decreases the water activity of a food. Examples of spoilage of frozen foods are: frozen oysters by pink yeasts (McCormack, 1950); frozen strawberries packed in 50% sucrose by a pink yeast (Berry and Magoon, 1934); and frozen peas (Mulcock, 1955). The spoilage of frozen oysters and frozen strawberries have been mentioned and discussed previously.

Mulcock (1955) recovered a strain of *Rhodotorula glutinis* from pack-ages of spoiled frozen peas. Some peas which had been held in storage for several months at $-17·8°$ were transferred to other storage, and as a result were held near freezing point; after a time spoilage was observed. Initially the peas were pink in colour and had an unpleasant odour. After the packages of spoiled peas were held at room temperature for a few days, the pink colour faded and became brownish-yellow. Isolates from the spoiled peas caused spoilage of frozen peas held near 0°, but not at the normal storage temperature of about − 17°. A second strain of *Rh. glutinis*, isolated from dust in the packaging plant, could not grow at 4°. The source of the yeasts in the peas was not established.

Ingram (1951) has stated that moulds and yeasts generally resist freezing better than bacteria do. McFarlane (1940, 1941, 1942) has investigated some of the factors influencing the survival of yeasts at subfreezing temperatures. He concluded that the more acid the suspending medium, the more rapid was the killing rate. He also did some experiments in which he suspended cells of a *Torula* sp., *Sacch. cerevisiae* and *Sacch. ellipsoideus* in unsweetened loganberry juice, in loganberry juice-sucrose solutions containing 20%, 30%, 40%, 50% and 60% solids, and in loganberry juice solutions to which citric acid had been added so that the total acidity of each medium approximated that of the unsweetened juice. Greatest destruction of yeasts occurred in unsweetened juice held at − 17·8°. The numbers of survivors tended to vary as the concentration of sucrose varied; but 50% and 60% solids did not always give the greatest protection. When temperature was the variable, greater destruction was observed, after several weeks storage, at − 10° than at − 20°.

XIV. Stored Grains

Under certain conditions yeasts may contribute to the spoilage of stored grains. Lund (1956) has reported the occurrence of considerable numbers of yeast cells on grains of barley before harvest but the occurrence of few yeast cells on samples of threshed barley. Teunisson (1954) found that during storage of freshly combined, naturally moist rough rice under essentially anaerobic conditions bacteria survived while moulds decreased and yeasts proliferated within a few days. This type of spoilage was not accompanied by heating, but a sour odour and discoloration of the kernels occurred. There was also a loss in seed viability.

Types of yeasts and yeast-like organisms isolated from rice stored for 34 days were *Endomycopsis chodati*, *Candida krusei*, *Oospora lactis*, *Candida tropicalis* and *Hansenula anomala*. After storage for seven months, the yeast population consisted mainly of *E. chodati*, *H. anomala* and *Pi. farinosa*. The largest populations of these yeasts occurred in the surface layers of the rice which, according to Teunisson, was due to a translocation of moisture to this area and a subsequent increase in relative humidity in that locale. The rice had a moisture content of 19·3%.

Ordinarily, yeasts are relatively unimportant in stored grains, according to Del Prado and Christensen (1952). On rice held at room temperature and 16·3% moisture, *Aspergillus glaucus* comprised 100% of the population. Similarly Bottomley *et al.* (1950) found *C. pseudotropicalis* to be most prevalent in yellow corn (maize) stored in an atmosphere of about 0·1% oxygen and relative humidities of 90% and higher; otherwise species of aspergilli and penicillia predominated.

XV. Concluding Remarks

Reference has been made in the previous discussion to various compounds that have been used to inhibit the growth of yeasts in foods. No attempt has been made to cover this subject in detail, but a few brief comments at this point seem worthwhile. Compounds frequently mentioned and accepted by food officials as food additives are benzoic acid and its salts, and sulphur dioxide and its salts. A number of other compounds are effective and have been used both experimentally and commercially.

According to Borgstrom (1954) 0·15% sodium benzoate is effective in most cases for checking the growth of yeasts in fruit juices. Nonionized benzoic acid is the effective preserving agent, which means that benzoates are most effective in acid foods. Borgstrom also states that the maximum amount of this preservative permitted by law in most countries is 0·1%, which will not always guarantee safe preservation. The amount of benzoic acid required to prevent spoilage varies with temperature, acidity and numbers of yeasts in the food product.

Sulphurous acid is more effective against mould spores and bacteria than yeast (Borgstrom, 1954). Sulphurous acid may be added to a food as sodium or potassium metabisulphite, calcium bisulphite, or as gaseous sulphur dioxide. Much of the added sulphurous acid combines to form bisulphite compounds with sugar, aldehydes and other reactive compounds; this combined sulphurous acid is ineffective against microorganisms. Apparently the non-ionized sulphurous acid is the lethal agent and, as with benzoic acid, the pH level has a profound influence on the amount of chemical required for preservation.

Balatsouras and Polymenacos (1963) compared the effectiveness of benzoic acid, sulphurous acid, sorbic acid, salicylic acid and formic acid as inhibitors of yeast growth. They used yeasts isolated from unpasteurized citrus juice and from olive and cucumber brines. The isolates represented species from the genera *Hansenula*, *Pichia*, *Debaryomyces*, *Sporobolomyces*, *Candida*, *Torulopsis*, *Rhodotorula* and the yeast-like organism *Oospora lactis*. Their ranking of the effectiveness of these chemicals against spoilage yeasts was: salicylic acid, benzoic or sulphurous acid, sorbic acid and formic acid. They concluded that salicylic acid was an excellent yeast-inhibitor in acid media, and that a concentration of 500 p.p.m. was sufficient under commercial practices. With sorbic acid, none of the isolates grew when the concentration was as high as 1,000 p.p.m. The presence of sodium chloride enhanced the effect of sorbic acid. They stated that, although formic acid is permitted in soft drinks in Greece, and in fruit juices and comfitures in Switzerland, Germany and other European countries, it is ineffective against yeast. Resistance to 1,000 p.p.m. was common, and levels of 1,600 p.p.m. were required to suppress growth of the whole group of yeasts. Smith *et al.* (1962), in their comparison, found potassium sorbate exhibited a greater inhibitory effect than did sodium benzoate over a pH range of 6·0 to 4·0 and a range of concentrations of 0·025% to 0·1%. Pimaricin, a fungistat, has been effective in controlling yeasts in dressed poultry, dairy products, berries and fruit juices at levels of 10 to 100 p.p.m. (Clark *et al.*, 1964). Diethylpyrocarbonate has been recommended for use in controlling yeasts in juices. This compound, which reacts with water to form ethanol and carbon dioxide, is effective against spoilage organisms of various beverages in a concentration of 30 to 200 p.p.m. (Genth, 1964). All substances which have been tested have not been mentioned, but the above are several that are now commonly used or show promise for future use.

The spoilage of foods by yeasts is not as prevalent as spoilage by bacteria and moulds, but can be a serious problem under certain circumstances. Control of spoilage of foods by yeasts has been achieved in some instances by good sanitary practices, which control initial contamination, by proper storage with controlled temperature and moisture levels, and by judicious use of heat-treatment and preservative agents. With an increased knowledge of the characteristics of spoilage yeasts and of the factors which will inhibit their growth, further improvement in control of yeasts in foods will be realized.

References

Allen, L. A. and Thornley, B. D. (1929). *Ann. appl. Biol.* **16**, 578–595.
Amerine, M. A. and Cruess, W. V. (1960). "The Technology of Wine Making". Avi Publishing Co., Westport, Connecticut.
Andersen, A. J. C. and Williams, P. N. (1965). "Margarine", 2nd rev. ed. Pergamon Press, Oxford.
Ayers, S. H. (1926). *Glass Container* **5** (4), 13–14, 20, 26.
Ayers, S. H., Barnby, H. A. and Voigt, E. L. (1930). *Fd. Inds* **2**, 61–64.
Ayres, J. C. (1955). *Adv. Fd Res.* **6**, 109–161.
Ayres, J. C. (1960). *Fd Res.* **25**, 1–18.

Ayres, J. C., Walker, H. W., Fanelli, M. J., King, A. W. and Thomas, F. (1956). *Fd Technol.*, *Champaign* **10**, 563–568.

Baker, E. E. and Mrak, E. M. (1938). *J. Bact.* **36**, 317–318.

Balatsouras, G. D. and Polymenacos, N. G. (1963). *J. Fd Sci.* **28**, 267–275.

Barnes, E. M. and Shrimpton, D. H. (1958). *J. appl. Bact.* **21**, 313–329.

Beech, F. W. (1958a). *In* "Recent Studies in Yeast and Their Significance in Industry", pp. 37–51. Society of Chemical Industry Monograph No. 3. MacMillan Co., New York.

Beech, F. W. (1958b). *J. appl. Bact.* **21**, 257–266.

Beech, F. W., Kieser, M. E. and Pollard, A. (1963). *Long Ashton Agric. Hort. Res. Sta. Ann. Rep.* pp. 147–149.

Bell, T. A. and Etchells, J. L. (1952). *Fd Technol.*, *Champaign* **6**, 468–472.

Bell, T. A. and Etchells, J. L. (1956). *Appl. Microbiol.* **4**, 196–201.

Bell, T. A., Etchells, J. L. and Borg, A. F. (1959). *J. Bact.* **77**, 573–580.

Bell, T. A., Etchells, J. L. and Jones, I. D. (1950). *Fd Technol.*, *Champaign* **4**, 157–163.

Bell, T. A., Etchells, J. L. and Jones, I. D. (1951). *Archs Biochem. Biophys.* **31**, 431–441.

Berry, J. A. (1934). *Science*, *N.Y.* **80**, 341.

Berry, J. A. and Magoon, C. A. (1934). *Phytopathology* **24**, 780–796.

Berry, J. M., Witter, L. D. and Folinazzo, J. F. (1956). *Fd Technol.*, *Champaign* **10**, 553–556.

Bioletti, F. T. (1911). *California Exp. Sta. Bull. No. 213.*

Borg, A. F., Etchells, J. L. and Bell, T. A. (1955). *Bact. Proc.*, 19.

Borgström, G. (1954). *In* "Chemistry and Technology of Fruit and Vegetable Juices". (D. K. Tressler and M. A. Joslyn, eds.), pp. 180–218. Avi Publishing Co., New York.

Bottomley, R. A., Christensen, C. M. and Geddes, W. F. (1950). *Cereal Chem.* **27**, 271–296.

Bouthilet, R. J., Neilson, N. E., Mrak, E. M. and Phaff, H. J. (1949). *J. gen. Microbiol.* **3**, 282–289.

Bowen, J. F., Strachan, C. C. and Davis, C. W. (1953). *Fd Technol.*, *Champaign* **7**, 102–105.

Boyd, J. W. and Tarr, H. L. A. (1955). *Fd Technol.*, *Champaign* **9**, 411–412.

Braverman, J. B. S. (1949). "Citrus Products". Interscience Publishers, New York.

Browne, C. A. (1918). *J. ind. Engng Chem.* **10**, 178, 198.

Brunkow, O. R., Peterson, W. H. and Fred, E. B. (1921). *J. Am. chem. Soc.* **43**, 2244–2245.

Bunker, H. J. (1961). *Progress ind. Microbiol.* **3**, 3–41.

Butjagin, B. (1904). *Zentbl. Bakt. ParasitKde (Abt. II)* **11**, 540–551.

Caldis, P. D. (1930). *J. agric. Res.* **40**, 1031–1051.

Challinor, S. W. (1955). *J. appl. Bact.* **18**, 212–223.

Challinor, S. W., Kieser, M. E. and Pollard, A. (1948). *Nature*, *Lond.* **161**, 1023–1025.

Chinn, S. H. F. and Nelson, F. E. (1946). *J. Dairy Sci.* **29**, 507.

Church, M. B., Paine, H. S. and Hamilton, J. (1927). *J. ind. Engng Chem.* **19**, 353–355.

Clague, J. A. and Fellers, C. R. (1933). *Arch. Mikrobiol.* **4**, 419–426.

Clark, D. S., Wallace, R. H. and David, J. J. (1954). *Can. J. Microbiol.* **1**, 145–149.

Clark, W. L., Shirk, R. J. and Kline, E. F. (1964). *In* "Microbial Inhibitors in Food", (N. Molin and A. Erichsen, eds.), pp. 167–184. Almquist and Wiksell, Stockholm.

Cole, G. M. (1955). *Fd Technol., Champaign* **9**, 38–45.

Continental Can Co. (1945). *Fd Pckr* **25**(3), 32–33.

Cordes, W. A. and Hammer, B. W. (1927). *J. Dairy Sci.* **10**, 210–218.

Cosbie, A. J. C., Tošic, J. and Walker, T. K. (1942). *J. Inst. Brew.* **48**, 156–160.

Costilow, R. N. (1957). *Fd Technol., Champaign* **11**, 591–595.

Costilow, R. N. and Fabian, F. W. (1953). *Appl. Microbiol.* **1**, 314–319.

Costilow, R. N., Coughlin, F. M., Robbins, E. K. and Hsu, W. T. (1957). *Appl. Microbiol.* **5**, 373–379.

Costilow, R. N., Etchells, J. L. and Blumer, T. N. (1954). *Appl. Microbiol.* **2**, 300–302.

Costilow, R. N., Ferguson, W. E. and Ray, S. (1955). *Appl. Microbiol.* **3**, 341–345.

Cruess, W. V. (1943). *Adv. Enzymol.* **3**, 349–386.

Cruess, W. V. and Irish, J. H. (1923). *Univ. California Agric. Exp. Sta. Bull. No. 359.*

Dakin, J. C. and Day, P. M. (1958). *J. appl. Bact.* **21**, 94–96.

Davey, A. E. and Smith, R. E. (1933). *Hilgardia* **7**, 523–551.

Deane, D. D., Nelson, F. E. and Baughman, R. W. (1954). *Am. Milk Rev.* **16**(5), 60–66.

De Camargo, R. and Phaff, H. J. (1957). *Fd Res.* **22**, 367–372.

Del Prado, F. A. and Christensen, C. M. (1952). *Cereal Chem.* **29**, 456–462.

Desrosier, N. W. (1963). "The Technology of Food Preservation". Rev. ed. Avi Publishing Co., Westport, Connecticut.

Do, J. Y. and Salunkhe, D. K. (1964). *Fd Technol., Champaign* **18**, 584–586.

Drake, S. D., Evans, J. B. and Niven, C. F., Jr. (1958). *Fd Res.* **23**, 291–296.

Drake, S. D., Evans, J. B. and Niven, C. F., Jr. (1959). *Fd Res.* **24**, 243–246.

Dyett, E. J. and Shelley, D. (1965). *In* "Food Science and Technology", (J. M. Leitch, ed.) Vol. 2, pp. 393–403. Gordon and Breach Science Publishers, Ltd., London.

Edson, H. A. (1912). *Vermont Agric. Exp. Sta. Bull.* No. 167.

Edwards, S. F. (1913). *Zentbl. Bakt. ParasitKde (Abt. II)* **39**, 449–455.

Eklund, M. W., Spinelli, J., Miyauchi, D. and Dassow, J. (1966). *J. Fd Sci.* **31**, 424–431.

Eklund, M. W., Spinelli, J., Miyauchi, D. and Groniger, H. (1965). *Appl. Microbiol.* **13**, 985–990.

Emard, L. O. and Vaughn, R. H. (1952). *J. Bact.* **63**, 487–494.

Empey, W. A. and Scott, W. J. (1939a). *Aust. Coun. scient. ind. Res. Bull.* No. 126.

Empey, W. A. and Scott, W. J. (1939b). *Aust. Coun. scient. ind. Res. Bull.* No. 129.

English, M. P. (1951). *Nature, Lond.* **168**, 391.

English, M. P. (1953). *J. gen. Microbiol.* **9**, 15–25.

Erickson, R. J. and Fabian, F. W. (1942). *Fd Res.* **7**, 68–79.

Esau, P. and Cruess, W. V. (1933). *Fruit. Prod. J.* **12**, 144–147.

Etchells, J. L. (1941). *Fd Res.* **6**, 95–104.

Etchells, J. L. and Bell, T. A. (1950a). *Farlowia* **5**, 87–112.

Etchells, J. L. and Bell, T. A. (1950b). *Fd Technol., Champaign* **4**, 77–83.

Etchells, J. L. and Jones, I. D. (1941). *Fruit Prod. J.* **20**, 370, 381.

Etchells, J. L., Bell, T. A. and Jones, I. D. (1953). *Farlowia* **4**, 265–304.

Etchells, J. L., Borg, A. F. and Bell, T. A. (1961). *Appl. Microbiol.* **9**, 139–144.

Etchells, J. L., Costilow, R. N. and Bell, T. A. (1952). *Farlowia* **4**, 249–264.

Fabian, F. W. and Bloom, E. F. (1942). *Fruit Prod. J.* **21**, 292–296.

Fabian, F. W. and Hall, H. H. (1933). *Zentbl. Bakt. ParasitKde (Abt. II)* **89**, 31–47.

Fabian, F. W. and Marshall, R. E. (1935). *Michigan Agric. Exp. Sta. Bull.* No. 98.

Fabian, F. W. and Quinet, R. I. (1928). *Michigan Agric. Exp. Sta. Tech. Bull.* No. 92.

Fabian, F. W. and Wadsworth, C. W. (1939). *Fd Res.* **4**, 511–519.

Fabian, F. W. and Wethington, M. C. (1950a). *Fd Res.* **15**, 135–137.

Fabian, F. W. and Wethington, M. C. (1950b). *Fd Res.* **15**, 138–145.

Farrell, L. and Lochhead, A. G. (1931). *Can. J. Res.* **5**, 539–543.

Faville, L. W. and Hill, E. C. (1951). *Fd Technol., Champaign* **5**, 423–425.

Fellers, C. R. (1929). *Fruit Prod. J.* **9**, 113.

Fellers, C. R. and Clague, J. A. (1932). *J. Bact.* **23**, 63.

Fellers, C. R. and Clague, J. A. (1942). *Fruit Prod. J.* **21**, 326–327, 347.

Ferguson, W. E. and Powrie, W. D. (1957). *Appl. Microbiol.* **5**, 41–43.

Fields, M. L. (1962). *Fd Technol., Champaign* **16**(8), 98–100.

Fields, M. L. (1964). *Fd Technol., Champaign* **18**, 114.

Foter, M. J., Anderson, E. O. and Dowd, L. R. (1941). *J. Dairy Sci.* **24**, 544–545.

Frank, H. A. and Willits, C. O. (1961). *Fd Technol., Champaign* **15**, 1–3.

Frank, H. A., Naghski, J., Reed, L. L. and Willits, C. O. (1959). *Appl. Microbiol.* **7**, 152–155.

Frazier, W. C. (1967). "Food Microbiology". 2nd ed. McGraw-Hill Book Co., New York.

Fred, E. B. and Peterson, W. H. (1922). *J. Bact.* **7**, 257–269.

Geminder, J. J. (1959). *Fd Technol., Champaign* **13**, 459–461.

Genth, H. (1964). *In* "Microbial Inhibitors in Food" (N. Molin and A. Erichsen, eds.), pp. 77–85. Almquist and Wiksell, Stockholm.

Gibbons, N. E. and Rose, D. (1950). *Can. J. Res. F.* **28**, 438–450.

Gilliland, R. B. and Harrison, G. A. F. (1966). *J. appl. Bact.* **29**, 244–252.

Grimes, M. and Doherty, J. (1929). *Sci. Proc. R. Dubl. Soc.* **19**, 261–267.

Haines, R. B. (1931). *J. Soc. chem. Ind., Lond.* **50**, 223T–227T.

Hall, H. H., James, L. H. and Nelson, E. K. (1937). *J. Bact.* **33**, 577–585.

Hammer, B. W. (1919). *Iowa Agric. Exp. Sta. Res. Bull.* No. 54.

Hammer, B. W. (1926). *J. Dairy Sci.* **9**, 507–511.

Hammer, B. W. (1948). "Dairy Bacteriology". John Wiley and Sons, Inc., New York.

Hammer, B. W. and Babel, F. J. (1957). "Dairy Bacteriology". 4th edn. John Wiley and Sons, Inc., New York.

Hammer, B. W. and Cordes, W. A. (1920). *Iowa Agric. Exp. Sta. Res. Bull.* No. 61.

Harding, H. A., Rogers, L. A. and Smith, G. A. (1900). *New York Agric. Exp. Sta. Bull.* No. 183.

Harrison, F. C. (1902). *Ontario Dept. Agric. Bull.* No. 120.

Hastings, E. G. (1906). *23rd Ann. Rep. Wisconsin Agric. Exp. Sta.*, pp. 107–115.

Hiscox, E. R. (1923). *Ann. appl. Biol.* **10**, 370–377.

Holck, A. A. and Fields, M. L. (1965). *J. Fd Sci.* **30**, 604–609.

Hood, E. G. and White, A. H. (1931). *Dom. Canada Dept. Agric. Bull.* No. 146.

Hood, E. G., Gibson, C. A. and Smith, K. N. (1952). *Scient. Agric.* **32**, 638–644.

Hope, G. W. (1963). *Fd Technol., Champaign* **17**, 123–124.

Hunter, A. C. (1920). *U.S. Dept. Agric. Bull.* No. 819.

Hunter, O. W. (1918). *J. Bact.* **3**, 293–300.

Ingram, M. (1949a). *Fd Mf.* **24**, 77–81.

Ingram, M. (1949b). *Fd Mf.* **24**, 121–124.

Ingram, M. (1950). *J. gen. Microbiol.* **4**, ix.

Ingram, M. (1951). *Proc. Soc. appl. Bact.* **14**, 243–260.

Ingram, M. (1955). "An Introduction to the Biology of Yeasts". Pitman, London.

Ingram, M. (1957). *In* "Microbial Ecology", pp. 90–133. Cambridge University Press.

Ingram, M. (1958). *In* "Chemistry and Biology of Yeasts", (A. H. Cook, ed.), pp. 603–633. Academic Press. Inc., New York.

Ingram, M. (1959). *Rev. Ferm. Ind. Aliment.* **14**, 23–33.

Ingram, M. (1960). *Acta microbiol. hung.* **7**, 95–105.

Ingram, M. (1962). *In* "Recent Advances in Food Science", (J. Hawthorne and J. M. Leitch, eds.) Vol. 2, pp. 272–284. Butterworths, London.

Johns, C. K. and Katznelson, H. (1941). *Can. J. Res. C*, **19**, 49–58.

Jones, A. and Harper, G. S. (1952). *Fd Technol., Champaign* **6**, 304–308.

Jones, C. H. (1912). *Vermont Agric. Exp. Sta. Bull.* No. 167.

Jones, I. D., Etchells, J. L., Veerhoff, O. and Veldhuis, M. K. (1941). *Fruit Prod. J.* **20**, 202–206, 219, 220.

Joslyn, M. A. and Cruess, W. V. (1929). *Hilgardia* **4**, 210–240.

Kroemer, K. and Krumbholz, G. (1931). *Arch. Mikrobiol.* **2**, 352–410.

Kroemer, K. and Krumbholz, G. (1932). *Arch. Mikrobiol.* **3**, 384–396.

Kühl, H. (1910). *Zentbl. Bakt. ParasitKde (Abt. I)* **54**, 5–16.

Lawrence, N. L., Wilson, D. C. and Pederson, C. S. (1959). *Appl. Microbiol.* **7**, 7–11.

Lea, C. H. (1931a). *J. Soc. chem. Ind., Lond.* **50**, 207T–213T.

Lea, C. H. (1931b). *J. Soc. chem. Ind., Lond.* **50**, 215T–220T.

Levine, M. (1940). *Fruit Prod. J.* **20**, 104–107.

Levine, M. and Toulouse, J. H. (1934). *Am. Bottlers of Carbonated Beverages Educational Bull.* No. 7.

Lochhead, A. G. and Farrell, L. (1930a). *Can. J. Res.* **3**, 51–64.

Lochhead, A. G. and Farrell, L. (1930b). *Can. J. Res.* **3**, 95–103.

Lochhead, A. G. and Farrell, L. (1931a). *Can. J. Res.* **5**, 529–538.

Lochhead, A. G. and Farrell, L. (1931b). *Can. J. Res.* **5**, 665–672.

Lochhead, A. G. and Farrell, L. (1936). *Fd Res.* **1**, 517–524.

Lochhead, A. G. and Heron, D. A. (1929). *Dom. Canada Dept. Agric. Bull.* No. 116 (new series).

Lochhead, A. G. and Landerkin, G. B. (1942). *J. Bact.* **44**, 343–351.

Lochhead, A. G. and McMaster, N. B. (1930–31). *Scient. Agric.* **11**, 351–360.

Lodder, J. and Kreger-van Rij, N. J. W. (1952). "The Yeasts". North-Holland Publishing Co., Amsterdam.

Lowings, P. H. (1956). *Appl. Microbiol.* **4**, 84–88.

Luh, B. S. and Phaff, H. J. (1951). *Archs Biochem. Biophys.* **33**, 212–227.

Lund, A. (1951). *J. Inst. Brew.* **57**, 36–41.

Lund, A. (1956). *Friesia* **5**, 297–302.

Lund, A. (1958). *In* "Chemistry and Biology of Yeasts", (A. H. Cooke, ed.), pp. 63–91. Academic Press, Inc., New York.

Lüthi, H. (1959). *Adv. Fd Res.* **9**, 221–284.

Macy, H., Coulter, S. T. and Combs, W. B. (1932). *Minnesota Agric. Exp. Sta. Tech. Bull.* No. 82.

Marshall, C. R. and Walkley, V. T. (1951). *Fd Res.* **16**, 448–456.

Marshall, C. R. and Walkley, V. T. (1952). *Fd Res.* **17**, 307–314.

Marten, E. A., Peterson, W. H. and Fred, E. B. (1929). *J. agric. Res.* **39**, 285–292.

McCormack, G. (1950). *Comml Fish. Rev.* **12**(11a), 128.

McDonald, V. R. (1963). *J. Fd Sci.* **28**, 135–139.

McFarlane, V. H. (1940). *Fd Res.* **5**, 43–57.

McFarlane, V. H. (1941). *Fd Res.* **6**, 481–492.

McFarlane, V.H. (1942). *Fd Res.* **7**, 509–518.

McKelvey, C. E. (1926). *J. Bact.* **11**, 98–99.

Michener, H. D. and Elliott, R. P. (1964). *Adv. Fd Res.* **13**, 349–396.

Miklik, E. (1953). *Milchwissenschaft* **8**, 23–26.

Miller, M. W. and Mrak, E. M. (1953). *Appl. Microbiol.* **1**, 174–178.

Miller, M. W. and Phaff, H. J. (1962). *Appl. Microbiol.* **10**, 394–400.

Morgan, M. E., Anderson, E. O., Hankin, L. and Dowd, L. R. (1952). *Connecticut (Storrs) Agric. Exp. Sta. Bull.* No. 284.

Morgan, R. H. (1938). "Beverage Manufacture (Non-Alcoholic)". Attwood and Colk, Ltd., London.

Morris, E. O. and Ross, S. S. (1965). *In* "Food Science and Technology", (J. M. Leitch, ed.) Vol. 2, pp. 385–392. Gordon and Breach Science Publishing, Ltd., London.

Morse, R. E., Fellers, C. R. and Levine, A. S. (1948). *J. Milk Fd Technol.* **11**, 346–351.

Mossel, D. A. A. (1952). *Antonie van Leeuwenhoek* **17**, 146–152.

Mossel, D. A. A. and Ingram, M. (1955). *J. appl. Bact.* **18**, 232–268.

Mrak, E. M. (1941). *Fruit Prod. J.* **20**, 267–276, 293.

Mrak, E. M. and Baker, E. E. (1940). *Proc. Third Int. Cong. Microbiol.*, pp. 707–708.

Mrak, E. M. and Bonar, L. (1938). *Fd Res.* **3**, 615–618.

Mrak, E. M. and Bonar, L. (1939). *Zentbl. Bakt. ParasitKde (Abt. II)* **100**, 289–294.

Mrak, E. M. and McClung, L. S. (1940). *J. Bact.* **40**, 395–407.

Mrak, E. M. and Phaff, H. J. (1948). *A. Rev. Microbiol.*, **2**, 1–46.

Mrak, E. M., Phaff, H. J. and Vaughn, R. H. (1942a). *J. Bact.* **43**, 689–700.

Mrak, E. M., Phaff, H. J., Vaughn, R. H. and Hansen, H. N. (1942b). *J. Bact.* **44**, 441–450.

Mrak, E. M., Vaughn, R. H., Miller, M. W. and Phaff, H. J. (1956). *Fd Technol., Champaign* **10**, 416–419.

Mulcock, A. P. (1955). *N.Z. Jl Sci. Technol.* B **37**, 15–19.

Murdock, D. I. (1964). *J. Fd Sci.* **29**, 354–359.

Naghski, J., Reed, L. L. and Willits, C. O. (1957). *Fd Res.* **22**, 176–181.

Naghski, J. and Willits, C. O. (1953). *Fd Technol., Champaign* **7**, 81–83.

Naghski, J. and Willits, C. O. (1957). *Fd Res.* **22**, 567–571.

Natarajan, C. P., Chari, C. N. and Mrak, E. M. (1948). *Fruit Prod. J.* **27**, 242–243, 267.

Nelson, J. A. (1928). *J. Dairy Sci.* **11**, 397–400.

Nickerson, W. J. (1943). *Mycologia* **35**, 66.

Nissen, W. (1930). *Milchw. Forsch.* **10**, 30–67.

Njoku-Obi, A. N., Spencer, J. V., Sauter, E. A. and Eklund, M. W. (1957). *Appl. Microbiol.* **5**, 319–321.

Nolte, A. J. and Loesecke, H. W. von (1940). *Fd Res.* **5**, 73–81.

Ogilvy, W. S. and Ayres, J. C. (1953). *Fd Res.* **18**, 121–130.

Oliver, M. (1962). *In* "Recent Advances in Food Science", (J. Hawthorn and J. M. Leitch, eds.) Vol. 2, pp. 265–271. Butterworths, London.

Olson, H. C. and Hammer, B. W. (1935). *Iowa St. Coll. J. Sci.* **10**, 37–43.

Olson, H. C. and Hammer, B. W. (1937). *Iowa St. Coll. J. Sci.* **11**, 207–213.

Ōnishi, H. (1963). *Adv. Fd Res.* **12**, 53–94.

Orla-Jensen, S. (1931). "Dairy Bacteriology", 2nd English Ed. The Blakiston Division, McGraw-Hill Book Co., Inc., New York.

Owen, W. L. (1913). *Zentbl. Bakt. ParasitKde (Abt. II)* **39**, 468–482.

Owen, W. L. (1918). *Louisiana Agric. Exp. Sta. Bull.* No. 162.

Owen, W. L. (1949). "The Microbiology of Sugars, Syrups, and Molasses". Burgess Publishing Co., Minneapolis.

Paine, H. S., Birckner, V. and Hamilton, J. (1927). *J. ind. Engng Chem.* **19**, 358–363.

Papadakis, J. A. (1960). *Annls Inst. Pasteur, Lille* **11**, 85–97.

Payne, D. B. and Perigo, J. A. (1960). *Annls Inst. Pasteur, Lille* **11**, 99–106.

Pearce, E. B. and Parker, R. T. P. (1908). *J. agric. Sci.* **3**, 55–79.

Pederson, C. S. (1931). *New York State Agric. Exp. Sta. Bull.* No. 595.

Pederson, C. S. (1936a). *Fd Res.* **1**, 9–27.

Pederson, C. S. (1936b). *Fd Res.* **1**, 301–305.

Pederson, C. S. and Breed, R. S. (1926). *New York State Agric. Exp. Sta. (Geneva) Bull.* No. 538.

Pederson, C. S. and Breed, R. S. (1929). *New York State Agric. Exp. Sta. (Geneva) Bull.* No. 570.

Pederson, C. S. and Kelly, C. D. (1938). *Fd Res.* **3**, 583–588.

Pederson, C. S., Albury, M. N., Wilson, D. C. and Lawrence, N. L. (1959a). *Appl. Microbiol.* **7**, 1–6.

Pederson, C. S., Wilson, D. C. and Lawrence, N. L. (1959b). *Appl. Microbiol.* **7**, 12–15.

Pederson, C. S., Albury, M. N. and Christensen, M. D. (1961). *Appl. Microbiol.* **9**, 162–167.

Pethybridge, G. H. (1906). *Econ. Proc. R. Dubl. Soc.* **1**, 306–320.

Peynaud, E. and Domercq, S. (1955). *Annls Inst. Pasteur, Paris* **89**, 346–347.

Phaff, H. J. and Douglas, H. C. (1944). *Fruit Prod. J.* **23**, 332–335.

Phaff, H. J., Mrak, E. M. and Williams, O. B. (1952). *Mycologia* **44**, 431–451.

Phaff, H. J., Miller, M. W. and Mrak, E. M. (1966). "The Life of Yeasts". Harvard University Press, Cambridge, Mass.

Phillips, G. F. and Mundt, J. O. (1950). *Fd Technol., Champaign* **4**, 291–293.

Poe, C. F. and Field, J. T. (1947). *Fruit Prod. J.* **27**, 112–116.

Pouncey, A. E. and Summers, B. C. E. (1939). *J. Soc. chem. Ind., Lond.* **58**, 162–165.

Prevot, A. R. and Thouvenot, H. (1960). *Annls Inst. Pasteur, Lille* **11**, 39–42.

Proszt, G. and Vas, K. (1962). *Proc. Third Cong. Hungarian Assoc. Microbiol.*, p. 254.

Rand, F. V. and Pierce, W. D. (1920). *Phytopathology* **10**, 189–231.

Recca, J. and Mrak, E. M. (1952). *Fd Technol., Champaign* **6**, 450–454.

Rice, F. E. and Markley, A. L. (1921). *New York State Agric. Exp. Sta. Bull.* No. 44.

Richter, A. A. (1912). *Mycol. Zentbl.* **1**, 67–76.

Robinson, J. F. and Hills, C. H. (1959). *Fd Technol., Champaign* **3**, 251–253.

Roelofsen, P. A. (1953). *Biochim. biophys. Acta* **10**, 410–413.

Ross, S. S. and Morris, E. O. (1965). *J. appl. Bact.* **28**, 224–234.

Rushing, N. B. and Senn V. J. (1962). *Fd Technol., Champaign* **16**(2), 77–79.

Russell, H. L. and Hastings, E. G. (1905). *Wisconsin Exp. Sta. Bull.* No. 128.

Savage, W. G. and Hunwicke, R. F. (1923). *Great Britain Dept. Sci. Ind. Res. Food Investigation Board Special Rep.* No. 13.

Scarr, M. P. (1951). *J. gen. Microbiol* 5, 704–713.

Scarr, M. P. (1953). *J. appl. Bact.* 16, 119–127.

Scheffer, W. R. and Mrak, E. M. (1951). *Mycopath. Mycol. appl.* 5, 236–249.

Schneider, I. S., Frank, H. A. and Willits, C. O. (1960). *Fd Res.* 25, 654–662.

Scott, W. J. (1936). *J. Coun. scient. ind. Res. Aust.* 9, 177–190.

Scott, W. J. (1937). *J. Coun. scient. ind. Res. Aust.* 10, 339–350.

Scott, W. J. (1957). *Adv. Fd Res.* 7, 83–127.

Sharf, J. M. (1960). *Annls Inst. Pasteur, Lille* 11, 117–132.

Shehata, A. M. E. (1960). *Appl. Microbiol.* 8, 73–75.

Sheneman, J. M. and Costilow, R. N. (1955). *Appl. Microbiol.* 3, 186–189.

Shrader, J. H. and Johnson, A. H. (1934). *Ind. Engng Chem.* 26, 869–874.

Shutt, D. B. (1925). *Scient. Agric.* 6(4), 118–119.

Silvestrini, D. A., Anderson, G. W. and Snyder, C. S. (1958). *Poult. Sci.* 37, 179–185.

Simpson, K. L., Nagel, C. W., Ng, H., Vaughn, R. H. and Stewart, G. F. (1959). *Fd Technol., Champaign* 13, 153–154.

Smith, E. S., Bowen, J. F. and MacGregor, D. R. (1962). *Fd Technol., Champaign* 16(3), 93–95.

Smith, R. E. and Hansen, H. N. (1931). *California Exp. Sta. Bull.* No. 506.

Snow, J. E. and Beard, P. J. (1939). *Fd Res.* 4, 563–585.

Soulides, D. A. (1955). *Appl. Microbiol.* 3, 129–131.

Soulides, D. A. (1956). *Appl. Microbiol.* 4, 274–276.

Steele, W. F. and Yang, H. Y. (1960). *Fd Technol., Champaign* 14, 121–126.

Sturges, W. S. (1923). *Abstr. Bact.* 7, 11.

Subba Rao, M. S., Soumithri, T. C., Johar, D. S., Sreenivasan, A. and Subrahmanyan, V. (1965). *In* "Food Science and Technology", (J. M. Leitch, ed.) Vol. 2, pp. 227–238. Gordon and Breach Science Publishers, Ltd., London.

Tanner, F. W. (1944). "The Microbiology of Foods". Garard Press, Champaign, Illinois.

Tanner, F. W. and Strauch, L. B. (1926). *Proc. Soc. exp. Biol. Med.* 23, 449–450.

Tarkow, L., Fellers, C. R. and Levine, A. S. (1942). *J. Bact.* 44, 367–372.

Tarr, H. L. A., Southcott, B. A. and Bissett, H. M. (1952). *Fd Technol., Champaign* 6, 363–366.

Teunisson, D. J. (1954). *Appl. Microbiol.* 2, 215–220.

Thatcher, F. S. and Loit, A. (1961). *Appl. Microbiol.* 9, 39–45.

Tilbury, R. H. (1966). *In* "Microbiological Deterioration in the Tropics", S.C.I. Monograph No. 23, pp. 63–79. Gordon and Breach, Science Publishers, Inc., New York.

Van Slyke, L. L. and Price, W. V. (1938). "Cheese". Orange Judd Publishing Co., New York.

Vaughn, R. H. (1954). *In* "Industrial Fermentations", (L. A. Underkofler and R. J. Hickey, eds.), pp. 417–478. Chemical Publishing Co., New York.

Vaughn, R. H., Douglas, H. C. and Gilliland, J. R. (1943). *California Agric. Exp. Sta. Bull.* No. 678.

Vaughn, R. H., Murdock, D. I. and Brokaw, C. H. (1957). *Fd Technol., Champaign* 11, 92–95.

Vickery, J. R. (1936a). *J. Coun. scient. ind. Res. Aust.* 9, 107–112.

Vickery, J. R. (1936b). *J. Coun. scient. ind. Res. Aust.* 9, 196–202.

Von Schelhorn, M. (1950a). *Adv. Fd Res.* **3**, 429–482.
Von Schelhorn, M. (1950b). *Z. Lebensmittelunters. u.-Forsch.* **91**, 117–124.
Vorbeck, M. L. and Cone, J. F. (1963). *Appl. Microbiol.* **11**, 23–27.
Wales, C. S. and Harmon, L. G. (1957). *Fd Res.* **22**, 170–175.
Walker, H. W. and Ayres, J. C. (1956). *Appl. Microbiol.* **4**, 345–349.
Walker, H. W. and Ayres, J. C. (1959). *Appl. Microbiol.* **7**, 251–255.
Weaver, C. A., Robinson, J. F. and Hills, C. H. (1957). *Fd. Technol., Champaign* **11**, 667–669.
Wehmer, C. (1905). *Zentbl. Bakt. ParasitKde (Abt. II)* **14**, 682–713.
Weinzirl, J. (1922). *J. Bact.* **7**, 599–603.
Weiser, H. H. (1942). *J. Bact.* **43**, 46.
Wells, F. G. and Stadelman, W. J. (1958). *Appl. Microbiol.* **6**, 420–422.
Whitehill, A. R. (1957). *Bact. Proc.* 20.
Wickerham, L. J. (1957). *J. Bact.* **74**, 832–833.
Wickerham, L. J. and Duprat, E. (1945). *J. Bact.* **50**, 597–607.
Williams, A. J., Wallace, R. H. and Clark, D. S. (1956). *Can. J. Microbiol.* **2**, 645–648.
Williams, O. B. and Mrak, E. M. (1949). *Fruit Prod. J.* **28**, 141, 153.
Wilson, H. F. and Marvin, G. E. (1929). *J. econ. Ent.* **22**, 513–517.
Witter, L. D., Berry, J. M. and Folinazzo, J. F. (1958). *Fd Res.* **23**, 133–142.
Worrell, L. (1951). *In* "The Chemistry and Technology of Food and Food Products", (M. B. Jacobs, ed.) Vol. 2, pp. 1706–1762. Interscience Publishers, New York.
Yacowitz, H., Pansy, F., Wind, S., Stander, H., Sassaman, H. L., Pagano, F. and Trejo, W. H. (1957). *Poult. Sci.* **36**, 843–849.
Yang, H. Y., Gomez, L. P., Rasulpuri, M. L. and Michalek, J. G. (1962). *Fd Technol., Champaign* **16**(4), 109–111.
Yokotsuka, T. (1954). *J. agric. Chem. Soc. Japan* **28**, 114.
Zein, G. N., Shehata, A. M. E. and Sedky, A. (1956). *Fd Technol., Champaign* **10**, 405–407.
Zieglor, F. and Stadelman, W. J. (1955). *Fd Technol., Champaign* **9**, 107–108.
Zuccaro, J. B., Powers, J. J., Morse, R. E. and Mills, W. C. (1951). *Fd Res.* **16**, 30–38.

Chapter 10

Miscellaneous Products from Yeast

J. S. HARRISON

*Research Department, Distillers Company (Yeast) Limited,
Epsom, England*

I. Introduction

Although taxonomically yeasts occupy a small niche in the evolutionary pattern of living organisms, and the species which are in common use can almost be counted on the fingers of one hand, their industrial applications vastly outweigh those of all other microorganisms. The commonly employed species include *Saccharomyces cerevisiae, Sacch. carlsbergensis* and *Candida utilis*.

Besides the well-known applications discussed in earlier chapters for the production of ethanol-containing beverages and industrial spirit and for food, either directly as a protein source or for baking, there are numerous other uses to which yeasts have been put. In order to carry out its basic function of maintaining life, it is necessary that chemical reactions operate continually within the living cells. The individual reactions, mainly enzyme controlled, are often reversible and are simple, complex molecules being synthesized in a series of steps. During the course of these reactions, compounds ranging from water and carbon dioxide to highly complex proteins, polysaccharides and nucleotides are formed. By mutation, in which the function of a single enzyme is destroyed, a break may occur in a sequence of reactions with the result that a useful compound is excreted in quantity, or artificial environmental conditions may be applied with similar results.

By various means based on these principles compounds falling into several groups can be derived from yeasts. These are divided, for convenience, into: (i) whole cells and structural components; (ii) intracellular constituents; (iii) excretion products; and (iv) compounds made by special reactions. The separate products are discussed in varying degrees of detail according to their industrial importance.

II. Whole Cells and Structural Components

The production and uses of yeasts for baking and for consumption directly as food for humans and animals are described elsewhere (Chapter 7, p. 349; Chapter 8, p, 421). A further commercial application of baker's yeast (*Sacch. cerevisiae*) and wine yeast (*Sacch. cerevisiae* var. *ellipsoideus*) is as a starter in vinegar fermentations (Vaughn, 1954). The yeast starter is added at a proportion of 2–10% by volume to a chosen solution of carbohydrate, which is thereby converted to ethanol prior to acetic acid formation by a slow secondary anaerobic fermentation by *Acetobacter*.

When live yeast cells are ingested, they have been shown to stay alive, in the main, and under certain conditions to remove assimilable nutrients, including vitamins, from the contents of the digestive tract, rather than to contribute additional food requirements (Parsons *et al.*, 1945). On the other hand, dead whole cells provide a pattern of compounds which is broadly similar to that present in animal and plant materials. The cell wall material, consisting largely of glucan and mannan (Trevelyan, 1958; Vol. 2, Chapter 5), is not easily digestible by most mammals. Certain micro-organisms, however, particularly moulds, can break down this material, and its nutritive status can be changed by their action. For the above reason whole autolysates and extracts, from which insoluble solids have been removed, are often used for human nutrition rather than dried intact cells. Brewer's yeast, for instance, can be debittered to remove the hop resins and is used in the extracted form (Chapter 8, p. 421). Such products are used as nutrient supplements, and for flavouring and, on a smaller scale, are important as constituents of media for the growth of micro-organisms. The yeasts employed commercially for these purposes are almost completely confined to the species *Sacch. cerevisiae* and *C. utilis*. In countries where the supply of nitrogenous feed-stuffs and fertilizers is scarce, dried yeast made from waste products such as wood waste may be used on the land as well as for feeding animals. Whole yeast is also present in the solid residues obtained in the final stages of distillery fermentations and distillations, and is used directly after separation or is dried prior to sale as a feed supplement (Chapter 6, p. 283). Yeast preparations and extracts have been recommended in the past for use medicinally in many specific diseases, but it would appear with present knowledge that the benefits were mainly due to the relief of vitamin deficiencies which are now treated more directly with the appropriate factor. Applications for the preparation of adhesives, plastics materials, and so on have been superseded by modern technology.

III. Cellular Constituents

A. ENZYMES

Yeasts contain many enzymes, all of which catalyse apparently simple changes in molecular structure, such as transfer of atoms or radicals (e.g. hydrogen, oxygen, hydroxyl, phosphate, amino and acetyl groups); or the linking of molecular units such as simple sugars to form polysaccharides, amino acids to produce peptides and proteins, and the cleavage of the complexes back to the simple constituents. De Becze (1960) lists over a hundred enzymes that have been found in yeasts, but only a small number of these are of commercial importance. A short review is also given by Prescott and Dunn (1959).

The enzymes can be divided for convenience into three categories, those which: (i) under natural conditions operate only in association with internal structures, and often spatially near other similar enzymes; (ii) are freely miscible with aqueous cell contents; and (iii) exist in extracellular form, or are sufficiently near the cell surface that, in effect, they behave as such in the sense that substances in the surrounding medium are chemically transformed (see Chapter 8, p. 421).

1. *Intracellular*

Enzymes in the first group, such as those contained in yeast mitochondria which catalyse the chain of reactions that comprise oxidative phosphorylation, while being essential to the functioning of the living cell, are biochemically still mainly of academic interest as their useful activities can only be elicited with great difficulty when detached from their natural environment. In the second category are a number of enzymes that can be separated by various means from the liquid cell contents. The commercially available examples include hexokinase, the initial enzyme in the series of reactions by which glucose is assimilated, thus:

$$\text{glucose} + \text{ATP} \longrightarrow \text{glucose 6-phosphate} + \text{ADP}$$

glucose 6-phosphate dehydrogenase, which is concerned in the reversible conversion:

$$\text{glucose 6-phosphate} + \text{NADP} \longrightarrow$$
$$\text{6-phospho-}\delta\text{-gluconolactone} + \text{NADPH}_2;$$

alcohol dehydrogenase which, as the final stage in the fermentation cycle in yeast, converts acetaldehyde to ethanol

$$\text{CH}_3\text{CHO} + \text{NADH}_2 \longrightarrow \text{CH}_3\text{CH}_2\text{OH} + \text{NAD}$$

and glutathione reductase. These enzymes are manufactured on a relatively small laboratory or pilot plant scale, and are used for research and analytical purposes. In the main they are free from other enzymes and can be used for their specific activities. Although some of the many enzymes involved in intracellular reactions are available commercially for similar purposes, they are prepared from sources other than yeast.

2. Invertase

The enzymes of the third group have evolved in such a way as to provide a mechanism for breaking down compounds in the extracellular environment which cannot enter the cell and be used directly as metabolic substrates. The most important of these are the carbohydrates. Terminal fructose groups in complexes are specifically hydrolysed by the invertase of yeast, a β-fructofuranosidase. In particular, sucrose is converted to fructose and glucose by its action, and raffinose is hydrolysed to fructose and melibiose. Both these reactions are important as steps in the conversion of common carbohydrate sources to assimilable sugars. The subject has been reviewed by Neuberg and Roberts (1946).

Invertase is present in many yeasts. Davies (1963) lists *Candida utilis, Debaryomyces dekkeri, Endomycopsis fibuliger, Hansenula jardinii, Saccharomyces ashbyi, Sacch. carlsbergensis* Y 267 and Y 572,

Sacch. cerevisiae, Sacch. fragilis, Sacch. kluyveri, Sacch. pastorianus, Sacch. uvarum and *Torulopsis colliculosa*. The enzyme is always closely associated with the mannan of the yeast cell wall (Fischer *et al.*, 1951). The invertase activity of many strains of yeast can be increased by aerobic growth under suitable conditions in the presence of sucrose (Pyke, 1958), particularly by the use of a slow growth rate for 8–10 h (Cochrane, 1961), and by the addition of mannose during propagation (Suomalainen *et al.*, 1967). The yeast is then separated and plasmolysed at a somewhat elevated temperature by the addition of toluene (Cochrane, 1961) or other agents (Pyke, 1958). When the yeast has liquefied papain is added, digestion is allowed to proceed for 48 h, and the cell debris is removed by a filter press. After chilling, the pH value of the liquid is adjusted to 4·5 and the invertase is precipitated by the addition of an equal volume of cold ethanol (industrial spirit). The precipitated "invertase gum" is dissolved in 55% glycerol to produce the salable commercial product. Neuberg and Roberts (1946) and Pyke (1958) review alternative procedures for making invertase products in liquid form (see also Chapter 8, p. 421).

Dry preparations are also available (Peppler and Thorn, 1960). Other patented processes include the absorption of concentrated liquid preparations on glucose or lactose (Wallerstein, 1932) and adsorption on calcium phosphate followed by washing of the precipitate and drying at low temperature (Neuberg, 1944).

The activity of invertase preparations can be standardized by any of a number of methods (Cochrane, 1961). In general either the turn-over rate of sucrose substrate per unit quantity of enzyme preparation, or the rate of hydrolysis of sucrose under precisely defined conditions, is measured. In Great Britain an arbitrary unit is used in which the positive optical rotation of the original sucrose solution is reduced to zero by the action of the enzyme; this represents a 75·9% turnover of a 10% sucrose solution (Cochrane, 1961). The official American method depends on the fact that the hydrolysis of sucrose follows the kinetics of a first order reaction (Johnston *et al.*, 1935). Standardizations carried out by these procedures, however, require careful interpretation in practice, as the exact conditions under which the enzyme is used affect its efficiency to a considerable extent.

The applications of invertase include the inversion of sucrose solutions as an alternative to acid hydrolysis, in order to avoid the production of undesirable impurities such as furfural, the production of fructose in confections in order to make use of its humectant properties (Gabel, 1950), and the conversion of originally solid sucrose mixtures to fluid consistency after covering with chocolate or other suitable coating to produce soft-centred confections.

3. Amylases

Certain yeasts are able to utilize starches present in the fermentation medium. *Endomycopsis fibuliger*, in particular, possesses strong α- and β-amylase activities (Tveit, 1967), and α-amylase has been prepared therefrom (Davies, 1963). The same enzyme has been observed in the crude filtrates from other species of yeast (Wickerham *et al.*, 1944; Davies, 1963). The amylolytic properties have been applied in a Swedish process (Tveit, 1967) in which *C. utilis* and *E. fibuligera* are used symbiotically to ferment starch-containing liquors. Starch suspensions are adjusted to pH 5, assimilable nitrogen is added in the form of urea, and an inoculum of equal quantities of the two yeasts is given. After 24 h the populations reach a maximum, typically about 1.5×10^9 *C. utilis*/ml and 1.75×10^8 *E. fibuliger*/ml. Five percent of the resultant mixture of yeasts, the bulk of which is *Candida*, is reserved as a suspension to which an appropriate amount of *E. fibuliger* is added to provide the inoculum for the next batch. The remainder is separated and dried for use as a supplement for animal feeds. The process described is designed to reduce the biochemical oxygen demand of starchy effluents while providing a useful by-product. There appears to be no commercial process by which amylases prepared from yeast are used in relatively pure form.

4. Lactase

Lactase, or β-D-galactosidase, is another enzyme for which patents relating to manufacture from yeast have been filed (Morgan, 1955; Meyers, 1956; Meyers and Stimpson, 1956; Young and Healey, 1957; Chapter 8, p. 421). Its function is to catalyse the hydrolysis of lactose to glucose and galactose.

Suitable sources of lactase are *Saccharomyces fragilis*, *Sacch. lactis*, *C. pseudotropicalis* and certain strains of *C. utilis* (Prescott and Dunn, 1959). According to the method patented by Meyers (1956) the chosen yeast is grown on whey obtained during cheese manufacture, from which protein has been removed by adjusting to pH 4.5 and heating to 85–104° and filtering. After adding 0.1% of ammonia the medium is cooled to 30°, inoculated with actively growing yeast, and held for 24 h with aeration. The yeast is then separated, washed and rapidly cooled to −18°. This treatment has the effect of destroying the fermentative system while retaining lactase activity. The vacuum-dried product provides a source of lactase. Various other methods of preparation also provide similar products containing a complex mixture of cell material (Prescott and Dunn, 1959).

Lactase preparations can be employed to improve the consistency of ice cream bases, frozen milk and concentrated milk by removing lactose, which tends to crystallize in hard particles, and it can be added to animal feeds (Pomeranz, 1964). Another valuable, but still small scale, application of lactase is in cases of deficiency in the ability of certain persons, notably Asians (Davis and Bolin, 1967; Flatz *et al.*, 1969) to metabolize milk products containing lactose, the result being digestive disorders, the medical cause of which can be elusive.

5. *Polygalacturonase*

The preparation and properties of this extracellular enzyme, which is produced by strains of *Sacch. fragilis*, have been described in a series of papers by Phaff and coworkers (Luh and Phaff, 1954; Demain and Phaff, 1954a). Polygalacturonase can be used to prepare pure mono- and di- (Phaff and Luh, 1952) and tri- and tetragalacturonic acids (Demain and Phaff, 1954b) from pectic acid. Yeasts possessing these properties have been observed to be associated with the undesirable softening of cucumber salt stock in the pickling industry (Prescott and Dunn, 1959).

B. COENZYMES

The coenzymes are, in the main, fairly simple chemical compounds whose purpose is often to act as carriers of atoms or radicals such as hydrogen, hydroxyl, phosphate, acetyl and the amino group. Although the coenzymes are in some cases present in yeast cells in relatively high concentrations, and are of vital importance to the function of intracellular enzymes systems, the majority have limited use outside the laboratory. In spite of being comparatively small molecules by cellular standards, they will not, in general, pass through the walls of living cells, and thus are not excreted. Similarly they cannot pass inwards when applied externally to cells, for instance *via* the blood stream of animals, in spite of a possible medical requirement on a biochemical basis. The former difficulty can be overcome by killing the cells before extraction, but the latter is normally insuperable, as the coenzymes must be formed within the cells, although a simpler permeable precursor of the coenzyme may provide a means of intracellular synthesis. Thus many components of coenzymes, such as certain growth factors (see Section III.F, p. 538), are freely taken up by cells.

Coenzymes have uses, however, in experimental work in isolated systems, and a number can conveniently be prepared from yeast. Those present in the highest amounts in yeast are the hydrogen-carriers nicotinamide adenine dinucleotide in the oxidized and reduced forms (NAD, $NADH_2$) and the corresponding phosphates (NADP, $NADPH_2$), the phosphate-transferring coenzymes which include the phosphates of uridine, cytidine and guanosine, but most important the di- and tri-phosphates of adenosine (ADP, ATP), and coenzyme A (CoA) which acts as a carrier of the acetyl, succinyl and malonyl radicals in intra-cellular synthetic reactions. Most of the modern techniques for pre-paring these compounds involve chromatography. For instance, in the method described by Munden et al. (1963) ATP and similar compounds which are in equilibrium with it in the living cell (Mann et al., 1958) are extracted from the yeast with hot water, freeing the solution of pre-cipitated protein and adsorbing the required compounds on activated carbon. The concentrated eluate is adsorbed on ion-exchange resin and the nucleotides are removed by elution at defined pH values and re-adsorbed on carbon. Final recovery of individual compounds is by rechromatography using ion-exchange resins.

CoA can be prepared from yeast (Reece et al., 1959) by extracting with cold water, adsorbing the soluble cell constituents on carbon, eluting with pyridine from which an acetone powder is prepared. Co-precipitation as a double cuprous complex with glutathione is then carried out and the glutathione is removed by means of an ion-exchange resin.

C. METABOLIC INTERMEDIATES

Within the living yeast cell many reactions proceed to provide energy and structural components. About twelve successive derivatives from the original carbohydrate source take part in the normal fermentation cycle, including glucose and fructose phosphates. Some of these com-pounds occur in relatively high concentrations, and can be prepared by methods similar to those used for the coenzymes, but they are mainly of academic interest and only manufactured in small amounts. Com-pounds that result from other reactions become building blocks for proteins and nucleic acids, and can in many cases be separated from the cell constituents, but it is often more economical to break down the final complexes under controlled conditions to obtain the simpler units such as amino acids, purines and pyrimidines.

Yeast nucleic acids, which are compounds containing adenine, guanine cytosine and uracil, constitute about 8% of the dry weight and can be extracted by mild alkaline treatment at 37° (Pyke, 1958). By immediate neutralization and filtration, followed by precipitation by acid from the cooled filtrate, washing and reprecipitating, the free nucleic acids are obtained. These can be converted to copper, iron or magnesium salts. By more severe treatment the individual purines and pyrimidines can be liberated; these can then be separated by chromatographic techniques. Ribose has also been prepared from nucleic acids (Pyke, 1958). Yeasts have been used in Japan to produce disodium guanylate, disodium inosinate and inosinic acid in some quantity (Peppler, 1967), and adenylic acid can be obtained in 63% yield by the action of yeast on adenosine (Anon., 1964). The latter four compounds are used as flavouring agents.

In similar fashion to the case of nucleic acids, many free amino acid intermediates in protein synthesis can be isolated from yeast cell contents, but their separation in quantity is difficult. Glutathione, a tripeptide, can however be extracted in commercial quantities (Pyke, 1958). A mixture of some twenty amino acids can also be obtained by hydrolysis of yeast protein but the same problem of isolation in the pure form exists. Although yeast mutants are known which produce reasonably high yields of amino acids, they have not found great commercial use. An alternative method, by which the intracellular concentration of specific amino acids is increased, is by supplying suitable chemical precursors. This is a useful technique when the complete chemical synthesis of the required acid is difficult, and has the virtue that the amino acid produced is in the *laevo*, or naturally assimilable, form.

Lysine can be made by supplying α-aminoadipic or α-ketoadipic acid to *C. utilis* in the presence of glucose and oxygen, when yields equivalent to 60% of the precursor can be obtained. In a similar way 5-formyl-2-oxovaleric acid, when fed to certain strains of *Saccharomyces*, will raise the lysine content of the yeast dry matter to 20%, and the lysine can fairly easily be recovered (Dulaney, 1967). Alanine can be obtained by supplying aspartic acid (Chibata *et al.*, 1965), and glutamic acid from α-ketoglutaric or citric acid (Otsuka *et al.*, 1957). Methionine is obtained by the use of γ-methyl mercapto-α-hydroxybutyric acid and *T. lactis* (Sumitomo Chemical Co., 1963), and *Rhodotorula gracilis* can be employed to produce cystine and methionine in good yield (Chiao and Peterson, 1953). By supplying anthranilic acid, tryptophan can be synthesized by strains of *Candida* and *Hansenula* (Terui and Enatsu, 1962). In spite, however, of much research and a large number of patents on this subject, yeasts are not used to a great extent for the commercial production of amino acids. Also, while mutants of various micro-organisms have been produced which excrete individual amino acids in commercially useful quantities, yeasts have not proved especially satisfactory for this purpose.

D. CARBOHYDRATES

The main carbohydrates of the yeast cell, other than the structural polysaccharides glucan and mannan, are glycogen and trehalose (Trevelyan, 1958). Although these occur in high proportions if yeast is grown under suitable conditions, no large-scale production is carried out. However, methods for preparing zymosan, a polysaccharide which has clinical uses, have been published (Ferranto *et al.*, 1965), and phosphomannans have been made from yeasts of the genus *Hansenula* (Slodki *et al.*, 1961).

E. LIPIDS

The lipids in yeasts mainly comprise glycerides of fatty acids, phospholipids and sterols (Eddy, 1958). The fatty acids include myristic, palmitic, stearic, hexadecanoic, oleic, linoleic, linolenic and many others (Prescott and Dunn, 1959; see Chapter 6, p. 283). Although many yeasts propagated by the normal processes contain only a small quantity of fats, certain species, when grown under special conditions, may reach as high a proportion as 63% of the dry weight. Prescott and Dunn (1959) list *Candida pulcherrima, C. reukaufii, C. utilis, Torulopsis lipofera, Rhodotorula gracilis* and *Trichosporon pullularis* as capable of producing high amounts of lipids. *Rhodotorula gracilis* has probably been studied more than most yeasts for this purpose. Sugars such as glucose, fructose, sucrose, invert sugar and blackstrap molasses are used at concentrations which can vary from 4% to 20%. The medium is made nitrogen-deficient to limit the production of protein, about 0·1% ammonium sulphate being a suitable level. A pH value of 4·6 has been employed (Pan *et al.*, 1949), at a temperature typically of about 28°. with aeration. Similar propagation systems are used for other species, although the pH and temperature optima may differ (Chapter 8, p. 421).

To extract the lipids, the cells are broken by treatment with acids, by autolysis or by grinding, for example with sand, and ether can be employed in the final stage of recovery. If autolysis is used, the yeast is left for two or three days at 50°. It is necessary to exclude oxygen during storage of the product. The main applications of yeast for the production of fats has been in cases where the normal sources are not available, particularly in war time (Dunn, 1952).

Spencer *et al.* (1962) have shown that certain hydroxy fatty acid glycerides are produced extracellularly by *T. magnoliae*, and that the yields can be increased by the addition to the medium of fatty acid esters, hydrocarbons and glycerides.

F. GROWTH FACTORS

Processes for the manufacture of the oil-soluble vitamin D, calciferol, have been developed using yeast as a source of ergosterol, which is then converted to the vitamin. The types of yeast most suitable for the production of ergosterol were found by Dulaney *et al.* (1954) to belong to the genus *Saccharomyces*, particularly brewer's yeasts of the species *Sacch. cerevisiae*, although strains of *Sacch. carlsbergensis* and *Sacch.*

oviformis were satisfactory. The selected strain can be grown in a molasses medium by a process designed to give the highest ergosterol content, and after being dehydrated by treatment, for example, with methanol, the dry yeast is extracted with ethanol or a mixture of lower alcohols, and the sterol esters in the extract are saponified (Petzoldt *et al.*, 1967a). The conversion to calciferol is carried out by dissolving the ergosterol in an organic solvent and passing through a quartz tube under ultraviolet illumination (Rosenberg, 1942). Dehydroergosterol and zymosterol can also be obtained from yeast (Petzoldt *et al.*, 1967b; see also Chapter 8, p. 421).

The water-soluble vitamins of yeast are discussed at some length in Chapter 8 (p. 421). These are normally administered medicinally as dried yeast, without isolation of the individual factors, although a process for manufacturing riboflavin by the use of yeast has been published (Hanson, 1967). A special case is that of vitamin B_1, thiamin, the content of which can be increased in yeast considerably. One commercial method involves adding to the propagation medium relatively high concentrations of two chemically synthesized moieties, 2-methyl-4-amino-5-hydroxymethyl pyrimidine and 4-methyl-5-(2-hydroxymethyl) thiazole, which are normally produced in small quantities within the living cell (Van Lanen, 1954). The synthesis is completed through the agency of the intracellular enzyme thiaminase. The thiamin content of yeast can be increased to as high a value as 1,000 µg/g dry weight by this method.

IV. Excretion Products

A. MONOHYDROXY ALCOHOLS

The production of ethanol in its many aspects has been fully described in earlier chapters. The mechanism of the formation of higher alcohols has also been discussed in connection with flavour-producing compounds in relation to several products. Fusel oils, the mixture of higher alcohols obtained during the distillation of fermented carbohydrate media, are collected during the preparation of pure ethanol and other spirits, and have commercial applications. A list of the higher alcohols found in distillates from spirit fermentations is given in Chapter 6 (p. 283).

Although some of the individual alcohols which are present in the highest amounts, for example 3-methyl propanol, 2-methyl butanol and 3-methyl butanol, can fairly easily be separated by fractional distillation, if required, the demand for these compounds does not normally warrant their isolation. Fusel oil is, however, used in the crude form as a lacquer solvent and for similar purposes.

B. POLYOLS

The formation and excretion of glycerol takes place as a normal side reaction during the anaerobic fermentation of sugars to ethanol and carbon dioxide (see Chapter 6, p. 283), but in the usual method of spirit production the amounts produced are small (1–2% of the carbohydrate metabolized). By two modifications, commonly known as Neuberg's second and third forms of fermentation (Neuberg *et al.*, 1920), the proportion of glycerol can be considerably increased. (Neuberg's first form is the normal fermentation system.) In the second form, sulphite added to the fermentation medium reacts with acetaldehyde produced in the fermentation reactions, with the consequence that glyceraldehyde phosphate formed earlier in the reaction sequence must act as hydrogen acceptor in place of acetaldehyde (see Chapter 6, p. 283), with the result that most of the sugar supplied is converted to acetaldehyde, glycerol and carbon dioxide according to the overall equation:

$$C_6H_{12}O_6 + \text{sulphite} \longrightarrow$$
$$CH_3CHO\text{-sulphite complex} + CO_2 + C_3H_8O_3$$

Alternatively, glycerol can be produced in high yield by the third Neuberg form of fermentation which is carried out under alkaline conditions. By this mechanism, the acetaldehyde is converted to acetic acid and ethanol to give an overall reaction which is largely represented as:

$$2\,C_6H_{12}O_6 + H_2O \longrightarrow 2\,CO_2 + CH_3COOH + C_2H_5OH + 2\,C_3H_8O_3$$

Several processes have been employed to exploit these two systems: these have been reviewed by Underkofler (1954). The German process of Connstein and Lüdecke (1921, 1924) was used in Germany during the first World War on a large scale, although recovery of pure glycerol was small compared with the amount produced. Beet sugar with added nutrient salts and sodium sulphite was inoculated with yeast and fermentation was allowed to continue for 48–60 h at 30°. The separated clear liquid contained 2–3% glycerol, along with 2–3% ethanol and 1% acetaldehyde. After distilling off the alcohol and acetaldehyde, the residual solution was further concentrated to provide crude glycerol. An alternative process of Fulmer *et al.* (1947), in which ammonium sulphite replaced the sodium salt and the fermentation was carried out at a controlled pH value of 6·8, was never used on a large scale. An alkaline method (Eoff, 1918) using sodium carbonate was also not used commercially. Further patents were taken out by Schade (1947) and Schade and Färber (1947) for processes in which the volatile compounds evolved during the fermentation were removed by passing air or other gases through the alkaline medium. More recently "insoluble-sulphite" processes have been investigated (Underkofler *et al.*, 1951a, 1951b), and a continuous fermentation system has been devised by Harris and Hajny (1960).

The principal difficulty in the commercial exploitation of all these methods appears always to have been the separation of the glycerol from other materials in the fermented medium. For this reason the production of glycerol by the use of yeasts has not been practised to any great extent, although according to the World-Wide Survey of Fermentation Industries 1963 (1966) 50,000 tons were made in Czechoslovakia in that year.

Several other polyols, including arabitol (Graham, 1961), mannitol and erythritol (Prescott and Dunn, 1959), can be produced in pure form and in good yield by fermentation methods using strains of *Hansenula* and *Endomycopsis*. The work of Spencer and Sallans (1956) has shown that osmophilic yeasts are particularly suitable for this purpose. It appears that glycerol is always formed along with the other polyols, but a study of the conditions (Spencer *et al.*, 1957) showed that high arabitol and low glycerol yields were obtained in the absence of supplementary nitrogen, and that increased aeration gave more glycerol but no improvement in the amount of arabitol formed. These workers concluded that about 60% of the sugar supplied could be converted to a mixture of the polyols. As arabitol, for example, is a solid at ordinary temperatures, the problem of separation is by no means as difficult as for glycerol. *Saccharomyces rouxii* strain P_3a was one of the osmophilic yeasts studied by Spencer and Shu (1957) for this purpose. In spite of the relative ease with which some of these compounds can be made by fermentation processes, no significant commercial applications are used.

V. Special Reactions

A. EPHEDRINE

Ephedrine, which is used in medicine and has in the past always been made from botanical sources, can be synthesized by a special fermentation process. The procedure, patented by Hildebrandt and Klavehn (1934), depends on the observation that benzaldehyde can be converted by certain yeasts to 1(1-phenyl-1-hydroxyl)-2-propane. The latter compound can then be condensed chemically with monomethylamine to give ephedrine. The benzaldehyde is added to a medium containing about 12·5 times its weight of glucose syrup and a heavy inoculum of yeast is added, after which fermentation proceeds for three days (Pyke, 1958). A study of the reactions involved has been made by Smith and Hendlin (1953).

B. OXIDATIONS AND REDUCTIONS

Many workers have investigated the oxidative and reductive abilities of yeasts, particularly in Germany between the two World Wars. The reduction of a large number of aldehydes and ketones has been recorded and such compounds as acetic acid can be oxidized. Mays *et al.* (1962) have patented a method of preparing hydroxycarboxylic acids by the reduction of the corresponding oxo-substituted acids by strains of *Saccharomyces* and *Candida*. At present none of these systems is used on a very large scale.

C. HYDROCARBON DERIVATIVES

Since, in recent years, interest has been concentrated on the metabolism of hydrocarbons by micro-organisms (McKenna and Kallio, 1965), efforts have been made to apply the enzymes of the living organisms to produce commercially useful transformations of hydrocarbon chemicals. The first stage in the assimilation of *n*-alkanes in the presence of atmospheric oxygen is commonly terminal and di-terminal oxidation (Atkinson and Newth, 1968). Among the many genera that have been reported to utilize hydrocarbons, various yeasts, including *Candida*, *Citeromyces*, *Debaryomyces*, *Endomycopsis*, *Hansenula* and *Torulopsis* are included (Quayle, 1968). However, most of the investigations concerned with the yeasts (particularly *Torulopsis candida*) have concentrated on the production of cell mass (see Chapter 8, p. 421) rather than extracellular chemical transformations, and therefore no industrial uses of this type have yet been fully developed. On the other hand, the composition of the yeast made in this way is similar to that of yeast grown on carbohydrates (Llewelyn, 1968); the same intermediates, amino acids and so on could therefore be extracted. There is also no reason why mutants of the same nature as those employed for the production of amino acids and other compounds as excretion products should not be developed but, as such systems are not used on any great scale for yeasts grown on carbohydrates, it is probable that this form of commercial application will not be common practice in the near future.

D. RADIOACTIVE BIOCHEMICALS

A specialized use of yeast for research purposes has been described by Laufer *et al.* (1964) in which *Candida utilis* is grown on sugar of high specific radioactivity prepared by the ingenious method of converting $^{14}CO_2$ into sucrose photosynthetically in the leaves of the *Canna* plant. The sugar is extracted and used for yeast propagation, when as much as 45% of the activity enters the cells, being incorporated into nucleotides, proteins, amino acids, polysaccharides and so on which may be extracted and used in biochemical research.

VI. Conclusions

In the previous pages of this volume, all the main commercial applications of yeasts have been described in some detail, and a somewhat arbitrary selection has been made in the present chapter of the minor uses, which are so numerous that the space does not allow of mention of the vast majority. The examples given are intended, however, to illustrate the potentialities of yeast as an industrial tool.

Although a wide variety of other micro-organisms are in extensive use for the production of certain products, notably organic acids such as acetic (vinegar), lactic, citric, gluconic, fumaric, itaconic and kojic, bacterial and fungal amylases and other enzymes, antibiotics, vitamins and hormones, and in waste disposal systems, the total industrial contribution is small in bulk compared with that of yeast. This is not to deny, of course, that applications such as in the manufacture of antibiotics are of vital importance in the modern world. One of the outstanding aspects of the main yeast product, ethanol, is the enormous revenue that it makes available to governments in the form of excise duty.

The reasons for this special role of the yeasts is possibly due, in the first place, to the ability of a limited number of species to become the predominant organisms under natural conditions in sugary solutions, combined with their harmless nature when ingested by humans. More research has almost certainly been carried out on a few species of yeasts than on any other single type of micro-organism, because of the early recognition of their vital function in alcoholic fermentation, and because of their availability in the form of baker's and brewer's yeasts. The result has been that many more useful applications have been discovered.

References

Anon. (1964). *Biotechnol. Bioengng* **6**, 63–64.
Atkinson, J. H. and Newth, F. H. (1968). *In* "Microbiology", (P. Hepple, ed.), pp. 35–45. Institute of Petroleum, London.
Becze, G. I. de (1960). *Wallerstein Labs Commun.* **23**, 99–124.
Chiao, J. S. and Peterson, W. H. (1953). *Agric. Fd Chem.* **1**, 1005–1008.
Chibata, I., Kakimoto, T. and Kato, J. (1965). *Appl. Microbiol.* **13**, 638–645.
Cochrane, A. L. (1961). Soc. Chem. Indust. Monograph No. 11, pp. 25–31.
Connstein, W. and Lüdecke, K. (1921). U.S. Patent 1,368,023.
Connstein, W. and Lüdecke, K. (1924). U.S. Patent 1,511,754.
Davies, R. (1963). *In* "Biochemistry of Industrial Micro-Organisms", (C. Rainbow and A. H. Rose, eds.), pp. 68–150. Academic Press, London.

Davis, A. E. and Bolin, T. (1967). *Nature, Lond.* **216**, 1244.

Demain, A. L. and Phaff, H. J. (1954a). *Nature, Lond.* **174**, 515.

Demain, A. L. and Phaff, H. J. (1954b). *Archs Biochem. Biophys.* **51**, 114–121.

Dulaney, E. L. (1967). *In* "Microbial Technology", (H. J. Peppler, ed.), pp. 308–343. Reinhold Publishing Corpn, New York.

Dulaney, E. L., Stapley, E. O. and Simpf, K. (1954). *Appl. Microbiol.* **2** (6), 371–379.

Dunn, C. G. (1952). *Wallerstein Labs Commun.* **15**, 61–81.

Eddy, A. A. (1958). *In* "The Chemistry and Biology of Yeasts", (A. H. Cook, ed.), pp. 157–249. Academic Press, New York.

Eoff, J. R. (1918). U.S. Patent 1,288,398.

Ferranto, A., Giuffrida, G. and Terranova, R. (1965). *Boll. Soc. Ital. Biol. Sper.* **41**, 1488–1491.

Fischer, E. H., Kohtès, L. and Fellig, J. (1951). *Helv. chim. Acta* **34**, 1132–1138.

Flatz, G., Saengudom, Ch. and Sanguanbhokhai, T. (1969). *Nature, Lond.* **221**, 758–759.

Fulmer, E. I., Underkofler, L. A. and Hickey, R. J. (1947). U.S. Patent 2,416,745.

Gabel, W. (1950). *Chem. Zbl.* II, 2984.

Graham, J. C. J. (1961). Brit. Patent 870,622.

Hanson, A. M. (1967). *In* "Microbial Technology", (H. J. Peppler, ed.), pp. 222–250. Reinhold Publishing Corpn, New York.

Harris, J. F. and Hajny, G. J. (1960). *Biochem. microbiol. technol. Engng* **2**, 9–24.

Hildebrandt, C. and Klavehn, W. (1934). U.S. Patent 1,956,950.

Johnston, W. R., Redfern, S. and Miller, G. E. (1935). *Ind. engng Chem., analyt. Edn* **7**, 82–86.

Laufer, L., Gutcho, S., Castro, T. and Grenner, R. (1964). *Biotechnol. Bioengng* **6**, 127–146.

Llewelyn, D. A. B. (1968). *In* "Microbiology", (P. Hepple, ed.), pp. 63–84. Institute of Petroleum, London.

Luh, B. S. and Phaff, H. J. (1954). *Archs Biochem. Biophys.* **48**, 23–37.

Mann, P. F. E., Trevelyan, W. E. and Harrison, J. S. (1958). Soc. Chem. Indust. Monograph No. 3, pp. 68–85.

Mays, G. T., Ven, B. van der and Jonge, A. P. de (1962). Canadian Patent 648,917.

McKenna, E. J. and Kallio, R. E. (1965). *A. Rev. Microbiol.* **19**, 183–208.

Meyers, R. P. (1956). U.S. Patent 2,762,748.

Meyers, R. P. and Stimpson, E. G. (1956). U.S. Patent 2,762,749.

Morgan, E. R. (1955). U.S. Patent 2,715,601.

Munden, J. E., Crook, E. M. and Donald, M. B. (1963). *Biotechnol. Bioengng* **5**, 221–230.

Neuberg, C. (1944). U.S. Patent 2,361,315.

Neuberg, C. and Roberts, I. S. (1946). "Invertase". Sugar Res. Foundation Sci. Rep. No. 4. New York.

Neuberg, C., Hirsch, J. and Reinfurth, E. (1920). *Biochem. Z.* **105**, 307–336.

Otsuka, S., Yazaki, H., Nagase, H. and Sakaguchi, K-I (1957). *J. gen. appl. Microbiol., Tokyo* **3**, 35–53.

Pan, S. C., Andreasen, A. A. and Kolachov, P. (1949). *Archs Biochem.* **23**, 419–433.

Parsons, H. T., Williamson, A. and Johnson, M. L. (1945). *J. Nutr.* **29**, 373–381.

Peppler, H. J. (1967). *In* "Microbial Technology", (H. J. Peppler, ed.), pp. 145–171. Reinhold Publishing Corpn, New York.

Peppler, H. J. and Thorn, J. A. (1960). U.S. Patent 2,992,748.

Petzoldt, K., Kieslich, K. and Koch, H. J. (1967a). German Patent 1,252,674.

Petzoldt, K., Kieslich, K. and Koch, H. J. (1967b). German Patent 1,252,675.

Phaff, H. J. and Luh, B. S. (1952). *Archs Biochem. Biophys.* **36**, 231–232.

Pomeranz, Y. (1964). *Fd Technol., Champaign* **18**, 682, 690.

Prescott, S. C. and Dunn, C. G. (1959). "Industrial Microbiology", 3rd. Ed. McGraw-Hill, New York.

Pyke, M. (1958). *In* "The Chemistry and Biology of Yeasts", (A. H. Cook, ed.), pp. 536–586. Academic Press, New York.

Quayle, J. R. (1968). *In* "Microbiology", (P. Hepple, ed.), pp. 21–34. Institute of Petroleum, London.

Reece, M. C., Donald, M. B. and Crook, E. M. (1959). *Biochem. microbiol. technol. Engng* **1**, 217–228.

Rosenberg, H. R. (1942). "Chemistry and Physiology of the Vitamins", p. 341. Interscience Publishers, New York.

Schade, A. L. (1947). U.S. Patent 2,428,766.

Schade, A. L. and Färber, E. (1947). U.S. Patent 2,414,838.

Slodki, M. E., Wickerham, L. J. and Cadmus, M. C. (1961). *J. Bact.* **82**, 269–274.

Smith, P. F. and Hendlin, D. (1953). *J. Bact.* **65**, 440–445.

Spencer, J. F. T. and Sallans, H. R. (1956). *Can. J. Microbiol.* **2**, 72–79.

Spencer, J. F. T. and Shu, P. (1957). *Can. J. Microbiol.* **3**, 557–567.

Spencer, J. F. T., Roxburgh, J. M. and Sallans, H. R. (1957). *J. agric. Fd Chem.* **5** (1), 64–67.

Spencer, J. F. T., Tulloch, A. P. and Gorin, P. A. J. (1962). *Biotechnol. Bioengng* **4**, 271–279.

Sumitomo Chemical Co. (1963). French Patent 1,331,847.

Suomalainen, H., Christiansen, V. and Oura, E. (1967). *Suomen Kemistilehti* **B340**, 286–287.

Terui, G. and Enatsu, T. (1962). *Technol. Rept Osaka Univ.* **12**, 477–493.

Trevelyan, W. E. (1958). *In* "The Chemistry and Biology of Yeasts", (A. H. Cook, ed.), pp. 369–436. Academic Press, New York.

Tveit, M. (1967). *In* "Biology and the Manufacturing Industries", (M. Brook, ed.), pp. 3–10. Academic Press, London.

Underkofler, L. A. (1954). *In* "Industrial Fermentations", (L. A. Underkofler and R. J. Hickey, eds.) Vol. I, pp. 252–270. Chemical Publishing Co., New York.

Underkofler, L. A., Fulmer, E. I., Hickey, R. J. and Lees, J. M. (1951a). *Iowa State Coll. J. Sci.* **26**, 111–133.

Underkofler, L. A., Fulmer, E. I., Lees, J. M. and Figard, P. H. (1951b). *Iowa State Coll. J. Sci.* **26**, 135–147.

Van Lanen, J. M. (1954). *In* "Industrial Fermentations" (L. A. Underkofler and R. J. Hickey, eds.), Vol. II, pp. 191–216. Chemical Publishing Co., New York.

Vaughn, R. H. (1954). *In* "Industrial Fermentations", (L. A. Underkofler and R. J. Hickey, eds.) Vol. I, pp. 498–535. Chemical Publishing Co., New York.

Wallerstein L. (1932). U.S. Patent 1,855,591.

Wickerham, L. J., Lockard, L. B., Pettijohn, O. G. and Ward, G. E. (1944). *J. Bact.* **48**, 413–427.

World-Wide Survey of Fermentation Industries 1963 (1966). *Pure appl. Chem.* **13**, 405–417.

Young, H. and Healey, R. P. (1957). U.S. Patent 2,776,928.

Author Index

Italic numbers indicate pages on which a reference is listed

I

M

W

Subject Index

A

Acetaldehyde, in yeast metabolism, 169
production by yeasts, 14, 41, 217
Acetic acid bacteria, contaminants of brewer's yeast, 208
Acetic acid, effect on yeast fermentation, 31
production by yeasts, 14, 43
Acetobacter, 131
effects on yeast fermentation, 302
α-Acetolactate, 202, 203
Acetoin, as contaminant of cider, 119
formation in beer production, 199–204, 217
in wine, 46
production by yeasts, 14, 243
Acid content of wine, methods of reduction, 44
Acetyl-coenzyme A, 170, 171, 178
Acid production, as measure of infection in fermentations, 330
as measure of progress of fermentation, 331
Acids, method of analysis, 331
Acrolein formation during spirit production, 302
Actidione, use in bacteriological testing, 121
Adaptation of yeasts, to high ethanol concentrations, 41
to nitrogen sources, 189
Adenine, 234
Adenosine triphosphate, in anaerobic yeast growth, 289, 291
preparation from yeast, 446, 536
Adenosine triphosphatase, 166
Aeration, effect on diacetyl formation, 201

efficiency, 375, 376
in continuous fermentation systems, 213, 215
in yeast production, 374–380
rate control, 376
system, practical difficulties, 377
Aerobic propagation, of wine yeast starter, 25
of yeast for *saké* brewing, 262, 265
Agglutination tests for yeasts, 210
Aggregation of yeast cells in mixed cultures, 237
Agitation in relation to yeast growth, 363
Air, as vector in yeast distribution, 82
Air filters, methods of testing, 374
types, 374
Air filtration, theoretical considerations, 374
α-Alanine, 164, 239, 240
preparation by use of yeast, 537
β-Alanine, 170, 205
Alcohol, economic importance, 1, 3, 247, 249
dehydrogenase, 184
preparation from yeast, 446, 532
high concentration in *saké* mash, 240, 244–246
tolerance of yeast, 241, 244–246
Alcoholic beverages, spoilage by yeasts, 511–514
Alcohols, effect on yeast function, 299
Aldehydes, effect on yeast function, 299
reduction of, 180
Ale, 198, 199
n-Alkanes, action of yeasts on, 542
American beers, 199, 206
Amino acid, excretion by yeasts, 167
"pattern" of yeast protein, 432, 433
pool in yeast, 157, 164, 166
synthesis by yeast, 185

QR
151
R79

Rose, Anthony H   comp.
    The yeasts; edited by Anthony H. Rose and
J.S. Harrison.   London, Academic Press, 1969-
1970.
        3v.   illus.   24cm.

    Includes bibliographical references.
    Contents.-v.1. Biology of yeasts.-v.2. The
physiology and biochemistry of yeasts.-v.3. Yeast
technology.

.Yeast-Collected works.   I.Harrison, John Stuart, joint
author.   II.Title.